Springer Aerospace Technology

The series explores the technology and the science related to the aircraft and spacecraft including concept, design, assembly, control and maintenance. The topics cover aircraft, missiles, space vehicles, aircraft engines and propulsion units. The volumes of the series present the fundamentals, the applications and the advances in all the fields related to aerospace engineering, including:

- structural analysis,
- aerodynamics,
- aeroelasticity,
- aeroacoustics,
- flight mechanics and dynamics
- orbital maneuvers,
- avionics,
- systems design,
- materials technology,
- launch technology,
- payload and satellite technology,
- space industry, medicine and biology.

The series' scope includes monographs, professional books, advanced textbooks, as well as selected contributions from specialized conferences and workshops.

The volumes of the series are single-blind peer-reviewed.

To submit a proposal or request further information, please contact: Mr. Pierpaolo Riva at pierpaolo.riva@springer.com (Europe and Americas) Mr. Mengchu Huang at mengchu.huang@springer.com (China)

The series is indexed in Scopus and Compendex

Zishun Liu · Renfu Li · Xiaodong He ·
Zhenghong Zhu
Editors

Advances and Challenges in Advanced Unmanned Aerial Systems

Selected Contributions of ICAUAS 2023

Springer

Editors
Zishun Liu
International Center for Applied Mechanics
Xi'an Jiaotong University
Xi'an, Shaanxi, China

Renfu Li
School of Aerospace Engineering
Huazhong University of Science
and Technology
Wuhan, Hubei, China

Xiaodong He
School of Astronautics
Harbin Institute of Technology
Harbin, Heilongjiang, China

Zhenghong Zhu
Space Engineering Design Laboratory
York University
Toronto, ON, Canada

ISSN 1869-1730 ISSN 1869-1749 (electronic)
Springer Aerospace Technology
ISBN 978-981-99-8044-4 ISBN 978-981-99-8045-1 (eBook)
https://doi.org/10.1007/978-981-99-8045-1

This Springer imprint is published by the registered company Springer Nature Singapore Pte Ltd.
The registered company address is: 152 Beach Road, #21-01/04 Gateway East, Singapore 189721, Singapore

Conference Organization

ICAUAS 2023

Conference Chairs

Renfu Li, Huazhong University of Science and Technology, China
Xiaodong He, Harbin Institute of Technology, China
Zhenghong (George) Zhu, York University, Canada
Xiaohui Wei, Nanjing University of Aeronautics and Astronautics, China
Zishun Liu, Xi'an Jiaotong University, China
Diofantos G. Hadjimitsis, Cyprus University of Technology, Cyprus

Conference Co-Chair

Chao Wang, Harbin Institute of Technology, China

International Advisory Committee

Edwin Lughofer, Johannes Kepler University Linz, Austria
Boong Yeol Ryoo, College of Architecture|Texas A&M University, USA
Zabih (Fary) Ghassemlooy, Northumbria University, UK

Program Chairs

Ian McAndrew, Capitol Technology University, USA
Jiann-Shiou Yang, University of Minnesota Duluth, USA
Hazry Desa, Universiti Malaysia Perlis, Malaysia
Hugh H.T. Liu, University of Toronto Institute for Aerospace Studies, Canada
Qingyu Peng, Harbin Institute of Technology, China

Local Chair

Ju Liu, Malmö University, Sweden

Publication Chairs

Hamid R. Ghaeini, CISPA Helmholtz Center for Information Security, German
Weicheng Jiao, Harbin Institute of Technology, China
Lifeng Hao, Harbin Institute of Technology, China

Publicity Chair

Chao Sui, Harbin Institute of Technology, China

Finance Chairs

Yi Hong, Harbin Institute of Technology, China
Qamar Naith, University of Jeddah, Saudi Arabia
Arfauz A Rahman, Queen's University Belfast, The United Kingdom

Program Committee

Yuanyuan Zhang, Huazhong University of Science&Technology, China
Jiwei Tang, Shanghai Jiao Tong University, China
Wu S.-F, Shanghai Jiao Tong University, China
Vasiu Radu, University of Timisoara Senate, Romania
Jolanta Mizera-Pietraszko, Opole University, Poland

International Scientific Committee

Mahmoud Quweider, University of Texas, USA
Taeghyun Kang, University of Central Missouri, USA
Oleg A Yakimenko, Naval Postgraduate School, USA
Jose Martinez-Carranza, Instituto Nacional de Astrofisica Optica y Electronica, Mexico
Tenreiro Machado, Polytechnic of Porto, Portugal
Teodor Lucian Grigorie, Military Technical Academy "Ferdinand I" in Bucharest, Romania
Ana Catarina S. Carapito, Universidade da Beira Interior, Portugal
Dragos Isvoranu, Univ. Politehnica of Bucharest1, Romania
Janusz Wielki, Opole University of Technology, Poland
Javaid Butt, Faculty of Science & Technology, United Kingdom
João Calado, Instituto Politécnico de Lisboa, Portugal
Jonghwa Kim, University of Augsburg, Germany
Jose Luis Vazquez-Poletti, Universidad Complutense de Madrid, Spain
Kasper Hallenborg, University of Southern Denmark, Denmark
Leo Mršić, IN2data Ltd Data Science Company, Zagreb, Croatia
Marek Kulbacki, Polish-Japanese Academy of Information Technology, Poland
Michael Opoku Agyeman, The University of Northampton, UK
Michail Kalogiannakis, University of Crete, Greece
Petrisor Parvu, Polytechnic University of Bucharest, ROMANIA
Sofia Meacham, Bournemouth University, UK
Seppo Sirkemaa, University of Turku, Pori Unit, Finland
Danilo Pelusi, Università degli Studi di Teramo, Italy
Fabio Leccese, Università Roma Tre, Italy
Takfarinas Saber, University of Galway, Ireland
Hamid R. Ghaeini, CISPA Helmholtz Center for Information Security, German
João Manuel Ferreira Calado, Instituto Superior de Engenharia de Lisboa, Portugal
Arfauz A Rahman, Queen's University Belfast, UK
Sule Yildirim Yayilgan, Norwegian University of Science and Technology, Norway
Vaios Lappas, University of Athens, Greece
Aqiang Lin, Northwestern Polytechnical University, China
Hongliang Ren, The Chinese University of Hong Kong (CUHK), China
Ferhan Şengür, Eskisehir Technical University, Turkey
Nguyen Dinh Duc, Vietnam National University, Vietnam
Mehmet Karaata, Kuwait University, Kuwait
Ilhan Asilturk, Selçuk University, Turkey
Mohammed A. Akour, Yarmouk University, Jordan
Murat Karakaya, Atılım University, Turkey
Nahlah Shatnawi, Yarmouk University, Jordan
Sevenpri Candra, Bina Nusantara University, Indonesia
Vuong Mai, KAIST, South Korea

Sakihara Reika, Tottori University, Japan
Fan Gao, ChongQing ChuanYi Automation Co., Ltd, China
Xin Lu, Iwate University, Japan
Fatih Karpat, Bursa Uludag University, Turkey
Mohd Na'im Abdullah, Universiti Putra Malaysia, Malaysia
Xiuxian Li, Tongji University, China
Ozlem Sahin, Eskişehir Technical University, Turkey
Yoshifumi Manabe, Kogakuin University, Japan
Ryspek Usubamatov, Kyrgyz State Technical University, Kyrgyz
Zhen Li, Jianyan Testing Group Chongqing Co., Ltd, China
Wudhichai Assawinchaichote, King Mongkut' s University of Technology Thonburi, Tailand
Nickolay (Mykola) Zosimovych, Xi' an Jiaotong Liverpool University, China
Mingqiang Luo, Beihang University, China
Heyong Xu, Northwestern Polytechnical University, China
Yufei Zhang, Tsinghua University, China
Weimin Sang, Northwestern Polytechnical University, China
Qiulin Qu, Beihang University, China
Yunsong Gu, Nanjing University of Aeronautics and Astronautics, China
Liguo Sun, Beihang University, China
Chuwei Zhou, Nanjing University of Aeronautics and Astronautics, China
Chen Wang, Nanjing University of Aeronautics and Astronautics, China
Yingyu Wang, Nanjing University of Aeronautics and Astronautics, China
Yan Shi, Nanjing University of Aeronautics and Astronautics, China
Jinhui Jiang, Nanjing University of Aeronautics and Astronautics, China
Wenzhong Xie, Nanjing University of Aeronautics and Astronautics, China
Huacheng Yuan, Nanjing University of Aeronautics and Astronautics, China
Yaping Hu, Nanjing University of Aeronautics and Astronautics, China
Changguo Wang, Harbin Institute of Technology, China
Lei Hou, Harbin Institute of Technology, China
Yao Li, Harbin Institute of Technology, China
Jian Sun, Harbin Institute of Technology, China
Xinsheng Wang, Harbin Institute of Technology, China
Weiran Yao, Harbin Institute of Technology, China
Changsheng Gao, Harbin Institute of Technology, China
Wei Kang, Northwestern Polytechnical University, China
Bo Zhang, Northwestern Polytechnical University, China
Dong Zhang, Northwestern Polytechnical University, China
Linyuan Jia, Northwestern Polytechnical University, China
Jiang Ding, Guangxi University, China
Zhan Li, Harbin Institute of Technology, China
Yue Wang, Beihang University, China
Kedong Wang, Beihang University, China
Zebbiche Toufik, University of Blida 1, Algeria

Preface

ICAUAS 2023

On behalf of the organizing committee, it is my great honor to welcome you to the 2023 International Conference on Advanced Unmanned Aerial Systems, which was held on July 13-16, 2023, multi-city and hybrid, in Harbin, China and Toronto, Canada.

The success of ICAUAS 2023 was due to the collective efforts of conference committee, which offers a unique platform for scholars, researchers, professionals, industry experts and young researchers to exchange ideas, present research findings, and discuss the latest advancements in the field. Through thought-provoking keynote speeches, enlightening plenary sessions and interactive panel discussions, this conference provides a dynamic and inclusive environment for stimulating intellectual discourse and fostering innovation.

In its first edition, ICAUAS 2023 received 176 submissions in total and attracted more than 110 attendees. After rigorous review by related top experts and rebuttal process, 42 excellent papers were accepted and included in the proceedings. During the conference, participants explored a wide array of topics, including but not limited to Autonomy and Control, Flight Dynamic and Control, Communication and Navigation, etc.

Finally, I would like to express my deepest appreciation to the diligent and devoted organizing committee and the Technical Program Committee members for their hard work in reviewing and selecting papers. Their unwavering commitment to excellence ensures that every participant of ICAUAS 2023 has experienced a truly exceptional conference. The tireless efforts and expertise of the organizing committee have made an immense contribution to conduct an unforgettable event for us all.

ICAUAS 2023 Conference Chair

Wuhan, China Renfu Li

Contents

Simple Electroaerodynamic Aircraft Testing with Multistage Staggered Arrangement of the Electrodes

Yulong Ying, Zewei Xia, Tong Lin, Jianfeng Wang, Mingying Huo, and Naiming Qi

Abstract Electroaerodynamic (EAD) devices use corona-induced ionic winds for in-atmosphere propulsion, from which EAD aircraft are derived with the advantages of low mechanical fatigue, low noise and zero carbon emissions. High electric field ionization of air produces charged particles that accelerate and then collide with air molecules in momentum to generate ionic winds and thrust. The EAD aircrafts are a high thrust-to-power ratio propulsion method, but its thrust and thrust density need to be improved. In this study, electrode unit arrays and multistage experiments are conducted on a staggered distribution of electrode configurations with low aerodynamic drag. The higher increase in thrust than linear superposition of electrode units of this structure in parallel compared to the collinear distribution. Based on the structure of electrode unit arrays and multistage, the open-loop flight test of the two-stage simple EAD aircraft is realized.

Keywords EAD aircraft · Corona discharge · Arrays and multistage experiment · Open-loop flight test

---

Y. Ying · Z. Xia · T. Lin · M. Huo (✉) · N. Qi
Harbin Institute of Technology, Harbin, China
e-mail: huomingying@hit.edu.cn

Y. Ying
e-mail: 22S118193@stu.hit.edu.cn

Z. Xia
e-mail: 3265521145@qq.com

T. Lin
e-mail: 18245030428@163.com

N. Qi
e-mail: qinmok@163.com

J. Wang
Aerodynamic Research Institute, Harbin, China
e-mail: 13208479940@163.com

Z. Liu et al. (eds.), *Advances and Challenges in Advanced Unmanned Aerial Systems*, Springer Aerospace Technology, https://doi.org/10.1007/978-981-99-8045-1_1

1 Introduction

Since the beginning of the aviation era, traditional propeller driven and jet engine driven aircraft have used large amounts of fossil fuels that pollute the environment, and there has been a growing interest in electric drive technology. Ion thrusters have been employed in electric propulsion systems in the aerospace industry for many years, but recently, the prospect of substituting conventional propulsion techniques with electroaerodynamic (EAD) devices that can generate propulsion in the atmosphere has grown [1–7]. Electroaerodynamic devices, also known as an electrohydrodynamic devices, accelerate charged particles by the electrostatic force in an electric field. When charged particles collide with neutral molecules, they transfer momentum to the neutral molecules, generating ion winds and thrust. EAD aircraft utilized for aircraft propulsion are distinct from unmanned aerial vehicles with fixed wings or rotors. Due to the lack of moving parts like turbines or propellers, EAD aircraft have various benefits including low noise, low mechanical wear, and zero carbon emissions. EAD devices have been researched and used in the fields of fluid pumping [8], heat transfer enhancement [9], and flow control [10] in addition to aircraft propulsion.

The electrode unit is the essential part of an EAD propulsion device, which consists of two (or two sets of) electrodes. An emitter and a collector are spaced apart by an air gap, and coupled to a high voltage source. The collector typically has a bigger frontal area, such as a cylinder or airfoil, and the emitter often has a very tiny curvature radius. The most common EAD device is to apply a DC voltage across the electrodes. Scholars have also investigated other methods to improve propulsion and efficiency, such as dielectric blocking discharges using an AC power source as an ion source [11]. The strong electric field that surrounds the emitter causes ions to form nearby, creating non-equilibrium or cold plasma. These ions then drift under the influence of the Coulomb force, dispersing towards the collector in an area known as the "drift region". Examples of corona-discharge geometries include wire-to-cylinder [12, 13], wire-to-airfoil [2, 14], and pin-to-mesh [4]. Early research mostly focused on wire-to-cylinder, but switching to an airfoil will greatly minimize the drag brought on by flow [14]. Drop-shaped collectors equipped on electrohydrodynamic propulsion aircraft which has vertical takeoff capability was presented by Khomich and Rebrov [2]. In 2018, Xu et al. from MIT [3, 15] has completed a test flight of the world's first plasma propulsion drone and published an article on it in Nature. This successful physical experiment demonstrated the viability of using EAD for aircraft propulsion. The collectors used in their ion-propelled aircraft were NACA 0010 airfoils.

As shown in Fig. 1, there were primarily two electrode distribution modes in earlier research on performance optimization. They were used to study changes in electric parameters like voltage and polarity as well as geometric parameters like the distance d between the emitter and the collector and the spacing s between two parallel electrodes. The emitter is positioned above the collector in Fig. 1a, which is called collinear distribution used in the majority of literature [3]. As shown in Fig. 1b, another distribution strategy called staggered distribution places the emitter

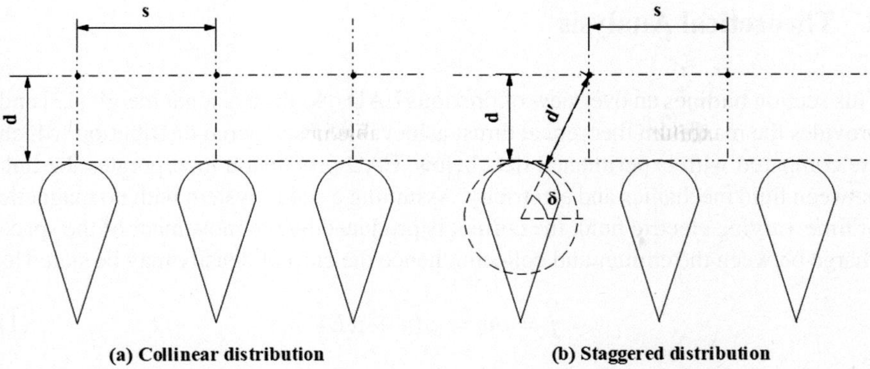

Fig. 1 Two main types of electrode distribution include collinear distribution (**a**) and staggered distribution (**b**). General description of electrode units and parameter definitions include the parameters used in the theoretical analysis in Sect. 2

in the middle of the succeeding collectors, which follows the same physical principles [2]. The literature data [16] indicates that the current-to-thrust conversion with such electrode distribution design is more efficient. At this point, the electrode unit consists of three electrodes, namely one emitter and two collectors. The emitter and collector will have a staggered layout when the electrode units are connected in parallel to form the arrays.

The performance of multistage and array arrangements has been studied in some detail in the literature [17], showing that when several electrode units are connected in parallel, the thrust decreases with the distance between the units. Collinear distributed wire-to-cylinder electrode configurations have been the main focus of previous multi-stage and array tests. The wire-to-airfoil and staggered electrode distribution multi-stage configuration may resemble the pattern found in reference [17], but has not been confirmed. However, from an EAD aircraft manufacturing perspective, this configuration may result in greater thrust and efficiency.

The aim of the present work is to provide a parametric analysis of multistage staggered arrangement of the electrodes with wire-to-airfoil. Applying the appropriate electrode unit geometry parameters and stage spacing results in the fabrication of a simple vertical takeoff and landing EAD aircraft. The flight parameters of the EAD aircraft are recorded with the change of voltage to verify the effect of voltage change on its flight state and acceleration.

The remainder of this paper is organized as follows. Theoretical analysis of staggered distribution of the electrodes is introduced in Sect. 2. Experimental setup for measuring thrust is described in Sect. 3. Furthermore, Experimental results of electrode unit arrays, multistage experiments and open-loop flight of the EAD aircraft are given in Sect. 4. Finally, the conclusions of this study are presented in Sect. 5.

2 Theoretical Analysis

This section outlines an overview of previous EAD one-dimensional theory [13] and provides the maximum theoretical thrust achievable in staggered distribution, which are compared with experimental data below. EAD propulsion incorporates the link between fluid mechanics and electricity. Assuming a stable system with no magnetic or time-varying electric field, the current is produced by the movement of the space charge between the emitter and collector, hence the current density may be stated as

$$j = \rho v_i = \rho(u + \mu E) \tag{1}$$

where j is the current density (current per unit area), ρ is the charge density, E is the electric field, and u is the neutral fluid velocity. Ion drift velocity term and convection term are added together to form the ion velocity v_i. The typical value of μE is at least 100 ms^{-1}, which is significantly higher than the velocity term for natural convection of air under the assumptions of an ionic mobility μ of $2 \times 10^{-4}\ \text{m}^2\text{V}^{-1}\ \text{s}^{-1}$ [18] and an average electric field intensity of $5 \times 10^5\ \text{Vm}^{-1}$.

For one-dimensional theory of wire-to-cylinder configuration for corona discharge geometry, total discharge current is

$$I = \int j\,dA = \rho\mu E A \tag{2}$$

where A is the cross-sectional area of corona discharge electrodes.

The EAD thrust is equal to the strength of the Coulomb force on the injected ions remaining in the gap between the two electrodes. And the EAD thrust per unit volume acting on the charged particles in the discharge volume can be expressed as

$$f = \rho E \tag{3}$$

Integrating over the entire discharge volume according to Eqs. (1) and (3), the total EAD thrust is obtained as

$$T = \int_V f\,dV = \int_V \frac{j}{\mu}\,dV = \frac{jAd}{\mu} = \frac{Id}{\mu} \tag{4}$$

where V is the entire discharge volume and d is the distance between the emitter and collector.

According to Townsend theory [19], regardless of the electric field distribution, the relationship between air corona discharge current and voltage can be simplified as

$$I = CV_a(V_a - V_0) \tag{5}$$

where V_a is the applied potential across the electrodes, V_0 is the corona inception voltage and C is a constant depending on the electrode geometry and ion mobility. Therefore, the relationship between thrust and voltage can be expressed as

$$T = \frac{CV_a(V_a - V_0)d}{\mu} \tag{6}$$

Furthermore, the thrust-to-power ratio is

$$\frac{T}{P} = \frac{d}{\mu V_a} \tag{7}$$

Thrust per unit area of electrode in the absence of resistance is

$$\frac{T}{A} = \frac{Id}{\mu A} \tag{8}$$

Taking into account the space charge and a number of assumptions [15], the maximum corona discharge current density in the one-dimensional situation, is described satisfactorily by the Mott–Gurney law:

$$j_{\max,1D} = \frac{9\varepsilon_0 \mu V_a^2}{8d^3} \tag{9}$$

where $\varepsilon_0 = 8.85 \times 10^{-12}$ F/m is the electric constant.

When the situation becomes staggered distribution as shown in Fig. 1b, we presume the distribution of the current density along the collector perimeter is similar to the wire-to-cylinder electrode arrangement, so introducing equivalent radius r_{eq} [2]

$$r_{eq} = \frac{C_a}{2\pi} \tag{10}$$

where C_a is airfoil electrode cross-sectional area perimeter.

Further correction of the maximum current density is necessary due to the change in staggered distribution. Assuming that the two collectors in the same electrode unit do not interfere with each other, the maximum theoretical current density in staggered distribution is

$$j_{\max,2D} = \frac{9\varepsilon_0 \mu V_a^2}{8d'^3} \tag{11}$$

where d' is the shortest connecting line distance between the emitter and the equivalent cylindrical collector surface if the emitter is considered to be a point.

As shown in Fig. 1b, the actual current density should vary with the δ angle [20], rather than Eq. (2) being uniformly distributed along the boundary of a cylinder or airfoil when the applied voltage and electrode gap are constant. If the emitter is considered to be a point, the shortest connecting line between the emitter and the equivalent cylindrical collector surface should be at the maximum current density value. j is a function of δ:

$$j(\delta) = j_{\max,\,2D} \cdot \left(\frac{E(\delta)}{E_{\max}}\right)^n = j_{\max,\,2D}\left(\frac{(1-\beta)^2}{1-2\beta\cos\delta+\beta^2}\right)^n \tag{12}$$

where $E(\delta)/E_{max}$ is the proportion of the equivalent cylinder's surface electric field strength to its highest value at $\delta = 0^o$, $\beta = \frac{r_{eq}}{r_{eq}+d'}$ and n is the exponent equal to 1.7 [2].

Integrate Eq. (12) to obtain the theoretical current per unit length of each airfoil collector:

$$I = 2\int_0^{\pi} j(\delta) \cdot r_{eq} d\delta \tag{13}$$

The results of Eq. (13) allow further derivation of the maximum theoretical thrust, thrust power ratio and thrust density in the case of staggered distribution.

3 Experimental Setup

An EAD electrode unit is composed of one 50 micron emitting tungsten wire and two airfoil collectors. The collectors are composed of balsa wood and aluminum foil wrapped around it. The selected collector geometry is consistent with Khomich (r_{eq} = 6.058 mm) [2]. This drop collector has also been tested in Belan [14].

The electrodes are supported by an exterior frame that makes it easier for them to connect. Balsa is used to make the frame and collectors because of its strong strength at low weight. In order to lessen the effects of sharp edges on aluminum foil, the effective length of the electrodes is $L = 18$ cm after the insulated fixed end of the collector is removed.

When electrode units are joined in parallel, the tungsten wire remaining straight is wound in a serpentine pattern through the pre-set holes on the balsa frame. Once the wire is slightly bent, it will create higher localized electric fields, which will change the uniformity of discharge and may lead to premature arc discharge. For multistage experiments the external frame height will be extended to accommodate the second stage electrode unit arrays.

The high voltage (HV) power supply *JMDC-P50-10* mA provides a positive voltage up to 50 kV and a current up to 10 mA. The output cable of high voltage power supply is fixed on a stable platform and carefully connected to the emitter to ensure that the weight and the stiffness of the cable do not affect the measurements. The collectors are grounded in all experiments. Due to the presence of leakage current [12] between electrodes, an additional ammeter is connected to ground for

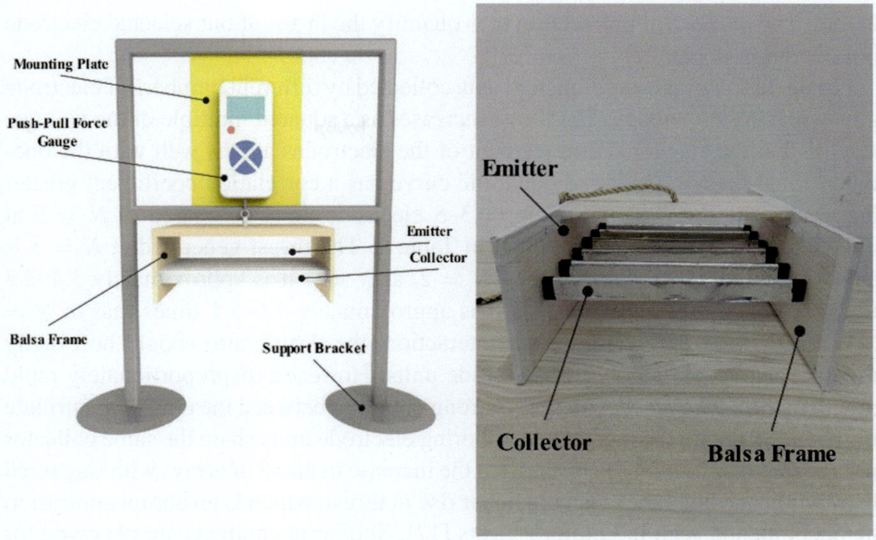

Fig. 2 Experimental setup for thrust measurements

current measurement. The resolution of the high-voltage output is 0.01 kV, and the resolution of the ammeter is 1 μA. Suspend the balsa frame under a push–pull force gauge (precision of $\pm 10^{-4}$ N) to measure the thrust generated by the EAD device. The measurement procedure is the following: high voltage is set to the desired value for at least 5 s, and then data are recorded by tension meter measurement software during 20 s at a sampling rate of 1 Hz before being averaged. All experiments are conducted at a temperature between 19 and 21 °C, an atmospheric pressure between 0.97 and 1.01, and a relative air humidity between 28 and 32%. The specific experimental setup details are shown in Fig. 2.

4 Result

4.1 Electrode Unit Arrays

The tests in this section require the determination of the geometric parameters of the electrode units. The results of both numerical simulations [21] and experiments [17] reveal that there is some optimal configuration for thrust for collinear distribution, with an optimal separation distance s ranging from $s = 1.1d$ to $s = 1.3d$. A smaller value of distance d at the same voltage results in increased thrust, but at higher voltages, it also causes an earlier spark breakdown. Based on the above conclusions, electrode unit geometry parameter is used in this paper with $s = 40$ mm and $d =$

34 mm. The purpose of this section is to quantify the thrust of our selected electrode unit operating in parallel.

Figure 3a shows the experimental data collected by different numbers of electrode units with varying voltage. The thrust increases as a squared multiple of the voltage, proving that the staggered arrangement of the electrodes agrees well with the one-dimensional theory. The fitted quadratic curve has a correlation coefficient greater than 0.99. The thrust ratios for N = 3–5 electrode units compared to $N = 2$ at 16 kV, 24 kV, and 30 kV are shown in Table 1. The thrust generated at $N = 3$ is approximately 2.0–2.3 times that of $N = 2$, at $N = 4$ it is approximately 3.4–3.8 times that of $N = 2$ and at $N = 5$ it is approximately 4.6–5.1 times that of $N = 2$. When there is no electrode unit interaction, the thrust ratio should be exactly proportionate to the number of electrode units. However, disproportionately rapid growth in thrust proves that there is a strong coupling between the units. We attribute this phenomenon to the fact that neighboring electrode units share the same collector in staggered distribution. It appears that the increase in thrust of arrays with staggered distribution can be higher than the linear rise in thrust, which is an abrupt contrast to the phenomenon seen in Gilmore's tests [17]. Similar phenomena are observed for other values of voltage, meaning that the thrust gain is greater for the larger number of electrode units ($N = 5$) chosen for the range of experiments tested.

The thrust density is defined in relation to propulsion system frontal area, For the EAD devices with staggered arrangement when N electrode units are considered, the thrust density per unit area is defined as

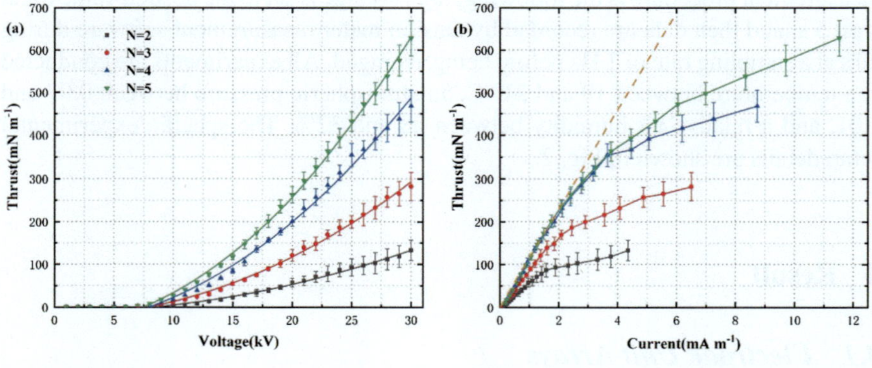

Fig. 3 Experimental data collected for the case of $s = 40$ mm and $d = 34$ mm with $N = 2$ to $N = 5$ electrode units: **a** thrust (mN·m^{-1}) versus voltage (kV) and **b** thrust versus current (mA·m^{-1})

Table 1 Thrust ratios for N = 3–5 electrode units compared to $N = 2$ at 16 kV, 24 kV, and 30 kV

Thrust ratio	$N = 3$	$N = 4$	$N = 5$
16 kV	2.168	3.636	5.037
24 kV	2.149	3.623	4.60
30 kV	2.125	3.523	4.727

$$\frac{T}{A} = \frac{T}{(Ns + b)} \quad (14)$$

where b is the width of individual collector. According to the Eqs. (4), (8) and (13), the unit thrust and thrust density at $N = 2$ is approximately 40%-44% and 53%-58% of the maximum theoretical value when the voltage is greater than 20 kV. The calculation of the ratio of thrust and thrust density relative to the theoretical maximum has some uncertainty, mainly due to errors in theoretical calculations such as the value of the ion mobility and the simplification of the way the current density is calculated and experimental errors.

As mentioned by Masuyama [13], thrust is no longer linear in current in Fig. 3b and a clear bilinear phenomenon can be observed. The thrust has a second linear state with a reduced gradient when the current exceeds a certain threshold. The orange line is an estimate of the slope when no performance degradation has occurred, where the ion mobility implied by the orange line is approximately $2.97 \times 10^{-4}\ \mathrm{m^2V^{-1}\ s^{-1}}$. We can attribute this phenomenon to the reverse current caused by the discharge of the ground electrode, resulting in an increase in current and a reduction in thrust given the opposing ion flow. There is a segmental drop in performance as the reverse currents start to be generated and these sharp degradation voltage values are consistent.

4.2 Multistage Testing

Figure 4 shows dimensionless thrust ratio T/T_{ss} and current ratio I/I_{ss} versus normalized stage spacing θ/d (θ is the distance between two stages), where T_{ss} and I_{ss} are the thrust and current of the single-stage array respectively. The single-stage array ($N = 5$) is evaluated in Sect. 4.1 with $s = 40$ mm and $d = 34$ mm. The ideal thrust of two-stage array should be twice that of single-stage array. We find that there is a similar trend between staggered distribution and collinear distribution in Fig. 4a. As θ/d increases, the overall thrust generated by two stages increases and gradually approaches its maximum value. For example, when $\theta/d \approx 0.59$ ($\theta = 20$ mm) and the applied voltage is 16 kV, the dimensionless thrust ratio is only 0.43, while when θ/d increases to close to 3, the ratio approaches 2. It has been confirmed that the reason for this trend is the generation of reverse ion wind [1]. We observed reverse spark breakdown under the condition of 20 kV when $\theta/d \approx 0.59$.

Due to the interactions between the collectors of the first stage and the emitters of the second stage, a discharge induced between stages increases the current. Figure 4b shows this trend. Similarly, the current generated by a single stage is quantified as 1. When θ/d is less than 2.0, as the θ/d increases, I/I_{ss} decreases sharply and tends to stabilize, approaching the ideal ratio of 2.0 when θ/d is greater than 2.0.

The thrust-to-power ratio will decrease with θ/d and the thrust density will increase with θ/d. We can choose superior performance while ensuring sufficient thrust by quantifying thrust-to-power and thrust density. The thrust density per unit volume is defined in relation to propulsion system total volume. For the EAD devices with

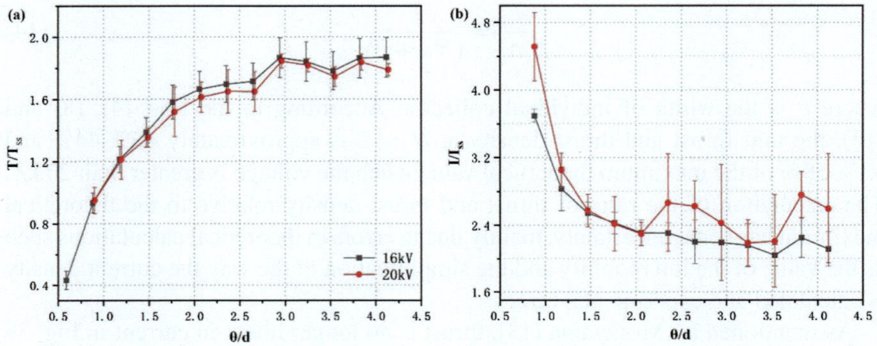

Fig. 4 Multistage dimensionless thrust T/T_{ss} (**a**) and current I/I_{ss} (**b**) as a function of normalized stage spacing θ/d

staggered distribution when K stages are considered in the EAD device, the thrust density per unit volume is defined as

$$\frac{T}{V} = \frac{T}{(Ns + b)[(K - 1)\theta + K(d + c)]} \tag{15}$$

where c is the chord length of individual collector. Substitute the parameters of our thruster configuration into the Eq. (15) to obtain the trend of thrust density change under multistage conditions in Fig. 5a. We find that the highest thrust per unit volume is achieved for a stage spacing between 50 and 100 mm, and record a small error. As the voltage increases, the peak thrust per unit volume significantly increases due to the increase in thrust. As the stage spacing θ continues to decrease, the reverse ion wind between stages has an increasing weakening effect on thrust, resulting in a decrease in thrust per unit volume. At a stage spacing of approximately $\theta/d >$ 3, the thrust per unit volume decreases given that the working volume considered increases, but thrust remains essentially the same. As expected, Fig. 5b shows that thrust-to-power increases with the gradual increase of θ/d, and the growth rate slows down and gradually stabilizes. The maximum thrust power ratio captured at 16 kV is 6.26 N/kW, with an uncertainty range of 6.05 to 6.49 N/kW.

4.3 Tethered Flight of the Simple EAD Aircraft

Our EAD aircraft uses a two-stage structure validated in Sect. 4.2. Positive high voltage is still provided by the high-voltage power supply, so it is necessary to connect and extend the lightweight wire to avoid the interference force and moment caused by wire pulling. Considering the mutual constraint relationship between thrust, thrust density per unit volume, and thrust-to-power, and taking thrust as the main evaluation index, the flight test stage spacing was chosen to be 90 mm. The primary goal of

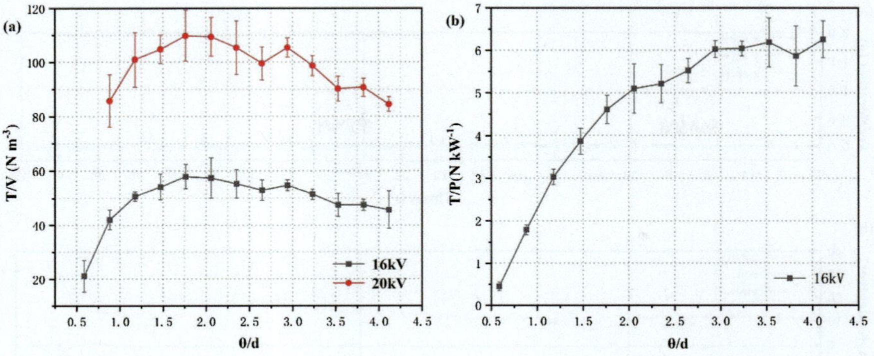

Fig. 5 Multistage thrust per unit volume T/V (**a**) and thrust to power ratio T/P (**b**) as a function of normalized stage spacing θ/d

this flight test is to control the weight of the aircraft and achieve vertical takeoff and landing flight. Hollow out the center of balsa wood with a density less than 80 kg/m^3 as a structural support for collector covered with aluminum foil. The structural shell of the aircraft is composed of two high-strength and lightweight foam plates which support 10 electrode units. All wire connections pass through the foam plates.

We place a 9 g aerial 9-axis commercial IMU away from the emitter to avoid unstable transient response or stoppage of IMU operation when high voltage is applied. Figure 6 demonstrates the attitude and acceleration data during takeoff transmitted by the on-board 9-axis IMU via Bluetooth. Figure 7 from left to right shows the images of the EAD aircraft during flight test. Liftoff occurs at approximately 24 kV, at which point the inferred takeoff weight based on multistage experimental records is approximately 20 g, which is less than the actual total weight of a simple EAD aircraft of 23.14 g. This finding is logical when the experimental uncertainty is taken into account.

The flight test process is as follows: gradually raise the voltage to 24 kV during the course of 0–10 s until a smooth takeoff by the EAD aircraft; 10–24 s, keep hovering at about 3–8 cm from the horizontal ground for observation; 24–40 s, rapidly alter voltage between 22–26 kV during this time to track the EAD aircraft's acceleration and deceleration; After 40 s, increase the voltage to spark breakdown and observe the transient effect of spark breakdown on the EAD aircraft.

Figure 6a displays the acceleration versus time curve during the 42-s experiment, whereas Fig. 6b displays the change in Euler angle (for ease of reading, the initial attitude angle of the sensor-mounted roll is set to -7.5 degrees). As can be seen, the EAD aircraft's status is stable because it does not depart the horizontal plane prior to 10 s. The aircraft displays slight changes in acceleration and attitude after hovering. Due to uncertainty moments such ion wind deflection, the acceleration swings dramatically and the yaw angle changes continually after altering the voltage. After the spark breakdown, the 3-axis acceleration rapidly shifts, and the aircraft immediately hits the ground, meaning the impact is fatal.

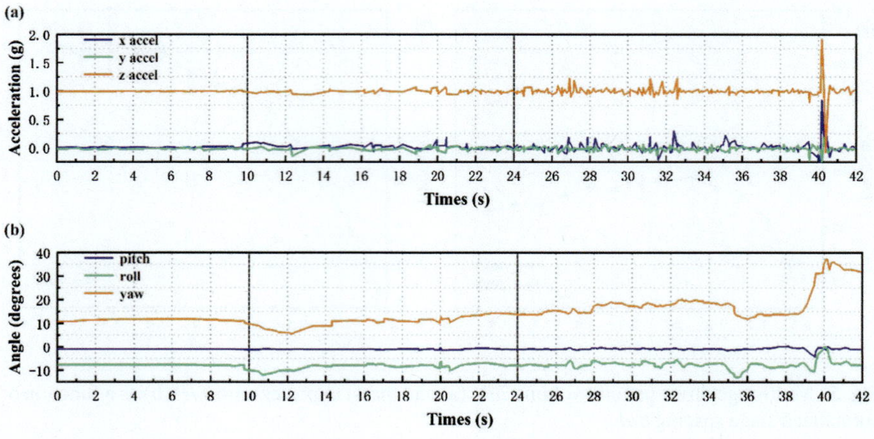

Fig. 6 Acceleration (**a**) and attitude angle (**b**) recordings of an EAD aircraft with a 9-axis commercial IMU load taking off under tethered, open-loop, uncontrolled conditions

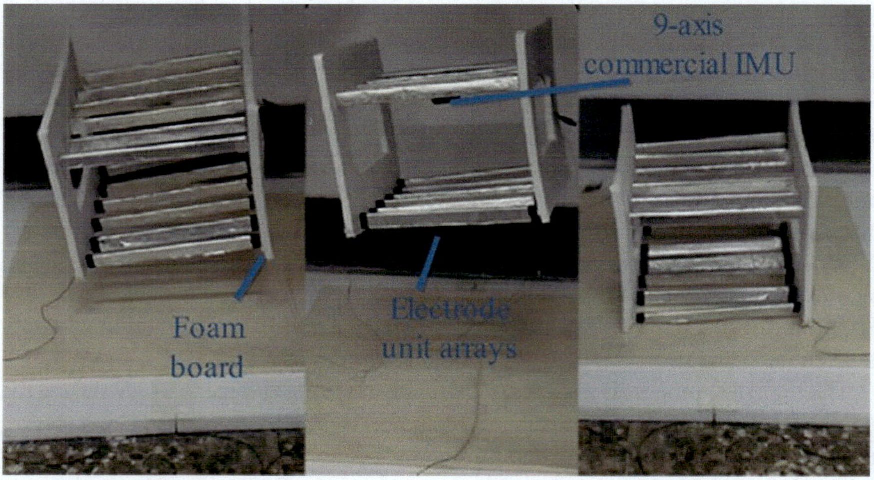

Fig. 7 Tethered flight test of the simple EAD aircraft (takeoff, hovering and landing from left to right)

5 Conclusion

Electroaerodynamic (EAD) device is a viable alternative to atmospheric propulsion. In this paper, we verify the performance of thrust and thrust density of propulsion units in parallel and multistage cases for the staggered arrangement of the electrodes. This distribution not only has less aerodynamic drag than the collinear distribution, but also produces superior thrust gains when connected in parallel. An optimal range of stage spacing exists for a multistage thruster, and the final stage spacing for the two-stage EAD aircraft was determined to be 90 mm. The EAD aircraft successfully completed a tethered, open-loop flight in the optimal configuration, displaying erratic acceleration during voltage changes and sudden changes in flying state during spark discharge.

Presently, EAD propulsion's low thrust and low thrust density have made it difficult for the technology to advance to practical applications. Future study should concentrate on developing new types of EAD propulsion and confirming the models through experiments or in-depth computational simulations to expect an order of magnitude improvement in performance. Another area for future research is how to modify the flight attitude by quantitative control of voltage. The efficacy of EAD devices is weakened by the free-stream velocity of air, according to preliminary investigations, but more thorough research is still required.

References

1. Wilson J, Perkins HD, Thompson WK (2009) An investigation of ionic wind propulsion. National Aeronautics and Space Administration (2009)
2. Khomich VY, Rebrov IE (2018) In-atmosphere electrohydrodynamic propulsion aircraft with wireless supply onboard. J Electrostat 95:1–12
3. Xu H et al (2018) Flight of an aeroplane with solid-state propulsion. Nature 563(7732):532–539
4. Drew DS et al (2018) Toward controlled flight of the ionocraft: a flying microrobot using electrohydrodynamic thrust with onboard sensing and no moving parts. IEEE Robot Autom Lett 3(4):2807–2813
5. Ieta A, Chirita M (2019) Electrohydrodynamic propeller for in-atmosphere propulsion. rotational device first flight. J Electrostatics 100
6. Chirita M, Ieta A (2022) Toroidal counter electrode for ionic propulsion. Scientific Reports 12(1)
7. Gomez-Vega N et al (2022) Model of multistaged ducted thrusters for high-thrust-density electroaerodynamic propulsion. AIAA J 61(2):767–779
8. Lai FC (2020) EHD gas pumping—a concise review of recent development. J Electrost 106
9. Franke ME, Hogue LE (1991) Electrostatic cooling of a horizontal cylinder. J Heat Transfer 113(3):544–548
10. Roth JR (2003) Aerodynamic flow acceleration using paraelectric and peristaltic electrohydrodynamic effects of a one atmosphere uniform glow discharge plasma. Phys Plasmas 10(5):2117–2126
11. Xu H, He Y, Barrett SRH (2019) A dielectric barrier discharge ion source increases thrust and efficiency of electroaerodynamic propulsion. Appl Phys Lett 114(25)
12. Monrolin N, Plouraboue F, Praud O (2017) Electrohydrodynamic thrust for in-atmosphere propulsion. AIAA J 55(12):4296–4305
13. Moreau E et al (2013) Electrohydrodynamic force produced by a wire-to-cylinder dc corona discharge in air at atmospheric pressure. J Phys D-Appl Phys 46(47)
14. Belan M et al (2021) A parametric study of electrodes geometries for atmospheric electrohydrodynamic propulsion. J Electrost 113
15. Masuyama K, Barrett SRH (2013) On the performance of electrohydrodynamic propulsion. Proc Royal Soc A 469
16. Moreau E et al (2015) Electrohydrodynamic force produced by a corona discharge between a wire active electrode and several cylinder electrodes—application to electric propulsion. J Electrostat 76:194–200

17. Gilmore CK, Barrett SRH (2015) Electrohydrodynamic thrust density using positive corona-induced ionic winds for in-atmosphere propulsion. Proceedings of the royal society a-mathematical physical and engineering sciences 471(2175)
18. Tyndall AM, Grindley GC (1926) The mobility of ions in air. Part I. Negative ions in moist air. Proc R Soc Lond A 110:341–364
19. Loeb LB (1965) Electrical coronas, their basic physics mechanisms. University of California Press. Berkeley, CA
20. Morrison RD, Hopstock DM (1979) The distribution of current in wire-to-cylinder corona. J Electrost 6:349–360
21. Coseru S, Fabre D, Plouraboue F (2021) Numerical study of ElectroAeroDynamic force and current resulting from ionic wind in emitter/collector systems. J Appl Phys 129(10)

Multi-RISs Assisted UAV-Enabled Wireless Sensor Network

Bin Duo, Aijia Shen, Yuankai Ning, Mingqian Shao, Jin Ning, Junsong Luo, and Zibin Wang

Abstract To fully take advantage of unmanned aerial vehicle (UAV)'s flight freedom in three-dimensional (3D) space, and enhance the capability of reconfigurable intelligent surface (RIS) for providing better communication services with multiple ground sensors, we design a multi-RISs assisted UAV-supported wireless sensor network (WSN). The target is to achieve the highest minimum average data collection rate by optimizing the communication scheduling of ground sensors, multi-RISs' phase-shift coefficients, and the UAV's 3D trajectory. This paper offers an iterative theorem that effectively obtains the high-quality suboptimal solution to the non-convex problem by utilizing successive convex approximation (SCA), alternating optimization (AO), and semidefinite relaxation (SDR) techniques. The results verify that the designed algorithm has significant improvement in data collection rate, and the practical significance of deploying multi-RISs and designing 3D UAV trajectory.

Keywords UAV communication · Wireless sensor network · Reconfigurable intelligent surface · Trajectory design · Passive beamforming

B. Duo (✉) · A. Shen · Y. Ning · M. Shao · J. Ning · J. Luo
College of Computer Science and Cyber Security, Chengdu University of Technology, Chengdu, China
e-mail: duobin@cdut.edu.cn

A. Shen
e-mail: aijiashen@stu.cdut.edu.cn

Y. Ning
e-mail: ningyuankai@stu.cdut.edu.cn

M. Shao
e-mail: mingqianshaw@stu.cdut.edu.cn

J. Ning
e-mail: ningjin@cdut.edu.cn

J. Luo
e-mail: luojuns@cdut.edu.cn

Z. Wang
National Key Laboratory of Science and Technology On Information System Security, Beijing, China

Z. Liu et al. (eds.), *Advances and Challenges in Advanced Unmanned Aerial Systems*, Springer Aerospace Technology, https://doi.org/10.1007/978-981-99-8045-1_2

1 Introduction

Nowadays, large-scale Internet of Things (IoT) systems are rapidly developing, and many sensors are incorporated into vehicles, industrial manufacturing, cities, and other scenarios to provide smart living [1]. Wireless sensor network (WSN) is considered to be an integral part of bringing the IoTs into reality and are already playing a key role in several applications, such as environmental monitoring, transportation, agricultural planting, and so on [2–6]. In [5], the authors used optical WSNs to detect foreign objects to ensure the safety of facilities in airports, and [6] has found more suitable routing protocols to improve WSN efficiency in agricultural scenarios. In addition, due to the technological progress in unmanned aerial vehicle (UAV) manufacturing, UAVs have been commonly applied in many fields including aerial photography, emergency rescue, and military [7]. Especially in the WSN, UAVs can provide strong assistance with enlarged wireless coverage, enhanced network throughput, and extended network lifetime, thanks to their benefits like high maneuverability and high-quality line-of-sight (LoS) communication link.

Recently, reconfigurable intelligent surface (RIS) as an advanced wireless communication technology that can reshape the environment of radio signal propagation, has received extensive attention. In short, RIS is a planar structure consisting of numerous passive reflecting units made from electromagnetic materials. Each unit possesses the capability to autonomously manipulate the amplitude and phase of incident signals [8]. By properly utilizing RIS in WSN and intelligently adjusting its reflection coefficients, the signal environment between transmitters and receivers can be flexibly reconfigured to significantly enhance communication performance and reduce power consumption [9].

According to the existing research, the application of UAVs and RISs in WSN to support the transmission of sensor information can improve the communication service quality of WSN, increase network coverage, and prolong the network life cycle. Therefore, it is natural to think that these two technologies can be combined and applied to WSN to better enhance network performance. In [10], the authors considered a RIS-assisted UAV-enabled WSN to effectively collect messages from multiple sensor nodes, and jointly optimized the RIS phase-shift and the UAV trajectory based on successive convex approximation (SCA) and block coordinate descent (BCD). Literature [11] employed RIS to aid UAVs in mitigating the detrimental effects caused by blockages in WSN and ensuring the transmission of fresh and up-to-date information. The deployment and trajectory of the UAV were jointly optimized in [12], to improve the energy efficiency of IoT data collection. Literature [13] considered urban environments and jointly designed UAV trajectory, communication scheduling, and RIS phase-shift, which indicated the importance of the elevation-distance trade-off for RIS-assisted UAV-supported WSN.

Although some related works considered the RIS-aided UAV-supported WSN, there are still some limitations and challenges that need further investigation: (1) The majority of works utilized the inaccurate LoS channel model (LCM), which ignores the practical effects of shadow fading and obstacle blocking, causing it more suitable for scenes such as rural areas; (2) The communication between ground sensors and the UAV is always blocked in urban areas by obstacles such as tall buildings, resulting in Non-LoS (NLoS) communication links. The current 2D UAV trajectory design cannot fully exploit the UAV's freedom of flight in space, failing to cooperate with RIS for effectively enhancing communication performance; (3) In a UAV-enabled WSN, the sensors are distributed over a wide area, so deploying only one RIS cannot cover all the sensors for providing reliable communications. The deployment of multiple RISs is expected to importantly enhance the performance of the UAV-enabled WSN. Nevertheless, since the phase-shift parameter setting of multiple RISs is highly related to the 3D trajectory of the UAV, their joint design is complicated to solve.

Motivated by the above limitations, this paper designs a multi-RISs aided UAV-supported WSN, to maximize the minimum average data collection rate, we cooperatively optimize the communication scheduling of sensor nodes, the phase-shift coefficients of multi-RISs, and the UAV 3D trajectory. But owing to the constructed non-convex optimization problem, an efficient iterative formula is planned, by applying alternating optimization (AO), semidefinite relaxation (SDR), and SCA techniques, to solve its high-quality suboptimal solution. The results verify that the algorithm proposed has a significant data collection rate improvement.

2 System Model

This paper considers a multi-RISs aided UAV-enabled WSN in urban environments. We employ a UAV to gather the sensing messages from the multiple ground sensors during its mission time. To efficiently serve all ground sensors and strengthen the data collection rate of the UAV, multiple RISs are placed to assist the ground-to-air transmission. This paper think over a 3D cartesian coordinate system, where the horizontal coordinates of the ground sensor nodes are denoted as $\mathbf{w}_k = [x_k, y_k]^T \in \mathbb{R}^{2\times1}, \forall k \in \mathcal{K} \triangleq \{1, \cdots, K\}$. Furthermore, RISs are installed at a fixed height z_R, and the corresponding horizontal coordinate is denoted as $\mathbf{w}_r = [x_r, y_r]^T \in \mathbb{R}^{2\times1}, \forall r \in \mathcal{R} \triangleq \{1, \cdots, R\}$, with each RIS having $M = M_{row} \times M_{col}$ passive reflective units with adjustable phases.

2.1 3D Trajectory Model of the UAV

Considered that the UAV communicates with the ground sensor nodes in different time slots for data collection, and the UAV mission time T can be divided into X adequately small time slots δ_t expressed as $T = \delta_t \mathrm{X}$, and $\mathbb{N} = \{1, \cdots, \mathrm{X}\}$ describe the set of these discrete time slots. Therefore, the UAV's 3D trajectory is denoted by $\{\mathbf{q}[\chi], z[\chi]\}, \chi \in \mathbb{N}$, where $\mathbf{q}[\chi] = [x[\chi], y[\chi]]^T$ represents the horizontal coordinates of the UAV in the χ th time slot, and $z[\chi]$ represents the UAV's vertical coordinates in the χ th time slot. To conduct periodic data collection, the UAV will fly back to the original position after accomplishing its mission and wait for the next. Therefore, UAV activities are subject to

$$\mathbf{q}[X] = \mathbf{q}[1], ||\mathbf{q}[\chi] - \mathbf{q}[\chi - 1]|| \leq S_{xy} \tag{1}$$

$$z[X] = z[1], |z[\chi] - z[\chi - 1]| \leq S_z \tag{2}$$

$$H_{\min} \leq z[\chi] \leq H_{\max} \tag{3}$$

where $\mathbf{q}[1]$ and $z[1]$ represent the UAV's initial level and perpendicular coordinates, $S_{xy} = V_{xy}\delta_t$ and $S_z = V_z\delta_t$ describe the UAV's maximum level and perpendicular distances in the χ th time slot, V_{xy} and V_z represent the UAV's maximum level and perpendicular speeds, $H_{\min}$ and $H_{\max}$ represent UAV's minimum and maximum altitudes, respectively.

2.2 Channel Modeling

Since the easy LoS channel model cannot precisely describe the shadow fading and path loss in the urban area, we use the probabilistic LoS channel model (PLCM) to represent the channel between the ground sensor and the UAV. Therefore, the LoS probability between ground sensor k and the UAV at the χ th time slot is defined by

$$P_{kU}^{L}[\chi] = (1 + c_1 \exp(-c_2(\psi_{kU}[\chi] - c_1)))^{-1} \tag{4}$$

where c_1 and c_2 depending on the environmental parameters, and

$$\psi_{kU}[\chi] = \frac{180}{\pi} arctan(\frac{z[\chi]}{||\mathbf{q}[\chi] - \mathbf{w}_k||}) \tag{5}$$

which represents the elevation angle from sensor node k to the UAV. Correspondingly, the NLoS probability from the UAV to ground sensor k at time slot χ can is denoted as $P_{kU}^{N}[\chi] = 1 - P_{kU}^{L}[\chi]$.

Thus, the channel gain between ground sensor k and the UAV at the χ th time slot in LoS is defined as $h_{kU}^{L}[\chi] = \sqrt{\rho_0 d_{kU}^{-\alpha_{kU}^{L}}}[\chi]$. Moreover, the channel gain in NLoS is defined as $h_{kU}^{N}[\chi] = \sqrt{\mu\rho_0 d_{kU}^{-\alpha_{kU}^{N}}}[\chi]$, where $d_{kU}[\chi] = \sqrt{||\mathbf{q}[\chi] - \mathbf{w}_k||^2 + (z[\chi])^2}$ represents the distance from the UAV to sensor node k at time slot χ, α_{kU}^{L} and α_{kU}^{N} represent the path-fading exponential factor in LoS and NLoS, respectively, ρ_0 and $\mu < 1$ represent the average channel gain and the additional signal attenuation coefficient in NLoS, respectively.

For the communication links between each ground sensor and different RISs, we assume that they follow Rayleigh fading, then the channel gain from sensor node k to RIS r at time slot χ is denoted as $g_{kr} = \sqrt{\rho_0 d_{kr}^{-\kappa}}\tilde{h}, \forall k \in \mathcal{K}, r \in \mathcal{R}$, where $d_{kr} = \sqrt{||\mathbf{w}_k - \mathbf{w}_r||^2 + z_r^2}$ represents the distance from RIS r to ground sensor k, $\tilde{h}$ represents the circularly symmetric complex gaussian (CSCG) random variable with zero mean and unit variance, and κ represents the path-fading exponential factor for Rayleigh fading. Therefore, the channel model between sensor node k and RIS r is denoted by

$$\begin{aligned}\mathbf{h}_{kr} = g_{kr}&\left[1, e^{-j\frac{2\pi d}{\lambda}\cos\phi_{kr}\sin\varphi_{kr}}, \cdots, e^{-j\frac{2\pi d(M_{row}-1)}{\lambda}\cos\phi_{kr}\sin\varphi_{kr}}\right]^T \\ &\otimes\left[1, e^{-j\frac{2\pi d}{\lambda}\sin\phi_{kr}\sin\varphi_{kr}}, \cdots, e^{-j\frac{2\pi d(M_{col}-1)}{\lambda}\sin\phi_{kr}\sin\varphi_{kr}}\right]^T\end{aligned} \tag{6}$$

where d and λ represent the spacing between the reflective units of the RIS and the signal carrier wavelength, respectively. Moreover $\cos\phi_{kr}\sin\varphi_{kr} = \frac{x_r - x_k}{d_{kr}}$ and $\sin\phi_{kr}\sin\varphi_{kr} = \frac{z_r}{d_{kr}}$, where ϕ_{kr} and φ_{kr} represent the angles of arrival (AoA) azimuth angle and elevation angle of the signal between ground sensor k and RIS r at time slot χ, respectively.

For the ground communication link between the RISs and the UAV, we also adopt the PLCM, whose LoS probability is denoted as

$$P_{rU}^L[\chi] = (1 + c_1\exp(-c_2(\psi_{rU}[\chi] - c_1)))^{-1} \tag{7}$$

where $\psi_{rU}[\chi]$ is the elevation angle from the UAV to RIS r in the χ th time slot, which is denoted as

$$\psi_{rU}[\chi] = \frac{180}{\pi}arctan(\frac{z[\chi] - z_r}{||\mathbf{q}[\chi] - \mathbf{w}_r||}) \tag{8}$$

Similarly, its NLoS probability is given by $P_{rU}^N[\chi] = 1 - P_{rU}^L[\chi]$. Therefore, the channel gain from the UAV to RIS r at the time slot χ for the LoS state is expressed as $\mathbf{h}_{rU}^L[\chi] = \sqrt{\rho_0 d_{rU}^{-\alpha_{rU}^L}[\chi]}\mathbf{h}_{rU}^{LoS}[\chi]$, where $d_{rU}[\chi] = \sqrt{||\mathbf{q}[\chi] - \mathbf{w}_r||^2 + (z[\chi] - z_r)^2}$ represents the distance from the UAV to RIS r at the χ th time slot, α_{rU}^L represents the corresponding path-fading exponential factor, and

$$\begin{aligned}\mathbf{h}_{rU}^{LoS}[\chi] = &\left[1, e^{-j\frac{2\pi d}{\lambda}\cos\phi_{rU}[\chi]\sin\varphi_{rU}[\chi]}, \cdots, e^{-j\frac{2\pi d(M_{row}-1)}{\lambda}\cos\phi_{rU}[\chi]\sin\varphi_{rU}[\chi]}\right]^T \\ &\otimes\left[1, e^{-j\frac{2\pi d}{\lambda}\sin\phi_{rU}[\chi]\sin\varphi_{rU}[\chi]}, \cdots, e^{-j\frac{2\pi d(M_{col}-1)}{\lambda}\sin\phi_{rU}[\chi]\sin\varphi_{rU}[\chi]}\right]^T\end{aligned} \tag{9}$$

where $\cos\phi_{rU}[\chi]\sin\varphi_{rU}[\chi] = \frac{x[\chi]-x_r}{d_{rU}[\chi]}$, $\sin\phi_{rU}[\chi]\sin\varphi_{rU}[\chi] = \frac{z[\chi]-z_r}{d_{rU}[\chi]}$, $\varphi_{rU}[\chi]$ and $\phi_{rU}[\chi]$ are the angles of departure (AoD) azimuth and elevation angles of the signal from the UAV to RIS r at time slot χ, respectively. Because of the multipath fading effect, the channel gain for the NLoS state from the UAV to RIS r is modeled by $\mathbf{h}_{rU}^{N}[\chi] = \sqrt{\rho_0 d_{rU}^{-\alpha_{rU}^{N}}[\chi]}\mathbf{h}_{ss}$, where α_{rU}^{N} denotes the corresponding path-fading exponential factor and $\mathbf{h}_{ss} \in \mathbb{C}^{M\times 1}$ is denoted as a CSCG random vector.

2.3 Data Collection Modeling

For the communication between ground sensors and the UAV, we use the Time Division Multiple Access (TDMA) protocol, which satisfies the constraints as follows:

$$\sum_{k=1}^{K} a_k[\chi] \leq 1 \tag{10}$$

$$a_k[\chi] \in \{0, 1\} \tag{11}$$

when $a_k[\chi] = 1$, sensor node k communicates with the UAV at time slot χ, otherwise $a_k[\chi] = 0$. The diagonal phase-shift matrix of RIS r at the time slot χ is defined as $\mathbf{\Theta}^r[\chi] = \text{diag}\{\beta_1^r e^{j\theta_1^r[\chi]}, \cdots, \beta_M^r e^{j\theta_M^r[\chi]}\}, \forall r \in \mathcal{R}, \chi \in \mathbb{N}$, where $\beta_m^r \in [0, 1], m \in \mathcal{M} = \{1, \cdots, M\}$ and $\theta_m^r[\chi] \in [0, 2\pi)$ denote the reflection phase and the reflection amplitude coefficient of the m th unit of RIS r, respectively. To obtain the maximum reflection amplitude gain, we let $|\beta_m^r| = 1, \forall m, r$. Due to the large path loss, only the signals reflected by the RIS for the first time are considered, and those reflected twice or more times are ignored [9]. Thus, the achievable rate for the UAV data collection in the χ th time slot for LoS and NLoS states is written as

$$R_k^C[\chi] = \log_2\left(1 + \gamma_0 \left| h_{kU}^{f}[\chi] + \sum_{r=1}^{R} (\mathbf{h}_{rU}^{o_r}[\chi])^H \mathbf{\Theta}^r[\chi]\mathbf{h}_{kr} \right|^2\right), f, o_r \in \{L, N\} \tag{12}$$

where $C = f o_1 \cdots o_R$ whose number of state combinations is 2^{R+1} represents the sequence in this time, and $\gamma_0 = P/\sigma^2$, where σ^2 represents the Additive White Gaussian Noise (AWGN) power of the receiver and P denotes the maximum transmission power of the sensor node.

Thus, the expected achievable rate between sensor node k and the UAV in LoS and NLoS at time slot χ is calculated by

$$\overline{R}_k[\chi] = \sum P_{kU}^{f}[\chi](\prod_{r=1}^{R} P_{rU}^{o_r}[\chi]) R_k^C[\chi], f, o_r \in \{L, N\} \tag{13}$$

3 Problem Formulation

By cooperatively optimizing the communication schedule $\mathbf{A} \triangleq \{a_k[\chi]\}$, the phase-shifts $\boldsymbol{\Theta} \triangleq \{\theta_m^r[\chi]\}$ for multi-RISs, the UAV horizontal trajectory $\mathbf{Q} \triangleq \{\mathbf{q}[\chi]\}$, and the UAV vertical trajectory $\mathbf{Z} \triangleq \{\mathbf{z}[\chi]\}$,where $m \in \mathcal{M}, r \in \mathcal{R}, k \in \mathcal{K}, \chi \in \mathbb{N}$, to maximize the minimum average data collection rate, which is formulated as

$$\max_{\mathbf{A},\boldsymbol{\Theta},\mathbf{Q},\mathbf{Z},\eta} \quad \eta \tag{14}$$

$$\text{s.t.} \quad \frac{1}{X}\sum_{\chi=1}^{X} a_k[\chi]\overline{R}_k[\chi] \geq \eta \tag{15}$$

$$0 \leq \theta_m^r[\chi] < 2\pi \tag{16}$$

$$(1)\text{–}(3), (5), (8), (10), (11)$$

Although the constraints (1)–(3), (10), and (16) are convex, the optimization problem is difficult to obtain because of the following reasons: (1) The optimization problem (14) is not joint convex with respect to $\mathbf{A}$, $\boldsymbol{\Theta}$, $\mathbf{Q}$, and $\mathbf{Z}$; (2) Eq. (15) is a highly complex constraint, including the RIS phase-shift $\boldsymbol{\Theta}$, the LoS probability of (4) and (7), and the achievable rate in LoS and NLoS states, which are related to $\mathbf{Q}$ and $\mathbf{Z}$; (3) Eqs. (5) and (8) are non-affine; (4) The main diagonal element of the RIS phase-shift matrix $\boldsymbol{\Theta}$ is the unit modulus constraint; (5) Eqs. (10) and (11) lead to problem (14) being a mixed-integer non-convex optimization problem. Therefore, an algorithm is proposed to solve this difficulty in the following section.

4 Proposed Solution

Based on AO, SDR, and SCA techniques, an efficient iterative theorem is proposed to get the sub-optimal solution to (14) in this part.

4.1 Optimization of Sensor Node Communication Scheduling

Given any feasible $\boldsymbol{\Theta}$,$\mathbf{Q}$ and $\mathbf{Z}$, the binary variable $a_k[\chi]$ in (11) can be relaxed to a continuous variable, and problem (14) can be expressed as

$$\max_{\mathbf{A},\eta} \quad \eta \tag{17}$$

$$\text{s.t.} \quad 0 \leq a_k[\chi] \leq 1 \tag{18}$$

$$(10), (15)$$

After such a transformation, problem (17) is a standard linear programming problem, by using CVX, it is efficiently solved [14].

4.2 Optimization of Multi-RISs Phase-Shift Matrix

Given $\mathbf{A}$ obtained from problem (17) and any feasible $\mathbf{Q}$ and $\mathbf{Z}$, the optimization problem of the RIS phase-shift matrix $\boldsymbol{\Theta}$ is reconstructed as

$$\max_{\boldsymbol{\Theta},\eta} \quad \eta \tag{19}$$

$$\text{s.t.} \quad (15), (16)$$

(19) is a non-convex optimization problem that is challenging to compute. Therefore, an AO algorithm is introduced to solve it. We decompose $\boldsymbol{\Theta}$ into different RIS phase-shift matrix variables which is given by $\boldsymbol{\Theta} = \{\boldsymbol{\Theta}^1, \cdots, \boldsymbol{\Theta}^r, \cdots, \boldsymbol{\Theta}^R, \forall r \in \mathcal{R}\}$. Because the different $\boldsymbol{\Theta}^r$ are independent of each other without coupling, the optimization problem (19) is equivalent to solving the optimization problem for a single RIS.

Therefore, according to [13], (19) is rewritten as

$$\max_{\boldsymbol{\Xi},\eta} \quad \eta \tag{20}$$

$$\begin{aligned} \text{s.t.} \quad & \frac{1}{X}\sum_{\chi=1}^{X} \alpha_k[\chi]\Big(P_{kU}^L[\chi]P_{RU}^L[\chi]\tilde{R}_k^{LL}[\chi] + P_{kU}^L[\chi]P_{RU}^N[\chi]\tilde{R}_k^{LN}[\chi] \\ & + P_{kU}^N[\chi]P_{RU}^L[\chi]\tilde{R}_k^{NL}[\chi] + P_{kU}^N[\chi]P_{RU}^N[\chi]\tilde{R}_k^{NN}[\chi]\Big) \geq \eta \end{aligned} \tag{21}$$

$$\boldsymbol{\Xi}[\chi] \succcurlyeq 0 \tag{22}$$

$$\boldsymbol{\Xi}_{r,r}[\chi] = 1, r = 1, 2, \cdots, M+1 \tag{23}$$

where $\tilde{R}_k^{LL}[\chi] = \log_2(1 + \gamma_0 \mathrm{Tr}(\boldsymbol{\Xi}[\chi]\mathbf{A}_1[\chi]))$, $\tilde{R}_k^{LN}[\chi] = \log_2(1 + \gamma_0 \mathrm{Tr}(\boldsymbol{\Xi}[\chi]\mathbf{A}_2[\chi]))\tilde{R}_k^{NL}[\chi] = \log_2(1+\gamma_0 \mathrm{Tr}(\boldsymbol{\Xi}[\chi]\mathbf{A}_3[\chi]))$, and $\tilde{R}_k^{NN}[\chi] = \log_2(1+\gamma_0 \mathrm{Tr}(\boldsymbol{\Xi}[\chi]\mathbf{A}_4[\chi]))$.

Therefore, the optimization problem (20) is a standard Semidefinite Program (SDP) problem, so we use CVX to obtain the optimal solution. In general, the relaxed optimization problem (20) may not get a solution with a rank equal to 1, which means that its optimal solution is only an upper bound. Therefore, we use a Gaussian randomization method to construct $\boldsymbol{\xi}[\chi]$ from high rank $\boldsymbol{\Xi}[\chi]$[15].

4.3 Optimization of UAV Horizontal Trajectory

Given $\mathbf{A}$ obtained from problem (17), $\boldsymbol{\Theta}$ obtained from problem (20), and any feasible $\mathbf{Z}$, the optimization problem of UAV horizontal trajectory is expressed as

$$\max_{\mathbf{Q},\eta} \quad \eta \tag{24}$$

$$\text{s.t.} \ (1), (5), (8), (15)$$

Due to the difficulties of solving problem (24), we introduce a relaxation transformation as follows [16]:

$$\alpha \log \zeta + \beta \leq \log(1+\zeta) \tag{25}$$

when $\zeta = \zeta_0$, the inequality above is strictly tightened and the approximation constants α and β are denoted as denoted as $\alpha = \frac{\zeta_0}{1+\zeta_0}$, $\beta = \log(1+\zeta_0) - \frac{\zeta_0}{1+\zeta_0}\log \zeta_0$.

According to (25), the lower bound of $R_k^C[\chi]$ can be obtained

$$\begin{aligned} &\log_2\left(1+\gamma_0\left|h_{kU}^f[\chi] + \sum_{r=1}^{R} (\mathbf{h}_{rU}^{o_r}[\chi])^H \boldsymbol{\Theta}^r[\chi]\mathbf{h}_{kr}\right|^2\right) \\ &\geq \alpha[\chi]\log_2\left(\gamma_0\left|h_{kU}^f[\chi] + \sum_{r=1}^{R} (\mathbf{h}_{rU}^{o_r}[\chi])^H \boldsymbol{\Theta}^r[\chi]\mathbf{h}_{kr}\right|^2\right) + \beta[\chi] \end{aligned} \tag{26}$$

where $\alpha[\chi]$ and $\beta[\chi]$ are denoted as $\alpha[\chi] = \frac{\zeta_0[\chi]}{1+\zeta_0[\chi]}$, $\beta[\chi] = \log(1+\zeta_0[\chi]) - \frac{\zeta_0[\chi]}{1+\zeta_0[\chi]}\log \zeta_0[\chi]$, and $\zeta[\chi] = \gamma_0\left|h_{kU}^f[\chi] + \sum_{r=1}^{R} (\mathbf{h}_{rU}^{o_r}[\chi])^H \boldsymbol{\Theta}^r[\chi]\mathbf{h}_{kr}\right|^2$.

Similarly, the lower bound of (12) is expressed as

$$\breve{R}_k^C[\chi] \triangleq 2\alpha[\chi]S^C[\chi] + \alpha[\chi]\log_2(\gamma_0) + \beta[\chi] \tag{27}$$

where $S^C[\chi] = \log_2\left(h_{kU}^f[\chi] + \sum_{r=1}^{R} (\mathbf{h}_{rU}^{o_r}[\chi])^H \boldsymbol{\Theta}^r[\chi]\mathbf{h}_{kr}\right)$.

To solve the complex non-convex variables $\mathbf{h}_{rU}^{LoS}[\chi]$ in (9), the $(l-1)$th iterative trajectory is used to approximate the trajectory of the l th iteration, and obtain an approximate $\mathbf{h}_{rU}^{LoS,l-1}[\chi]$. Then, $S^C[\chi]$ can be reconstructed as

$$S^C[\chi] = \log_2\left(\frac{B_{kU}^f}{d_{kU}^{\alpha_{kU}^f/2}[\chi]} + \sum_{r=1}^{R} \frac{B_{krU}^{o_r}[\chi]}{d_{rU}^{\alpha_{rU}^{o_r}/2}[\chi]}\right), f, o_r \in \{L, N\} \tag{28}$$

where $B_{kU}^{L} = \sqrt{\rho_0}$, $B_{krU}^{L}[\chi] = \sqrt{\rho_0}(\mathbf{h}_{rU}^{LoS,l-1}[\chi])^H \mathbf{\Theta}^r[\chi]\mathbf{h}_{kr}$, $B_{kU}^{N} = \sqrt{\mu\rho_0}$ and $B_{krU}^{N}[\chi] = \sqrt{\rho_0}(\mathbf{h}_{ss})^H \mathbf{\Theta}^r[\chi]\mathbf{h}_{kr}$.

Furthermore, $\overline{R}_k[\chi]$ can be defined as

$$\overline{R}_k[\chi] \triangleq \sum \Omega^C[\chi] \tag{29}$$

where $\Omega^C[\chi] = \frac{1}{g^f[\chi]} \prod_{r=1}^{R} \frac{1}{t_r^{o_r}[\chi]} \breve{R}_k^C[\chi]$, $f, o_r \in \{L, N\}$, $g^L[\chi] = 1 + c_1 \exp(-c_2(\psi_{kU}[\chi] - c_1))$, $g^N[\chi] = 1 + c_1^{-1} \exp(c_2(\psi_{kU}[\chi] - c_1))$, $t_r^L[\chi] = 1 + c_1 \exp(-c_2(\psi_{rU}[\chi] - c_1))$, and $t_r^N[\chi] = 1 + c_1^{-1} \exp(c_2(\psi_{rU}[\chi] - c_1))$.

Therefore, (24) is reformulated as

$$\max_{\mathbf{Q},\eta} \quad \eta \tag{30}$$

$$\text{s.t.} \quad \frac{1}{\mathrm{X}} \sum_{\chi=1}^{\mathrm{X}} \alpha_k[\chi] \sum \Omega^C[\chi] \geq \eta \tag{31}$$

$$(1), (5), (8)$$

By introducing the relaxation variables $\mathbf{g} \triangleq \{g^L[\chi], g^N[\chi]\}$, $\mathbf{t} \triangleq \{t_r^L[\chi], t_r^N[\chi]\}$, $\mathbf{u} \triangleq \{u[\chi]\}$ and $\mathbf{v} \triangleq \{v[\chi]\}$, the optimization problem (30) is reconstructed as

$$\max_{\mathbf{Q},\mathbf{\Omega},\mathbf{\Lambda},\mathbf{\Psi}_{kU},\mathbf{\Psi}_{rU},\mathbf{S},\mathbf{g},\mathbf{t},\mathbf{u},\mathbf{v},\eta} \quad \eta \tag{32}$$

$$\text{s.t.} \quad \Omega^C[\chi] \leq \frac{1}{g^f[\chi]} \prod_{r=1}^{R} \frac{1}{t_r^{o_r}[\chi]} \Lambda^C[\chi] \tag{33}$$

$$\Lambda^C[\chi] \leq 2\alpha[\chi] S^C[\chi] + \alpha[\chi] \log_2(\gamma_0) + \beta[\chi] \tag{34}$$

$$S^C[\chi] \leq \log_2\left(\frac{B_{kU}^f}{u^{\alpha_{kU}^f/2}[\chi]} + \sum_{r=1}^{R} \frac{B_{krU}^{o_r}[\chi]}{v^{\alpha_{rU}^{o_r}/2}[\chi]} \right) \tag{35}$$

$$g^L[\chi] \geq 1 + c_1 \exp(-c_2(\psi_{kU}^L[\chi] - c_1)) \tag{36}$$

$$g^N[\chi] \geq 1 + \frac{1}{c_1} \exp(c_2(\psi_{kU}^N[\chi] - c_1)) \tag{37}$$

$$t_r^L[\chi] \geq 1 + c_1 \exp(-c_2(\psi_{rU}^L[\chi] - c_1)) \tag{38}$$

$$t_r^N[\chi] \geq 1 + \frac{1}{c_1} \exp(c_2(\psi_{rU}^N[\chi] - c_1)) \tag{39}$$

$$\psi_{kU}^{L}[\chi] \leq \frac{180}{\pi}\tan^{-1}\left(\frac{z[\chi]}{||\mathbf{q}[\chi]-\mathbf{w}_k||}\right) \tag{40}$$

$$\psi_{kU}^{N}[\chi] \geq \frac{180}{\pi}\tan^{-1}\left(\frac{z[\chi]}{||\mathbf{q}[\chi]-\mathbf{w}_k||}\right) \tag{41}$$

$$\psi_{rU}^{L}[\chi] \leq \frac{180}{\pi}\tan^{-1}\left(\frac{z[\chi]-z_r}{||\mathbf{q}[\chi]-\mathbf{w}_r||}\right) \tag{42}$$

$$\psi_{rU}^{N}[\chi] \geq \frac{180}{\pi}\tan^{-1}\left(\frac{z[\chi]-z_r}{||\mathbf{q}[\chi]-\mathbf{w}_r||}\right) \tag{43}$$

$$d_{kU}[\chi] \leq u[\chi] \tag{44}$$

$$d_{rU}[\chi] \leq v[\chi] \tag{45}$$

(1), (31).

where $\mathbf{\Psi}_{kU}[\chi] = \{\psi_{kU}^{L}[\chi], \psi_{kU}^{N}[\chi]\}$ and $\mathbf{\Psi}_{rU}[\chi] = \{\psi_{rU}^{L}[\chi], \psi_{rU}^{N}[\chi]\}$ denote elevation-related auxiliary variables, $\mathbf{\Lambda} = \{\Lambda^{C}[\chi]\}$, $\mathbf{\Omega} = \{\Omega^{C}[\chi]\}$ and $\mathbf{S} = \{S^{C}[\chi]\}$ denote rate-related auxiliary variables.

Using the monotonic property of the logarithmic function, (33) can be transformed into

$$\ln(\Lambda^{C}[\chi]) \geq \sum_{r=1}^{R}\ln(t_r^{o_r}[\chi]) + \ln(g^{f}[\chi]) + \ln(\Omega^{C}[\chi]) \tag{46}$$

Although (46) is still non-convex, each term on its left is concave with its corresponding variable. As a result, for the local points $\mathbf{\Omega}^{l} = \{\Omega^{C,l}[\chi]\}$, $\mathbf{g}^{l} = \{g^{f,l}[\chi]\}$ and $\mathbf{t}^{l} = \{t^{o_r,l}[\chi]\}$ at the l th iteration, where $o_r, f \in \{L, N\}$, using a first-order Taylor expansion can obtain their corresponding upper bounds, as follows

$$\begin{aligned}\ln(\Lambda^{C}[\chi]) \geq & \sum_{r=1}^{R}\left(\ln(t^{o_r,l}[\chi]) + \tfrac{1}{t^{o_r,l}[\chi]}(t^{o_r}[\chi]-t^{o_r,l}[\chi])\right) \\ & + \ln(g^{f,l}[\chi]) + \tfrac{1}{g^{f,l}[\chi]}(g^{f}[\chi]-g^{f,l}[\chi]) \\ & + \ln(\Omega^{C,l}[\chi]) + \tfrac{1}{\Omega^{C,l}[\chi]}(\Omega^{C}[\chi]-\Omega^{C,l}[\chi])\end{aligned} \tag{47}$$

Note that since $A_0 > 0$, $A_{r0} > 0$, $x > 0$ and $y_r > 0$, $f(g(x, y_1, \cdots, y_R)) = \log_2\left(\frac{A_0}{x^{\alpha/2}} + \sum_{r=1}^{R}\frac{A_{r0}}{y_r^{\beta/2}}\right)$ is convex, the SCA technique is used to approximate (35) to obtain its lower bound. Through the first-order Taylor expansion at local points $\mathbf{v}^{l} = \{v^{l}[\chi]\}$ and $\mathbf{u}^{l} = \{u^{l}[\chi]\}$, i.e.,

$$S^C[\chi] \le \log_2 A^{C,l}[\chi] + \frac{B^{f,l}[\chi]}{A^{C,l}[\chi]\ln 2}(u[\chi] - u^l[\chi]) + \sum_{r=1}^{R}\left(\frac{C^{o_r,l}[\chi]}{A^{C,l}[\chi]\ln 2}(v[\chi] - v^l[\chi])\right) \tag{48}$$

where $A^{C,l}[\chi] = \frac{B^f_{kU}}{(u^l[\chi])^{\alpha^f_{kU}/2}} + \sum_{r=1}^{R} \frac{B^{o_r}_{krU}[\chi]}{(v^l[\chi])^{\alpha^{o_r}_{rU}/2}}$, $B^{f,l}[\chi] = -\frac{\alpha^f_{kU} B^f_{kU}}{2(u^l[\chi])^{(\alpha^f_{kU}+2)/2}}$, $C^{o_r,l}_r[\chi] = -\frac{\alpha^{o_r}_{rU} B^{o_r}_{krU}}{2(v^l[\chi])^{(\alpha^{o_r}_{rU}+2)/2}}$.

Since the right terms of (40) and (42) are convex with respect to $||\mathbf{q}[\chi] - \mathbf{w}_k||$ and $||\mathbf{q}[\chi] - \mathbf{w}_r||$ respectively and the right terms of (41) and (43) are concave with respect to $\mathbf{q}[\chi]$, using first-order Taylor expansion at the local point $\mathbf{Q}^l = \{\mathbf{q}^l[\chi]\}$, the following constrain transformation can be obtained

$$\psi^L_{kU}[\chi] \le D^l_1[\chi] - D^l_2[\chi](||\mathbf{q}[\chi] - \mathbf{w}_k|| - ||\mathbf{q}^l[\chi] - \mathbf{w}_k||) \tag{49}$$

$$\psi^N_{kU}[\chi] \ge D^l_1[\chi] - \frac{D^l_2[\chi]}{||\mathbf{q}^l[\chi] - \mathbf{w}_k||}(\mathbf{q}^l[\chi] - \mathbf{w}_k)^T(\mathbf{q}[\chi] - \mathbf{q}^l[\chi]) \tag{50}$$

$$\psi^L_{rU}[\chi] \le E^l_1[\chi] - E^l_2[\chi](||\mathbf{q}[\chi] - \mathbf{w}_r|| - ||\mathbf{q}^l[\chi] - \mathbf{w}_r||) \tag{51}$$

$$\psi^N_{rU}[\chi] \ge E^l_1[\chi] - \frac{E^l_2[\chi]}{||\mathbf{q}^l[\chi] - \mathbf{w}_r||}(\mathbf{q}^l[\chi] - \mathbf{w}_r)^T(\mathbf{q}[\chi] - \mathbf{q}^l[\chi]) \tag{52}$$

where $D^l_1[\chi] = \frac{180}{\pi}\tan^{-1}(\frac{z[\chi]}{||\mathbf{q}^l[\chi]-\mathbf{w}_k||})$, $D^l_2[\chi] = \frac{180}{\pi}\frac{z[\chi]}{||\mathbf{q}^l[\chi]-\mathbf{w}_k||^2+(z[\chi])^2}$, $E^l_1[\chi] = \frac{180}{\pi}\tan^{-1}(\frac{z[\chi]-z_r}{||\mathbf{q}^l[\chi]-\mathbf{w}_r||})$, and $E^l_2[\chi] = \frac{180}{\pi}\frac{z[\chi]-z_r}{||\mathbf{q}^l[\chi]-\mathbf{w}_r||^2+(z[\chi]-z_r)^2}$.

Similarly, at local points $\mathbf{v}^l = \{v^l[\chi]\}$ and $\mathbf{u}^l = \{u^l[\chi]\}$, using the first-order Taylor expansion on $u^2[\chi]$ and $v^2[\chi]$, we can obtain $-u^2[\chi] \le (u^l[\chi])^2 - 2u^l[\chi]u[\chi]$, $-v^2[\chi] \le (v^l[\chi])^2 - 2v^l[\chi]v[\chi]$.

Thus, (32) is converted to the convex problem as follows

$$\max_{\mathbf{Q},\mathbf{\Omega},\mathbf{\Lambda},\mathbf{\Psi}_{kU},\mathbf{\Psi}_{rU},\mathbf{S},\mathbf{g},\mathbf{t},\mathbf{u},\mathbf{v},\eta} \quad \eta \tag{53}$$

$$\text{s.t.} \quad (d_{kU}[\chi])^2 + (u^l[\chi])^2 - 2u^l[\chi]u[\chi] \le 0 \tag{54}$$

$$(d_{rU}[\chi])^2 + (v^l[\chi])^2 - 2v^l[\chi]v[\chi] \le 0 \tag{55}$$

$$(1), (31), (36)–(39), (47), (48), (49)–(52)$$

(53) is a standard convex optimization problem, using CVX, it is effectively solved.

4.4 Optimization of UAV Vertical Trajectory

Given **A**, **Θ**, **Q** obtained from (53), the optimization problem of the UAV vertical trajectory **Z** is given by

$$\max_{\mathbf{Z},\eta} \quad \eta \tag{56}$$

$$\text{s.t. } (2), (3), (5), (8), (15)$$

Since this problem has a similar form to the optimization problem of the UAV horizontal trajectory, following the process, this UAV vertical trajectory optimization problem can be solved. Due to space limitations, we omit this part of the derivation.

5 Numerical Results

To verify the proposed data collection system based on the joint design of multi-RISs and UAV 3D trajectories under the PLCM (marked as JT&mRIS-PLCM), we compare it with the following benchmark schemes: (1) The joint design scheme of the multi-RISs and UAV 3D trajectory based on the LoS channel model (marked as JT&mRIS-LCM); (2) The joint design scheme of the multi-RISs and UAV horizontal trajectory based on the PLCM, where the UAV's height is fixed as the lowest height (marked as J(2D)T&mRIS-PLCM); (3) The joint design scheme of UAV 3D trajectory based on the PLCM, which is no RIS assist communication (marked as T/RIS-PLCM). We consider a 200 × 300 m^2 rectangular area in the simulation, where $K = 4$ ground sensors are randomly deployed. Table 1 list other main simulation parameters.

We consider the convergence performance of Algorithm 1 under different T, as shown in Fig. 1. The max–min rate enlarges with the number of iterations until convergence after about 25 iterations. Moreover, as T increases, the performance of maximizing the minimum rate has a significant improvement, because the UAV can obtain a longer communication time with each sensor node.

Figure 2 compares the UAV trajectories and max–min rate performance with different scenarios when $T = 120$, $R = 3$ and $M = 500$. In the T/RIS-PLCM, the UAV will fly to each sensor and hover at its lowest altitude. The reason is that it can not only maintain the maximum elevation angle between the UAV and the sensor node to increase the LoS probability, but also reduce the path fading loss. Without consideration of the elevation angle, in the JT&mRIS-LCM scheme, the UAV flies close to the RIS or hovers at a certain position between the sensor node and the RIS at the lowest altitude. The reason is that the path loss between the sensor nodes and the UAV can be kept minimum when the UAV flies at the minimum altitude. Furthermore, because the RISs can provide the cascade channel gain, the UAV will hover at the optimal hovering location to gain from direct transmission links and cascade channels simultaneously. Compared with the JT&mRIS-LCM scheme, in

Table 1 Simulation parameters

Symbol	Physical meaning	Simulation value
$V_{xy},\ V_z$	UAV maximum horizontal and vertical velocity	15 m/s, 10 m/s
δ_t	Time slot length	1 s
$H_{min},\ H_{max}$	UAV minimum and maximum height	60 m, 150 m
z_R	RIS deployment height	30 m
$\alpha_{rU}^L,\ \alpha_{rU}^N$	The path-fading exponential factor in LoS and NLoS between RISs and UAV	2.2, 3.2
$\alpha_{kU}^L,\ \alpha_{kU}^N$	The path-fading exponential factor in LoS and NLoS between sensor nodes and the UAV	2.7, 3.7
ρ_0	Average channel gain	−30 dB
c_1, c_2	Environment constants	11.95, 0.14
κ	The path loss exponent for Rayleigh fading channels	2.2
σ^2	Additive white Gaussian noise	−80 dBm
P	Transmit power of sensor nodes	0.01 W
d	The reflective unit interval of RIS	$\lambda/2$
ε	Iteration accuracy	0.0001
μ	The additional signal attenuation factor	−20 dB

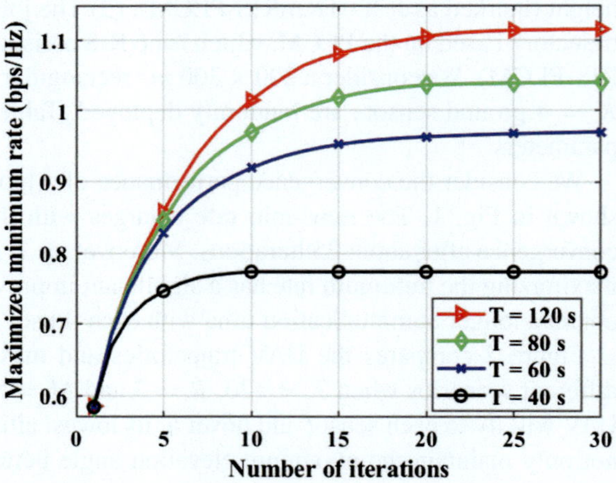

Fig. 1 Max–min rate versus the number of iterations

the J(2D)T&mRIS-PLCM, the UAV hovers very close to the RISs, the reason is that the reflective channel gain between the RISs and the UAV is more dominant, and the UAV hovering over the RISs can achieve the maximum elevation angle and the shortest path with the RISs, resulting in better communication quality. For the JT&mRIS-PLCM scheme, in Fig. 2b, the UAV makes full use of the degrees of

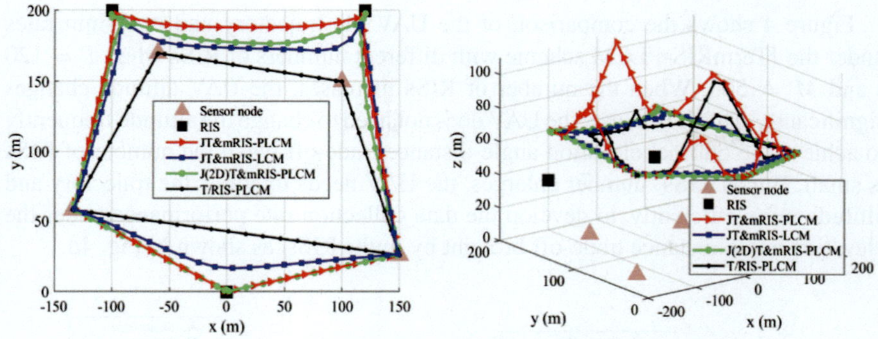

(a) 2D trajectory of the UAV (b) 3D trajectory of the UAV

Fig. 2 Comparison of the max–min rates and UAV trajectories under various schemes

freedom in the 3D space, and adjusts its vertical trajectory more flexibly and makes a trade-off between elevation and distance, leading to significant performance gains.

As shown in Fig. 3, the JT&mRIS-PLCM scheme proposed is better than other benchmark scenarios. Specifically, the data collection rate of the JT&mRIS-LCM scheme is the worst, which means that in multi-RIS aided UAV-enabled WSN, it is practically significant to adopt a more accurate channel model. The J(2D)T&mRIS-PLCM scheme has superior performance to the T/RIS-PLCM scheme, which indicates that the RISs can provide important assistance for UAV data collection. Finally, the JT&mRIS-PLCM scheme has better performance than the J(2D)T&mRIS-PLCM scheme. The reason is that the (3D)Multi-RIS-PLCM scheme can better trade-off the elevation angle and distance, thereby improving the communication quality between UAV and sensor nodes.

Fig. 3 The max–min rate versus T under different schemes

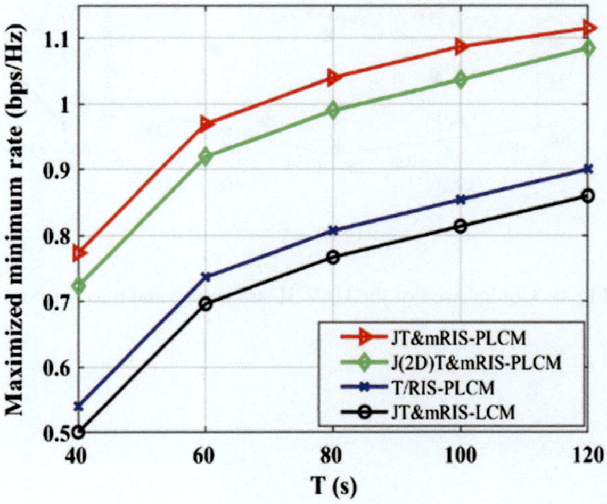

Figure 4 shows the comparison of the UAV 3D trajectory and max–min rates under the JT&mRIS-PLCM scheme with different numbers of RISs when $T = 120$ s and $M = 500$. When the number of RISs increases, the UAV altitude changes significantly. This is because, the UAV does not need to change the altitude frequently to achieve the optimal elevation angle-distance trade-off when the number of RISs is small. But as RISs' number enlarges, the UAV needs to adjust the trajectory and altitude more efficiently, to develop the data collection rate performance under the elevation angle-distance trade-off brought by multi-RISs, as shown in Fig. 4d.

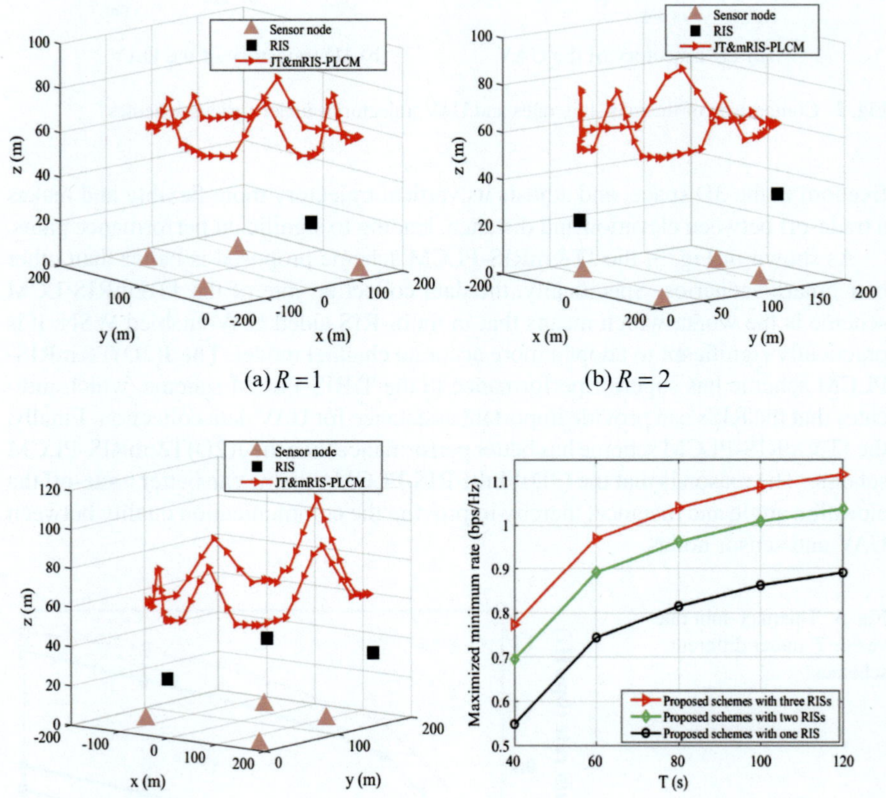

(a) $R = 1$ (b) $R = 2$

(c) $R = 3$ (d) Max-min rate versus T

Fig. 4 Comparison of the UAV 3D trajectory and max–min rates with various numbers of RIS

6 Conclusion

We construct a multi-RISs aided UAV-supported WSN in this paper. By jointly optimizing the phase-shift coefficient of multi-RISs, sensor node communication scheduling, and the UAV 3D trajectory, the minimum average data collection rate among all sensor nodes is maximized. The numerical results prove that our proposed solution can successfully enhance the data collection rate when compared to other benchmark schemes, showing the necessity and importance of deploying multi-RISs in UAV-enabled WSN. Although this paper only discusses the case of a single UAV, to be considered for more complex scenarios, the algorithm in this paper can be further investigated in depth in the case of multiple UAVs.

References

1. Zhang L, Liang YC, Niyato D (2019) 6G visions: mobile ultra-broadband, super internet-of-things, and artificial intelligence. China Commun 16(8):14
2. Palazzi V, Gelati F, Vaglioni U et al (2019) Leaf-compatible autonomous RFID-based wireless temperature sensors for precision agriculture. 2019 IEEE topical conference on wireless sensors and sensor networks (WiSNet), Orlando, FL, USA, 1–4
3. Madhumathi R, Arumuganathan T, Shruthi R (2020) Soil NPK and moisture analysis using wireless sensor networks. 2020 11th international conference on computing, communication and networking technologies (ICCCNT), Kharagpur, India, 1–6
4. Liu J, Zhao Z, Ji J et al (2020) Research and application of wireless sensor network technology in power transmission and distribution system. Intell Converged Netw 1(2):199–220
5. Cagnetti M, Leccisi M, Leccese F (2020) Simulation of a WSN routing protocol for airport runway application. 2020 IEEE 7th international workshop on metrology for aerospace (MetroAeroSpace). IEEE, 522–528
6. Cagnetti M, Leccisi M, Leccese F (2020) Reliability comparison of routing protocols for WSNs in wide agriculture scenarios by means of ηl index. SENSORNETS 2020-proceedings of the 9th international conference on sensor networks, 169–176
7. Zeng Y, Wu Q, Zhang R (2019) Accessing from the Sky: a Tutorial on UAV communications for 5G and beyond. Proc IEEE 107(12):2327–2375
8. Wu Q, Zhang R (2020) Towards smart and reconfigurable environment: intelligent reflecting surface aided wireless network. IEEE Commun Mag 58(1):106–112
9. Wu Q, Zhang S, Zheng B et al (2021) Intelligent reflecting surface-aided wireless communications: a tutorial. IEEE Trans Commun 69(5):3313–3351
10. You C, Kang Z, Zeng Y et al (2021) Enabling smart reflection in integrated air-ground wireless network: IRS meets UAV. IEEE Wirel Commun 28(6):138–144
11. Fan X, Liu M, Chen Y et al (2023) RIS-assisted UAV for fresh data collection in 3d urban environments: a deep reinforcement learning approach. IEEE Trans Veh Technol 72(1):632–647
12. Dong L, Liu Z, Jiang F et al (2022) Joint optimization of deployment and trajectory in UAV and IRS-assisted IoT data collection system. IEEE Internet Things J 9(21):21583–21593
13. Shao M, Liul Y, Duo B, et al (2022) Joint passive beamforming and elevation angle-dependent trajectory design for RIS-aided UAV-enabled wireless sensor networks. 2022 19th annual IEEE international conference on sensing, communication, and networking (SECON). IEEE, 488–496
14. Grant M, Boyd S (2020) CVX: matlab software for disciplined convex programming, version 2.2 [EB/OL]. http://cvxr.com/cvx

15. Wu Q, Zhang R (2019) Intelligent reflecting surface enhanced wireless network via joint active and passive beamforming. IEEE Trans Wirel Commun 18(11):5394–5409
16. Papandriopoulos J, Evans JS (2009) SCALE: a low-complexity distributed protocol for spectrum balancing in multiuser DSL networks. IEEE Trans Inf Theory 55(8):3711–3724

Research on Landing Attitude Control of Terrain Adaptive UAVs

Xinlei Ni, Qiaozhi Yin, Peilin Zhong, Xiaohui Wei, and Hao Sun

Abstract When landing on unstructured terrains in the field, UAVs often cannot go-around as they tilt on the ground, and even face the risk of roll-over with the attitude angles increasing. In this paper, a terrain adaptive UAV with adjustable landing legs is designed aiming at the stable landing and going-around requirements of multirotor UAVs under unstructured terrain conditions. a drop dynamic model of UAV is described in detail and drop simulation on flat ground is carried out in MATLAB/Simulink to verify the landing performance of UAV. The VMC method is used to establish the UAV's landing attitude control loop, which adjusts the landing legs actively through the feedback of attitude angle to suppress the variation of landing attitude caused by the terrains. The results of UAV's drop simulation under the terrain with height difference show that the attitude angle converges quickly to a range suitable for going-around during the landing process.

Keywords Attitude control · Virtual model control · UAV · Terrain adaptive · Drop simulation

1 Introduction

Multirotor Unmanned Aerial Vehicles (UAVs) are widely used in the field environment due to their characteristics like flexibility, maneuverability and vertical taking-off and landing. Most multirotor UAVs need to land on flat ground with their trussed

---

X. Ni · Q. Yin · P. Zhong · H. Sun
Key Laboratory of Fundamental Science for National Defense - Advanced Design Technology of Flight Vehicle, Nanjing University of Aeronautics and Astronautics, Nanjing 210016, China

Q. Yin
National Key Laboratory of Helicopter Aeromechanics, Nanjing University of Aeronautics and Astronautics, Nanjing 210016, China

X. Wei (✉)
State Key Laboratory of Mechanics and Control for Aerospace Structures, Nanjing University of Aeronautics and Astronautics, Nanjing 210016, China
e-mail: wei_xiaohui@nuaa.edu.cn

Z. Liu et al. (eds.), *Advances and Challenges in Advanced Unmanned Aerial Systems*, Springer Aerospace Technology, https://doi.org/10.1007/978-981-99-8045-1_3

landing gear to ensure that the lift provided by rotors does not produce horizontal component forces and rotating torque during going-around. When landing on uneven or sloping terrains in the field, UAVs often cannot go-around as they tilt on the ground, and even face the risk of roll-over with the attitude angles increasing. Therefore, new configurations of terrain adaptive UAVs and corresponding landing attitude control technology are needed to meet the stable landing and going-around requirement under unstructured terrain conditions of multirotor UAVs.

A variety of innovative aircraft configurations based on multirotor UAV were proposed for the adaptation between UAVs' structure and different terrains. The Skolkovo Innovation Center [1] in Russia designed a quadcopter UAV with a robotic landing gear, which can estimate and adapt to the landing terrain by the torque transducers and inertial measurement units. Liu [2] proposed a quadcopter with a bionic finger structure, which curls the "fingers" to adapt to the ground through a single motor and lands on different unstructured terrains. Roderick [3] proposed a new quadcopter UAV with bird-like aerial grasper and closed-loop balance controller, which can perch on a wider range of surface geometries and catch objects during flight. Hang [4] was also inspired by birds and proposed a modularized framework of UAV which can adapt to specific structures through different contact models. As terrain adaptive UAVs based on multirotor UAVs are mainly small quadcopters at present, it is necessary to propose a configuration of terrain adaptive UAV with large load capacity to enrich potential uses.

The attitude control of Vertical Taking-off and Landing (VTOL) aircrafts mainly focuses on flight attitude control before landing. Wang [5] proposed a method for accurate control of rocket's velocity, attitude angle and landing point in the return landing section, which ensures the rocket achieve an ideal landing state when landing on the offshore platform. Yan [6] proposed a landing control method for the descent process of a flexible asteroid lander, which changes the contact conditions and enhances the landing stability of the lander. The jumping-landing process of the robot has a certain similarity to the landing process of the terrain adaptive UAV's deformable structure, and there are three types of common control algorithms: Central pattern generator (CPG) [7], Zero Moment Point (ZMP) [8] and Intuitionistic control method [9]. Therefore, it is necessary to combine the unique mechanism of terrain adaptive UAV and control methods in the field of robotics to actively adjust the attitude of the landing process and get rid of the uncontrollability of multirotor UAV's Landing- Stabilizing/Overturning process.

In this paper, a terrain adaptive UAV configuration and a corresponding landing attitude controller are proposed, aiming at the stable landing and going-around requirements of multirotor UAV under unstructured terrain conditions. In Sect. 2, the design process and parameters of terrain adaptive UAV are described. In Sect. 3, a drop dynamic model of UAV is described and drop simulation on flat ground is carried out in MATLAB/Simulink to verify the landing performance. In Sect. 4, the landing position and attitude controller and improved controller based on Virtual Model Control (VMC) method are designed, and the effectiveness of the controllers are verified by drop simulation under the terrain with height difference. Conclusions are drawn in Sect. 5.

2 Conceptual Design of Terrain Adaptive UAV

The main design indicators of traditional multirotor UAVs include energy consumption, agility, control robustness [10], structural weight ratio, flight speed and other parameters. In order to realize the active adaptability of terrain adaptive UAVs for different terrains, part of fuselage needs to be designed as deformable structures, resulting in problems such as increased weight proportion and increased energy consumption.

In this paper, UAVs are simply divided into two systems for design iteration: propulsion system (including rotors, batteries and so on) and mission system (including adaptive landing structures, airborne sensors and so on). In the mission system, the four-legged bionic structure scheme with the best landing stability and weight [11] was selected for the preliminary design of the taking-off and landing structure, and common mission loads such as camera and lidar are also taken into consideration. Based on the load weight and flight performance requirements of the mission system, the propulsion system component selection [12] and analysis of structural weight are carried out to evaluate the reliability of the taking-off and landing structure. After several rounds of iteration, the overall structure of the drone is shown in Fig. 1 and part of parameters are shown in Table 1.

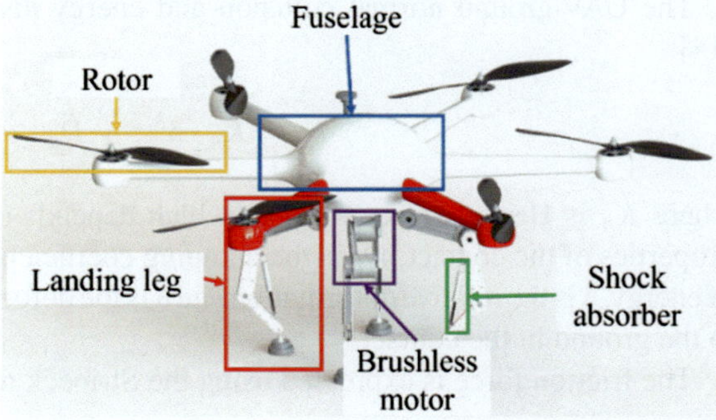

Fig. 1 Diagram of terrain adaptive UAV's structure

Table 1 Main parameters of terrain adaptive UAV

Parameter	Value	Unit
Structural weight	35	kg
Maximum take-off weight	60	kg
Maximum cruising speed	36	km/h
Blade	32	inch
Rotor wheelbase	1.9	m
Maximum adaptative height difference	0.3	m

3 Mathematical Modeling and Simulation

3.1 Dynamic Modeling of Terrain Adaptive UAV

The body coordinate system $\{O_b\}$ takes the center of mass as the origin; OX_b axis points forward in the symmetric plane of fuselage; OZ_b axis is perpendicular to the OX_b axis in symmetric plane and points down; OY_b axis is perpendicular to the symmetric plane and points to the right. The attitude angle of the aircraft is obtained in the order of yaw, pitch, and roll. The force acting on UAV during landing can be expressed as [13]:

$$\begin{bmatrix} \tau \\ F \end{bmatrix} = \begin{bmatrix} \sum_{i=1}^{4} r_i \times F_i \\ G + \sum_{i=1}^{4} F_i \end{bmatrix} \tag{1}$$

where r_i is a three-dimensional vector pointing from the origin of $\{O_b\}$ to the footpad (which is the point contacting with ground), $G = \begin{bmatrix} 0 \ 0 \ mg \end{bmatrix}^T$ is the gravity vector, and $F_i = \begin{bmatrix} F_{\mu x} \ F_{\mu y} \ F_n \end{bmatrix}^T$ are the tangential friction along the flat ground and normal impact perpendicular to the flat ground.

The UAV-ground normal collision and energy dissipation can be expressed as [14]:

$$F_n = K_n \cdot \delta^{3/2} + D_n \cdot \dot{\delta} \tag{2}$$

where K_n is Hertz contact stiffness which depends on the geometry and material properties of the contact, D_n is the damping coefficient which means the dissipation of energy, δ is the relative deformation and $\dot{\delta}$ is the deformation velocity perpendicular to the ground in the contact.

The friction force is expressed using the Stribeck model:

$$\begin{cases} F_\mu = F_n \cdot \mu(v) \\ \mu(v) = \mu_s \cdot \mathrm{sgn}(v) - k_1 \cdot v + k_2 \cdot v^3 \end{cases} \tag{3}$$

where F_μ is friction force in the opposite direction of the relative motion between footpad and ground, F_n is the normal collision force from Eq. (2), v is a non-dimensional relative velocity between footpad and ground, k_1 and k_2 are constants related to material properties and friction coefficients.

The cushioning force of the shock absorber can be expressed by a spring-damping system as:

$$F_m = K_m \cdot \delta + D_m \cdot \dot{\delta} \tag{4}$$

where K_m is the axial equivalent stiffness coefficient, D_m is the axial equivalent damping coefficient, δ is the cushion stroke and $\dot{\delta}$ is the relative cushion speed of the compression-only shock absorber.

3.2 Drop Simulation

The drop simulation of UAV under flat terrain condition is carried out according to the mathematical model in Sect. 3.1. UAV starts free fall from the air with no initial attitude angle and lands at a speed of 1.5 m/s on the ground. All landing legs are locked during landing. The displacement of center of mass is defined to be positive vertically downward. The normal force is defined to be positive vertically upward along the ground plane. The compression of shock absorber is seen as the positive direction of cushion stroke. The initial height of free fall is related to the landing speed and acceleration of gravity according to the equation $h = v^2/(2 \cdot g)$.

The simulation results under flat terrain condition are shown in Fig. 2. UAV starts free fall at 0 s and touches the ground at about 0.15 s. The vertical displacement of UAV reaches a peak of 155 mm at 0.22 s, drops to 110 mm at 0.35 s and then quickly converges to 125 mm, as shown in Fig. 2a. Figure 2b shows that the cushion stroke reaches a peak of 11.8 mm at about 0.22 s and drops to about 0.5 mm at 0.35 s, then stabilize to 4 mm, as shown in. The normal collision force reaches the first peak of 450N at 0.15 s, reaches the second peak of 245N at 0.2 s, and gradually stabilize to about 80N after dropping to about 10N at 0.35 s according to Fig. 2c.

It can be seen in combination with Fig. 2a–c that the generation of normal collision force after landing at 0.15 s drives the shock absorber to compress to an extreme value, and the center of mass is pressed down to the maximum value. The UAV bounds back after 0.22 s and reaches a limit at 0.35, but does not completely detach from the ground. Then the movement of UAV converges quickly and stabilize on the ground. The simulation results show that the UAV designed in this paper has good landing performance on the conventional landing surface.

4 Design of Position and Attitude Controller

The interaction process between the terrain adaptive UAV and the landing ground is mainly adjusted by the action of landing legs. Before landing, UAV needs to identify terrains by lidar and camera, and the UAV moves the footpads to match the ground terrain through the position controller. During the landing process, the roll angle and pitch angle of fuselage can be adjusted by curling and straightening the landing legs to prevent the fuselage rolling over or the rotor hitting the ground. Before going-around, it is necessary to adjust the attitude to make the rotor horizontal and lift direction vertical to ensure the safety of going-around. Therefore, it is necessary to

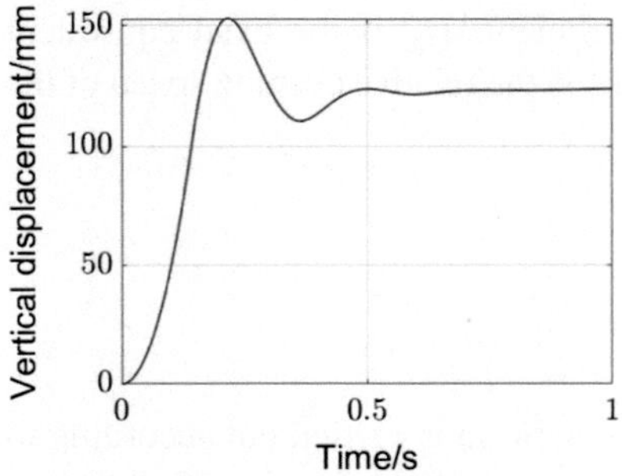

(a) Vertical displacement of the center of mass.

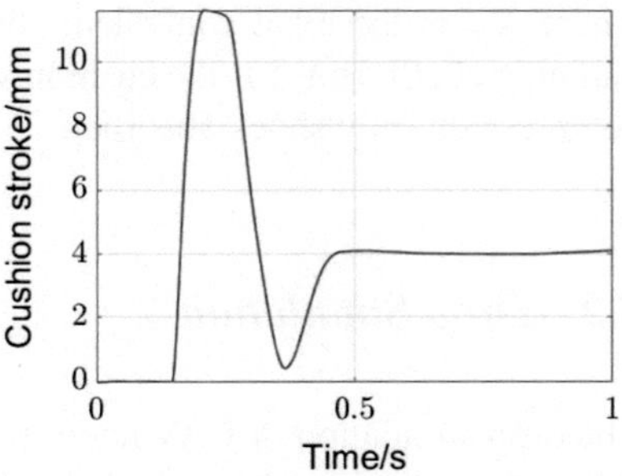

(b) Cushion stroke of shock absorber.

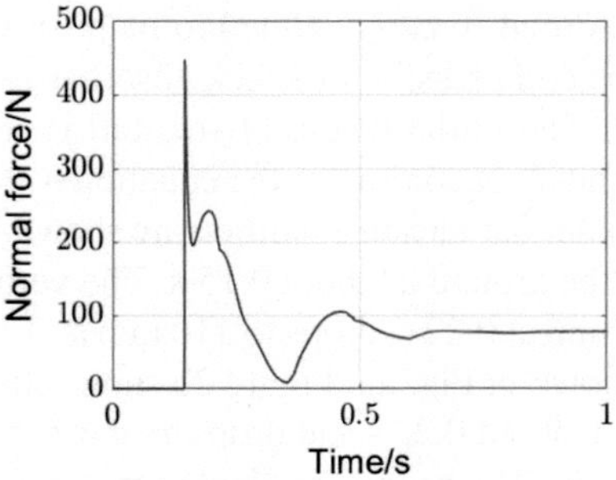

(c) Normal collision force.

Fig. 2 Results of drop simulation under flat terrain condition

design a set of position and attitude controller for the landing process of UAV based on the dynamic model.

4.1 Position and Attitude Controller Based on VMC Method

The Virtual Model Control (VMC) method is an intuitive control method and has been proved to be stable by the Lyapunov method [15]. The control principle is to generate a virtual force through imaginary virtual components (such as springs, dampers, etc.) between the feature points of the controlled object and the desired motion trajectory, and "drives" the subject to desired state by the virtual force. This method does not involve inverse kinematics solving and joint trajectory tracking, so it enjoys simple calculation and high efficiency [16].

As shown in Fig. 3, when the footpad position $\left(X_i^b, Y_i^b, Z_i^b\right)$ of landing leg No.i in the body coordinate system $\{O_b\}$ deviates from the desired position $\left(X_{id}^b, Y_{id}^b, Z_{id}^b\right)$, the virtual force along the body coordinate system $\{O_b\}$ generated by the virtual spring-damping system can be expressed as:

$$\begin{bmatrix} F_{i\_virtual\_X} \\ F_{i\_virtual\_Y} \\ F_{i\_virtual\_Z} \end{bmatrix} = \begin{bmatrix} K_X & 0 & 0 \\ 0 & K_Y & 0 \\ 0 & 0 & K_Z \end{bmatrix} \begin{bmatrix} X_{id}^b - X_i^b \\ Y_{id}^b - Y_i^b \\ Z_{id}^b - Z_i^b \end{bmatrix} + \begin{bmatrix} B_X & 0 & 0 \\ 0 & B_Y & 0 \\ 0 & 0 & B_Z \end{bmatrix} \begin{bmatrix} \dot{X}_{id}^b - \dot{X}_i^b \\ \dot{Y}_{id}^b - \dot{Y}_i^b \\ \dot{Z}_{id}^b - \dot{Z}_i^b \end{bmatrix} \tag{5}$$

where K_X, K_Y, K_Z are the stiffness coefficient of the virtual spring and B_X, B_Y, B_Z are the damping coefficient of the virtual damper.

As shown in Fig. 3, a virtual torque can also be constructed by imagining a torsion spring and a rotary damper when the attitude angle of fuselage deviates from the expected value. In view of the needs of UAV landing and going-around on unstructured terrains, the attention on roll angle and pitch angle is obviously greater than on yaw angle, so the virtual moment established along the body coordinate system $\{O_b\}$ is:

$$\begin{bmatrix} \Gamma_{b\_virtual\_\varphi} \\ \Gamma_{b\_virtual\_\theta} \end{bmatrix} = \begin{bmatrix} K_\varphi & 0 \\ 0 & K_\theta \end{bmatrix} \begin{bmatrix} \varphi_d - \varphi \\ \theta_d - \theta \end{bmatrix} + \begin{bmatrix} B_\varphi & 0 \\ 0 & B_\theta \end{bmatrix} \begin{bmatrix} \dot{\varphi}_d - \dot{\varphi} \\ \dot{\theta}_d - \dot{\theta} \end{bmatrix} \tag{6}$$

where K_φ, K_θ are stiffness coefficient of virtual torsion spring, B_φ, B_θ are the damping coefficient of virtual rotary damper, φ_d, θ_d are the expected roll angle and pitch angle and φ, θ are the actual roll angle and pitch angle.

The virtual moment can be directly decomposed into virtual driving torques of horizontally placed brushless motors, which may lead unpredictable movement of landing leg's footpad along OX_b and OY_b axis of $\{O_b\}$ and make the landing legs

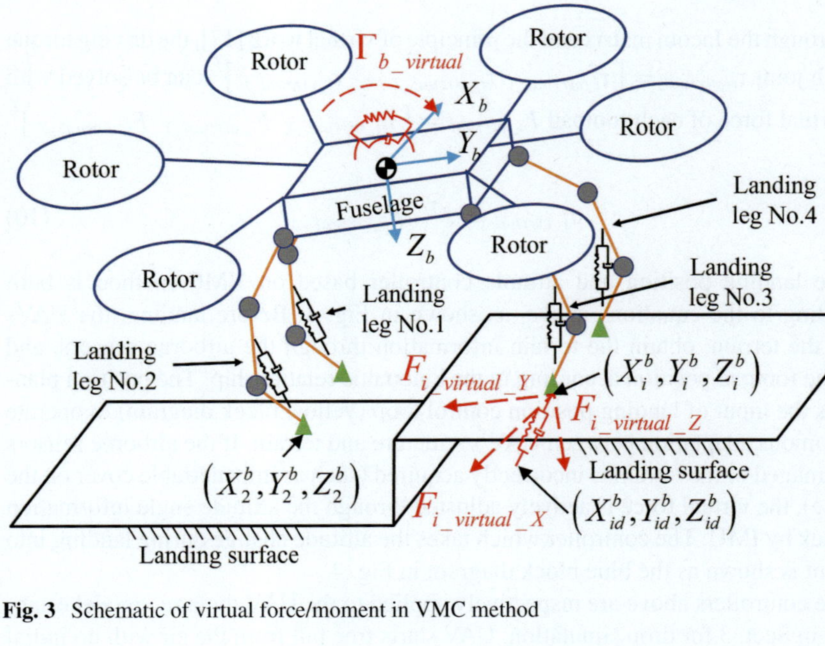

Fig. 3 Schematic of virtual force/moment in VMC method

hit unknown obstacles. Therefore, the virtual moment is decomposed into the virtual force along OZ_b axis of $\{O_b\}$ acting on each footpad (The relationship between the action force and the reaction force should be considered):

$$F_{i\_virtual\_Z} = -\frac{\Gamma_{b\_virtual\_\varphi}}{2 \cdot Y_{id}^b} + \frac{\Gamma_{b\_virtual\_\theta}}{2 \cdot X_{id}^b} \tag{7}$$

The virtual force generated by these imaginary components does not actually act on the UAV, so it is necessary to convert the virtual force/moment in the Task Space into the joint torque in the Joint Space of landing legs and drive the brushless motors by these input joint torques. Based on the forward kinematics relationship of the UAV landing legs, the coordinates of each footpad in the body coordinate system $\{O_b\}$ can be expressed as:

$$P_i^b = f(q) \tag{8}$$

where $P_i^b = \left[X_i^b \; Y_i^b \; Z_i^b \right]^T$ is the footpad's position vector of the landing leg No.i in $\{O_b\}$, and $q = \left[q_1 \; q_2 \; \cdots \; q_n \right]^T$ is the joint angle vector of each landing leg.

Deriving the Eq. (8) gives the corresponding Jacobian matrix as:

$$J = \begin{bmatrix} \frac{\partial X_i^b}{\partial q_1} & \cdots & \frac{\partial X_i^b}{\partial q_n} \\ \frac{\partial Y_i^b}{\partial q_1} & \cdots & \frac{\partial Y_i^b}{\partial q_n} \\ \frac{\partial Z_i^b}{\partial q_1} & \cdots & \frac{\partial Z_i^b}{\partial q_n} \end{bmatrix} \tag{9}$$

Through the Jacobi matrix and the principle of virtual work [17], the driving torque of each joint $\tau_{i\_virtual} = \left[\tau_{i\_virtual\_1} \; \tau_{i\_virtual\_2} \; \cdots \; \tau_{i\_virtual\_n} \right]^T$ can be solved with the virtual force of each footpad $F_{i\_virtual} = \left[F_{i\_virtual\_X} \; F_{i\_virtual\_Y} \; F_{i\_virtual\_Z} \right]^T$ as:

$$\tau_{i\_virtual} = J^T \cdot F_{i\_virtual} \tag{10}$$

The landing position and attitude controller based on VMC method is built according to the equations above, as shown in Fig. 4. Before landing, the UAVs sense the terrain, obtain the terrain information through the airborne sensors, and plan the footpad position according to the kinematic relationship. The position planning is the input of landing position control loop (yellow block diagram) to operate autonomous adaptation between UAV's structure and terrain. If the airborne sensors are damaged or the terrain is incorrectly acquired (such as impenetrable cover on the surface), the virtual force is actively adjusted through the attitude angle information fed back by IMU. The controller which takes the attitude change during landing into account is shown as the blue block diagram in Fig. 4.

The controllers above are respectively applied to the UAV dynamic model established in Sect. 3 for drop simulation. UAV starts free fall from the air with no initial

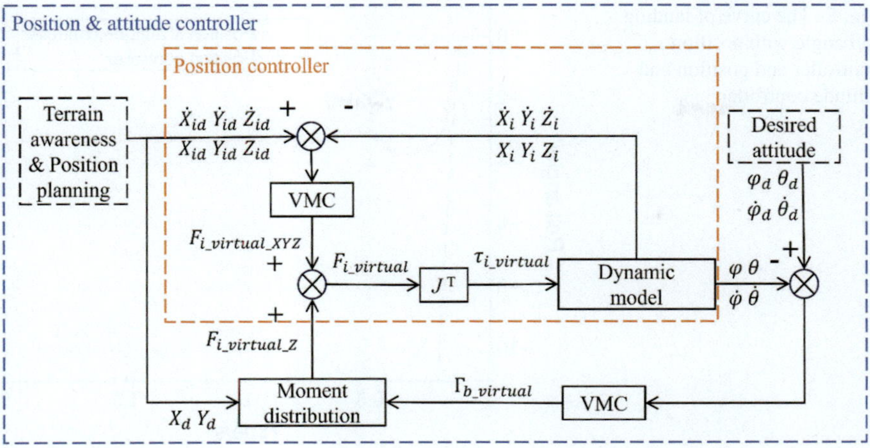

Fig. 4 Block diagram of the position and attitude controller based on VMC

attitude angle and lands at a speed of 1.5 m/s on the ground. The landing surface is a step ground with a height difference of 150 mm. The landing leg No.1 and 2 land in the area of lower plane and landing leg No.3 and 4 land in the area of higher plane, as shown in Fig. 3. The expected values for both controllers are shown in Table 2 and Table 3.

The results of the drop simulation are shown in Fig. 5, where the blue dashed line is the result of position controller, and the red solid line is the result of position and attitude controller. The roll angle of UAV using only position controller increases rapidly after touching the ground at about 0.15 s and exceeds 10° at 0.2 s. This puts the UAV at risk of roll-over and rotors hitting the ground. When using position and attitude controller, the roll angle increases slowly after touching the ground to a peak value (about 3.8°) at 0.4 s, and converges to a steady-state value (about 3.2°).

When the UAV lands on the ground with height difference of 150 mm, the landing leg No.3 and 4 contact the ground before the other two legs and generate a normal collision force. When only the position controller is used, the controller controls the

Table 2 Expectations of position and velocity

	X_{id}^b	Y_{id}^b	Z_{id}^b	$\dot{X}_{id}^b$	$\dot{Y}_{id}^b$	$\dot{Z}_{id}^b$
Unit	m	m	m	m/s	m/s	m/s
Value	±0.2	±0.3	0.4	0	0	0

Table 3 Expectations of attitude angle and angular velocity

	φ_d	θ_d	$\dot{\varphi}_d$	$\dot{\theta}_d$
Unit	rad	rad	rad/s	rad/s
Value	0	0	0	0

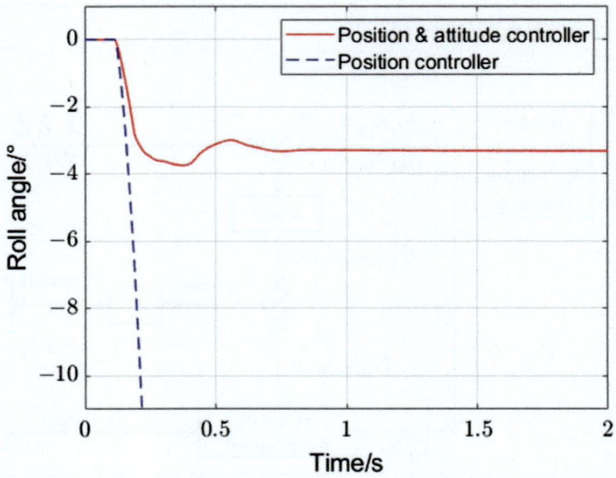

Fig. 5 The curve of landing roll angle with position controller and position and attitude controller

position of each footpad to maintain the expected value, and the contact force acts on the fuselage as a rolling moment which causes a large roll angular velocity. The position and attitude controller curls up the landing leg No.3 and 4, and extends down the landing leg No.1 and 2 to contact the ground according to the virtual moment generated by a negative roll angle. The contact force generates a rolling moment opposite to the roll angle and angular velocity to reduce the rotation. This allows the attitude of UAV to remain stable during landing.

The VMC method uses a virtual spring-damping system to convert the attitude error into a control output, so it always has a steady-state error and cannot achieve accurate attitude control after compensating for the constant disturbance caused by the terrain height difference (just like a spring-damping system under gravity cannot recover to zero). The steady-state error can be minimized to make the control effect approach to the target state by modifying the stiffness coefficient and damping coefficient in the equations.

4.2 *Improved Position and Attitude Controller*

Since the desired attitude angle and angular velocity of UAV are generally 0 during landing, the position and attitude controller in Sect. 4.1 can be regarded as an extension of PD controller in classical PID theory. Obviously, the imaginary spring element of the pose controller can be regarded as the proportion part of PID, and the imaginary damping element can be regarded as the differentiation part of PID. An addition of integration part can output a control signal to eliminate steady-state errors based on the accumulation of past errors according to the classical PID theory. The block diagram of improved controller is shown in Fig. 6.

The drop simulation of the improved controller is carried out with the same initial conditions in Sect. 4.1. The results are shown in Fig. 7, where the blue dashed

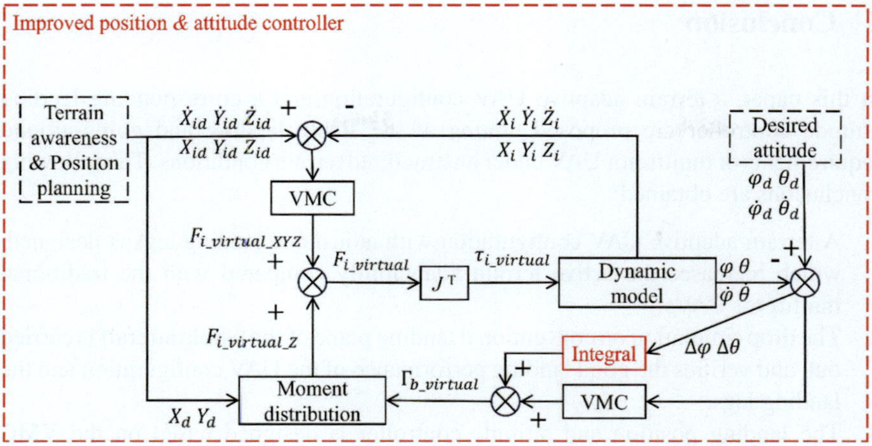

Fig. 6 Block diagram of the improved position and attitude controller

line is the simulation result of the original position and attitude controller, and the solid red line is the simulation result of the improved controller. The peak of roll angle after landing is reduced from 3.8º to 3.3º and then the roll angle gradually decreases to 0.16º by increasing the integration link to the controller. The improved position and attitude controller eliminates the steady-state error of attitude angle on the basis of ensuring a stable landing on uneven terrain, and meets the requirements of going-around as the rotor is horizontal and lift direction is vertical.

Fig. 7 The curve of landing roll angle with original and improved position and attitude controller

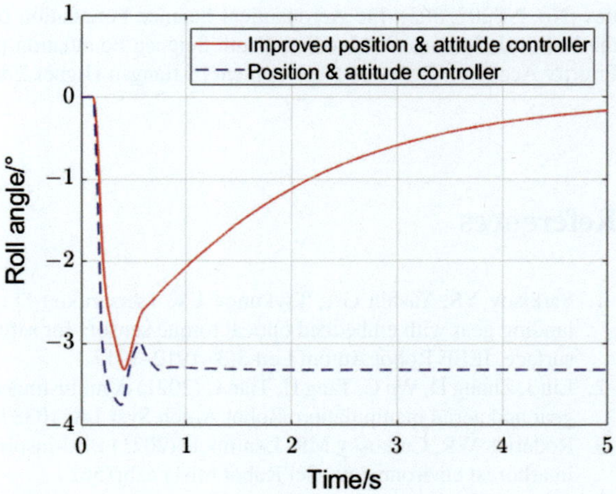

5 Conclusion

In this paper, a terrain adaptive UAV configuration and a corresponding landing attitude controller are proposed aiming at the stable landing and going-around requirements of multirotor UAV under unstructured terrain conditions. The following conclusions are obtained:

1. A terrain adaptive UAV configuration with adjustable landing legs is designed, which increases the active terrain adaptability compared with the traditional multirotor UAVs.
2. The drop simulation on conventional landing plane of the whole aircraft is carried out, and verifies the good landing performance of the UAV configuration and the landing legs.
3. The landing position and attitude controller is designed based on the VMC method, which maintains the peak of attitude angle within -4° when landing on a step ground with a height difference of 150 mm, and prevents the fuselage rolling over or the rotor hitting the ground.
4. The integration link is introduced into the landing attitude controller to eliminate the influence of steady-state error in traditional VMC method. The peak of the landing attitude angle reduces to 3.3° and converges to 0.16° in 5 s, which meets the attitude requirement of the UAV's going-around.

Acknowledgements This study was supported by the Fund of Prospective Layout of Scientific Research for NUAA (Nanjing University of Aeronautics and Astronautics), the National Natural Science Foundation of China (No. 52275114), the China Postdoctoral Science Foundation Funded Project (No. 2021M691565), the Fundamental Research Funds for the Central Universities (No. NS2023005), the Aeronautical Science Foundation of China (No. 202000410520002), the National Defense Outstanding Youth Science Foundation (No. 2018-JCJQ-ZQ-053), and the Priority Academic Program Development of Jiangsu Higher Education Institutions.

References

1. Sarkisov YS, Yashin GA, Tsykunov EV, Tsetserukou D (2018) Dronegear: a novel robotic landing gear with embedded optical torque sensors for safe multicopter landing on an uneven surface. IEEE Robot Autom Lett 3(3):1912–1917
2. Liu J, Zhang D, Wu C, Tang H, Tian C (2021) A multi-finger robot system for adaptive landing gear and aerial manipulation. Robot Auton Syst 146:103878
3. Roderick WR, Cutkosky MR, Lentink D (2021) Bird-inspired dynamic grasping and perching in arboreal environments. Sci Robot 6(61):eabj7562
4. Hang K, Lyu X, Song H, Stork JA, Dollar AM, Kragic D, Zhang F (2019) Perching and resting—a paradigm for UAV maneuvering with modularized landing gears. Sci Robot 4(28):eaau6637
5. Wang H (2018) Research on the landing mechanism and landing dynamics simulation of reusable launch vehicle. Nanjing Univ Aeronaut Astronaut. (in chinese)
6. Yan W, Feng R, Baoyin H (2022) Stability of a flexible asteroid lander with landing control. Aerospace 9(11):719

7. Raibert M, Blankespoor K, Nelson G, Playter R (2008) Bigdog, the rough-terrain quadruped robot. IFAC Proc Volumes 41(2):10822–10825
8. Vukobratović M, Borovac B (2004) Zero-moment point—thirty five years of its life. Int J Humanoid Rob 1(01):157–173
9. Raibert MH (1986) Legged robots that balance. MIT Press
10. Liscouët J, Pollet F, Jézégou J, Budinger M, Delbecq S, Moschetta JM (2022) A methodology to integrate reliability into the conceptual design of safety-critical multirotor unmanned aerial vehicles. Aerosp Sci Technol 127:107681
11. Black RJ (1964) Quadrupedal landing gear systems for spacecraft. J Spacecr Rocket 1(2):196–203
12. Dai X, Quan Q, Ren J, Cai KY (2019) An analytical design-optimization method for electric propulsion systems of multicopter UAVs with desired hovering endurance. IEEE/ASME Trans Mechatron 24(1):228–239
13. Liu H, Liu X, Ren J, Wang J, Dong J, Fang Z (2023) Design and Control Algorithm for Hexapod Landing Gear of Unmanned Helicopter. China Mech Eng 34(4). (in chinese)
14. Ni X, Yin Q, Wei X, Zhong P, Nie H (2022) Research on landing stability of four-legged adaptive landing gear for multirotor UAVs. Aerospace 9(12):776
15. Pratt JE (1995) Virtual model control of a biped walking robot. Massachusetts Institute of Technology
16. Zhang G, Rong X, Li Y, Chai H, Li B (2016) Control of the quadrupedal trotting based on virtual model. Robot 38(1):64–74 (in chinese)
17. Xie H (2015) Intuitive control based on virtual model for quadruped robot trot gait. National University of Defense Technology. (in chinese)

Optimal Event-Triggered Control of Quadrotor with Low Power Consumption and High Stability

Hanbing Zou, Junjie Zeng, Jiahao Huang, Pan Wang, Linbo Chen, and Yaosong Long

Abstract As one of unmanned aerial vehicles (UAV), the quadrotor has invariably been troubled by low battery energy density, high controller power consumption, and low blade efficiency, which are responsible for the short endurance of the quadrotor. To provide the quadrotor with low controller power consumption under the condition of accurate response, an optimal event-triggered control method (OECM) is proposed in this paper. In OECM, a frequency optimization function for event-triggered control of quadrotors is presented to regulate the system states of the quadrotor with minimum energy and fast response. The altitude and position tracking stability of the desired dynamics in finite time are proved by Lyapunov technique. Finally, the simulation results illustrate the efficiency and applicability of the designed control procedure in comparison with other control techniques.

Keywords Quadrotor event-triggered control · Optimal update frequency · Finite-time stability

1 Introduction

Recently, quadrotors have been widely used in many fields owing to their advantages of vertical take-off and landing, ease of operation, low cost, and flexible maneuverability [1–4]. The applications of quadrotors have extended in many areas such as environmental monitoring, power line inspection, emergency rescue, and plant protection [1]. Effective control of the quadrotors is crucial to meet high-performance mission requirements. In reality, three issues exist in quadrotor control, including multi-input multi-output (MIMO) systems, unspecified characteristics, and time-varying states and delays [5], which determine the control of quadrotors to become a complex and difficult segment for quadrotors. Therefore, it's necessary to improve control efficiency [6].

H. Zou · J. Zeng · J. Huang · P. Wang · L. Chen · Y. Long (✉)
School of Aerospace Engineering, Huazhong University of Science and Technology, Wuhan, China
e-mail: longyaosong@hust.edu.cn

Z. Liu et al. (eds.), *Advances and Challenges in Advanced Unmanned Aerial Systems*, Springer Aerospace Technology, https://doi.org/10.1007/978-981-99-8045-1_4

Famous for the ease of analysis, the time-triggered control system is the most common method of ordinary control theory in quadrotors [7]. Such a control strategy uses a fixed update frequency, which is always high so that the quadrotor can perform rapidly, but it causes a lot of energy consumption. To solve this problem, an event-triggered controller has been used in the control of quadrotors in recent years. The main feature is aperiodic sampling to save communication resources and reduce computation [8–10]. In [11], an event-triggered H2 controller is designed for attitude control of a quadrotor, the event-triggered mechanism saves communication network resources, and the performance of the system is guaranteed by the H2 controller. In [12], a hyperbolic tangent function-based fixed-time event-triggered controller is designed for solving the prescribed performance trajectory tracking problem of quadrotor aircraft with six degrees of freedom. In [13], under the directed topology, a fast-formation event-triggered controller is designed for the communication noise interference between quadrotors. Most of the existing event-triggered controllers can improve the efficiency of the quadrotors to a certain extent, but none of them can maximize it [14].

In event-triggered control system, the triggering threshold plays an important role, which determines the computation and energy consumption of the quadrotor. Most of the popular event-triggered control theories at present use a fixed threshold. Zhu et al. presented an event-triggered controller that solves the problem of trajectory tracking for a quadrotor with thrust actuation in a single body-fixed direction and full angular velocity actuation [2]. An experiment was conducted to verify that the quadrotor was stable, and the inter-event time was 0.068 s, reasonably efficient compared with the time-triggered control system. But for the fixed triggering threshold, the efficiency of the quadrotor can be improved. Therefore, the dynamic threshold was brought up. Wang et al. designed an event-triggered control strategy with dynamic threshold parameters to deal with the formation control problem for a group of quadrotor unmanned aerial vehicles [3]. The dynamic threshold can further increase the performance of the quadrotor, but the existing literature is incapable of showing the most appropriate one.

In this paper, motivated by the aforementioned discussions, a quadrotor optimal event-triggered control method (OECM) is designed to further improve efficiency. This method optimizes the update frequency of the controller compared with existing event-triggered control method. Different from the traditional sliding mode control method, the efficiency of the quadrotors' control system reaches its maximum and the flight performance is improved by Hamilton function [15, 16].

The organizational structure of this paper is presented as follows. Section 2 presents the structure and the dynamic model of the quadrotor, Sect. 3 describes the design method of the controller, where stability is proved and control parameters are discussed in detail. In Sect. 4, simulation results are presented.

2 Mathematical Model of the System

Figure 1 shows the dynamics of the quadrotor in the inertial and body coordinate systems.

The XYZ axes in this Fig. 1 represent the inertial coordinate system, and the $X_C Y_C Z_C$ axes represent the body coordinate system [17].

2.1 Rotational Dynamics Analysis

The angular velocity of the quadrotor is represented by the Euler's angle and the body angular velocity, respectively, where the Euler's angle is $\mathbf{\Omega} = (\phi, \theta, \psi)^{\mathrm{T}}$ and the body angular is $\mathbf{R} = \left(\mathrm{r_x}, \mathrm{r_y}, \mathrm{r_z}\right)^{\mathrm{T}}$. By analyzing the kinematics of the quadrotor, the rotation matrix $\mathbf{Q}$ can be obtained [18].

$$\dot{\Omega} = \mathbf{Q}(\Omega)\mathrm{R} \tag{1}$$

$$\mathbf{Q}(\Omega) = \begin{pmatrix} 1 & \tan\theta\sin\phi & \tan\theta\cos\phi \\ 0 & \cos\phi & -\sin\phi \\ 0 & \sin\phi/\cos\theta & \cos\phi/\cos\theta \end{pmatrix} \tag{2}$$

Define ω_i as the rotation speed of the i_{th} rotor. Based on the relationship between the aerodynamic forces and the speed of the blade in aerodynamics, the lift L generated by the quadrotor can be obtained

$$\mathrm{L} = \sum_{i=1}^{4} \mathrm{L}_i = k_t\left(\omega_1^2 + \omega_2^2 + \omega_3^2 + \omega_4^2\right) \tag{3}$$

where $\mathrm{k_t}$ is the aerodynamic coefficient.

Fig. 1 The reference coordinate system of an UAV

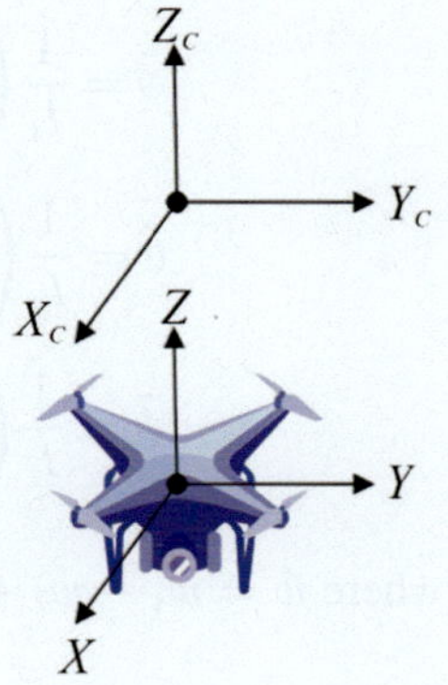

According to the dynamics [1], the torque of the lift generated by the quadrotor propeller to the XYZ axis of the body frame can be obtained.

$$\mathbf{M} = \begin{pmatrix} \mathrm{T}_x \\ \mathrm{T_y} \\ \mathrm{T_z} \end{pmatrix} = \begin{pmatrix} \frac{\sqrt{2}}{2} l(-\mathrm{L}_1 + \mathrm{L}_2 + \mathrm{L}_3 - \mathrm{L}_4) \\ \frac{\sqrt{2}}{2} l(\mathrm{L}_1 - \mathrm{L}_2 + \mathrm{L}_3 - \mathrm{L}_4) \\ \frac{k_d}{k_t}(\mathrm{L}_1 + \mathrm{L}_2 - \mathrm{L}_3 - \mathrm{L}_4) \end{pmatrix} = \begin{pmatrix} \frac{\sqrt{2}}{2} l k_t \left(-\omega_1^2 + \omega_2^2 + \omega_3^2 - \omega_4^2\right) \\ \frac{\sqrt{2}}{2} l k_t \left(\omega_1^2 - \omega_2^2 + \omega_3^2 - \omega_4^2\right) \\ k_d \left(\omega_1^2 + \omega_2^2 - \omega_3^2 - \omega_4^2\right) \end{pmatrix} \tag{4}$$

where T_* is the moment to the *-axis, l is half the wheelbase of the quadrotor, and k_d is the aerodynamic coefficient related to the moment.

Through the above analysis, we can obtain the rotational dynamics model of the quadrotor. This model can be expressed as the following equation.

$$\begin{cases} \dot{\Omega} = \mathbf{Q}(\Omega)\mathrm{R} \\ \mathbf{IR} = -\mathrm{R} \times (\mathrm{IR}) + \mathbf{G} + \mathbf{M} + \mathbf{f} + \mathbf{d} \end{cases} \tag{5}$$

where **I** characterizes the moment of inertia of the quadrotor to the body frame. And the symmetry of the quadrotor leads to $\mathbf{I} = \begin{bmatrix} \mathrm{I_X} & 0 & 0 \\ 0 & \mathrm{I_Y} & 0 \\ 0 & 0 & \mathrm{I_Z} \end{bmatrix}$ being a diagonal matrix. **G** characterizes the gyroscopic moment and $\mathbf{f} = f\left[\dot{x}\ \dot{y}\ \dot{z}\right]^T$ characterizes the air friction. **d** denotes the external disturbance and **d** can be written as a vector containing three components in the Euler angular direction $\mathbf{d} = \left[\mathrm{d}_\varphi\ \mathrm{d}_\theta\ \mathrm{d}_\psi\right]^{\mathrm{T}}$. The value of **d** is small compared to other values, which ensures that the stability of the control system is satisfied even when **d** is unknown.

When hovering, the change of quadrotor angle is relatively small, so the trigonometric function in the rotation matrix can be linearized [19]. In turn, the angular acceleration expression can be obtained:

$$\begin{aligned} \ddot{\phi} &= \frac{1}{I_x}\left((I_y - I_z) r_y r_z + T_x - \frac{\sqrt{2} f l}{2}\dot{\phi} + J_r r_y \hat{w} - d_\phi\right) \\ \ddot{\theta} &= \frac{1}{I_y}\left((I_z - I_x) r_x r_z + T_y - \frac{\sqrt{2} f l}{2}\dot{\theta} - J_r r_x \hat{w} - d_\theta\right) \\ \ddot{\psi} &= \frac{1}{I_z}\left((I_x - I_y) r_x r_y + T_z - \frac{\sqrt{2} f l}{2}\dot{\psi} - d_\psi\right) \end{aligned} \tag{6}$$

where $\hat{w} = w_1 - w_2 + w_3 - w_4$ and J_r is the rotor's rotational inertia.

2.2 Translational Dynamics Analysis

According to Newton's second law, the center of mass of the quadrotor under the body coordinate system satisfies the following equation of motion [20].

$$m\mathbf{a} = \mathbf{G} - \mathbf{L} - \mathbf{f} \tag{7}$$

where m is the mass of the quadrotor and $\mathbf{a} = \left[\ddot{x}\ \ddot{y}\ z\right]^T$ is the acceleration of the quadrotor center of mass.

Dividing both sides of Eq. (6) by the same mass m then yields the expression for acceleration a.

$$\mathbf{a} = \mathbf{g} - \frac{\mathbf{L}}{\mathbf{m}} - \frac{\mathbf{f}}{\mathbf{m}} \tag{8}$$

where $\mathbf{g} = \left[0\ 0\ g\right]^T$ is the acceleration of gravity.

In order to convert the kinetic equations in the body coordinate system to the inertial coordinate system, it is necessary to multiply the rotation matrix $\mathbf{Q}$ to both left and right sides of Eq. (6), and then obtain the following equation.

$$\mathbf{a} = \mathbf{g} - \mathbf{Q}\frac{\mathbf{L}}{\mathbf{m}} - \frac{\mathbf{f}}{\mathbf{m}} \tag{9}$$

Substituting Eq. (2) into the above equation [21], the equation for the advection dynamics for a quadrotor can be obtained as

$$\begin{cases} \ddot{x} = \frac{-L}{m}(\cos\phi\sin\theta\cos\psi + \sin\phi\sin\psi) - \frac{f\dot{x}}{m} \\ \ddot{y} = \frac{-L}{m}(\cos\phi\sin\theta\cos\psi + \sin\phi\sin\psi) - \frac{f\dot{y}}{m} \\ \ddot{z} = g - \frac{L}{m}(\cos\theta\cos\phi) - \frac{f\dot{z}}{m} \end{cases} \tag{10}$$

2.3 Additive Parameter Uncertainties

For quadrotors, the theoretical parameters used for calculations and the actual parameters obtained from measurements are often different. In order to enhance the applicability of the control law, additive parameter uncertainties for quadrotors will be considered in this paper [1].

$\Delta\alpha$ denotes the difference between the theoretical value α and the actual value $\mathring{\alpha}$.

$$\Delta\alpha = \alpha - \mathring{\alpha} \tag{11}$$

Define tracking error as

$$\begin{cases} e_\phi = \phi - \phi_d \\ e_\theta = \theta - \theta_d \\ e_\psi = \psi - \psi_d \end{cases} \tag{12}$$

where ϕ_d, θ_d, ψ_a are reference values and they are second order derivable.

Considering additive parameter uncertainties and tracking error into Eq. (6), we can obtain the error equation.

$$\begin{aligned} \mathop{I}\limits^{\circ}_{x}\ddot{e}_\phi &= \left(\mathop{I}\limits^{\circ}_{y} - \mathop{I}\limits^{\circ}_{z}\right) r_y r_z + T_x - \frac{\sqrt{2} f \mathop{l}\limits^{\circ}}{2}\dot{\phi} - \mathop{I}\limits^{\circ}_{x}\ddot{\phi}_d - \delta_\phi \\ \mathop{I}\limits^{\circ}_{y}\ddot{e}_\theta &= \left(\mathop{I}\limits^{\circ}_{z} - \mathop{I}\limits^{\circ}_{x}\right) r_x r_z + T_y - \frac{\sqrt{2} f \mathop{l}\limits^{\circ}}{2}\dot{\theta} - \mathop{I}\limits^{\circ}_{y}\ddot{\theta}_d - \delta_\theta \\ \mathop{I}\limits^{\circ}_{z}\ddot{e}_\varphi &= \left(\mathop{I}\limits^{\circ}_{x} - \mathop{I}\limits^{\circ}_{y}\right) r_x r_y + T_z - f \mathop{l}\limits^{\circ}\dot{\varphi} - \mathop{I}\limits^{\circ}_{z}\ddot{\phi}_d - \delta_\varphi \end{aligned} \tag{13}$$

where

$$\begin{aligned} \delta_\phi &= \Delta I_x \ddot{\phi} - \Delta\left(I_y - I_x\right) r_y r_z + \frac{\sqrt{2} f}{2} \Delta l \dot{\phi} - J_r r_y \hat{w} + d_\phi \\ \delta_\theta &= \Delta I_y \ddot{\theta} - \Delta(I_z - I_x) r_x r_z + \frac{\sqrt{2} f}{2} \Delta l \dot{\theta} + J_r r_y \hat{w} + d_\theta \\ \delta_\psi &= \Delta I_x \ddot{\psi} - \Delta\left(I_x - I_y\right) r_x r_y + f \Delta l \dot{\psi} + d_\psi \end{aligned} \tag{14}$$

In addition, the lumped uncertainty of the system satisfies the condition:

$$\begin{cases} \left|\frac{1}{\mathop{x}\limits^{\circ}_{x}}\delta_\phi\right| < N_{\phi 1} + N_{\phi 2}|\phi| + N_{\phi 3}|\dot{\phi}| \\ \left|\frac{1}{\circ}\delta_\theta\right| < N_{\theta 1} + N_{\theta 2}|\theta| + N_{\theta 3}|\dot{\theta}| \\ \left|\frac{1}{I_y}\delta_\psi\right| < N_{\psi 1} + N_{\psi 2}|\psi| + N_{\psi 3}|\dot{\psi}| \end{cases} \tag{15}$$

where $N_{*_1}, N_{*_2}, N_{*_3}$ are all positive numbers representing the different factors affecting the uncertainty of the system.

For the convenience of later derivation, Eq. (6) is rewritten in the following form, taking A as an example.

$$\begin{cases} \dot{\phi}_1 = \phi_2 \\ \dot{\phi}_2 = \frac{1}{I_x}\left(\left(I_y - I_z\right) r_y r_x + J_r r_y \hat{\omega} - d_\phi\right) + \frac{T_x}{I_x} - \frac{\sqrt{2}}{2}\frac{fl}{I_x}\phi_2 \end{cases} \tag{16}$$

3 Controller Design

In this section, a quadrotor optimal event-triggered control method (OECM) will be designed to ensure the stability of the quadrotor in the presence of disturbances and uncertainties. An exposition of the event-triggered control law, encompassing both ideal and practical scenarios, is presented and its stability is rigorously established. Subsequently, the determination of the optimal update frequency for the event-triggered control law is elucidated.

Define the sliding mode surface as follows [22]:

$$\sigma = \dot{s} + ks \tag{17}$$

where

$$s = \dot{e} + \lambda e \tag{18}$$

where s and σ are both positive parameters.

The sliding mode surface $\sigma = \left[\sigma_\phi \; \sigma_\theta \; \sigma_\psi\right]^T$ in each direction are as follows:

$$\begin{aligned}
\sigma_\phi &= \ddot{e}_\phi + (\lambda + k)\dot{e}_\phi + k\lambda e_\phi \\
\sigma_\theta &= \ddot{e}_\theta + (\lambda + k)\dot{e}_\theta + k\lambda e_\theta \\
\sigma_\psi &= \ddot{e}_\psi + (\lambda + k)\dot{e}_\psi + k\lambda e_\psi
\end{aligned} \tag{19}$$

Consider the ideal case of linear surfaces. In the ideal case, without considering UAV interference and parameter measurement error, the linear sliding mode surface satisfy the following condition:

$$\sigma = 0 \tag{20}$$

From Eqs. (6), (12), and (19), the expression for the moment T_x in the x axis direction can be obtained as follows:

$$T_x = \mathscr{H}\cos\theta + \mathscr{L}\cos\phi + \mathscr{F}\cos^2\phi + \mathscr{A} I_x + \mathscr{K} \tag{21}$$

where $\mathscr{H}$, $\mathscr{L}$, $\mathscr{F}$, $\mathscr{A}$, and $\mathscr{K}$ are all functions related to the state of the system.

The specific expressions of these functions are shown below:

$$\begin{aligned}
\mathscr{H} &= -J_r\hat{\omega}\dot{\phi}\sin\phi + \left(I_y - I_z\right)\dot{\psi}\dot{\phi} \\
\mathscr{L} &= \left(I_y - I_z\right)\dot{\phi}^2\sin^3\phi\cos^2\theta - J_r\hat{\omega}\dot{\phi} \\
\mathscr{F} &= -2\left(I_y - I_z\right)\dot{\psi}\dot{\phi}\cos\theta \\
\mathscr{A} &= \lambda k e_\phi - (\lambda + k)\dot{e}_\phi + \ddot{\phi}_d \\
\mathscr{K} &= \frac{\sqrt{2}}{2} fl\phi + d_\phi
\end{aligned} \tag{22}$$

The Lyapunov function of (21) is defined as:

$$V = \frac{1}{2}\sigma^2 \tag{23}$$

For ideal case, From Eq. (20), $\sigma = 0$. Thus $_s\ \dot{V} = \sigma\dot{\sigma} = 0$. Thus Lyapunov stability is proven.

In event-triggered systems, the actual T_x tends to differ from the ideal value. In event-triggered systems, sampling is often periodic. During the sampling period, the real T_x remains constant and is equal to the ideal one at the previous sampling moment. The difference between the actual T_x and the ideal one is:

$$\eta(t) = T_{xr}(t) - T_x(t) \tag{24}$$

where T_{xr} means the real T_x, $t \in \left[t_k, t_{k+1}\right)$.

Theorem 1 Consider a quadrotor that fits the dynamical model (18). The system will be stable as long as the following conditions are satisfied:

$$\|\eta(t)\| = \|T_{xr}(t) - T_x(t)\| \in [0, \Lambda] \tag{25}$$

where

$$\Lambda = \left|\frac{1}{3(k+\lambda)}\left(\mathbb{Z}\lambda + \mathbb{R}I_x + \mathbb{N}I_x\lambda^2 + \mathbb{Q}\right)\right| \tag{26}$$

where $\mathbb{Z}$, $\mathbb{R}$, $\mathbb{N}$, and $\mathbb{Q}$ are all functions related to the system state.

These functions contain information not only about the quadrotor parameters, but also about the quadrotor flight state. Their specific expressions are as follows:

$$\begin{aligned}
&\mathbb{Z} = \left((\phi_d - \phi)k^2 + 3k\left(\dot{\phi}_d - \dot{\phi}\right) + 2\dot{\phi}_d - 2r_y r_z\right)I_x + \sqrt{2}fl\dot{\phi} - 2\left(T_x - \delta_\phi\right)\\
&\mathbb{R} = \left(-\dot{\phi} - \dot{\phi}_d\right)k^2 + (2k-1)\ddot{\phi}_d - \ddot{\phi}\\
&\mathbb{N} = (\phi_d - \phi)k - \dot{\phi} + \dot{\phi}_d\\
&\mathbb{Q} = 2k\left(-\left(I_y - I_z\right)r_y r_z + \frac{\sqrt{2}}{2}fl\dot{\phi} - T_x + \delta_\varphi\right)
\end{aligned} \tag{27}$$

Proof The Lyapunov function in the real case is:

$$\dot{V}_2 = \sigma\cdot\dot{\sigma} = (\dot{s} + ks)(\ddot{s} + k\dot{s}) \tag{28}$$

From (6), (13), (14), (17), it follows that:

$$V_2 = \left[\frac{1}{I_x}\mathbb{C} + (k+\lambda)\dot{e}_\phi + k\lambda e_\phi\right]\left[\frac{\lambda+k}{I_x}\mathbb{C} + k\lambda\dot{e}_\phi + e_\phi^{(3)}\right] \tag{29}$$

where

$$\mathbb{C} = \left(I_y - I_z\right)r_y r_z + I_x - \frac{\sqrt{2}}{2}fl\dot{\phi} - I_x\dot{\phi}_d - \delta_\phi + \eta \tag{30}$$

Deriving the left and right sides of Eq. (29) simultaneously, we can obtain the expression for V_2:

$$\dot{V}_2 = A\eta^2 + B\eta \tag{31}$$

where A and B are only related to the state of the system and unrelated to the control error. The specific expressions of A and B are:

$$\begin{aligned}
A &= \frac{3(k+\lambda)}{T^2} \\
B &= \frac{2(k+\lambda)}{T^2}\cdot\left(\left(I_y - I_z\right)\left(r_y r_z - \dot{\phi}\right) + T_x - I_x\ddot{\phi}_d - \delta_\phi\right) \\
&+ \left(\frac{k\lambda}{T_x} + (\lambda+k)^2\right)\dot{e}_\phi + e_\phi^{(3)} + (\lambda+k)k\lambda e_\phi
\end{aligned} \tag{32}$$

It is easy to see that $\dot{V}_2$ is a quadratic function of η.Since the quadratic term coefficient A of $\dot{V}_2$ is positive, therefore $\dot{V}_2$ is less than zero when η is located between two intersections of $\dot{V}_2$ and the x axis. It can be known that when η equals 0, $\dot{V}_2$ equals 0. Therefore 0 is one of the intersection points.

When $\eta = \Lambda$

$$\begin{aligned}
\dot{V}_2 =& A\Lambda^2 + B\Lambda = \frac{1}{3(k+\lambda)^2}\left(\mathbb{Z}\lambda + \mathbb{R}I_x + \mathbb{N}I_x\lambda^2 + \mathbb{Q}\right)^2 \\
&+ B\frac{1}{3(k+\lambda)}\left(\mathbb{Z}\lambda + \mathbb{R}I_x + \mathbb{N}I_x\lambda^2 + \mathbb{Q}\right) = 0
\end{aligned} \tag{33}$$

The above analysis shows that when $\|\eta(t)\| \in [0, \Lambda]$, $V_2 \le 0$. It indicates that the system can still satisfy Lyapunov stability even in the presence of external disturbances and uncertainties. This completes the proof.

Remark 1

The analysis above demonstrates the Lyapunov stability of this control law but also gives an upper bound on error η. When the input error η reaches Λ, the event-triggered system must update the control law, thus ensuring stability. This upper bound Λ is also the critical parameter for the subsequent optimal update frequency analysis.

The control law for the quadrotor in ideal and non-ideal conditions is derived above. The event-triggered control samples the system state at irregular intervals.

Our goal is: to define the optimal control frequency such that the sampling interval is maximized when the system performance function [2] is minimized. This section will use the max–min optimization problem to achieve this goal.

The nonlinear system with an event-based control strategy (21) given by (16) can be rewritten as:

$$\dot{\phi}_2(t) = f(\phi_2) + \boldsymbol{g}(\phi_2)T_x(t) \tag{34}$$

where $f(\phi_2) = \frac{1}{I_x}\left((I_y - I_z)r_y r_x + J_r r_y \hat{\omega} - d_\phi\right) - \frac{\sqrt{2}}{2}\frac{fl}{I_x}\phi_2$, $\boldsymbol{g}(\phi_2) = \frac{1}{T_x(x)}$

Define the performance function ζ of the system as [2]:

$$\|\varsigma(t)\|^2 = K(\phi(t)) + ST_x^2(t) \tag{35}$$

where $K(*)$ is a positive definite function satisfying $K(0) = 0$ and S is a positive number. The function $K(*)$ and the positive number S penalize the state and the control policy, respectively.

In order to optimize the event trigger interval, the input error η must be maximized. Using Eq. (35) to define the infinite horizon performance index:

$$R(\phi, \eta, T_x) = \int_t^\infty \|\varsigma(\tau)\|^2 - \beta^2\eta^2\big]d\tau \tag{36}$$

where $\sigma > 0$ represents the attenuation constant [3].

The above analysis shows that the problem is a max–min optimization problem based on a two-person zero-sum game. The control input T_x of the quadrotor is the minimum participant and the input error η is the maximum participant. The objective of the problem is to find a saddle-point solution, which satisfies the following condition:

$$V^*(T_x(t)) = \min_{T_x}\max_{\eta} R(T_x, \eta) = \max_{\eta}\min_{T_x} J(T_x, \eta) \tag{37}$$

For a reachable zero-state observation system [3] $Q(x) = C^2(x)$,, the nonlinear mapping $C(x)$, there exists a minimum positive definite solution to the optimal value function (37), when $\beta > \beta^*$ and β^* is the H_∞ gain [3].

Based on the performance function defined in (33), the Hamilton function of the system (6) can be given as:

$$H(\phi_2, \eta, T_x) = K(\phi_2) + aT_x^2 - \beta^2\eta^2 + V_{\phi_2}^*[f(\phi_2) + \dot{\boldsymbol{g}}(\phi_2)T_x(t) + \boldsymbol{g}(\phi_2)\eta(t)] \tag{38}$$

where $V_{\phi_2}^* = \frac{\partial V^*}{\partial \phi_2}$, $V_{\phi_2}^*$ is the optimal function determined by Eq. (37) and a is a positive parameter [4].

The optimal control law T_x^* can be derived from the optimal condition equation $\frac{\partial H(\phi, T_x, \eta)}{\partial T_x} = 0$.

$$T_x^* = -\frac{1}{2I_x a} V_{\phi_2}^* \tag{39}$$

Similarly, the control error in the case of (37) can be found by $\frac{\partial H(\phi, T_x, \eta)}{\partial \eta} = 0$.

$$\eta^* = \frac{1}{2I_x \beta^2} V_{\phi_2}^* \tag{40}$$

In order to ensure the stability and high efficiency of the quadrotor, η must meet the following conditions: $|\eta| \leq \min\left\{\Lambda, \frac{V_\phi^*}{2\beta^2 I_x}\right\}$. Once $|\eta|$ meets Λ or $\frac{V_\phi^*}{2\beta^2 I_x}$, the control law of the quad must update.

The above analysis shows that the optimal control strategy T_x^* and the control error η^* can be written as a function of $V_{\phi_2}^*$.. Therefore, the problem of solving T_x^* and η^* becomes the problem of solving $V_{\phi_2}^*$.. To solve $V_{\phi_2}^*$., the two Eqs. (39) (40) are substituted into Eq. (38) to obtain the following HJI equation:

$$H = K(\phi_2) + V_{\phi_2}^* f - \frac{1}{4\alpha} V_{\phi_2}^2 g^2 + \frac{1}{4\beta^2} V_{\phi_2}^2 g^2 \tag{41}$$

The solution of the above HJI equation is the optimal function $V_{\phi_2}^*$..

As discussed earlier, the closed solution of the HJI equation is difficult to express in an analytic form [4, 5]. In order to solve $V_{\phi_2}^*$., a neural network-based method is introduced to solve the approximate solution.

The Bellman equation in integral form can be obtained by integrating Eq. (40) over the interval t_k to t_{k+1}.

$$V^*(t_{k+1}) - V^*(t_k) = \int_{t_k}^{t_{k+1}} \left[-K(\phi_2) - aT_x^2 + \beta^2\eta^2\right] d\tau \tag{42}$$

Assuming that the solution of the HJI equation is smooth, a two-layer linear parametric neural network with fixed input layers can be used to approximate the optimal value function. It means that the value function in parametric form can be expressed in terms of the constant weight matrix y and ξ as:

$$V^*(\phi(\tau)) = w^T \cdot \xi(\phi(\tau)) + \varepsilon(\phi(\tau)) \tag{43}$$

where $\xi(\phi(\tau)) = \left[\phi_1\ \phi_2\ \phi_1\phi_2\ \phi_1^2\ \phi_2^2\ \phi_1^2\phi_2^2\ \phi_1^3\right]^T$ is the basis vector for the HJI SOLA implementation [23]. The function $\varepsilon(*)$ is the reconstruction error, and w^T is weight matrix.

Substituting Eq. (41) into Eq. (42), the following equation is obtained:

$$w^T \Delta\xi(\phi) + \Delta\varepsilon = \int_{t_\varepsilon}^{t_{k+1}} \left[-K(\phi_2) - aT_x^2 + \beta^2\eta^2\right] d\tau \tag{44}$$

where $\Delta\xi(\phi) = \xi(\phi(t_{k+1})) - \xi(\phi(t_k))$, $\Delta\varepsilon = \varepsilon(\phi(t_{k+1})) - \varepsilon(\phi(t_k))$.

Denote $\mathscr{X}(\tau) = \begin{cases} \int_{t_{i-1}}^{t_k} \left(K(\phi_2) + aT_x^2 - \beta^2\eta^2\right) d\tau + \hat{V}(t_k) - \hat{V}(t_{k-1}), \tau = t_k \\ \int_{t_k}^{\tau} \left(K(\phi_2) + aT_x^2 - \beta^2\eta^2\right) d\tau + \hat{V}(\tau) - \hat{V}(t_k), \tau \in (t_k, t_{k+1}) \end{cases}$.

That means $X_{k+1} = V_{t_{k+1}}^* - \hat{V}_{t_{k+1}} + V_{t_k}^* - \hat{V}_{t_k}$.Since the exact solution of the HJI equation cannot be calculated, the estimate of the V^*, $\hat{V}$, is used for the iterative calculation. The expression of $\hat{V}$ is given as follows.

$$\hat{V} = \hat{w}^T \xi(\phi) \tag{45}$$

According to [7], the following NN weight adaptation rule is selected.

$$\dot{\hat{W}} = \begin{cases} -\alpha \frac{[\Delta\xi(\phi(\tau))]}{\left(1+[\Delta\xi(\phi(\tau))]^T[\Delta\xi(\phi(\tau))]\right)^2} X^T(\tau), \tau = t_k \\ -\alpha \frac{[\Delta\xi(\phi(\tau))]}{\left(1+[\Delta\xi(\phi(\tau))]^T[\Delta\xi(\phi(\tau))]\right)^2} X^T(\tau), \tau \in (t_k, t_{k+1}). \end{cases} \tag{46}$$

The above equation yields an estimate of the weight matrix $\hat{W}$. Using it, we can calculate $\hat{V}$. This gives us the estimate of the optimal control law $T_x^* T_x^*$

$$\hat{T}_x(t) = -\frac{\hat{V}_\phi}{2aI_x} = -\frac{[\nabla\xi(\phi)]^T}{2aI_x}\hat{w} \tag{47}$$

$$\hat{T}_e(t) = -\frac{[\nabla\xi(\phi_e)]^T}{2aI_x}\hat{w} \tag{48}$$

4 Simulation Results

In this section, the control effect of OECM is verified by simulation, and the power consumption and control stability of OECM are compared with those of traditional time-triggered update frequency methods.

4.1 Simulation Setup

The quadrotor dynamics model given in (49) is applied as follows:

$$\begin{cases} \dot{\phi}_1 = \phi_2 \\ \dot{\phi}_2 = \frac{1}{I_x}\big((I_y - I_z)r_y r_x + J_r r_y \hat{\omega} - d_\phi\big) + \frac{T_x}{I_x} - \frac{\sqrt{2}}{2}\frac{fl}{I_x}\phi_2 \end{cases} \tag{49}$$

where $I_x = 0.9$; $I_y = 0.7$; $I_z = 0.4$; T_x is the input, given by (21);

In determining the system update frequency, OECM uses formula (25) to obtain.

The parameters are selected as follows: $a = 0.5, k = 1, \lambda = 0.1, \beta = 0.7$. The traditional time-triggered control method is set to a fixed value $f = 100Hz$.

4.2 Interpretation of Result

Figure 2 shows the OECM control attitude Angle ϕ_t given in this paper and the ϕ_d obtained by sensor sampling; It can be seen that the fitting degree of ϕ_t and ϕ_d increases with time. Figure 3 shows the attitude angle ϕ of traditional time-triggered control method and the curve ϕ_d obtained by sensor sampling.

By comparison, it is found that the control error of OECM is quite close to that of the traditional control, and both are at a low level.

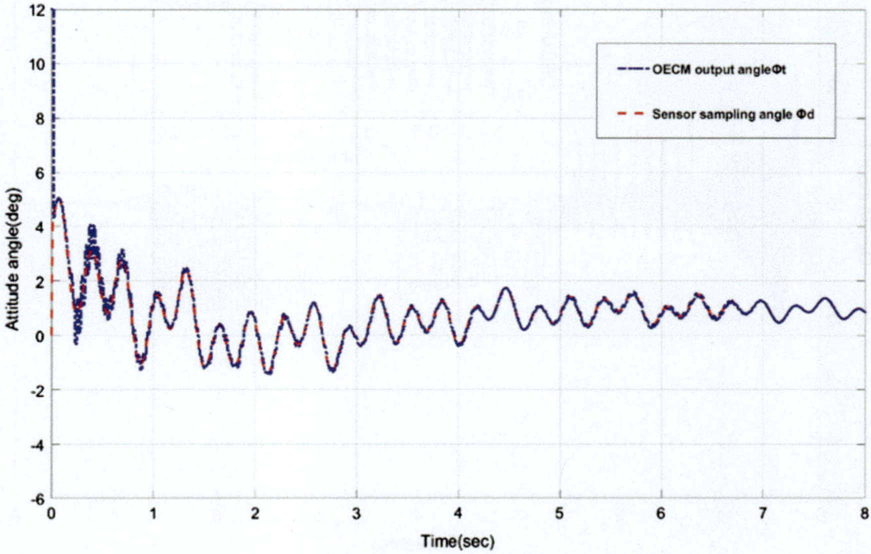

Fig. 2 The output effect of OECM

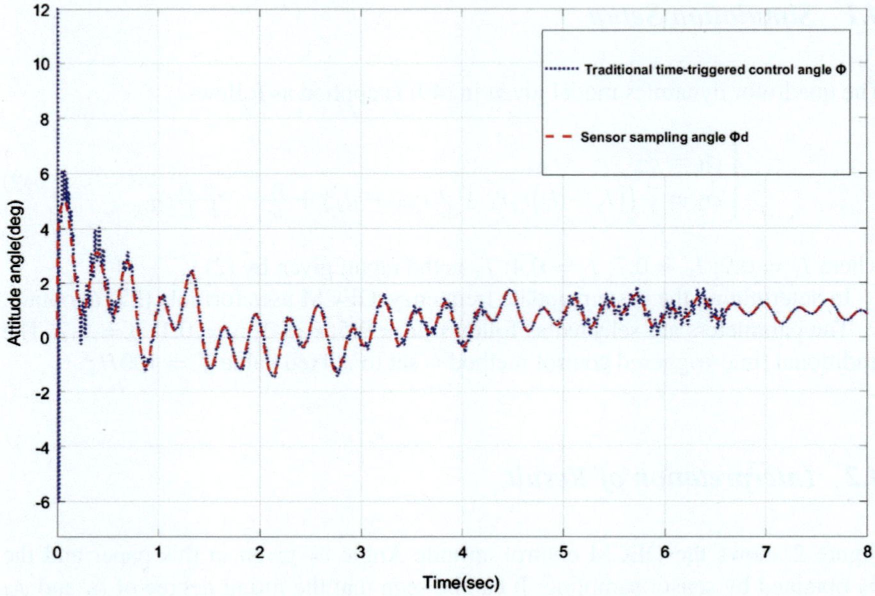

Fig. 3 The output effect of traditional time-triggered control method

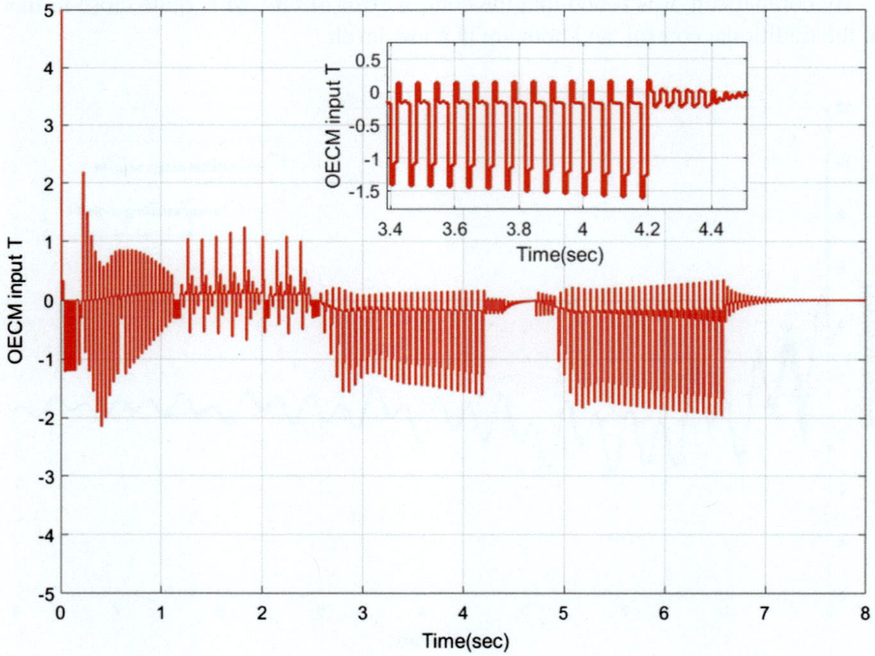

Fig. 4 OECM input value

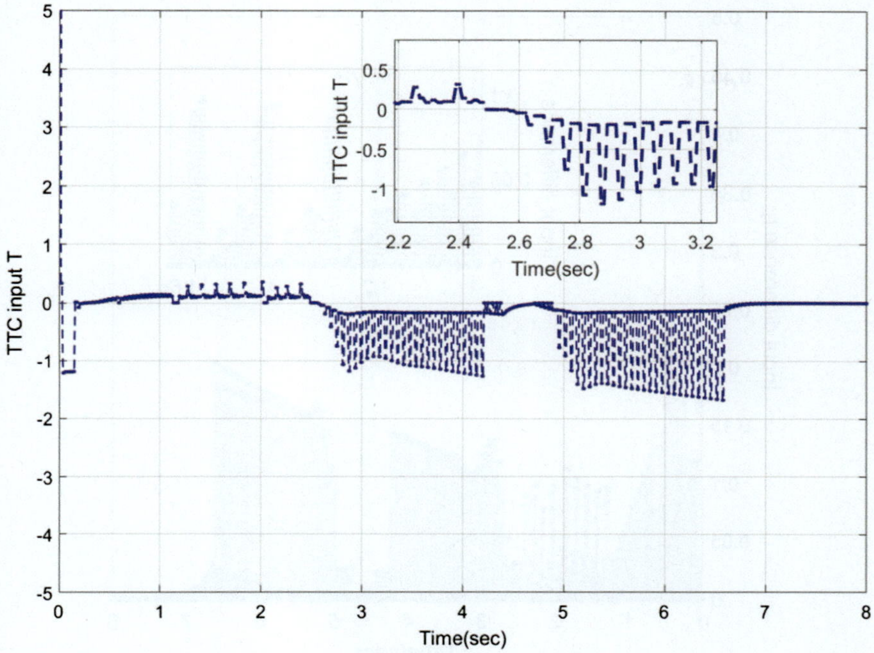

Fig. 5 Traditional time-triggered control method input value

Figures 4 and 5 depict the variation of input value T of OECM and the traditional time-triggered control method with time respectively. Figure 6 shows the input difference η of the two methods. In this simulation, the time when η is zero is the update time of OECM.

By comparison, it can be found that the difference between OECM and the traditional control methods in input is kept in a very low range, not only that, the number of OECM updates in the experiment is less than 50% of the traditional control. In other words, OECM can reduce the power consumption of the control system to less than 50% of the traditional control method while ensuring that the system has a high degree of fit in the control attitude angle.

Figure 7 shows the curve of the optimal update threshold. It can be seen that the value of Λ is always kept below 1.5 and decreases continuously until it is finally kept in a stable range, which ensures that OECM can timely eliminate errors while utilizing errors. And Λ also reflects the internal mechanism of the dynamic change of OECM update frequency over time. It not only reduces the power consumption of OCEM control to the optimal value but also ensures the stability of OECM control.

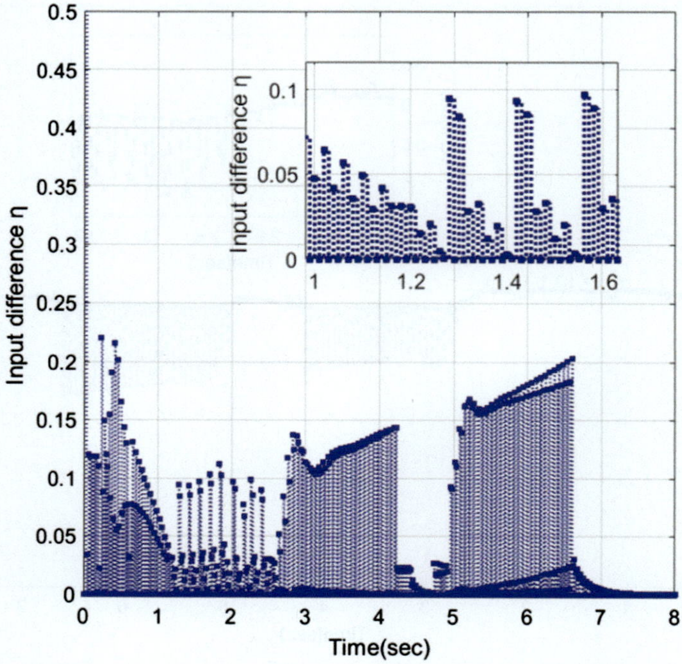

Fig. 6 Input difference

Fig. 7 The optimal update threshold

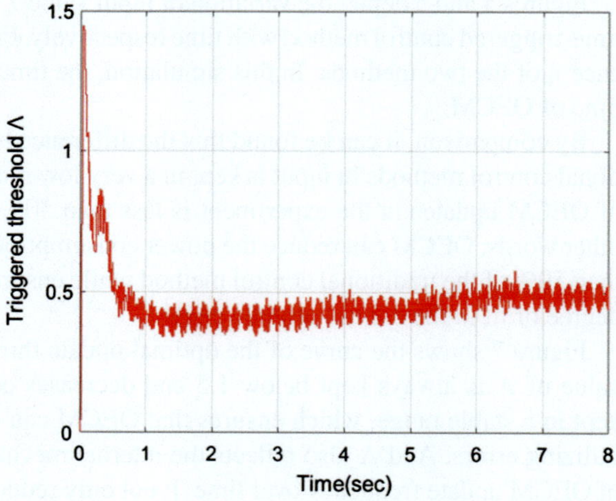

5 Conclusions

In this paper, an optimal event-triggered control method (OECM) of the quadrotor is proposed. Simulation results show that the proposed controller converges to the switching plane in the presence of external disturbances. Next, the optimal update frequency of this event-triggered controller is computed by solving the max–min optimization problem. Numerical simulations demonstrate that this control rate can reduce the power consumption of the quadrotor flight control. The experimental testing of this algorithm in practice is a focus of future work. In addition, the application of the algorithm in quadrotor cluster formation will be investigated in the future.

References

1. Lian S, Meng W, Lin Z et al (2021) Adaptive attitude control of a quadrotor using fast nonsingular terminal sliding mode. IEEE Trans Industr Electron 69(2):1597–1607
2. Narayanan V, Sahoo A, Jagannathan S (2018) Optimal event-triggered control of nonlinear systems: a min-max approach. 2018 Annual American control conference (ACC). IEEE, 3441–3446
3. Basar T, Bernhard PH (1991) [∞]-optimal control and related minimax design problems : a dynamic game approach. Autom Control IEEE Trans 41(9):1397
4. Molin A, Hirche S (2012) On the optimality of certainty equivalence for event-triggered control systems. IEEE Trans Autom Control 58(2):470–474
5. Lewis FW, Jagannathan S, Yesildirak A (1998) Neural network control of robot manipulators and non-linear systems. CRC Press
6. Chen P, Jiang Y, Dang Y et al (2022) Real-time efficient trajectory planning for quadrotor based on hard constraints. J Intell Rob Syst 105(3):52
7. Narayanan V, Jagannathan S (2016) Approximate optimal distributed control of uncertain nonlinear interconnected systems with event-sampled feedback. 2016 IEEE 55th conference on decision and control (CDC). IEEE, 5827–5832
8. Pouzesh M, Mobayen S (2022) Event-triggered fractional-order sliding mode control technique for stabilization of disturbed quadrotor unmanned aerial vehicles. Aerosp Sci Technol 121:107337
9. Tabuada P (2007) Event-triggered real-time scheduling of stabilizing control tasks. IEEE Trans Autom Control 52(9):1680–1685
10. Feng J, Hao F (2021) Event-triggered sliding mode control for time-delay uncertain systems. Asian J Control 23(3):1407–1418
11. Jung H, Han S, Lee S (2020) Event-Triggered H 2 attitude controller design for 3 DOF hover systems. IEMEK J Embedded Syst Appl 15(3):139–148
12. Wang J, Wang P, Tian B (2022) Hyperbolic tangent function-based fixed-time event-triggered control for quadrotor aircraft with prescribed performance. J Franklin Inst 359(12):6267–6285
13. Dou L, Cai S, Zhang X et al (2022) Event-triggered-based adaptive dynamic programming for distributed formation control of multi-UAV. J Franklin Inst 359(8):3671–3691
14. Aranda-Escolástico E, Rodríguez C, Guinaldo M et al (2018) Asynchronous periodic event-triggered control with dynamical controllers. J Franklin Inst 355(8):3455–3469
15. Warren D (2014) Optimal adaptive control and differential games by reinforcement learning principles. J Guidance Control Dyn
16. Agrawal OP, Baleanu D (2007) A Hamiltonian formulation and a direct numerical scheme for fractional optimal control problems. J Vib Control 13(9–10):1269–1281

17. Beard R (2008) Quadrotor dynamics and control rev 0.1
18. Saviolo A, Loianno G (2023) Learning quadrotor dynamics for precise, safe, and agile flight control. Ann Rev Control
19. Maaruf M, El Ferik S, Saif AWA (2022) Robust hovering control of a quadrotor. 2022 19th international multi-conference on systems, signals & devices (SSD). IEEE, 2180–2185
20. Fernando H, De Silva ATA, De Zoysa MDC, et al. (2013) Modelling, simulation and implementation of a quadrotor UAV. 2013 IEEE 8th international conference on industrial and information systems. IEEE, 207–212
21. Zhao B, Xian B, Zhang Y et al (2014) Nonlinear robust adaptive tracking control of a quadrotor UAV via immersion and invariance methodology. IEEE Trans Industr Electron 62(5):2891–2902
22. Besnard L, Shtessel YB, Landrum B (2012) Quadrotor vehicle control via sliding mode controller driven by sliding mode disturbance observer. J Franklin Inst 349(2):658–684
23. Dierks T, Jagannathan S (2010) Optimal control of affine nonlinear continuous-time systems using an online Hamilton-Jacobi-Isaacs formulation. 49th IEEE conference on decision and control (CDC). IEEE, 3048–3053

Research on the Intelligent Control System of Aerial Robots

Yuhan Yuan and Menghua Zhou

Abstract Intelligent aerial robots, featuring high stability, flexibility and adaptability, have a promising future in their wide applications in military and civilian fields, as automation, information technology and unmanned technology is booming. However, the current closed-source flight controllers fail to gain the underlying control, and thus the self-developed control algorithms cannot be verified; while the study of open-source flight controllers requires impossibly large effort due to the huge code architecture and may have many redundant functions. Therefore, to develop an intelligent flight control system independently becomes an important research topic. A whole process composing of mathematical model building, control law design and simulation experiment verification was completed in this paper, based on cascade control. The tests proved that the designed flight control system is highly accurate and stable with favorable dynamic performance, and is useful for the engineering application of UAV flight.

Keywords Aerial robot · Quadrotor unmanned aerial vehicle · Flight control · Modeling · Simulation verification

1 Introduction

Intelligent aerial robots refer to multi-degree-of-freedom (MDoF) aircrafts equipped with embedded on-board computers, inertial measurement units (IMUs), wireless communication devices and cameras, that can sense the environment and fly autonomously in the air and accomplish designated assignments under unmanned conditions [1]. Due to their high stability, flexibility and adaptability, intelligent aerial robots have a promising future in their wide application in military and civilian fields.

Y. Yuan
Jiangsu Meicun Senior High School, Wuxi, China

M. Zhou (✉)
Nanjing University of Aeronautics and Astronautics, Nanjing, China
e-mail: 1642567550@qq.com

Z. Liu et al. (eds.), *Advances and Challenges in Advanced Unmanned Aerial Systems*, Springer Aerospace Technology, https://doi.org/10.1007/978-981-99-8045-1_5

At present, quadrotors are the most popular in the market [2], which have fast response, can take off and land vertically and hover stably, can adapt to complex and changeable environment [3], and can also design built-in software and carry various load equipment according to demand [4].

Recently, the theory and practice of quadrotors technology have been developing steadily [5]. However, in the current research on flight control methods, the underlying control of closed-source flight control can not be obtained, and the open-source flight control source code architecture is huge and difficult to study, and there are many redundant functions. Therefore, the independent development of a set of intelligent flight control system is of great significance to verify the new control algorithm.

In this paper, quadrotors are used as the main research object. First, the accurate mathematical model is established, and then the cascade PID control scheme is used to build the flight control software subsystem, and the design, debugging and simulation of each loop are realized in the MATLAB/Simulink environment. The experimental results show that this paper has certain theoretical significance and engineering reference value for promoting the development of intelligent control system of quadrotors.

2 Dynamical and Kinematic Modeling of an Aerial Robot

An "x-shaped" quadrotor aerial robot was chosen as the research basis for the control law design. As in Fig. 1, the robot can roll, pitch, yaw and lift vertically by adjusting the four motor speeds to change the corresponding rotor speeds.

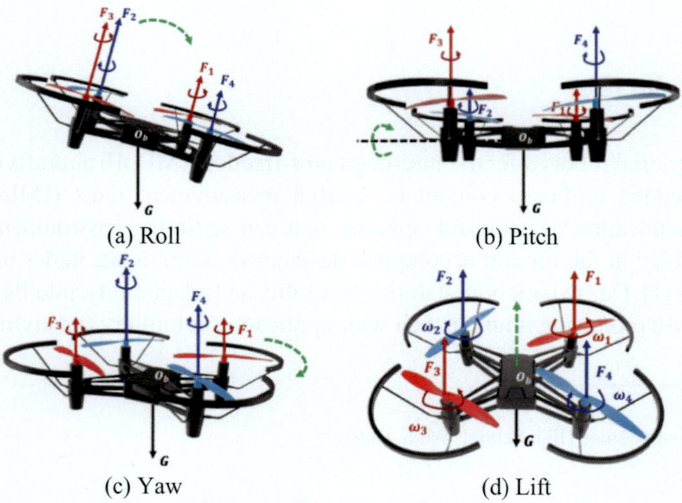

Fig. 1 Basic motions of an "X-shaped" quadrotor aerial robot

2.1 Dynamic Equations of the Quadrotor Aerial Robot

According to Newton's Second Law of Motion-Force and Acceleration:

$$\mathbf{F}_n = \begin{bmatrix} F_{xn} \\ F_{yn} \\ F_{zn} \end{bmatrix} = m\frac{d\mathbf{V}_n}{dt} = m\ddot{\boldsymbol{\xi}}_n = m\begin{bmatrix} \ddot{x} \\ \ddot{y} \\ \ddot{z} \end{bmatrix} \tag{1}$$

In the equation, $\mathbf{F}_n$ is the vector of resultant external forces on the robot, $\boldsymbol{\xi}_n$ is the absolute position and $\mathbf{V}_n$ is the robot's linear velocity.

The following formula can be obtained:

$$\begin{bmatrix} \ddot{x} \\ \ddot{y} \\ \ddot{z} \end{bmatrix} = \frac{1}{mR_n^B}\mathbf{F}_B = \frac{1}{m}R_B^n\mathbf{F}_B$$
$$= \begin{bmatrix} \frac{1}{m}(F_1 + F_2 + F_3 + F_4)(\cos\phi\sin\theta\cos\psi + \sin\phi\sin\psi) \\ \frac{1}{m}(F_1 + F_2 + F_3 + F_4)(\cos\phi\sin\theta\sin\psi - \sin\phi\cos\psi) \\ \frac{1}{m}(F_1 + F_2 + F_3 + F_4)(\cos\phi\cos\theta) - g \end{bmatrix} \tag{2}$$

In the equation, F_B is the vector sum of external forces on the robot in flight, the matrix R_n^B presents the transformation from the navigation coordinate system {n} to the body axis coordinate system {B}.

Based on Euler Equations [6], the following formula can be obtained:

$$\dot{\omega}_B = \begin{bmatrix} \dfrac{(F_2 + F_3 - F_1 - F_4)l - qr\left(J_{zz} - J_{yy}\right)}{J_{xx}} \\ \dfrac{(F_3 + F_4 - F_1 - F_2)l - pr(J_{xx} - J_{zz})}{J_{yy}} \\ \dfrac{(F_1 + F_3 - F_2 - F_4)l - qp\left(J_{yy} - J_{xx}\right)}{J_{zz}} \end{bmatrix} \tag{3}$$

In the equation, J is the rotational inertia under the body axis coordinate system, F_1, F_2, F_3 and F_4 is the lift of propellers corresponding to #1, #2, #3 and #4 motors.

In a nutshell, Eqs. (2) and (3) are dynamic equations of the quadrotor aerial robot.

2.2 Kinematic Equations of the Quadrotor Aerial Robot

Kinematic equations of the robot in translational and rotational motions are:

$$\dot{\xi}_n = \mathbf{V}_n \tag{4}$$

$$\dot{\eta} = \mathbf{T}\omega_B \tag{5}$$

In the equations, **T** is Euler Matrix [7], which can be expressed as:

$$\mathbf{T} = \begin{bmatrix} 1 & \sin\phi\tan\theta & \cos\phi\tan\theta \\ 0 & \cos\phi & -\sin\phi \\ 0 & \sin\phi\sec\theta & \cos\phi\sec\theta \end{bmatrix} \tag{6}$$

When the robot is in flight at a low speed, they can be approximated as:

$$\mathbf{T} = \begin{bmatrix} 1 & 0 & 0 \\ 0 & 1 & 0 \\ 0 & 0 & 1 \end{bmatrix} \tag{7}$$

$$\dot{\eta} = \boldsymbol{\omega}_B \tag{8}$$

3 Control Law Design of the Quadrotor Aerial Robot

Control algorithms are pivots for the quadrotor aerial robot to fly stably, including the attitude control loop, the horizontal position control loop and the altitude control loop. Control algorithms that have been studied and widely applied in academic and engineering fields include PID control, linear quadratic regulator (LQR) control, H∞ control, feedback linearization control and adaptive control. In this paper, PID control was adopted, but feedforward, restricted and filtering algorithms were added to some loops to avoid excessively frequent adjustments in the robot's pose and overcorrect fluctuations.

3.1 Principles of Single-Stage PID Control and Position PID Algorithm

PID controllers are widely employed in theoretical research and engineering practices, since they boast simple structures, favorable control effects and strong robustness, and they do not require accurate mathematical models of the controlled objects [8]. The basic structure of the PID control system is shown in Fig. 2 when disturbance noise disturbance signals are considered.

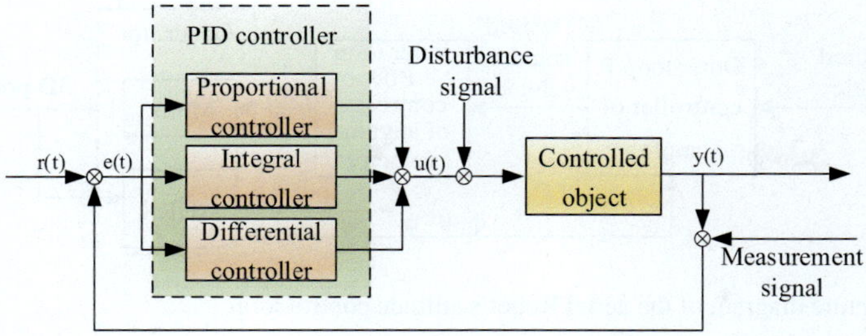

Fig. 2 Basic Structure of the PID control System with Disturbance Considered

3.2 *Realization of Cascade PID Algorithm*

Where a large nonlinear part of the controlled object is involved in the inner loop, its working point will move and destabilize the single-stage PID control. Therefore, the system will conspicuously oscillate and be less stable when there is wind or other disturbing factors [9]. However, when the inner loop of a cascade structure is disturbed or sees changes in condition, the entire system will still be stable. Thus, the closed-loop cascade PID control law was employed in this paper to enhance the robot system's stability and anti-disturbance capability, as presented in Fig. 3.

Attitude Control Loop. Figure tells that the attitude control loop and the horizontal position control loop works in cascade, and the outer loop control loop constantly adjusts the setting value of the inner loop controller according to disturbance or instruction changes. Therefore, the attitude control loop should improve the fast following performance, given that the loop is "fast, accurate and stable" [10].

The structure diagram of the attitude control loop is showcased in Fig. 4. The cascade PID control of the attitude loop can accurately and stably control both the attitude angle and the attitude angular velocity, thus laying a foundation for the stability of the horizontal position control loop of the system's outer loop [11].

Horizontal Position Control Loop. The horizontal position control loop adopts the closed-loop cascade PID control, realizing the conversion from the deviation in the

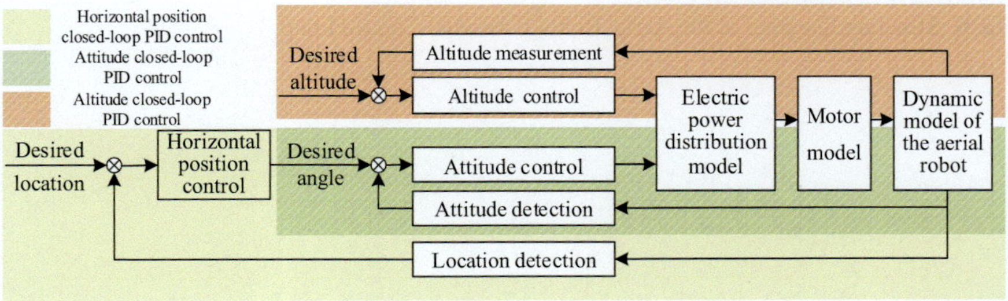

Fig. 3 Structure diagram of the Robot's closed-loop cascade PID control law

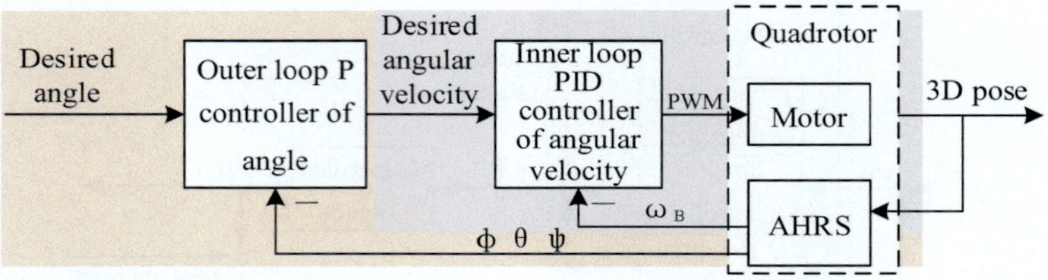

Fig. 4 Structure diagram of the aerial Robot's attitude control loop

current position to the desired attitude angle, as presented in Fig. 5. As the outer loop, the horizontal position controller constantly adjusts the desired value of the attitude loop of the inner loop according to varied loads and instructions, so as to enable the attitude to move or keep balance at a required horizontal position.

Altitude Control Loop. The altitude position control loop also adopts the closed-loop cascade PID control, realizing the conversion from the current altitude deviation to the median throttle value. The loop's structure diagram is shown in Fig. 6 [12].

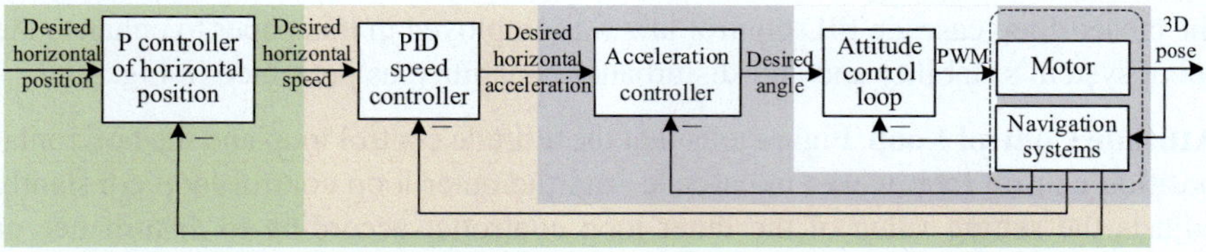

Fig. 5 Structure diagram of the aerial Robot's horizontal position control loop

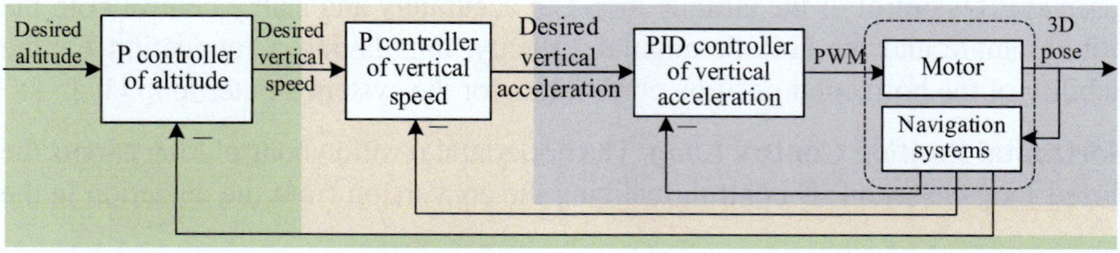

Fig. 6 Structure Diagram of the Aerial Robot's Altitude Control Loop

4 Flight Control Simulation

4.1 Cascade PID Control Simulation of All Loops

Based on the established module of the quadrotor aerial robot model, PID control loops of attitude, horizontal position and altitude were created, and were validated in Simulink respectively for their simulation effect.

(1) Attitude Control Loop

The desired Euler angle was set to be $\left[target\_roll\ target\_pitch\ target\_yaw \right] = \left[0.36\ 0.02t\ 0 \right]$ (unit: radian) and parameters of the attitude control loop were shown in Table 1.

Figure 7 is the simulation results of the attitude angle. The simulation results demonstrate that the step response of the roll angle in Fig. 7a witnesses the rise time of $t_r = 0.108s$, the regulation time of $t_s = 0.124s$, an overshoot of $\sigma\% = 1.01\%$ and a steady-state error of 0. That means the system responds quickly with moderate damping and favorable dynamic and stable performance; The ramp response of the angle of pitch in Fig. 7b can tell that there is a steady-state tracking error of $\varepsilon = 0.02 \times T = 0.00161 rad = 0.09°$ and the system is accurate in tracking. The yaw angle command input is zero in flight, so that the actual yaw angle in Fig. 7c denotes that the yaw angle fluctuates within 10^{-6} in the previous 1s, which can be neglected.

(2) Horizontal Position Control Loop

Based on the attitude control loop, parameters of the horizontal position control loop were selected to conduct the simulation validation of the PID control. The desired horizontal position was determined to be $\left[target\_pos\_x\ target\_pos\_y \right] = \left[0.4\ 0.6 \right]$ (unit: m) and parameters of the loop were set as in Table 2.

Figure 8 shows the simulation results. Analysis of the simulation results demonstrates that the step response towards the direction x in Fig. 8a witnesses $t_r = 1.25s$, $t_s = 1.39s$ and $\sigma\% = 0.15\%$; the step response towards the direction y witnesses

Table 1 Parameters of the attitude control loop

Parameter	Value
Body mass m/kg	1.50
Proportionality coefficient of the angular loop *p_angle*	$\left[20\ 20\ 20\right]$
Proportionality coefficient of the angular velocity loop *p_omega*	$\left[0.6\ 0.12\ 1\right]$
Integral coefficient of the angular velocity loop *i_omega*	$\left[0.002\ 0.0058\ 0.082\right]$
Differential coefficient of the angular velocity loop *d_omega*	$\left[0.0095\ 0.0095\ 0.004\right]$

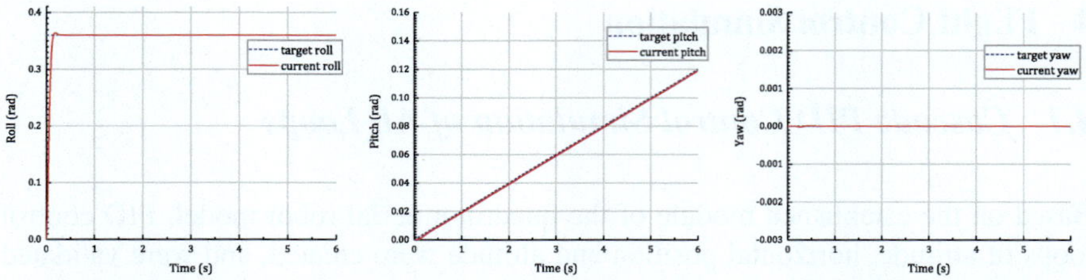

(a) Simulation Result of the Roll Angle

(b) Simulation Result of the Pitch Angle

(c) Simulation Result of the Yaw Angle

Fig. 7 Simulation results of the attitude control loop

Table 2 Parameters of the horizontal position control loop

Parameter	Value
Body mass m/kg	1.50
Proportionality coefficient of the horizontal position loop *p_pos*	$\begin{bmatrix} 1.68 & 1.28 \end{bmatrix}$
Proportionality coefficient of the horizontal velocity loop *p_vel*	$\begin{bmatrix} 3.9 & 3.9 \end{bmatrix}$
Integral coefficient of the horizontal velocity loop *i_vel*	$\begin{bmatrix} 0 & 0 \end{bmatrix}$
Differential coefficient of the horizontal velocity loop *d_vel*	$\begin{bmatrix} 0.003 & 0.003 \end{bmatrix}$

$t_r = 1.25s$, of $t_s = 1.46s$, $\sigma\% = 0.24\%$ and steady-state errors towards both directions of 0. That means the system responds quickly with moderate damping and favorable dynamic and stable performance.

(3) Altitude Control Loop

The desired horizontal position was designed to be $target\_alt = 0.4m$ and parameters of the altitude control loop were demonstrated in Table 3.

Figure 9 the simulation results of the altitude, vertical velocity and vertical acceleration respectively. Analysis of the simulation results indicates that the step response of altitude witnesses $t_r = 1.25s$, $t_s = 1.39s$, $\sigma\% = 1.28\%$ and a steady-state error of 0. That means quick responds and stable performance.

4.2 *Cascade PID Attitude Control Simulation*

Loop control models established were combined into a cascade PID control loop. Input and output modules were added to build a simulation platform model of the robot [13, 14]. Then the simulation validation of the cascade PID pose control could be conducted in Simulink.

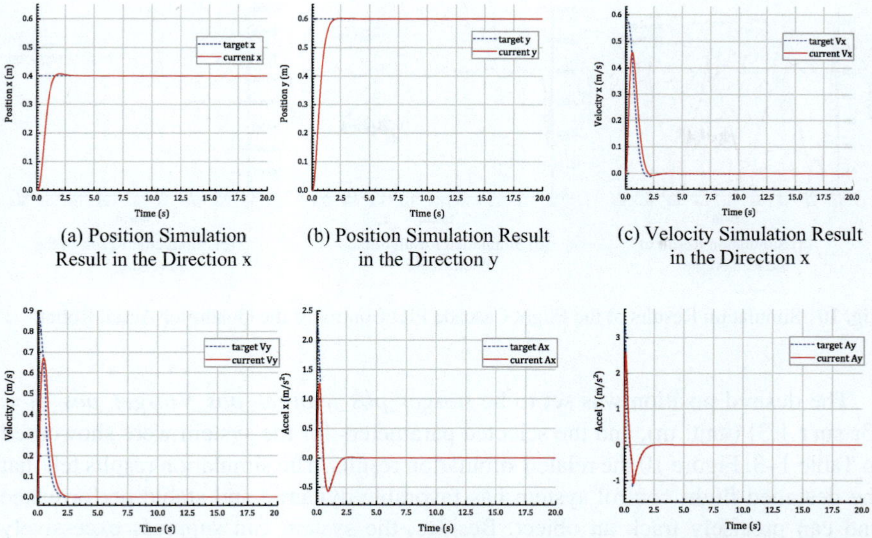

(a) Position Simulation Result in the Direction x

(b) Position Simulation Result in the Direction y

(c) Velocity Simulation Result in the Direction x

(d) Velocity Simulation Result in the Direction y

(e) Acceleration Simulation Result in the Direction x

(f) Acceleration Simulation Result in the Direction y

Fig. 8 Simulation results of the horizontal position control loop

Table 3 Parameters of the altitude control loop

Parameter	Value
Body mass m/kg	1.50
Proportionality coefficient of the vertical position loop p_alt	3.3
Proportionality coefficient of the vertical velocity loop p_vel_z	25
Proportionality coefficient of the vertical acceleration loop p_acc_z	16
Integral coefficient of the vertical acceleration loop i_acc_z	10
Differential coefficient of the vertical acceleration loop d_acc_z	0.006

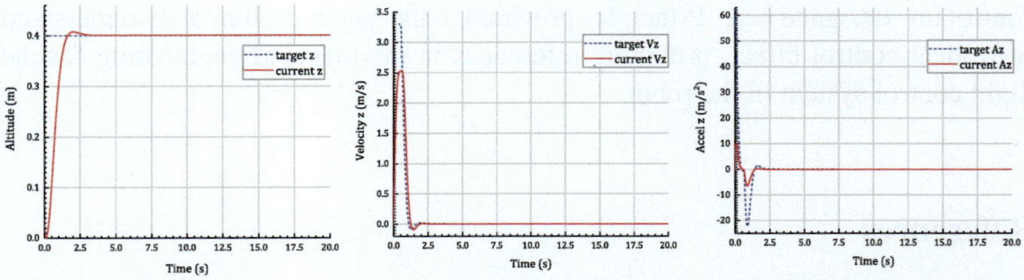

(a) Simulation Structure of Altitude

(b) Simulation Result of the Vertical Velocity

(c) Simulation Result of the Vertical Acceleration

Fig. 9 Simulation Results of the Altitude Control Loop

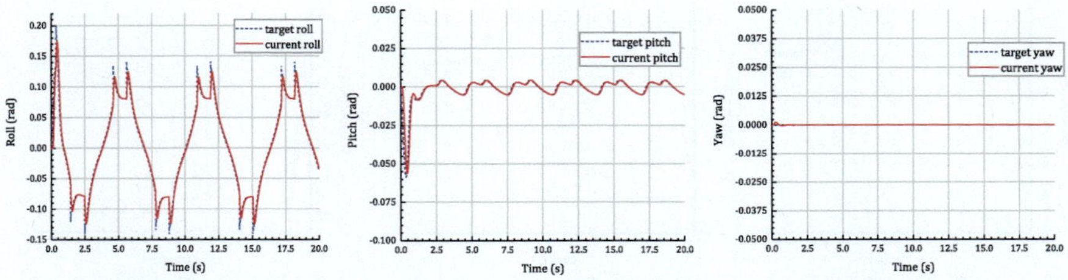

(a) Simulation Result of the Roll Angle (b) Simulation Result of the Pitch Angle (c) Simulation Result of the Yaw Angle

Fig. 10 Simulation Results of the Flight Cascade PID Control of the Quadrotor Aerial Robot

The desired position was set to be [*target_pos_x target_pos_y target_pos_z*] = [$3t$ $\sin t$ 1.3] (unit: m), and the selected parameters for the system were showcased in Table 1–3. Figure 10 the related simulation results. The simulation results tell that the designed flight control system has favorable dynamic and stable performance and can precisely track an object. Besides, the system can suppress excessively frequent adjustments and overcorrect fluctuations to some extent.

5 Conclusions

In this paper, the flight control system of a quadrotor aerial robot was studied. The systematic structure and flight principles of the robot were elaborated, based on which a six degrees of freedom dynamic model; cascade PID control and lowpass filtering were employed to realize accurate flight control, based on the hardware architecture of the system. Moreover, loops were designed, debugged and simulated in MATLAB/Simulink. By comparing theoretically calculated values and data provided by simulation experiments, the accuracy and stability of the motor model and the quadrotor aerial robot were validated, as well as the feasibility and practicality of the flight control law designed here. Principles provided in this paper can be easily understand with ideal control effect, providing references in theories and engineering for the flight control system of the robot.

References

1. Din AFU, Mir I, Gul F et al (2023) Robust flight control system design of a fixed wing UAV using optimal dynamic programming. Soft Comput 27:3053–3064. https://doi.org/10.1007/s00500-022-07484-z
2. Kim HJ, Shim DH (2003) A flight control system for aerial robots: algorithms and experiments. Control Eng Pract 11(12):1389–1400
3. Zhang G, Hsu LT (2018) Intelligent GNSS/INS integrated navigation system for a commercial UAV flight control system. Aerosp Sci Technol 80:368–380

4. Goupil P (2011) AIRBUS state of the art and practices on FDI and FTC in flight control system. Control Eng Pract 19(6):524–539
5. Kada B, Ghazzawi Y (2011) Robust PID controller design for an UAV flight control system. Proceedings of the world congress on engineering and computer science 2(1–6):1–6
6. Bouadi H, Mora-Camino F (2018) Modeling and adaptive flight control for quadrotor trajectory tracking. J Aircr 55(2):666–681
7. Olfati R (2001) Nonlinear control of underactuated mechanical systems with application to robotics and aerospace vehicles. Massachusetts Inst of Technol
8. Yu X, Jiang J (2011) Hybrid fault-tolerant flight control system design against partial actuator failures. IEEE Trans Control Syst Technol 20(4):871–886
9. Tischler MB (2018) System identification methods for aircraft flight control development and validation. Routledge
10. Jackson PB (2010) Overview of missile flight control systems. J Hopkins APL Tech Dig 29(1):9–24
11. Chi Y, Zhang Y, Liu Z (2015) A survey on technologies for automatic forest fire monitoring, detection, and fighting using unmanned aerial vehicles and remote sensing techniques. Can J For Res 45(7):150312143318009
12. Guo K, Jia J, Yu X et al (2020) Multiple observers based anti-disturbance control for a quadrotor UAV against payload and wind disturbances. Control Eng Pract 102:104560
13. Eliker K, Grounic S, Tadjine M et al (2020) Practical finite time adaptive robust flight controlsystem for quad-copter UAVs. Aerosp Sci Technol 98:105708
14. Goupil P (2010) Oscillatory failure case detection in the A380 electrical flight control system by analytical redundancy. Control Eng Pract 18(9):1110–1119

A Multi-Task Path Planning Algorithm of UAV Based on Improved RRT

Hongyu Deng and Hui Zhang

Abstract To address the multi-task path planning problem of UAVs in complex and realistic 3D geographic environments. We propose an improved RRT algorithm and use it for multi-task path planning. First, an equivalent 3D map is constructed based on the digital elevation model. Then, the RRT algorithm generates fewer effective sampling points and easily falls into local problems. Optimization of sampling space, introduction of variable step size and target guidance strategy are proposed to improve it. Finally, the paths are smoothed by cropping paths and B-sample curves. Experimental results show the improved algorithm significantly reduces path planning time and path length.

Keywords UAV · DEM · RRT algorithm · 3D path planning · Multi-task

1 Introduction

With the steady maturation of unmanned aerial vehicle (UAV) technology, the application of UAVs has been increasingly pervasive across a variety of industries, from industrial inspections to forest fire monitoring and pesticide spraying [1–3]. Path planning has emerged as a dominant area of focus in UAV research. Currently, 3D path planning is predominantly executed using a range of sophisticated methods, including the A* algorithm [4], D* algorithm [5], genetic algorithm (GA) [6], and rapidly exploring random tree (RRT) algorithm [7]. However, most of the path planning studies set up a single start point and end point, while less research has been done on multi-task point path planning. Multi-task point path planning involves finding a collision-free path from the start point through the task points and finally to the end point. The literature [8] proposed a multi-constraint optimization model for a multi-objective point path planning problem and solved it with a genetic algorithm. The literature [9] first clusters the multi-task objectives and performs path traversal of the clustered task points using an optimized genetic algorithm and a

H. Deng · H. Zhang (✉)
Inner Mongolia University, Hohhot 010021, China
e-mail: hui.zhang@imu.edu.cn

Z. Liu et al. (eds.), *Advances and Challenges in Advanced Unmanned Aerial Systems*, Springer Aerospace Technology, https://doi.org/10.1007/978-981-99-8045-1_6

neighborhood search algorithm. The literature [10] fuses the mathematical model of the Unmanned Surface Vessel (USV) with the RRT algorithm and uses a genetic algorithm to determine the order of multi-task point visits to generate the optimal path.

The RRT algorithm is a popular choice for solving path planning problems due to its simplicity. However, it has limitations, including a lack of search optimization, significant computational demands, and a tendency for path planning to fail in narrow environments. This paper has two main contributions. The first addresses the limitations of the RRT algorithm and the constraints of the UAV multitasking path planning and map model. We propose a multi-task path planning algorithm of UAV based on improved RRT. Secondly, we propose a B-spline curve trajectory planning method. It meets the requirements of smooth paths and having to pass through each task point.

2 Path Planning Problem Description

2.1 *Description of the UAV Multi-Task Point Path Planning Problem*

In this paper, considering the UAV's own constraints and the tasks to be performed, we establish a multi-condition constraint including obstacle avoidance, path length, flight altitude, angle and UAV multi-task point path planning objective function.

Multi-Condition Constraints

Maximum Range and Minimum Path Segment Length Constraints. The length between adjacent path points in the planned UAV path must be greater than or equal to the Euclidean distance between two points l_{min}. The sum of the length of all path segments after the UAV completes its task cannot exceed the farthest distance L_{max} flown by the UAV.

$$\forall l_i \geq l_{min}, \sum_{i=1}^{n} l_i \leq L_{max}, i = 1, 2, \cdots, n \tag{1}$$

l_i is the length between adjacent path points. n i s the number of segments in the path.

Maximum Turning Angle and Maximum Climbing Angle Constraints. Turn angle constraint means that the angle formed by three adjacent points in the UAV path point $p_{i-1}(x_{i-1}, y_{i-1}, z_{i-1})$, $p_i(x_i, y_i, z_i)$, $p_{i+1}(x_{i+1}, y_{i+1}, z_{i+1})$ vector $\overrightarrow{p_{i-1}p_i}$ and $\overrightarrow{p_i p_{i+1}}$ projection in the horizontal plane cannot exceed the maximum turn angle, and the maximum turn angle is set to 60° in this paper. The climb angle constraint means that the angle formed by two adjacent points in the vertical direction of the

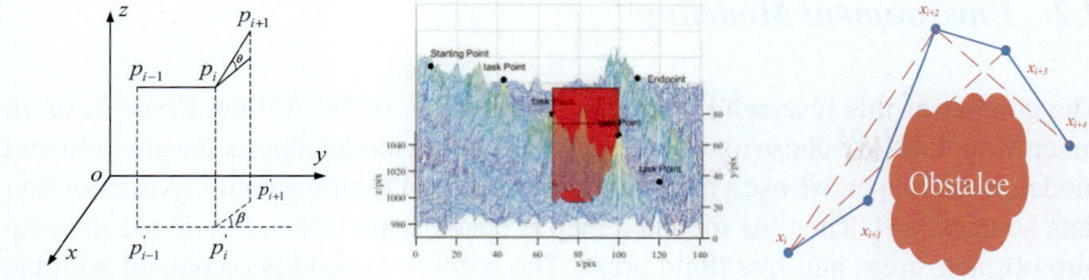

(a)Turning and climbing angle (b) Sample Space (c) Path cropping

Fig. 1 Three pairs of schematics

UAV path point cannot exceed the maximum climb angle, and the maximum climb angle is 90° in this paper. Figure 1a shows the diagram of the turn angle and climbs angle.

$$\theta = arctan\frac{|z_{i+1} - z_i|}{\sqrt{(x_{i+1} - x_i)^2 + (y_{i+1} - y_i)^2}}$$
$$\beta = arccos\frac{\overrightarrow{p_{i-1}'p_i'} \cdot \overrightarrow{p_i'p_{i+1}'}}{\left|\overrightarrow{p_{i-1}'p_i'}\right|\left|\overrightarrow{p_i'p_{i+1}'}\right|} \quad (2)$$

Flight Height Constraint. The photos taken by the drone on the ground during flight have high-quality requirements, and too high a flight height will lead to unclear photos with too low a resolution, and too low a flight height will easily lead to collisions. Therefore, considering the flight task and safety of the UAV, it is assumed that the height z_i of the path point $p_i(x_i, y_i, z_i)$ satisfies the following equation constraints.

$$H_{min} \le Z_i \le H_{max} \quad (3)$$

Objective Function. In this paper, the objective function of UAV path planning for multi-tasking is to generate a collision-free feasible trajectory from the starting point to the endpoint by passing through the task points sequentially. With $q(x, y, z, \theta, \beta)$ denoting the attitude of the UAV at a position in flight, (x, y, z) denoting the spatial coordinates of the UAV, θ denoting the horizontal turning angle of the UAV, and β denoting the vertical climbing angle of the UAV. With $L(p_1, \cdots, p_n)$ denotes the collision-free path from the starting point through the task point and finally reaching the endpoint. The objective function can be written in the following form.

$$q_1(x_1, y_1, z_1, \theta_1, \beta_1) \overset{L(p_1,\cdots,p_n)}{\rightarrow} q_n(x_n, y_n, z_n, \theta_n, \beta_n) \quad (4)$$

2.2 Environment Modeling

The context of this research is the UAV observation of the Yellow River Basin in Inner Mongolia. We chose to use the digital elevation model map as the environment model. In this paper, we use a regular grid model DEM for information extraction and data sources [11]. Then the modeled map is spatially partitioned to divide the map into obstacle areas and free flight areas. The specific method is combined with the flight height constraint of the UAV mentioned earlier, and the suitable flight height is $(\boldsymbol{H}_{min}, \boldsymbol{H}_{max})$. The accessible and inaccessible areas of the UAV constitute the 3D map $\boldsymbol{M}$.

$$M = \{m = (x, y, z)|x = 1, 2. \cdots, x_{max}, y = 1, 2, \cdots, y_{max}, z = h(x_i, y_i)\} \quad (5)$$

$$m(x, y, z) = \begin{cases} 1, z < h(x_i, y_i) \\ 0, z \geq h(x_i, y_i) \end{cases} \quad (6)$$

X_{max} and Y_{max} represent the scale of the selected map. $z = h(x_i, y_i)$ is the altitude value of the point obtained by reading the DEM data. $m(x, y, z) = 0$ is the area where the UAV can fly, and $m(x, y, z) = 1$ is the area where the UAV cannot pass.

3 Classical RRT Algorithm and Improved RRT Algorithm

3.1 Classical RRT Algorithm

Classical RRT algorithm is a sampling-based path planning algorithm. Randomly generate new points from a given starting point by a fixed step. If no collision occurs, the new point is added to the path until the new point is the endpoint.

3.2 Improved RRT Algorithm

Optimization of Sampling Space. Given the randomness of the sampling points generated by the RRT algorithm, we propose to reduce the sampling space to improve the effective sampling point generation rate. The sampling space is determined based on the coordinates of the starting point (x_1, y_1, z_1) and the endpoint (x_2, y_2, z_2) of the currently planned path segment according to the order of the starting and endpoints and task points. Firstly, the size of x_1, x_2, y_1, y_2 is compared to determine the value range of x and y, and the value range of z is the suitable flight height of the UAV. Figure 1b shows the reduced sampling space diagram.

The blind expansion of random points may create a large number of useless nodes and increase the planning time. To solve this problem, the idea of sampling point target guidance is proposed. A fixed value p in [0,1] is set, and a random number n in [0,1] is generated initially at each random point generation, and if $n \geq p$, the sampling space is sampled for the corresponding planning path segment, and assign the end point to the sampled point directly when n is less than p.

Variable Step Size Strategy. The step length in the traditional RRT algorithm is fixed, which increases the planning time when the nodes expand in the region with few obstacles; and when the nodes expand to the region with many obstacles, it leads to frequent collisions and getting into localization, resulting in planning failure. In this paper, we propose calculating the shortest distance between the current parent node and the current obstacle region each cycle, and selecting different step lengths based on comparing the shortest distance to the threshold. In the region with few obstacles, a large step size is used to pass the current region quickly; in the region with many obstacles, a small step size is used to reduce the possibility of collision.

$$step = \begin{cases} step1, d \leq thr \\ step2, d > thr \end{cases} \tag{7}$$

d is the shortest length between the parent node and the obstacle in the current area, the length of thr is $step2$, $step2$ is the large step size, and $step1$ is the small step size.

3.3 Path Trimming and Smoothing

The improved RRT algorithm using the above approach, the initial path is obtained by passing through all task points from the starting point and finally reaching the endpoint. Then the initial path is cropped to reduce the path length. Finally, the cropped path is smoothed with a B-sample curve to obtain a path that satisfies the performance of the UAV and can be practically flown.

Path Cropping. Figure 1c shows a diagram of path cropping, where the lines from the starting point x_i are connected to the next path points x_{i+1}, x_{i+2},and x_{i+3}, and these lines are checked for crossing obstacles. When $x_i x_{i+2}$ and $x_i x_{i+3}$ are detected as not crossing the obstacle, the path $x_i x_{i+1} x_{i+2} x_{i+3}$ is trimmed to $x_i x_{i+3}$, x_{i+1} and x_{i+2} are deleted, and x_{i+2} is kept if a collision occurs. Then start detecting the connection $x_{i+2} x_{i+4}$ until the extra points are removed.

Path Smoothing. For path point clipping, the path collapses and the direction and climb angle of some path points are too large to meet the practical flying demands of the UAV, so the path is smoothed using B spline curves. Treat the path plan as interpolated points, inverse performance control points, and then select points in each control path segment to generate a smoothed curve to ensure that the curve still passes through the task point. The expression of B spline is

$$P(u) = \sum_{i=0}^{n} P_i B_{i,k}(u) \tag{8}$$

$B_{i,k}(u)$ is the i th k th order B-sample basis function corresponding to the control point p_i. u are independent variables. We use the Triple B Spline curve to smooth the trajectory.

3.4 UAV Multi-Task Path Planning Algorithm Implementation

Figure 2 shows the process framework of the improved RRT-based UAV multi-task path planning algorithm in this paper.

4 Simulation Experiment and Result Analysis

To verify the feasibility of our proposed algorithm, it was contrasted with the classic RRT algorithm and the A* algorithm, and two equivalent simulation maps with complex terrain in the Yellow River basin of Inner Mongolia (data source [11]) were selected for 30 simulation experiments. The resolutions of the two maps are 30 m/pix. The average iterations, path length, and time data are derived from the sum of 30 experiments, divided by 30.

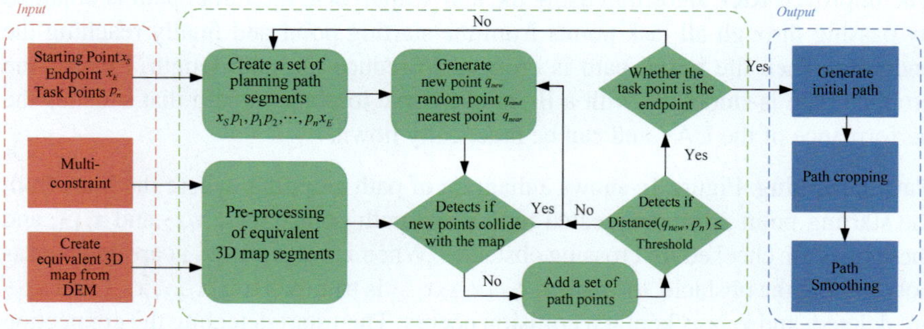

Fig. 2 A process framework for multi-task path planning algorithms for UAVs

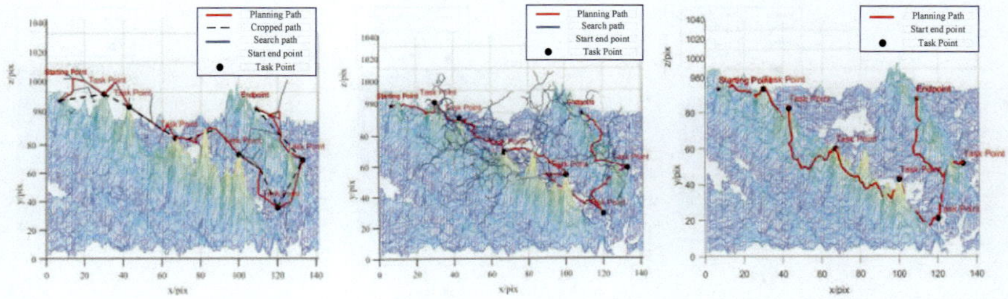

(a) Algorithm of this paper (b) RRT algorithm (c) A* Algorithm

Fig. 3 Small map simulation comparison experiments

Table 1 30 times comparison experiment of three algorithms

Planning algorithms	Average iterations	Average length/pix	Average time/s
Algorithm of this paper	**96**	**250.97**	**1.083**
RRT algorithm	1872	303.811	27.139
A* Algorithm	\	498.802	3.387

4.1 Experimental Comparison

Map 1 is small, the size is 141 × 91 × 100pix. set the starting point coordinates (7,85,995), task point coordinates (30,69,1020), (43,56,1024), (67,33,1025), (100,10,1036), (120,13,997), (123,39,997), (133,35,1006), the order of the task points listed is the order of the task points passed. The coordinates of the endpoint are (16,56,1010), the altitude range of the suitable airspace is [985,1050], and the size of the processed map is 141 × 91 × 65pix. Figure 3 shows the path planning results of the three algorithms. Table 1 shows the data comparison of 30 experiments.

The size of Map 2 is 562 × 316 × 150pix. Let the coordinates of the starting point be (99,292,1060), and the coordinates of the task points are (86,218,1045), (70,127,1030), (67,71,1025), (121,47,990), (195,62,990), (281,79,1031), (378,107,1015), (439,143,995). The endpoint coordinates are (505,216,985), the suitable altitude range is [940,1090]. Figure 4 shows the path planning results of the three algorithms. Table 2 shows the data comparison of 30 experiments.

4.2 Path Smoothing, Turning Angle, Climbing Angle Change

Figure 5 shows the results of a simulation experiment. Figure 5a is a schematic diagram of the smoothing of the cropped path using three B spline curves. It can be seen that after smoothing using three B spline curves, the result is good and it

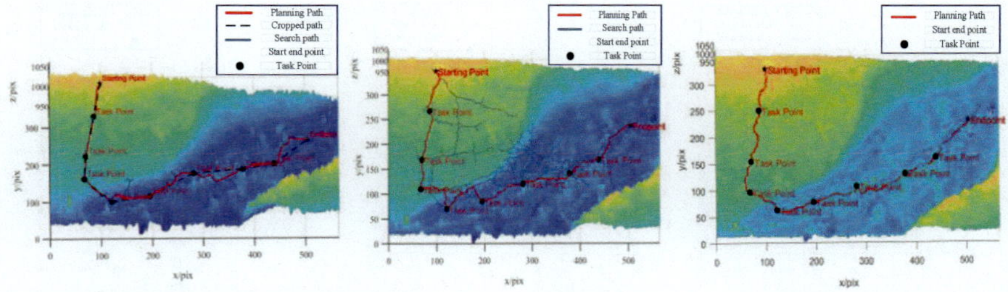

(a) Algorithm of this paper (b) RRT algorithm (c) A* Algorithm

Fig. 4 Large map simulation and comparison experiments

Table 2 30 times comparison experiment of three algorithms

Planning algorithms	Average iterations	Average length/pix	Average time/s
Algorithm of this paper	**276**	**752.129**	**3.825**
RRT algorithm	3206	926.061	42.096
A* Algorithm	\	1169.594	17.852

also goes through the task point. Figure 5b, c show the change in turn angle and climb angle after smoothing, respectively. The initial number of path points is 130, the number of path points after cropping is 22, and the number of path points after smoothing is 150. It can be seen that the turning angle of the smoothed path is greatly reduced contrasted with the previous path, and the climbing angle is also reduced after smoothing.

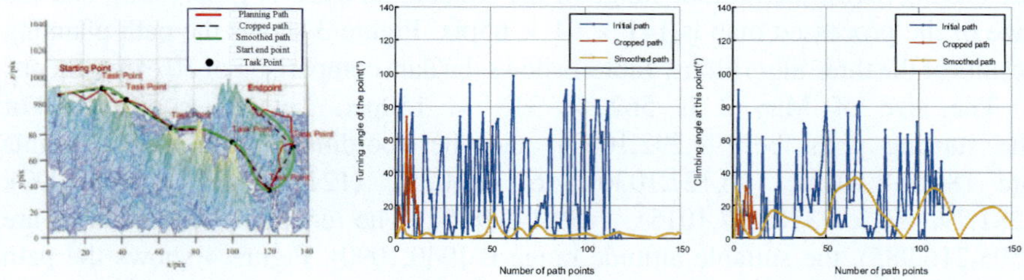

(a) Smoothed path (b) Turning angle change (c) Climbing angle change

Fig. 5 Results of a certain experiment

5 Conclusion

In this paper, in order to solve the UAV multi-task path planning issues in complex 3D space environment. First, the mathematical model of the problem is established, and then the RRT algorithm is improved for the shortcomings of the RRT algorithm. Then it is used for UAV multi-task point path planning. Finally, simulation experiments were conducted with real maps. The algorithm of this paper is contrasted with RRT algorithm and A* algorithm. The path length, planning time, number of iterations, turning angle and climbing angle indicators are evaluated to verify the feasibility of the algorithm in this paper. At the same time, the method proposed in this paper has limitations, including a lack of consideration for the dynamic and control constraints of the UAV. Additionally, there have been no physical experiments conducted to test the practical feasibility of the algorithm.

References

1. Breitenmoser A, Tâche F, Caprari G et al (2010) toward multi climbing robots for power plant inspection. In: Proceedings of the 9th international conference on autonomous agents and multiagent systems, 1713–1720
2. Merino L, Caballero F et al (2012) An unmanned aircraft system for automatic forest fire monitoring and measurement. J Intell Rob Syst 65:533–548
3. Li L et al (2010) A brief discussion on the application areas and prospects of drones. Geospatial Inf 8(05):7–9
4. Sun J, Tang J, Lao S (2017) Collision avoidance for cooperative UAVs with optimized artificial potential field algorithm. IEEE Access. 5:18382–18390
5. Cheng N et al (2020) A Dijkstra-based algorithm for intelligent vehicle trajectory planning under multiple constraints. Northwestern Polytechnic University 38:1284–1290
6. Yan S, Pan F (2019) Research on route planning of auv based on genetic algorithms. In: IEEE international conference on unmanned systems and artificial intelligence, 184–187
7. LaValle SM (1998) Rapidly-exploring random trees: a new tool for path planning
8. Xiao C, Zhou Y et al (2019) Multi-UAV multi-target point path planning in obstructed areas. Space Control Technol Appl 45(04):46–52
9. Liu W, Wang Y (2019) Research on the path planning problem of multi-UAV collaborative search for multiple targets. Electro-Optics Control 26(3):35–38+73
10. Liu J et al (2021) A multi-task point path planning method for surface unmanned boats. J Shanghai Maritime Univ 42(4):1–6
11. Geospatial Data Cloud Homepage, http://www.gscloud.cn. Accessed 1 March 2023

Effects of Geometric Nonlinearity on the Gust Responses of Near-Space Solar-Powered Vehicle Landing Phase

Jinyang Wu, Guanghui Wu, and Xiaohui Wei

Abstract The near-space solar-powered aircraft was taken as the research object in this study, which was characterized by lightweight, high flexibility and low landing speed. Based on the CFD/CSD loosely coupled analysis method, the grid velocity method was employed to introduce gust loads, and the effects of geometric nonlinearity on the static equilibrium state and gust load state during landing phase of near-space solar-powered aircraft were investigated. The influence of geometric nonlinearity on the structure and aerodynamic performance of the aircraft during landing phase was revealed. The results showed that after the gust response of the aircraft reached its peak, the lift coefficient, the structure bending moment of wing root, and wingtip displacement obtained without considering the geometric nonlinearity of the aircraft did not recover to the static equilibrium state, and the wingtip had a Z-direction displacement of −0.18 m and the maximum counter-twist of 3.3°. The calculation results considering geometric nonlinearity were more consistent with the actual situation of various aircraft parameters. Therefore, the geometric nonlinearity of the structure should be taken into account when analysing the gust response of near-space solar-powered aircraft during landing phase.

Keywords Near-space vehicle · Gust response · Landing phase · Geometric nonlinearity · CFD/CSD coupled

1 Introduction

Near-space solar-powered aircrafts typically cruise in the near-space region between 20 and 100 km. Compared to conventional aircraft, near-space solar-powered aircraft have advantages such as lightweight, high lift-to-drag ratio, long cruising time, low

J. Wu · X. Wei (✉)
College of Aerospace Engineering, Nanjing University of Aeronautics and Astronautics, Nanjing, China
e-mail: Wjysx2101140@nuaa.edu.cn

G. Wu
Commercial Aircraft Corporation of China, Shanghai, China

Z. Liu et al. (eds.), *Advances and Challenges in Advanced Unmanned Aerial Systems*, Springer Aerospace Technology, https://doi.org/10.1007/978-981-99-8045-1_7

energy consumption, and strong concealment, which make them widely applicable in both military and civilian fields [1]. In recent years, the research and development of near-space solar-powered aircraft has achieved rapid development [2]. However, a review of the work carried out domestically and internationally reveals that relatively little research has been conducted on the takeoff and landing processes of near-space solar-powered aircraft. Near-space solar-powered aircraft often has geometric nonlinear structural features such as large size, lightweight, long wingspan, and high body flexibility; high lift and low drag aerodynamic performance requirements; and extremely low flight speeds and high air density during landing compared to cruising conditions. These factors can easily cause large bending and torsional deformations of the wing under sudden airflow load conditions such as wind shear and turbulence, leading to changes in the wing's aerodynamic load distribution and, in turn, significant impacts on the aircraft's aerodynamic characteristics, even lead to aircraft disintegration and crash accidents. Therefore, studying the aerodynamic-elastic gust response of near-space solar-powered aircraft during takeoff and landing is of great significance to flight safety.

In the past, due to technical limitations, the gust response analysis of flexible aircraft mainly used aerodynamic-elastic spatial state models based on linear rigid-flexible coupling models. However, this method is inadequate for large flexible aircraft with significant deformation, as it is necessary to consider the gust response characteristics brought by geometric nonlinearity [3]. As a result, researchers at home and abroad have carried out a large amount of work, such as using nonlinear beam structure and high angle of attack aerodynamic model [4], quasi-modal method and unsteady vortex lattice method [5], quasi-modal method and Modified Theodorsen method combined with strip theory [6] to carry out structural and aerodynamic modeling, and discussed the changes of structure and aerodynamic characteristics of large flexible aircraft under different gust conditions. In recent years, it has become feasible and rapidly developing to couple CFD based on Euler equations or Navier–Stokes equations with CSD to solve the precise calculation of nonlinear aerodynamic-elastic problems [7]. In the study of geometrically nonlinear large flexible aircraft gust response, CFD/CSD coupling demonstrates good performance, accurately reflecting the dynamic characteristics of the aircraft under large displacement and large structural deformation conditions [3].

Based on CFD/CSD loose coupling analysis method, the grid velocity method is employed to introduce gust loads, and the influence of geometric nonlinearity on the static equilibrium state and gust response of near-space solar-powered aircraft during the initial landing phase are investigated. The effects of geometric nonlinearity on aircraft's structure and aerodynamic performance are founded.

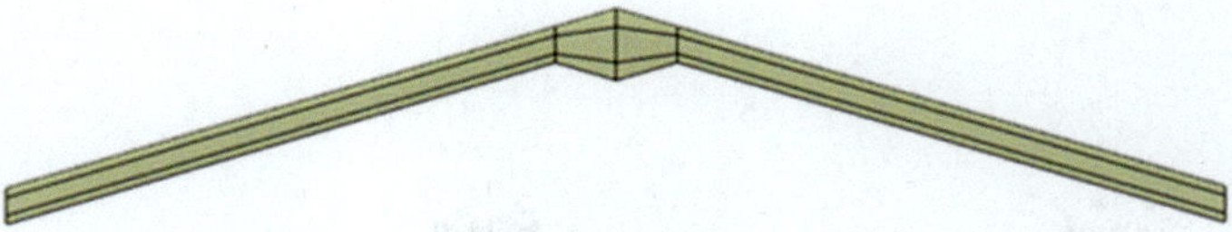

Fig. 1 The near-space solar-powered aircraft model

2 Model of Near-Space Solar-Powered Aircraft

2.1 *Subject Design*

Referring to the design of U.S. Facebook's Aquila unmanned aerial vehicle, the calculated near-space solar-powered aircraft model is shown in Fig. 1. The solar-powered aircraft adopts an inner and outer wing flying-wing layout; the wingspan of the aircraft is 48.00 m, the average aerodynamic chord length is 1.47 m, the aspect ratio is 32.65, and the total mass is 280 kg. The material parameters used in the aircraft are refer to the reference [8].

2.2 *Model Construction and Boundary Conditions*

The Ansys software was used for model construction. During modeling, considering the landing state, the aircraft has a certain angle of attack, and the incoming flow speed is very low, with air density close to sea level. Therefore, the boundary conditions were set with the aircraft's flight altitude at 500 m, flight speed at 40 km/h (11.1 m/s), an initial angle of attack at landing of 4°, and atmospheric density of 1.167 kg/m^3. Given the symmetry of the calculation object, to improve computational efficiency, only one side of the flow field was analyzed, hence a semi-solid model was chosen. To reduce the number of fluid grids, the fluid domain was constructed with the front and back planes parallel to the leading edge of the aircraft. The Ansys Mechanical was used for fluid and solid grid division, with refined processing on the fluid–solid coupling face grid. Figure 2 presents a schematic of part of the model grid.

2.3 *Setting up Geometric Nonlinearity*

Within the realm of elasticity mechanics, if the assumption is based on small deformation, i.e., the displacement and deformation are relatively small, the impact of this deformation on stiffness can be disregarded. Therefore, in general linear calculations, the establishment of equilibrium equations is based on the initial geometric relationship of the structure, and the stiffness matrix of the unit remains unchanged,

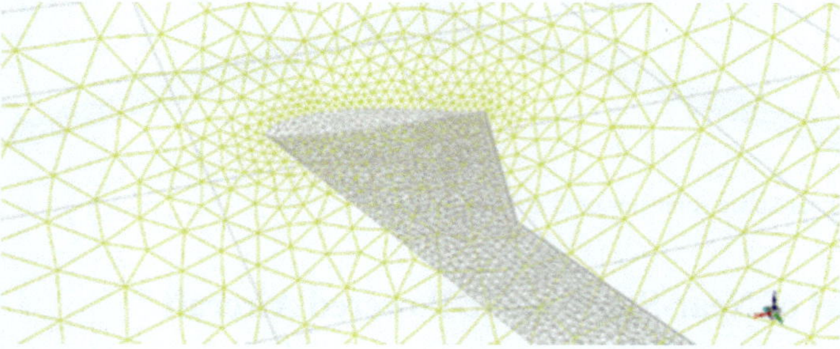

Fig. 2 Schematic diagram of model grid

that is, the matrix under the initial state. However, in some nonlinear problems, the effect of deformation on stiffness needs to be considered.

When using Ansys for simulation analysis, for geometric nonlinearity calculations, the "Large Deflection" analysis setting is selected. The stiffness changes caused by large deflections, large rotations, and large strains that lead to changes in shape and direction of the structure are taken into account. During each iteration process, a new stiffness matrix is regenerated based on the new geometric position coordinates of the structure, making the calculation more precise.

3 Calculation Method

Though the low Reynolds number effect is significant during the cruising stage of near-space solar-powered aircraft under study, the Reynolds number exceeds one million during the takeoff and landing processes, falling into the high Reynolds number range. Consequently, a loosely coupled CFD/CSD method is employed for solving. In the CFD section, the finite volume method based on the SST turbulence model is utilized to compute the compressible integral form of Reynolds-averaged Navier–Stokes (RANS) equations. During the spatial discretization process, the convective flux term adopts the second-order accuracy upwind scheme of Roe's format; the viscous flux term opts for the second-order accuracy central discretization method; the time advancement employs the LU-SGS implicit time advancement technique.

3.1 Aerodynamic Model

The SST turbulence model combines the standard k-ε model with the standard k-ω model through a blending function [9]. This model exhibits excellent numerical characteristics in near-wall regions and free shear layers and can effectively simulate flows with significant adverse pressure gradient features. The general governing equation for the SST turbulence model is:

$$\frac{\partial}{\partial t}(\rho k) + \frac{\partial}{\partial x_i}(\rho k \overline{u}_i) = \frac{\partial}{\partial x_j}\left(\Gamma_k \frac{\partial k}{\partial x_j}\right) + F_k - Y_k \tag{1}$$

$$\frac{\partial}{\partial t}(\rho \omega) + \frac{\partial}{\partial x_i}(\rho \omega \overline{u}_j) = \frac{\partial}{\partial x_i}\left(\Gamma_\omega \frac{\partial \omega}{\partial x_i}\right) + F_\omega - Y_\omega + D_\omega \tag{2}$$

where t represents time; ρ (kg/m^3) denotes density; k stands for turbulent kinetic energy; ω signifies specific turbulent dissipation; $\overline{u}_i$ and $\overline{u}_j$ are the mean values of turbulent velocities; x_i and x_j are coordinate components; Γ_k and Γ_ω are effective diffusion coefficients; F_k and F_ω represent turbulent generation terms; Y_k and Y_ω are the dissipation terms for k and ω, respectively; D_ω denote diffusion terms.

3.2 Grid Velocity Method

The gust model is introduced using the grid velocity method, and the flow field velocity in the fluid region can be expressed as:

$$\boldsymbol{V} = (u - x_t)\boldsymbol{i} + (v - y_t)\boldsymbol{j} + (w - z_t)\boldsymbol{k} \tag{3}$$

In the formula: u, v and w represent the velocity components in the Cartesian coordinate system; x_t, y_t and z_t are the grid velocity components in the three directions; $\boldsymbol{i}$, $\boldsymbol{j}$ and $\boldsymbol{k}$ represent the unit vectors in each direction of the Cartesian coordinate system.

When the fluid region is affected by an upward gust with a velocity of w^i, a sudden increase in velocity of w^i occurs in the Z-direction of the fluid region grid, while in reality, the fluid domain grid does not move. At this time, the fluid domain is equivalent to the inflow effect of $-w^i$ under the condition that the grid does not move. Based on this, the flow field velocity in the fluid domain under the gust effect can be expressed as:

$$\boldsymbol{V} = (u - x_t)\boldsymbol{i} + (v - y_t)\boldsymbol{j} + [w - (z_t - w^i)]\boldsymbol{k} \tag{4}$$

3.3 Gust Calculation Model

The gust pattern experienced by the aircraft is shown in Fig. 3. The commonly used 1-cos engineering gust model [10] is employed, where the entire aircraft is subjected to the gust influence in the vertical direction as expressed in Eq. (5):

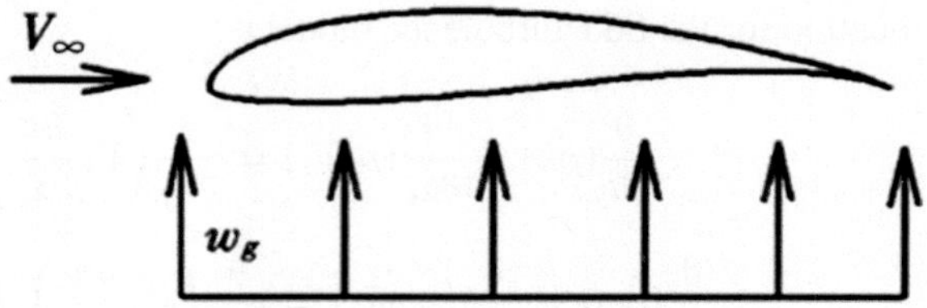

Fig. 3 The gust pattern of the aircraft [11]

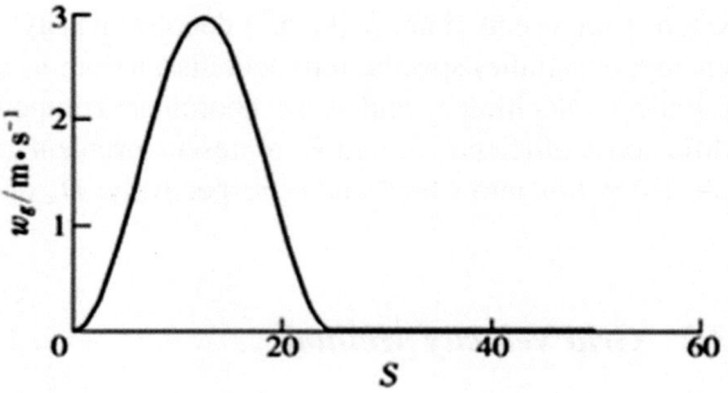

Fig. 4. 1-cos type gust velocity pattern [11]

$$w_g = 0.5W_0\left(1 - \cos\frac{2\pi x}{L_g}\right) \tag{5}$$

In the equation: W_0 represents the gust amplitude; L_g represents the gust length, and x is the spatial length. Figure 4 illustrates the 1-cos gust velocity pattern encountered by the aircraft. Assuming the average aerodynamic chord length of the aircraft is c, and defining the dimensionless time $S = 2V_\infty t/c$.

3.4 Calculation Process

In this study, even when the aircraft does not encounter gusts during landing, the wing itself is in a state of equilibrium with a certain deformation. Therefore, establish the computational process of the CFD/CSD loose coupling method as shown in Fig. 5.

4 Calculation Results and Analysis

4.1 Example Verification

In order to verify the accuracy of the computational method, a uniform cross-section wing with one end fixed and the other end free was selected as the validation case [12]. The wing profile adopts the NACA0012 airfoil, with an aspect ratio of 1:5, a

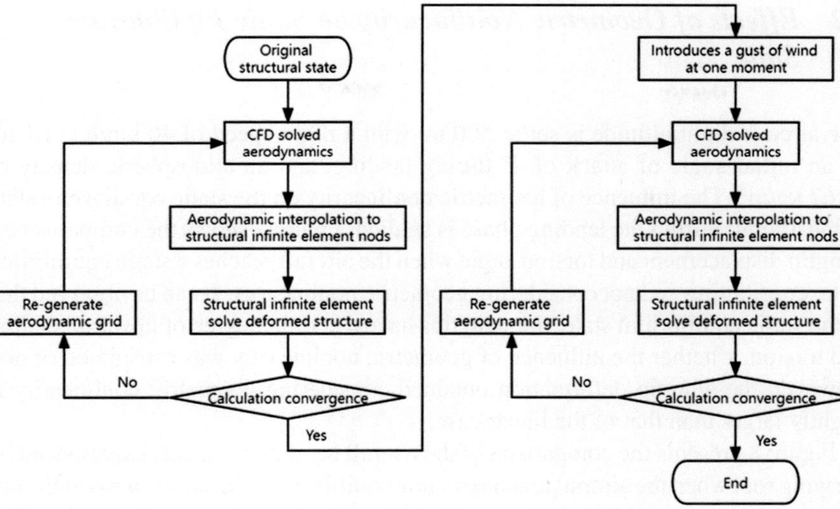

Fig. 5 Computing flowchart for loose coupling method

root chord length of 1 m, a bending stiffness EI = 100kN·m^2, and a unit length wing mass = 10 kg/m; the specific computational conditions can be found in reference [12]. Figure 6 shows the comparison between the calculated values and the literature values. As can be seen from the figure, except for the oscillation frequency being slightly larger than the literature value (the reason being that the literature only considers the first-order bending mode, while this study considers both bending and torsional modes), the calculated results are in good agreement with the literature values [12], demonstrating the effectiveness and accuracy of the computational method presented in this paper.

Fig. 6 Comparing the calculated value with the literature value

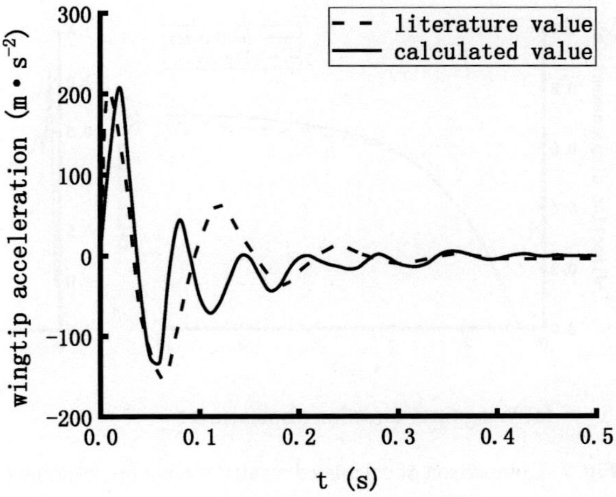

4.2 Effects of Geometric Nonlinearity on Static Equilibrium State

The aircraft flight altitude is set to 500 m, with a flight speed of 40 km/h (11.1 m/s), an initial angle of attack of 4° during landing, and an atmospheric density of 1.167 kg/m^3. The influence of geometric nonlinearity on the static equilibrium state of the aircraft during the landing phase is studied. Figure 7 shows the comparison of wingtip displacement and torsion angle when the aircraft reaches a static equilibrium state, considering and not considering geometric nonlinearity. It can be observed that in the static equilibrium state, the wingtips have a certain degree of upward bending and torsion, whether the influence of geometric nonlinearity was considered or not. However, the wingtip deformation obtained considering geometric nonlinearity is slightly larger than that of the linear case.

Figure 8 presents the comparison of the overall bending moments experienced by the wing root when the aircraft reaches a static equilibrium state under two conditions, with and without considering geometric nonlinearity. It can be seen that in the static equilibrium state, the bending moment obtained considering geometric nonlinearity is slightly larger than that of the linear case, and there is a periodic small-amplitude oscillation in the bending moment under linear conditions. This results in the wingtip deformation obtained considering geometric nonlinearity being slightly larger than that of the linear case.

Figure 9 displays the comparison of the overall lift coefficient of the aircraft when it reaches a static equilibrium state under two conditions, with and without considering geometric nonlinearity. It can be observed that in the static equilibrium state, the overall lift coefficient results obtained considering and not considering the effect of aircraft geometric nonlinearity show a relatively slight difference. This is due to the fact that the overall deformation of the aircraft in the static equilibrium state is quite small.

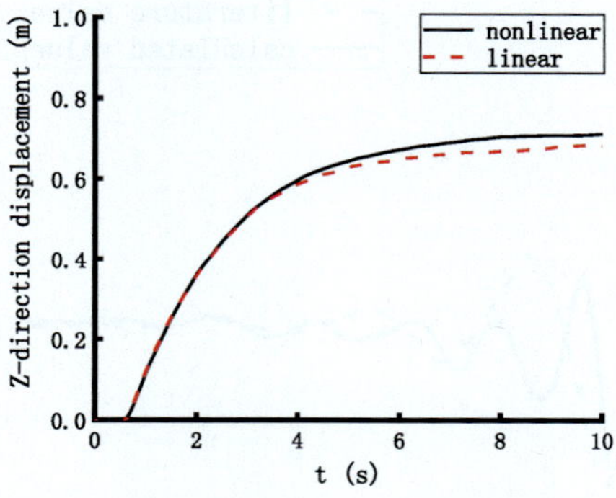

(a) Wing tip Z-direction displacement

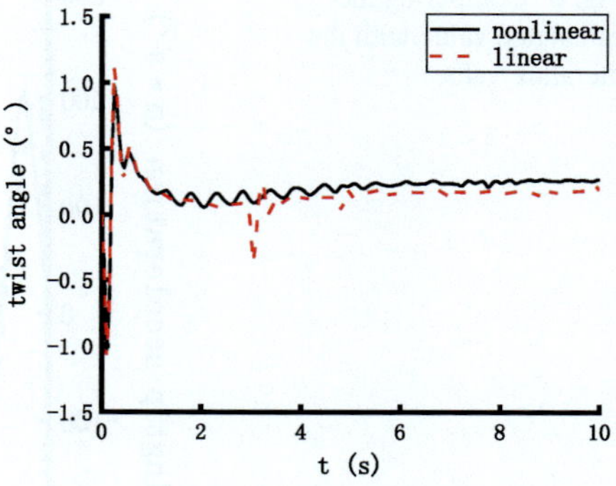

(b) Wing tip twist angle

Fig. 7 Comparison of calculated results of wing tip deformation in static equilibrium state

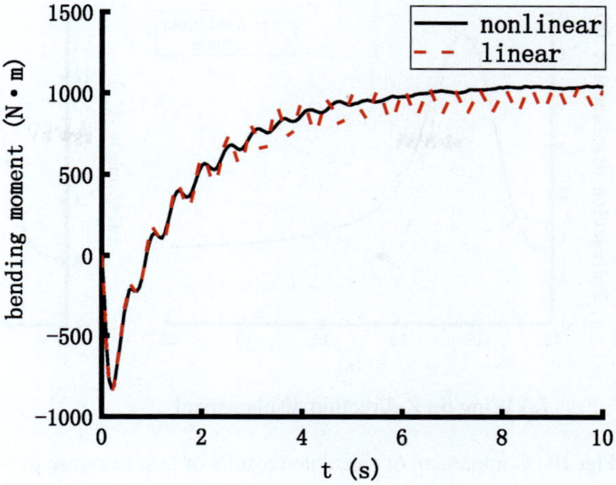

Fig. 8 Comparison of calculation results of bending moment in static equilibrium state

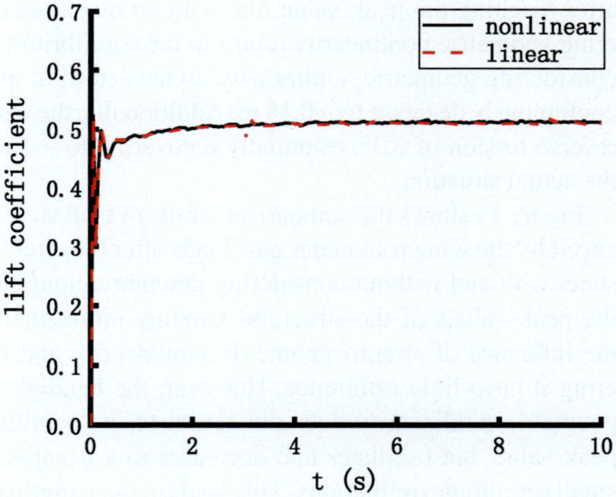

Fig. 9 Comparison of calculation results of aircraft lift coefficient in static equilibrium state

4.3 Effects of Geometric Nonlinearity on Gust Response

Based on the static equilibrium state, the gust load is applied at the 10th second. The gust amplitude is 6 m/s, and the gust length is set at 25 times the aircraft's average aerodynamic chord. The effect of geometric nonlinearity on the gust response of the aircraft during landing is investigated. Figure 10 shows the comparison of wingtip displacement and torsion angle under gust loads after the aircraft reaches a static equilibrium state, with and without considering geometric nonlinearity. It can be observed that both the wingtip deformation obtained considering the influence of aircraft geometric nonlinearity and that obtained without considering it exhibit large response peak values, with little difference between the peak values. However,

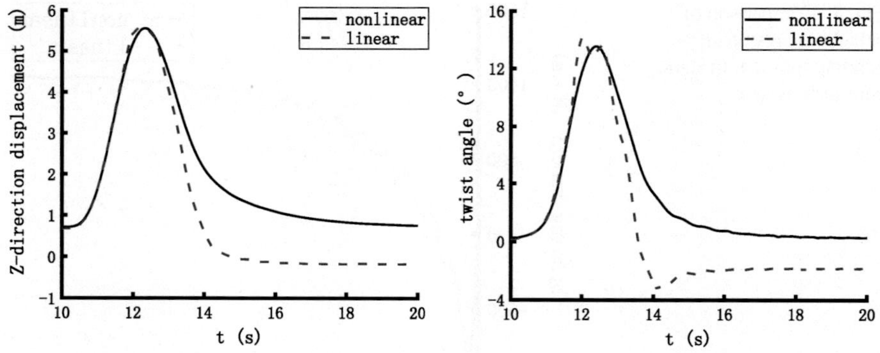

(a) Wing tip Z-direction displacement (b) Wing tip twist angle

Fig. 10 Comparison of calculated results of gust response to wingtip deformation

after reaching the peak value, the wingtip displacement and torsion angle considering geometric nonlinearity return to the equilibrium position, while those without considering geometric nonlinearity do not return to the static equilibrium state, but continuously decrease to −0.18 m. Additionally, the wingtip experiences a maximum reverse torsion of 3.3°, eventually recovering to −1.8°, which is inconsistent with the actual situation.

Figure 11 shows the comparison of the overall structural bending moment experienced by the wing root under gust loads after the aircraft reaches a static equilibrium state, with and without considering geometric nonlinearity. It can be observed that the peak values of the structural bending moment response obtained considering the influence of aircraft geometric nonlinearity and that obtained without considering it have little difference. However, the bending moment without considering geometric nonlinearity does not return to the equilibrium state after reaching the peak value, but oscillates and decreases to a negative value, followed by irregular small-amplitude oscillations. This leads to the wingtip deformation obtained without considering geometric nonlinearity decreasing to a negative value, resulting in an outcome inconsistent with reality.

Figure 12 shows the comparison of the overall lift coefficient under gust loads after the aircraft reaches a static equilibrium state, with and without considering geometric nonlinearity. It can be observed that the peak values of the overall lift coefficient response obtained considering and not considering the influence of aircraft geometric nonlinearity have little difference. However, the overall lift coefficient without considering geometric nonlinearity does not return to the equilibrium state after reaching the peak value, but oscillates and decreases to around 0.4. This is due to the fact that after being subjected to gust loads, the overall aircraft deformation without considering geometric nonlinearity does not recover to the static equilibrium position, but exhibits reverse torsion, resulting in a decrease in the aircraft attack angle.

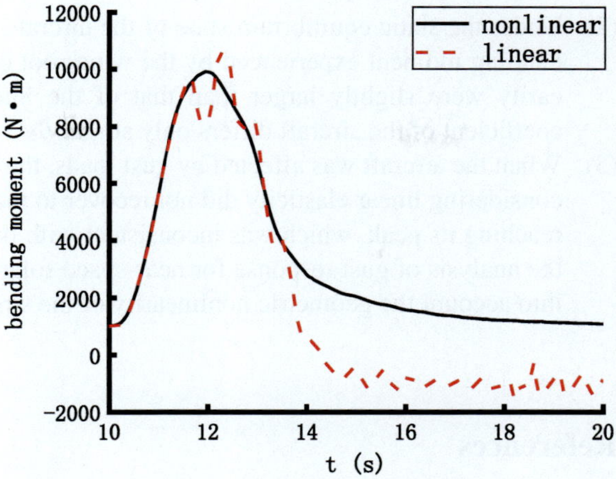

Fig. 11 Comparison of calculation results of gust response to bending moment

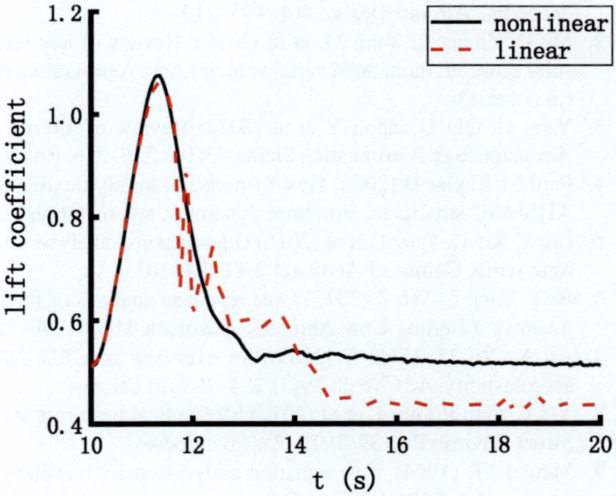

Fig. 12 Comparison of calculation results of gust response to aircraft lift coefficient

5 Conclusion

For the gust loads that near-space solar-powered aircraft might encounter during landing phase, the CFD/CSD loosely coupled analysis method was employed, and gust loads were introduced through the grid velocity method. The influence of geometric nonlinearity on the static equilibrium state and gust response during aircraft landing was analyzed and summarized. The following conclusions could be obtained from the study:

(1) Near-space solar aircraft are characterized by light structural weight, high body flexibility, and low descent speed. If there is gust loads during landing, the aircraft will exhibit significant peak gust responses.

(2) Under the static equilibrium state of the aircraft, the wingtip deformation and bending moment experienced by the wing root considering geometric nonlinearity were slightly larger than that of the linear case, but the overall lift coefficient of the aircraft differs only slightly.
(3) When the aircraft was affected by gust loads, the response result obtained only considering linear elasticity did not recover to the static equilibrium state after reaching its peak, which was inconsistent with the actual situation. Therefore, the analysis of gust response for near-space solar-powered aircraft should take into account the geometric nonlinearity of the structure.

References

1. Bolsunovsky AL, Buzoverya NP, Gurevich BI et al (2001) Flying wing—problems and decisions. Aircraft Design 4(4):193–219
2. Ma D, Zhang L, Yang M, et al (2020). Review of key technologies of ultra-long-endurance solar powered unmanned aerial vehicle. Acta Aeronautica et Astronautica Sicina 41(3):34–63. (in chinese)
3. Yang C, Qiu Q, Zhou Y et al (2022) Review of aircraft gust alleviation technology. Acta Aeronautica et Astronautica Sicina 43(10):216–256. (in chinese)
4. Patil M, Taylor D (2006) Gust response of highly flexible aircraft. 47th AIAA/ASME/ASCE/AHS/ASC structures, structural dynamics, and materials conference
5. Liu Y, Xie C, Yang C et al (2016) Gust response analysis and wind tunnel test for a high-aspect ratio wing. Chinese J Aeronaut 29(1):91–103
6. Bi Y, Yang C, Wu Z (2015) Gust response analysis of flexible aircraft with aerodynamic non linearity. J Beijing Univ Aeronaut Astronaut 41(7):1208–1214. (in chinese)
7. An X, Xu M, Chen S (2009) An overview of CFD/CSD coupled solution for nonlinear aeroelasticity. Adv Mech 39(3):284–298.(in chinese)
8. Yin C, Gao P, Guo J, et al (2012) Mechanical performance analysis on wing's Zephyr7 UAV. Struct Environ Eng 39(3):19–25. (in chinese)
9. Menter FR (1994). Two-equation eddy-viscosity turbulence models for engineering applications. AIAA J 32(8):1598–1605
10. Hoblit FM (1988) Gust loads on aircraft: concepts and applications. AIAA Educ Series, AIAA, Washington, D.C.
11. Zhao W, Huang J, Zhang S et al (2020) Research on gust responses characteristics of solar-powered UAV coupling with propeller slipstream. Flight Dyn 38(3):11–17. (in chinese)
12. Gu N, Lu Z, Zhang J (2011) CFD-based analysis for gust response of aircraft wing. Acta Aeronautica et Astronautica Sicina 32(5):785–791. (in chinese)

Research and Application of Intelligent Automatic Drone Port and Image Recognition in Precise Inspection of Pole Tower

Fan Gao and Zhongyi Long

Abstract Based on the characteristics of low cost, easy control and high flexibility of drone, the application of drone in inspection of pole tower and power lines had become the best tool for efficiently and safely completing inspection operations, and had been widely used in power inspection. Traditional drone inspection require long-term flight control operation of drone, "drone crash into tower" and other accidents were also increasing with the expansion of business, the industry urgently needs the more intelligent, safe and controllable inspection operation. From this, "automatic drone power fine inspection solution" was proposed, firstly, the overall architecture design of the solution was introduced, including software and hardware design scheme, network architecture and data transmission mode. Then, the design scheme of the flight route for the fine inspection of the pole and tower was introduced, the flight path of the drone from the automatic drone port flight to the target pole tower was planned, and the single pole tower shooting scheme and multi-group pole tower shooting scheme were designed. Then it focuses on solving the problem that the key components such as insulators, damper and link fitting account for a relatively small proportion in the shooting scheme taken by drones, FPN (region generation network) was used to build Faster R-CNN (Regional Convolutional Neural Network) detection model to improve the efficiency and accuracy of hidden danger detection of small components. In the practical application case, the Bolt missing pin and other hidden defects identified by image recognition method of automatic drone in the task of pole tower inspection were introduced. The efficiency, accuracy, timely response and cost savings of this solution had been verified in field cases. The demonstration application scenario of automatic drone was formed through the functions of on-site deployment of mission, automatic drone takes-off and landing, automatic patrol, remote control to perform tasks.

F. Gao (✉)
ChongQing ChuanYi Automation Co., Ltd, Chongqing, China
e-mail: gaofan_sic@sina.com

Z. Long
Hunan Valin Xiangtan Iron & Steel Co., Ltd, Xiangtan, China
e-mail: 279719963@qq.com

Z. Liu et al. (eds.), *Advances and Challenges in Advanced Unmanned Aerial Systems*, Springer Aerospace Technology, https://doi.org/10.1007/978-981-99-8045-1_8

Keywords Pole tower · Precise inspection · Automatic drone port · Faster R-CNN

1 Introduction

Nowadays, the application of drone is becoming more and more mature, with the development of artificial intelligence technology, computer vision technology, automatic control technology, sensor technology, the characteristics of low cost, easy control and high flexibility are becoming more acute. It can complete special tasks from the air without causing casualties [1, 2]. Compared with traditional manual inspection, using unmanned drone to conduct fine inspection on pole towers and power lines has advantages such as fast speed, low labor cost and low personnel risk, and has become the best tool to complete automatic inspection tasks quickly and efficiently.

The use of drone helps inspection team members improve inspection efficiency, and it has been widely used in power inspection. However, the new working mode also brings new challenges [3, 4]. Drone inspection requires long-term flight control operation of drone, which is not a low requirement for the operation skills of the grass-roots team personnel. The occurrence of accidents such as "drone crash into tower" were also increasing with the expansion of business. Under the premise of safe flight, the same route in expert and novice flying may also produce different inspection conclusions. The industry is in urgent need of a more intelligent, safer and more controllable inspection operation [5–7]. Lowering the threshold of drone operation, reducing the impact of personnel experience on inspection work, and further improving the efficiency of drone patrol operation will become a technical trend followed by the industry.

2 Design of Automatic Inspection System

2.1 Network Architecture and Data Transmission

In order to realize the fine inspection task of the autonomous pole tower of the intelligent networked drone, the automatic intelligent inspection system platform is customized to control the inspection behavior of the drone and process the data during the inspection. As shown in Fig. 1, remotely control the drone automatic takeoff and landing/automatic battery replacement through network, continuous task execution, real-time data return to the back end for analysis and processing, to achieve "automation" and "unmanned". At the same time, it can plan automatic flight inspection routes based on the drone port, so as to realize intelligent and efficient flight and liberate the labor force.

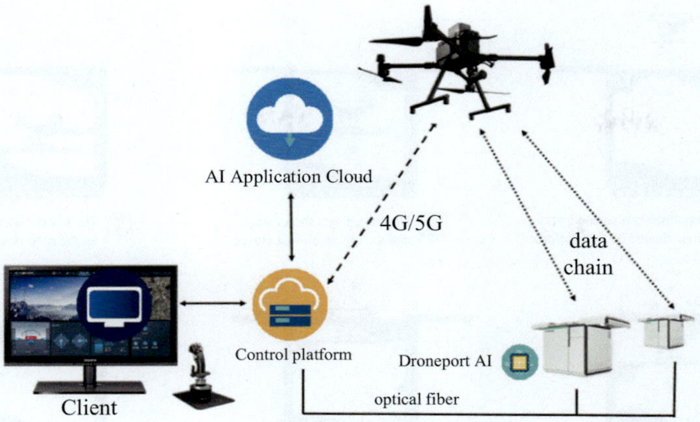

Fig. 1 Network architecture of automatic inspection system

2.2 *Automatic Drone Port*

Drone port refers to an drone port or hub designed for or dedicated to drones. The intelligent network drone automatic drone port is a ground automatic facility to assist the whole process operation of intelligent network drone, which can replace manual operation and greatly improve the automatic operation efficiency of drone.

The drone is stored in the automaton field. When there is a flight demand, the drone takes off from the drone port autonomously, and lands back to the automaton field automatically after completing the task. In the automated airfield, the drone can automatically charge the battery and automatically change the battery for the drone, ready for the next mission. With the automatic drone port as the basic guarantee, the drone can take off and land by itself without human intervention, replace the battery, and achieve fully automated operation.

As shown in Fig. 2, the automated drone port is responsible for storage, battery replacement and ground-to-air relay communication of the drone. It can realize the automatic recovery and energy supply of the drone, and facilitate the drone to complete various operations such as daily inspection and fine inspection by itself in the case of unattended.

2.3 *Functionality of Automatic Inspection System*

The automatic intelligent inspection system uses core technologies such as image recognition and AI flight to make autonomous decisions about flight path, attitude and shooting parameters, and obtain high-quality inspection data.

Real-time transmission of image data. When the drone is performing the inspection task, it can send the video back to the background monitoring center in real time,

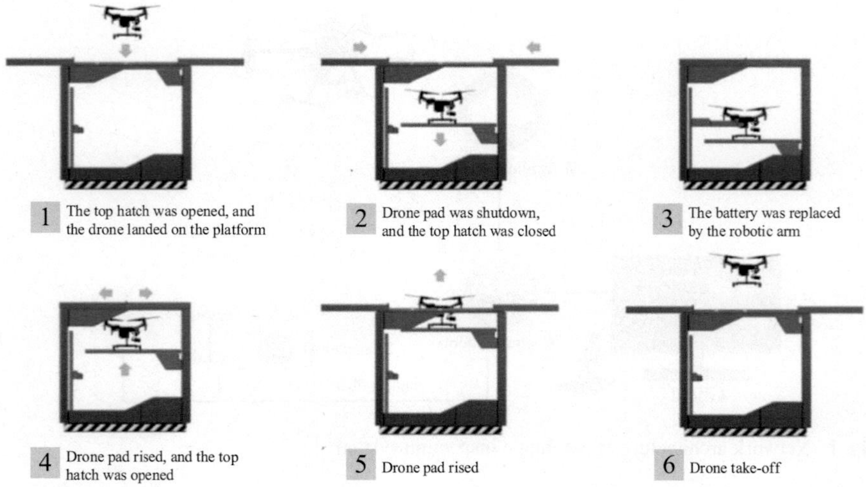

Fig. 2 Workflow diagram of drone port

view the drone picture in real time, and automatically save the inspection shooting or inspection video, and manage the inspection shooting files.

Intelligent Patrol. According to the planned route, the drone automatically takes off, performs tower inspection tasks, and automatically lands, which is intelligent in the whole process without manual control.

Automatic Alignment of Drone in Fine Inspection. The image control algorithm is used to control the drone head to point the airborne high-definition camera, automatically adjust the focal length to enlarge the area, and always lock the target in the center of the frame. At the same time, adjust the focal length to enlarge the area for detailed inspection and targeted fault diagnosis.

Automatic Defect Identification of Image. After the drone inspection is completed, the platform processing system can integrate and analyze the features of the images, realize the automatic positioning and identification of the insulator parts, conduct the corresponding defect analysis of the insulator parts according to the set defect identification rules, and realize the analysis and identification of common defects of pole tower insulators.

3 Drone Operation Flight Planning and Shooting Scheme

3.1 Fine Inspection Task

The automated drone port auxiliary network drone carries out pole and tower inspection. After the route planning is completed, the drone takes off from the on-site automatic drone port, flies automatically to the target pole and tower according to the planned route, uses the automatic inspection function to inspect the pole and tower autonomously, carries out inspection and shooting, and flies back to the automatic drone port after the inspection.

In the inspection process, the image control algorithm is used to control the drone head to point the airborne high-definition camera at the inspection area, automatically adjust the focal length to enlarge the area, and always lock the target in the center of the frame, while always lock the target in the middle of the frame, so as to carry out fine inspection and targeted inspection target shooting. After the inspection shooting, the platform data processing system carries out identification and analysis on the captured image data, realizes intelligent analysis and identification of common defects of pole tower insulators, and completes intelligent diagnosis of the health status of pole tower.

3.2 Drone Operation Flight Planning

In the complex working environment of steel mills, the inspection of the insulator for the pole and tower facilities has more stringent requirements for the shooting point and focusing of the drone, and the conventional route planning cannot meet the accuracy requirements, so it is necessary to solve the precision problems of the drone orientation, the specific implementation scheme is as follows:

Single Pole Power Shooting Scheme. The flight path of the drone from the automatic drone port to the target pole tower is planned by means of manual positioning. As shown in the figure below, the drone reaches the designated position and navigation point 5. Plan the shooting path for the 6 target components of the target tower (1, 2, 3, 4, 5, 6): 1→2→3→4; Waypoint 1 and 4 are hovering shooting points, and the drone executes tasks in sequence.

As shown in Fig. 3, after the drone reaches the shooting point 1 (waypoint 5), it drops to the appropriate shooting point 1 (waypoint 1), and uses the decoupling function of waypoint camera action and waypoint to realize the execution of multiple task actions at the single waypoint, tilt Angle of the head, and take pictures of the multi-pole tower components 1, 2, and 3 (as shown in the figure below). Then the drone rises to waypoint 2, flies in the waypoint order of 2→3→4, hovers at waypoint 4, and adds waypoint instructions to achieve shooting of pole and tower components 5, 6, 7.

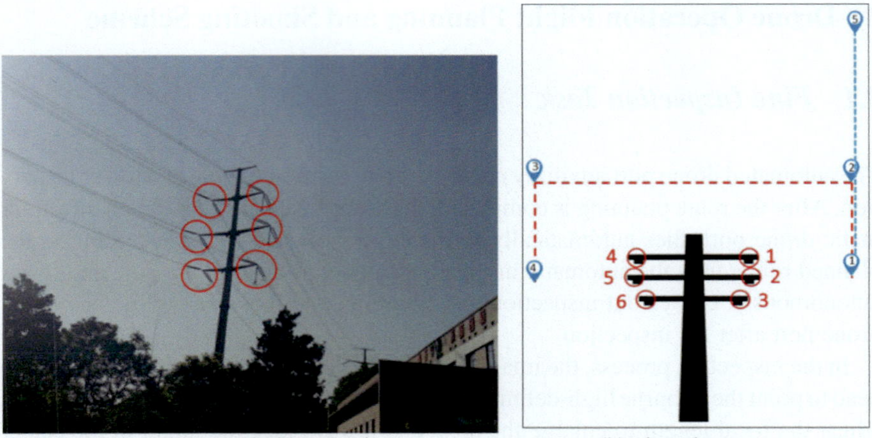

(a) actual photograph (b) elevation

Fig. 3 Flight path diagram of single pole power inspection

Multiple Pole and Tower Shooting Schemes. Without considering the optimal path and the optimal power consumption strategy, the drone operation route is planned by manual doting according to the arrangement order of the existing poles and towers. As shown in Fig. 4, each operation executes any route, the drone executes in order, and returns to the automatic drone port after shooting pole towers to send the photos back to the background. In the comparison system, the route ID can be used as a unique identity, and the operation order of each route is fixed, which can solve the problem of confusion in the comparison object.

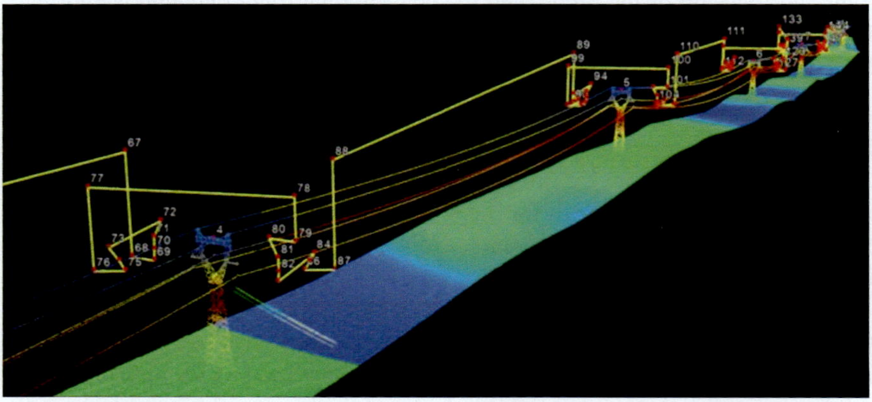

Fig. 4 Flight path diagram of multiple pole power inspection

4 Fast R-CNN Image Recognition Model and Practical Application Case

4.1 Image Recognition Model Based on Fast R-CNN

In recent years, deep learning has made a lot of achievements in the field of image recognition and intelligent detection [8–10]. In particular, R-CNN (Regional Convolutional Neural Network) algorithm has achieved remarkable results in image recognition detection. The advantages of convolutional neural networks in processing image morphological features make them widely used in image recognition and detection tasks. In order to reduce the time spent in extracting feature vectors using CNN model for R-CNN candidate regions, scientists proposed Fast R-CNN algorithm. In order to further improve efficiency, the Faster R-CNN algorithm introduces RPN (region generation network) to directly generate candidate regions, which is more efficient and lightweight than the traditional R-CNN algorithm and Fast R-CNN algorithm. The overall network structure based on Faster R-CNN was shown in Fig. 5, input of the networkis an image containing the hidden dangers of the tower pole itself (including seven key parts such as tower footing, tower mast, insulator, damper, link fitting, spacer, grading ring etc.). After detection and recognition, the probability score of abnormal hidden features in key parts and the boundary box of hidden parts are output by classification algorithm.

As shown in Fig. 6, FPN uses bottom-up path, top-down path and horizontal connection to combine the high-level features of low resolution and high semantic information with the low-level features of high resolution and low semantic information, so that the features at each scale have very rich semantic information.

FPN was used instead of single-scale feature maps to accommodate RPN. In the process of image recognition and prediction, the feature information needs to be classified. In the lightweight edge computing algorithm, it can be approximately regarded as an optimization problem of convex quadratic programming, whose solution can be obtained by constructing the Lagrange function:

$$L(w, a, b) = \frac{1}{2}w^T w - \sum_{i=1}^{n} a_i[y_i(w \bullet x_i + b) - 1] \quad (1)$$

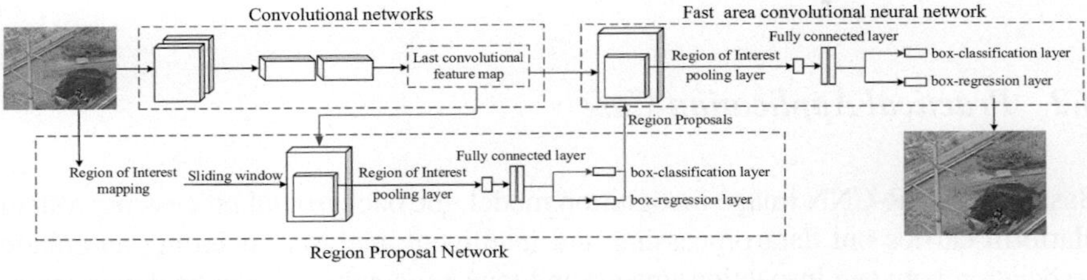

Fig. 5 Multiple pole and tower inspection flight trajectory diagram

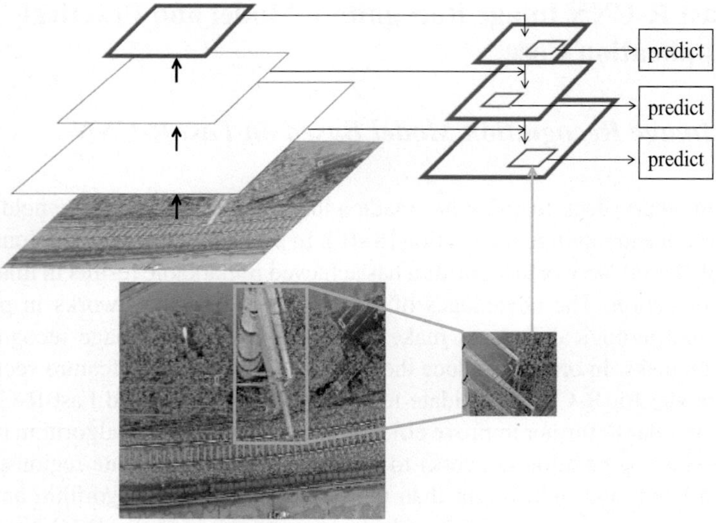

Fig. 6 Multiple pole and tower inspection flight trajectory diagram

in the formula (1): $a_i \geq 0$ is Lagrange multiplier.

Take the partial derivative of w and b respectively, and make them equal to 0, there is:

$$\begin{cases} \frac{\partial L(w,a,b)}{\partial w} = 0 \Rightarrow w = \sum_{i=1}^{n} a_i x_i y_i \\ \frac{\partial L(w,a,b)}{\partial b} = 0 \Rightarrow \sum_{i=1}^{n} a_i y_i = 0 \end{cases} \tag{2}$$

Bring it into formula (1), there is:

$$L(w, a, b) = \sum_{i=1}^{n} a_i - \sum_{i=1}^{n} \sum_{j=1}^{n} a_i a_j y_i y_j \bullet (x_i x_j) \tag{3}$$

Fast R-CNN uses RoI (Region of Interest) pooling layer to extract features, it need to assign different scales of RoI to pyramid level. If RoI is small, it should be mapped to a more granular level.

4.2 Practical Application Case

Based on Fast R-CNN image recognition model, the background processing system platform carries out data processing and intelligent analysis, including automatic association between inspection images and tower accounts, automatic defect annotation and information entry, and automatically records key data and defect analysis

Fig. 7 Bolt missing pin

results [11, 12]. The system automatically records the key data and defect analysis results, generates the pole tower inspection operation report, and locate the abnormal information of the insulator through the operation report, helping to eliminate safety hazards and ensure the safe use of electricity in the plant.

In the practical application, defect analysis results were shown as Figs. 7, 8, 9, 10 and 11. Figure 7 shows the identified bolt missing pin defect of different pole tower.

Figure 8 shows the identified loss of grading ring defect of different pole tower.

Figure 9 shows the identified ground wire corrosion defect of different pole tower.

Figure 10 shows the identified corrosion of small metal fittings defect. Figure 11 shows the identified tower corrosion defect of different pole tower.

The platform system summarizes and analyzes the analysis and identification results, and classifies them according to the component type, defect type and defect level, so as to facilitate classified query according to requirements. After obtaining the refined inspection images of insulators, the background processing system platform carries out data processing and intelligent analysis, including automatic association of inspection images with tower accounts, automatic defect labeling and information entry, automatically records key data and defect analysis results.

5 Summary

Through the research and application of intelligent automatic drone port and image recognition in precise inspection of pole tower, the research results are summarized as follows:

Fig. 8 Loss of grading ring

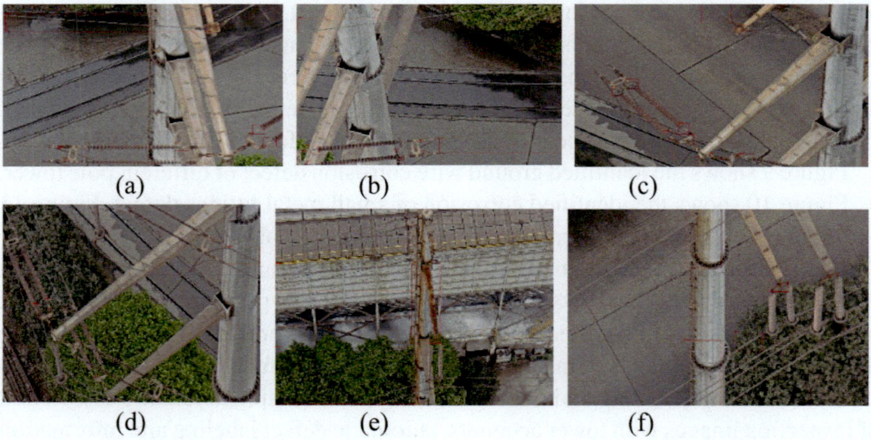

Fig. 9 Ground wire corrosion

(1) Automatic drone power fine inspection solution was formed through the functions of on-site deployment of mission, automatic drone takes-off and landing, automatic patrol, remote control to perform tasks;
(2) According to the planned route, the drone automatically takes off and landing, performs tower inspection tasks, which achieve the whole intelligent automatic process without manual control;

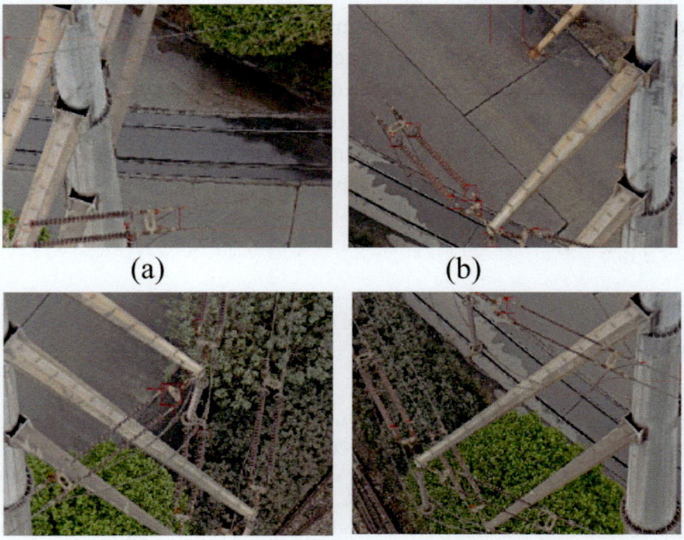

(a) (b)

(b) (d)

Fig. 10 Corrosion of small metal fittings

(a) (b)

(b) (d)

Fig. 11 Tower corrosion

(3) Based on Fast R-CNN image recognition model, the background processing system platform carries out data processing and intelligent analysis, all inspection navigation processes are strictly numerical, with strict process specifications and error control, and high accuracy consistency, avoiding the situation of insufficient inspection caused by manual inspection with experience.The efficiency, accuracy, timely response and cost savings of this solution had been verified in field cases.

References

1. Ruhai L et al (2021) Research on inclination monitoring technology of transmission tower based on drone. J Phys Conf Ser 2030(1):012076
2. Kumar PS et al (2020) Energy management system for small scale hybrid wind solar battery based microgrid. IEEE Access 8:8336–8345
3. Mannem NS et al (2021) Broadband active load-modulation power amplification using coupled-line baluns: a multifrequency role-exchange coupler doherty amplifier architecture. IEEE J Solid-State Circuits 56(10):3109–3122
4. Wang SL et al (2022) Semantic segmentation and defect detection of aerial insulators of transmission lines. J Phys Conf Ser 2185(1):436–440
5. Chen D-Q et al (2020) Safety distance analysis of 500kV transmission line tower UAV patrol inspection. IEEE Lett Electromag Comp Pract Appl 2(4):124–128
6. Fangcheng L et al (2021) Intelligent detection and parameter adjustment strategy of major electrical equipment based on optimized YOLOv4. Trans China Electrotechn Soc 36(22):4837–4848. (in chinese)
7. Shao G et al (2020) Research progress in unmanned aerial vehicle inspection technology on overhead transmission lines. High Voltage Eng 46(1):14–22. (in Chinese)
8. Guowei LIU et al (2021) Insulator fault detection method based on image recognition technology and intelligent algorithm. Insulators Surge Arresters 4:191–195 (in Chinese)
9. Wang Z et al (2021) Two stage insulator fault detection method based on collaborative deep learning. Trans China Electrotechn Soc 36(17):3594–3604. (in Chinese)
10. Zhang Y et al (2021). Research on UAV transmission line following method based on laser ranging and ROS system. Chinese J Electron Dev 44(4):935–940. (in Chinese)
11. Bai W et al (2021) E-FCNN based electric power inspection image enhancement. Electric Power 54(05):179–185. (in Chinese)
12. He N et al (2021) Research on infrared image missing insulator detection method based on deep learning. Power Syst Protect Control 49(12):132–140. (in Chinese)

Numerical Analysis of Low Reynolds Number Compressible Flows with Triangle Airfoil

Junli Wang, Chang Wang, Chen Liu, and Jian Wu

Abstract The Martian atmosphere is characterized by low density and low temperatures, resulting in Mars vehicles typically operating at low Reynolds numbers compressible flow. To obtain the difference in the prediction of such flows by different computational numerical models. Based on Reynolds-averaged Navier–Stokes (RANS) (including the S-A and SST k-ω turbulence models) and large eddy simulation (LES) models, a three-dimensional flow model of Mars environmental blades is established. And this paper numerically studied the triangular airfoil under compressible flow conditions with a low Reynolds number (Re = 3000, Mach = 0.5), considering various angles of attack (AOA). The results show that LES can predict lift coefficient (C_l) and drag coefficient (C_d) well. S-A turbulence model has high fidelity only at low angles of attack, and it is difficult to accurately predict flow separation at high AOA. A similar situation applies to the SST k-ω turbulence. When the AOA increases from 10 to 11°, the increase of lift coefficient is mainly dominated by the suction surface. Moreover, even at an AOA as low as 3°, flow separation was observed on the airfoil surface in the Martian environment.

Keywords Low Reynolds number · Compressible flow · LES · CFD

1 Introduction

As a Mars exploration tool, Mars unmanned aerial vehicle (UAV) can effectively make up for the shortcomings of rover, such as narrow detection range, slow moving speed and difficult detection of complex terrain. The successful flight of the "Ingenuity" on Mars has created another milestone in human spaceflight. Although Mars has an atmosphere that makes it suitable for flight, it is quite different from Earth's

J. Wang · C. Wang · J. Wu (✉)
School of Energy Science and Engineering, Harbin Institute of Technology, Harbin 150001, China
e-mail: jian.wu@hit.edu.cn

C. Liu
Laboratory for Space Environment and Physical Science, Harbin Institute of Technology, Harbin 150001, China

Z. Liu et al. (eds.), *Advances and Challenges in Advanced Unmanned Aerial Systems*, Springer Aerospace Technology, https://doi.org/10.1007/978-981-99-8045-1_9

atmosphere. The atmosphere of Mars is characterized by low density and low temperature [1]. Mars spacecraft usually operate at compressible flow conditions with a low Reynolds number [2]. At low Reynolds number, the viscous shear layer is not easy to transition and reattach, resulting in a decrease in lift drag ratio [3]. And laminar flow separation will cause rotor stall and low-frequency oscillation, which will significantly affect the aerodynamic performance of UAVs. The compressibility of the fluid enhances the stability of the shear layer. In addition, when the flight is in a high Mach number state, it is easy to produce air compression effect and shock wave oscillation [2].

At present, some scientific research institutions [4–7] have developed Mars simulation environment, conducted airfoil wind tunnel experiments or Mars UAV hover experiments, and evaluated the UAV performance. Mars environment simulation usually requires a low temperature and low pressure confined space, the experiment system is complex. Numerical simulation is an effective means to study such special environmental flow problems. Recently, Desert et al. [8] used numerical methods to study the effect of Mach number and Reynolds number on the performance of curved airfoil. Their results show that, at low Reynolds numbers, changes in aerodynamic performance caused by changes in Reynolds number and Mach number are of the same order of magnitude. The results of previous studies show that the airfoil suitable for Mars gas environment is quite different from the traditional airfoil. Among them, the triangular airfoil has attracted the attention of some scholars [9–12]. The numerical research is mainly based on direct numerical simulation (DNS). In addition, some scholars have used RANS model to study the low Reynolds numbers flow. Morizawa et al. [13] used the RANS model to numerically simulate the NACA0012-34 airfoil whose Reynolds number and Mach number were 4100 and 0.22, respectively. It was found that the C_{l} and C_{d} predicted numerically at a high AOA deviated from the experimental value. In addition, the numerical calculation results of Park et al. [14] show that there is a big difference between the prediction of surface pressure coefficient of low Reynolds number airfoil by RANS (SST $k-\omega$) and the calculation results of DNS.

To sum up, there are relatively few studies on numerical simulation of compressible flow conditions with a low Reynolds number in Mars environment, and transverse comparative analysis among numerical simulation methods is lacking. Therefore, in order to analyze the difference between different numerical simulation methods in predicting compressible flow with low Reynolds number, a three-dimensional large eddy simulation numerical simulation method was established in the Martian gas environment. A triangular airfoil with an incoming flow of Reynolds number 3000 Mach number 0.5 was numerically simulated, and the lift-drag characteristics of the airfoil at different AOA were obtained. The results of LES and RANS model are compared and analyzed, which aid in the development of unmanned airfoil for the Mars environment.

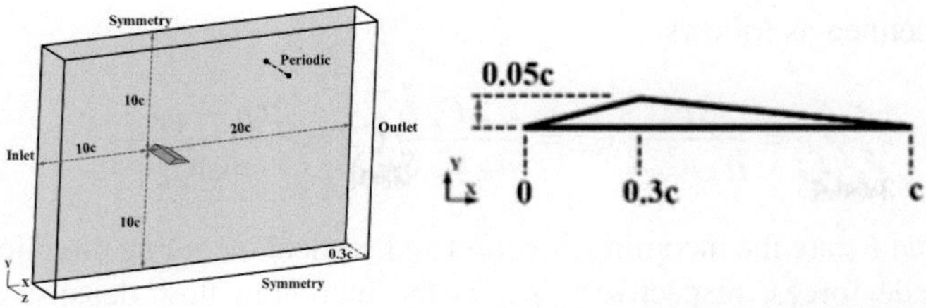

(a) Triangle airfoil (b) Computational domain

Fig. 1 Triangle airfoil and computational domain

2 Numerical Approach

2.1 Physical Model

For the current Mars UAV, its propellers operate at low Reynolds numbers compressible flow (Re = 10^3–10^4, M = 0.1–0.9) [15]. Therefore, Re = 3000 and M = 0.5 is selected in this paper for numerical study. Non-smooth airfoil has better adaptability to low Reynolds number flows [3, 16]. Therefore, this paper considers the flow around the triangular airfoil as shown in Fig. 1a under uniform flow. And Fig. 1b shows the schematic diagram of the computational domain and boundary conditions.

2.2 Governing Equation

For compressible Newtonian fluids, here are the dimensionless for N-S and energy equations:

$$\frac{\partial \rho}{\partial t}+\frac{\partial \rho u_i}{\partial x_i}=0\quad \frac{\partial \rho u_i}{\partial t}+\frac{\partial \rho u_i u_j}{\partial x_j}=-\frac{\partial p}{\partial x_i}+\frac{1}{Re}\frac{\partial}{\partial x_j}\left[\mu\left(\frac{\partial u_i}{\partial x_j}+\frac{\partial u_j}{\partial x_i}-\frac{2}{3}\frac{\partial u_k}{\partial x_k}\delta_{ij}\right)\right] \tag{1a}$$

$$\frac{\partial \rho e}{\partial t}+\frac{\partial(\rho e+p)u_j}{\partial x_j}=\frac{\partial}{\partial x_i}\left[\frac{Ma\mu}{RePr(\gamma-1)}\right]\frac{\partial T}{\partial x_i}+\frac{1}{Re}\frac{\partial}{\partial x_j}u_i\left[\mu\left(\frac{\partial u_i}{\partial x_j}+\frac{\partial u_j}{\partial x_i}-\frac{2}{3}\frac{\partial u_k}{\partial x_k}\delta_{ij}\right)\right] \tag{1b}$$

where, the total energy per unit mass $e = p/(\rho(\gamma - 1)) + u_i^2/2$, γ is the ratio of specific heats. μ is the molecular viscosity, which is a function of temperature and satisfies Sutherland's formula.

In addition, the lift coefficient, drag coefficient and pressure coefficient, which are the main aerodynamic parameters to measure the aerodynamic performance of airfoil, are defined as follows:

$$C_l = \frac{2F_y}{\rho_\infty u_\infty^2 S} C_d = \frac{2F_x}{\rho_\infty u_\infty^2 S} C_p = \frac{2(p - p_\infty)}{\rho_\infty u_\infty^2} \tag{2}$$

where, F_x and F_y are the incoming direction and vertical incoming direction components of blade forces, respectively. ρ_∞ is the incoming flow density, u_∞ is the incoming flow velocity, S is the characteristic area. In this paper, the product of blade chord length and wingspan is taken.

In this paper, Finite volume method is used to discretize the governing equations in space. The Coupled algorithm [17] is adopted to realize the pressure–velocity coupling for this study. The gradient term is discretized by least square method. The convection terms of LES model are discretized by bounded central difference scheme, and those of RANS model are discretized by second-order upwind scheme. An implicit second-order method is used to advance the equations. The dynamic Smagorinsky-Lily [18] subgrid model is used for LES calculation. The S-A and SST k-ω turbulence models can better deal with the transport of turbulent shear stress in the inverse pressure gradient and separated boundary layer, so these two models are selected in this paper for RANS numerical calculation.

2.3 Numerical Verification

A structured hexahedral grid was used for meshing, as shown in Fig. 2. Besides, it is guaranteed that the first layer of wing surface mesh y+ is less than 1. And Δx+ is less than 30 and Δz+ is less than 15 [19]. In the unsteady calculation process, ensure that the CFL number is less than 1 and the dimensionless time step is 5.6 $\times 10^{-3}$. Considering that large eddy simulation has higher grid requirements than Reynolds mean, this paper adopts large eddy simulation and selects different grid numbers to verify grid independence. The blade attack angle is 12°, a total of 120 convective times are calculated, and the last 60 convective times are selected for time average. Table 1 shows the C_l and C_d obtained by calculation of different examples. The results show that when the number of grids is 2.4 million, the change of C_l and C_d is less than 1%, so 2.4 million grids are selected for numerical calculation.

3 Results

In this paper, LES and RANS (S-A and SST k-ω turbulence models) are used to simulate a triangular airfoil with low Re compressible flow conditions (Re = 3000, Mach = 0.5) at different AOA. The lift coefficient, drag coefficient, pressure coefficient and flow separation characteristics are compared and analyzed. And the process of nonlinear lift generation is also analyzed.

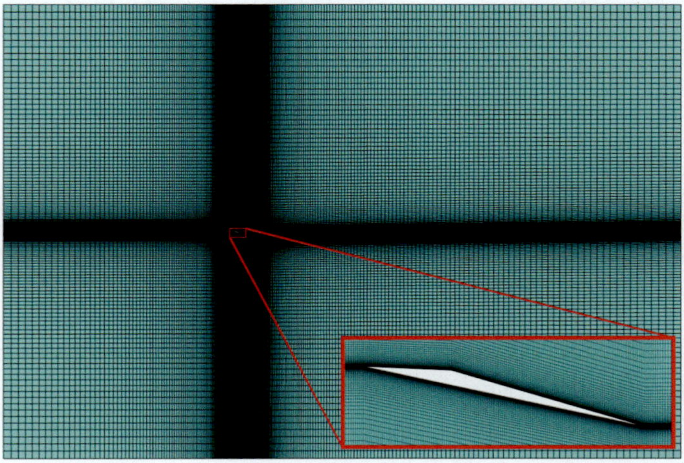

Fig. 2 Schematic of computing grid

Table 1 C_l and C_d with different grids

Case	Grid numbers ($\times 10^6$)	C_l	Error (%)	C_d	Error (%)
Case 1	0.6	0.966	18.4	0.234	12.7
Case 2	1.2	1.185	0.9	0.268	0.4
Case 3	2.4	1.196	0.9	0.269	0.7
Case 4	4.8	1.207	–	0.271	–

3.1 Lift Coefficient and Drag Coefficient

Figure 3a shows the C_l of airfoil at different AOA obtained by different models. The results show that only LES has less error in the overall parameter interval than DNS [9]. When the angle of attack is in the range of 0–10°, the Cl increases approximately linearly. However, between 10 and 11°, the C_l experiences a sharp increase, indicating nonlinear aerodynamic behavior. Subsequently, stall phenomenon occurs, and the C_l remains nearly constant in the range of 11–14° of the AOA. Based on S-A and SST k-ω turbulence models, the obtained C_l have similar trends with the AOA. When the AOA ranges from 0 to 6°, their predicted C_l are consistent with those predicted by LES and DNS. However, with the increase of the AOA, their predicted values gradually deviate from those of LES and DNS, and they did not predict the nonlinear lift and stall phenomena.

Figure 3b shows the airfoil C_d at different AOA calculated by different models. The results show that LES has less error and higher fidelity than DNS. When the AOA is in the range of 0–4°, the C_d exhibits a slow and low-level increase. In the range of 4–11°, the C_d experiences a rapid increase, followed by an approximately linear increase. The S-A and SST k-ω turbulence models predict similar trends in

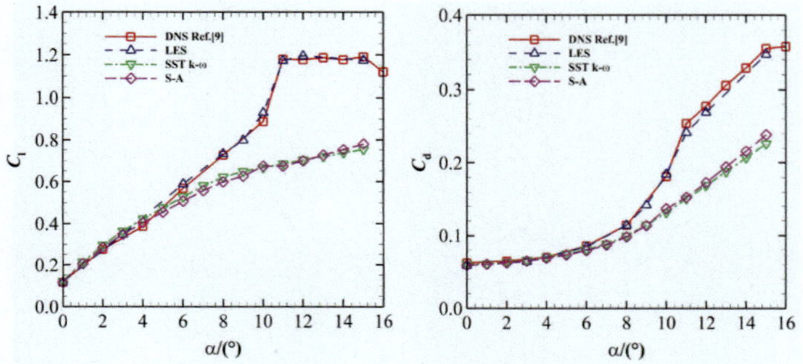

(a) C_l (b) C_d

Fig. 3 C_l and C_d of airfoil at different AOA under different calculation models

the variation of the C_l and C_d with AOA. And the increase trend is the same as that predicted by LES and DNS. The difference is that the predicted rapid increase of the resistance coefficient is in the range of 4–10°.

In conclusion, for such compressible flows with low Reynolds number, LES can better predict the C_l and C_d with high fidelity. The S-A and SST k-ω turbulence models have high fidelity only at low AOA.

3.2 Pressure Coefficient and Flow Field Characteristics

To further investigate the factors influencing the variation of the airfoil C_l and explore the disparities in predicted airfoil aerodynamic characteristics among the three calculation models, three specific AOA, namely 3, 8, and 12°, were chosen for analysis. Figure 4 shows the airfoil C_p calculated by three models at different attack angles. The results show that when the airfoil attack angle is 3°, the C_p predicted by the three models are basically the same. For the pressure surface, there is a large pressure gradient near the leading edge of the blade, and a small pressure gradient in the middle of the blade. For the suction surface, there is no inverse pressure region between the leading edge and the vertex of the blade except near the leading edge, and the downstream region after the vertex of the blade is all inverse pressure region, which indicates that the vertex can effectively inhibit the conduction of the inverse pressure region to the upstream when the AOA is kept low. When the airfoil attack angle is 8°, the C_p predicted by the three models are different. For the pressure surface, the difference is mainly near the trailing edge, and the pressure predicted by the LES model is smaller. For the suction surface, the C_p predicted by the three models is similar in distribution, but the values are different. At this time, the suction surface predicted by the three models all has a reverse pressure gradient, and the front part of the blade is strong. The C_p predicted by the S-A and SST k-ω turbulence model is generally large, so the predicted C_l is generally small. When the airfoil

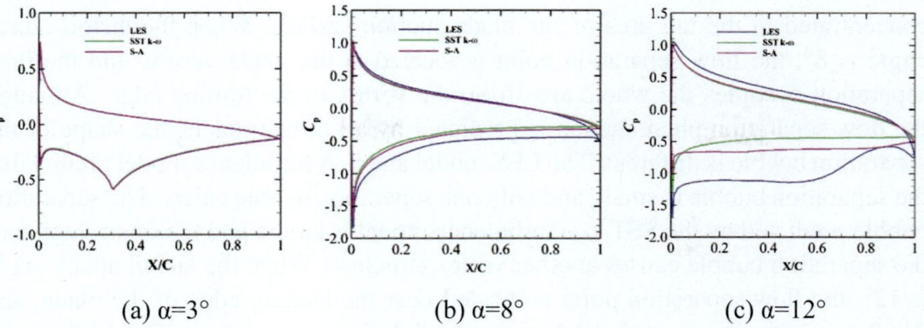

Fig. 4 C_p of airfoil at different AOA under different calculation models

attack angle is 12°, the C_p of pressure surface predicted by the three models predict are different greatly. And the C_p predicted by S-A and SST k-ω turbulence model are generally small. For the suction surface, the C_p predicted by the three models are significantly different. The distribution of C_p predicted by the S-A and SST k-ω turbulence model is similar, but there are obvious numerical differences, but they are both higher than those predicted by the LES model. The LES model predicts that there is a large inverse pressure gradient at the leading edge of the blade, and the pressure coefficient in the middle of the blade changes little. In addition, there is a positive pressure gradient near the trailing edge of the blade.

Figure 5 shows the vorticity and streamline distribution under different attack angles calculated by different models. The results show that when the airfoil attack angle is 3°, the flow fields calculated by the three models are similar, and flow separation still exists even at low attack angle, and the flow separation is mainly

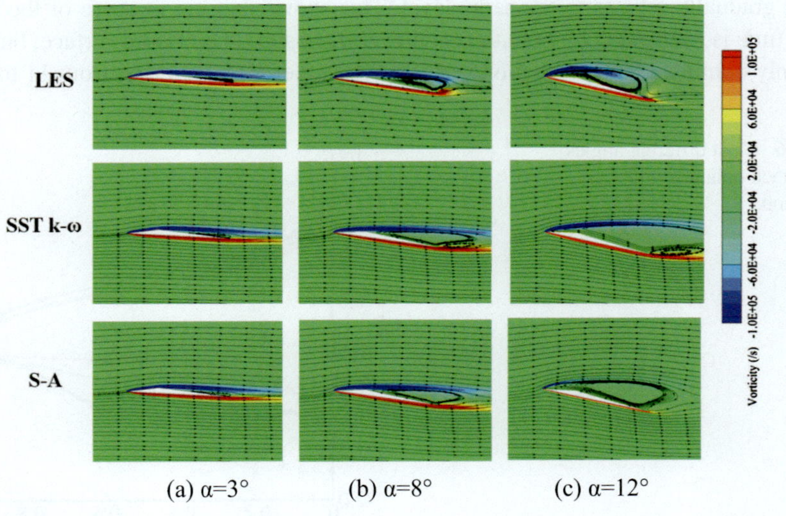

Fig. 5 Vorticity and streamline distribution at different angles of attack

concentrated in the tail area of the blade suction surface. When the airfoil attack angle is 8°, the flow separation point is located at the blade vertex, and the flow separation occupies the whole area from the vertex to the trailing edge. Although the flow separation phenomenon is predicted by all three models, the shape of the separation bubble is different. The LES model and S-A turbulence model predict that the separation bubble is small, and only one separation bubble exists. The separation bubble predicted by the SST k-ω turbulence model is larger, and the return region of the separation bubble causes another vortex structure. When the airfoil attack angle is 12°, the flow separation point is located near the leading edge of the blade, and the flow separation occupies a large part of the suction surface. The LES model and the S-A turbulence model also predict a single separation bubble, but the LES model predicts a small separation bubble size and the presence of reattachment flow separation reattachment at the tail of the blade. Two separation bubbles are predicted by the SST k-ω turbulence model, and the size of the separation bubbles is further increased.

According to the variation of lift coefficient with Angle mentioned above, it can be seen that the triangular airfoil has nonlinear lift phenomenon. In order to further analyze the causes of nonlinear lift of airfoil, airfoils with attack angles of 9, 10, 11 and 12° were selected. Figure 6 shows the pressure coefficients of airfoils with different attack angles calculated by LES model. The results show that when the AOA increases from 9 to 10°, the C_p of the pressure surface is basically unchanged, and the distribution law of the C_p of the suction surface is similar, but the value decreases, and it is mainly concentrated in the middle and lower reaches, which indicates that the increase of the C_l at this time mainly comes from the suction surface. When the AOA increases from 10 to 11°, the C_p of the pressure surface increases slightly, the distribution law of the pressure coefficient of the suction surface changes, the C_p in the middle and upper reaches of the blade decreases sharply, and the C_p in the tail area gradually returns to the same level. This shows that the increase of the Cl at this time is influenced by both the suction surface and the pressure surface, but it is mainly dominated by the suction surface. When the AOA increases from 11 to 12°,

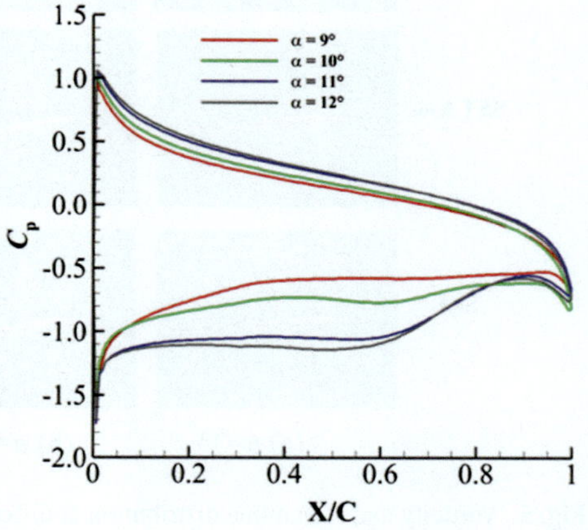

Fig. 6 C_p at different angles of attack obtained with LES method

the C_p of suction surface and pressure surface remain basically unchanged, and the C_l tends to be stable.

Figure 7 shows the vorticity and flow line distribution under different attack angles obtained by using LES. The research results show that when the attack angle of the airfoil increases from 9 to 10°, the flow separation point moves further forward, the flow separation area of the suction surface expands, and the separation bubble increases. This shows that the increase of the C_l is mainly due to the expansion of the flow separation region of the suction surface. When the AOA of the airfoil increases from 10 to 11°, the flow separation point moves further forward, but the change is not obvious, and flow separation and reattachment appear in the tail area of the blade, which indicates that the increase of C_l at this time is mainly from the flow separation and reattachment of the suction surface. When the airfoil attack angle increases from 11 to 12°, the flow pattern does not change. Figure 8 shows the Q-criterion isosurface (Q = 1) colored by dimensionless spanwise velocity. The results show that when the angle of attack is 10°, the spanwise velocity is very vicious and the flow field near the blade is approximately two-dimensional. However, when the angle of attack increases to 11°, the local spanwise velocity increases and becomes a three-dimensional flow. This shows that the three-dimensional flow structure can improve the lift performance of the blade.

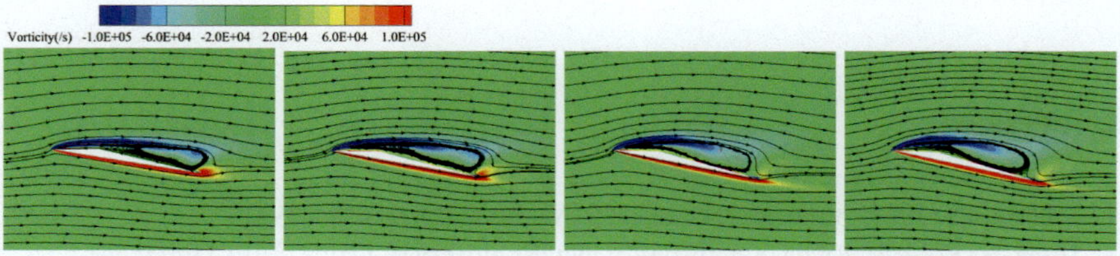

(a) α=9° (b) α=10° (c) α=11° (d) α=12°

Fig. 7 Vorticity and streamline distribution at different angles of attack obtained with LES method

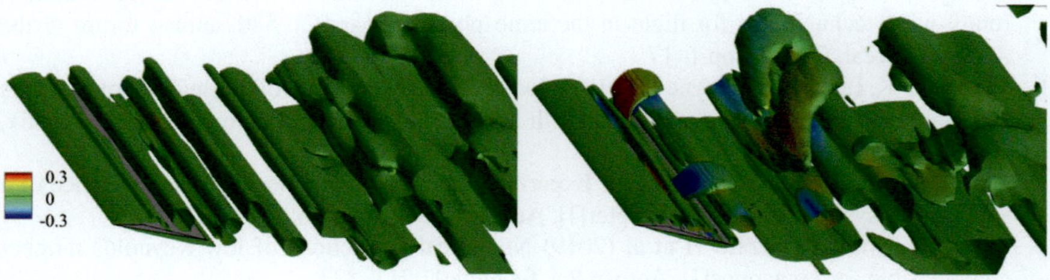

(a) α=10° (b) α=11°

Fig. 8 The Q-criterion isosurface (Q = 1) colored by dimensionless spanwise velocity

4 Conclusion

In this paper, LES model, RANS models (S-A and SST k-ω) are used to conduct three-dimensional numerical simulation of a triangular airfoil (Re = 3000, Mach = 0.5) with at different AOA. The fidelity of these three models for predicting compressible flows at low Reynolds number is studied, and the causes of nonlinear lift are analyzed. The main conclusions are as follows:

(1) LES can better predict C_l and C_d, with high fidelity. The S-A and SST k-ω turbulence models have high fidelity only in the range of 0–6° of AOA.
(2) Flow separation occurs on the surface of the airfoil even when the AOA is 3°. The difference between the three models in the prediction of pressure coefficient is mainly reflected in the suction surface, and the S-A and SST k-ω turbulence models are difficult to accurately predict the flow separation phenomenon at high AOA. When the AOA increases from 10 to 11°, the increase of lift coefficient is mainly dominated by the suction surface.

Acknowledgements This work is supported by the Open Research Fund of State Key Laboratory of Hydro-Power Equipment (No. SKLHE-ORF-202108) and the Fundamental Research Funds for the Central Universities (Grant No. AUGA9803500921).

References

1. Johnson FS (1948) Atmosphere of mars[J]. Nature 161(4094):594–594
2. Drela M (1992) Transonic low-Reynolds number airfoils[J]. J Aircr 29(6):1106–1113
3. Mueller TJ, DeLaurier JD (2003) Aerodynamics of small vehicles[J]. Annu Rev Fluid Mech 35(1):89–111
4. Anyoji M (2011) Development of a mars wind tunnel and its applications to low Reynolds number and high-subsonic airfoil testing[D]. Tohoku University, Sendai
5. Young L A, Aiken E, Derby M, et al (2002) Experimental investigation and demonstration of rotary-wing technologies for flight in the atmosphere of mars[C]. 58th annual forum of the AHS, Montreal, Canada, pp 1–17
6. Witold J F K, Ethan A R, Wayne J (2018) Low Reynolds number airfoil evaluation for the Mars helicopter rotor[C]. AHS International 74th annual forum and technology display, phoenix, United States, pp 1–17
7. Zhao P, Quan Q, Chen S et al (2019) Experimental investigation on hover performance of a single-rotor system for Mars helicopter[J]. Aerosp Sci Technol 86:582–591
8. Désert T, Jardin T, Bézard H et al (2019) Numerical predictions of low Reynolds number compressible aerodynamics[J]. Aerosp Sci Technol 92:211–223
9. Munday PM, Taira K, Suwa T et al (2015) Nonlinear lift on a triangular airfoil in low-Reynolds-number compressible flow[J]. J Aircr 52(3):924–931
10. Hidaka H, Okamoto M (2014) An experimental study of triangular airfoils for Mars airplane[J]. Aerosp Technol Jpn 12:Pk_21–Pk_27
11. Caros L, Buxton O et al (2022) Direct numerical simulation of flow over a triangular airfoil under Martian conditions[J]. AIAA J 60:3961–3972
12. Caros L, Buxton O, et al (2023) Optimization of triangular airfoils for Martian helicopters using direct numerical simulations[J]. AIAA J 1–11

13. Morizawa S, Nonomura T, Oyama A et al (2018) Effect of Mach number on Airfoil characteristics at Reynolds number of 3,000[J]. Trans Jpn Soc Aeronaut Space Sci 61(6):258–267
14. Park K, Jung J, Jeong S (2023) Multi-objective shape optimization of airfoils for mars exploration aircraft propellers[J]. Int J Aeronaut Space Sci 24(1):9–23
15. Koning WJ, Johnson W, Allan BG (2018) Generation of Mars helicopter rotor model for comprehensive analyses[R]. AHS Aeromechanics Des Transform Vert Flight
16. Okamoto M (2019) An experimental study of the dynamic aerodynamic characteristics of a yaw-oscillating wind turbine airfoil[J]. Phys Fluids 31(6):067102
17. Chen ZJ, Przekwas AJ (2010) A coupled pressure-based computational method for incompressible/compressible flows[J]. J Comput Phys 229(24):9150–9165
18. Lilly DK (1992) A proposed modification of the Germano subgrid-scale closure method[J]. Phys Fluids A 4(3):633–635
19. Piomelli U, Balaras E (2002) Wall-layer models for large-eddy simulations[J]. Annu Rev Fluid Mech 34(1):349–374

A Multi-stage Target Assignment Method Based on Improved Genetic Algorithm

Tianyan Zhou, Ruoming An, Changsheng Gao, and Yuqing Li

Abstract Aiming at the different importance of target assignment tasks in each stage of multi-stage air attack operations, an improved genetic algorithm is designed. The decimal one-dimensional multi-segment chromosome is used to represent the multi-stage target assignment scheme. By setting the differential crossover and mutation probability for each segment of the chromosome, the opportunity to find the optimal assignment scheme in the operation stage with higher assignment quality requirements is improved. At the same time, the oscillation of the assignment scheme in the secondary operation stage with lower assignment quality requirements is reduced, and the performance of the target assignment algorithm is effectively improved. Finally, the simulation example shows that the improved genetic algorithm can better solve the problem of target assignment in multi-stage air strike operations.

Keywords Multi-stage target assignment · Genetic algorithm · Dynamic target assignment · Crossover and mutation

1 Introduction

The target assignment problem is to explore how to reasonably assign equipment to each target, thereby improving the success rate of achieving mission goals. According to the different task objectives, the target assignment problem can be divided into the target assignment of the defender and the target assignment of the attacker.

T. Zhou · R. An · C. Gao · Y. Li (✉)
School of Astronautics, Harbin Institute of Technology Harbin, Harbin, China
e-mail: bradley@hit.edu.cn

T. Zhou
e-mail: 1030902504@qq.com

R. An
e-mail: anruoming@hit.edu.cn

C. Gao
e-mail: gaocs@hit.edu.cn

Z. Liu et al. (eds.), *Advances and Challenges in Advanced Unmanned Aerial Systems*, Springer Aerospace Technology, https://doi.org/10.1007/978-981-99-8045-1_10

From whether the problem involves time factors, the target assignment problem can be divided into static target assignment problem and dynamic target assignment problem.

At present, the solution of the target assignment problem mainly uses various heuristic algorithms [1–9] such as genetic algorithm, particle swarm optimization, ant colony algorithm, bee colony algorithm, simulated annealing algorithm and machine learning methods such as neural network and reinforcement learning [10]. Liu et al. [11] used the improved particle swarm optimization algorithm to solve the problem of multi-objective dynamic target assignment by considering the constraints such as weapon firing time window and discreting the whole task process into multiple stages. Wenhai Wu et al. [12] used an improved differential evolution algorithm to solve the dynamic target assignment problem by dispersing the dynamic target assignment problem into multiple static target assignment problems. Scholars have proposed many algorithms to solve the target assignment problem, and have achieved good results [13]. However, for the multi-stage target assignment problem, the task objectives of each stage are different, the importance is different, and the difficulty of the assignment operation is different. At present, there is not enough attention to above problems.

This paper mainly focuses on the multi-stage target assignment problem of the attacker, mathematically models the multi-stage target assignment problem, and analyzes the task objectives and importance of each task stage. Aiming at the multi-stage target assignment problem, the crossover and mutation operation of the general genetic algorithm is improved, so that the improved genetic algorithm can better adapt to the multi-stage target assignment problem, and improve the stability of the algorithm and the quality of the solution.

2 Multi-stage Target Assignment Mathematical Model

2.1 Problem Description and Mathematical Constraints

The multi-stage target assignment problem of the attacker is to assign his own weapons to each defensive unit that needs to be eliminated. Defence units that need to be destroyed usually include interception units, which can intercept incoming targets entering the range and reduce the hit rate of the attacker. Therefore, in this kind of multi-stage target assignment problem, the hit rate of the attacker to each target is not constant. As the stage progresses, the number of interception units of the defender gradually decreases, and the hit rate of the attacker will continue to rise.

In general, when calculating the hit rate of each target, the attacker needs to take into account the position of the target and the interception unit around the target and the range of the interception unit. In this model, in order to simplify the calculation, the interception matrix is directly used to represent the cover information of each interception unit to other units obtained by the attacker before the start of the operation.

Suppose that there are a total of targets, where the target is the interception unit, then the interception matrix I^{itc} is, which represents whether the target x can protect the target x, and can be protected when 1 is taken, and can not be protected when 0 is taken. The interception matrix x at each stage will change as the action continues. For the interceptor unit destroyed in each stage, in the subsequent stage, its corresponding parameters in the intercept matrix will be reset to 0.

Suppose there are a total of m targets, where the n $m(1 < n < m)$ target is the interception unit, then the interception matrix I^{itc} is $n \times m$, $S_{ij}^{p}(i = n, n+1, \cdots m, j = 1, 2, \cdots m)$ represents whether the target i can protect the target j, when 1 is able to protect, when 0 is not. The interception matrix I^{itc} at each stage will change as the action continues. For the interceptor unit destroyed in each stage, its corresponding parameters in the intercept matrix will be reset to 0 in the subsequent stage.

$$I^{itc} = \begin{bmatrix} S_{n1}^{p} & S_{n2}^{p} & \cdots & S_{nm}^{p} \\ S_{(n+1)1}^{p} & S_{(n+1)2}^{p} & \cdots & \vdots \\ \vdots & \vdots & \ddots & \vdots \\ S_{m1}^{p} & S_{m2}^{p} & \cdots & S_{mm}^{p} \end{bmatrix} \tag{1}$$

Assuming that the attacker has a total of St task stages, the number of equipment used in each stage is $W_{num}(num = 1, 2 \cdots, St)$, and x_{kl} equipment is assigned to the target $l(l = 1, 2 \cdots m)$ in stage $k(k = 1, 2 \cdots St)$, then the target assignment matrix T is:

$$T = \begin{bmatrix} x_{11} & x_{12} & \cdots & x_{1m} \\ x_{21} & x_{22} & \cdots & \vdots \\ \vdots & \vdots & \ddots & \vdots \\ x_{St1} & x_{St2} & \cdots & x_{Stm} \end{bmatrix} \quad s.t \begin{cases} x_{k1} + x_{k2} \cdots x_{km} \le W_k \\ x_{kl} \ge 0 \end{cases} \tag{2}$$

The main optimization direction of the attacker's target assignment problem is to cause as much damage to the target as possible, so the total damage to all targets after the end of all task stages is taken as the objective function. Assuming that the life value of each target is $H_j(j = 1, 2, \cdots m)$, the interception rate of the interception unit is $R_i(i = n, n+1 \cdots, m)$, the interception matrix of each stage is $I_k^{itc}(k = 1, 2 \cdots St)$, and the lethality of each equipment of the attacker is D, the objective function is:

$$HP_l(x) = \begin{cases} H_l & x > H_l \\ x & x \le H_l \end{cases} (l = 1, 2, \cdots m),$$

$$F = \sum_{k=1}^{St} \sum_{l=1}^{m} HP_l[D \cdot x_{kl} \prod_{i=n}^{m} (1 - R_i \cdot S_{il}^{P})] \tag{3}$$

where, in a stage, if the damage caused to a target exceeds his life value, the target is considered to be destroyed, and the score obtained is equal to the life value of the target, rather than the damage caused, preventing the calculation of invalid scores. The killing value of each stage to the target is equal to the sum of the product of lethality and hit rate.

2.2 Analysis of the Characteristics of Each Stage of Multi-stage Target Assignment Problem

The main characteristic of the multi-stage target assignment problem is that the action effect of the previous stage will affect the action effect of the next stage. According to the objective function (3), for the attacker, the successful elimination of the interception unit of the defender in the early stage can improve the hit rate in the follow-up stage, thus improving the killing value to the target in the follow-up stage and effectively improving the action effect in the follow-up stage.

Therefore, in the multi-stage target assignment problem of the offensive side, the target assignment in the early stage will be more important. If the defense of the other side can be effectively weakened, the opportunity to finally complete the task target can be improved. However, at the same time, in the early stage, the interception power is relatively strong, and the hit rate of each target is relatively low. It is very difficult to find a reasonable target assignment scheme in the early stage and obtain a better task effect in order to find a target with a relatively high hit rate in the early stage, so as to cause enough damage to it and destroy it, and at the same time, to destroy the interception unit that protects a large number of other targets, so as to weaken the defense ability as much as possible and pave the way for the follow-up action. In the later stage, because the number of remaining targets is much less than that of the previous stage, and the hit rate also increases a lot, the difficulty of target assignment has decreased significantly.

Therefore, in view of the characteristics of the problem, this paper improves the parameters of the general genetic algorithm, increases the search ability of the algorithm for the better target assignment scheme in the early stage, and gives up some search ability for the target assignment scheme in the later stage, so as to improve the stability of the whole algorithm.

3 Improved Genetic Algorithm for Multi-stage Target Assignment Problem

The chromosome used to record the target assignment scheme in this paper is a decimal one-dimensional chromosome. The number of equipment assigned to each target in each stage is arranged in turn, and then each stage is connected in turn. The length of the chromosome is $St \times m$.

3.1 Algorithm Improvement Ideas

Genetic algorithm (GA) is a search algorithm that imitates the process of biological genetic evolution. It has a good effect on solving NP-complete problems such as target assignment. The genetic algorithm mainly searches the solution space through crossover and mutation operations. The larger the crossover and mutation probability, the stronger the search ability for the solution space, but the excessive probability may cause the excellent solution fragment to be unable to retain, and cause the algorithm to oscillate or even diverge. In this paper, by adopting dynamic and differentiated crossover and mutation parameters, the probability of crossover and mutation in different parts of a chromosome is different. The pre-chromosome segment (which records the target assignment scheme in the pre-action stage) has larger crossover and mutation parameters, improves the probability of crossover and mutation, makes the search range of the pre-segment solution larger, and has more opportunities to find better solutions. The part with low search demand in the back segment uses a lower probability of crossover and mutation to prevent the instability of the population caused by excessive crossover and mutation parameters, as shown in Fig. 1. At the same time, the crossover and mutation parameters will gradually decrease with the increase of evolutionary algebra, which will reduce the convergence difficulty of the algorithm.

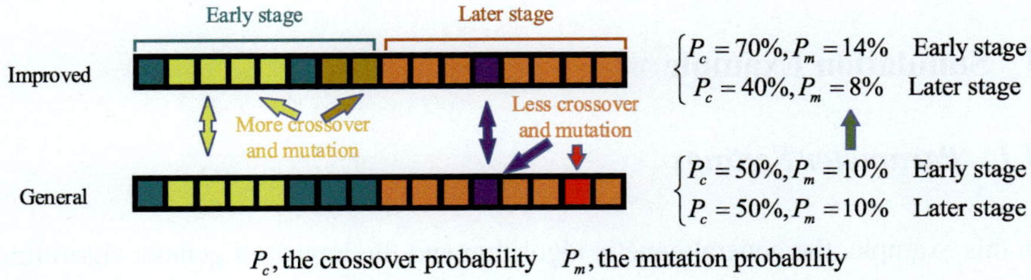

Fig. 1 Differential crossover and mutation probability

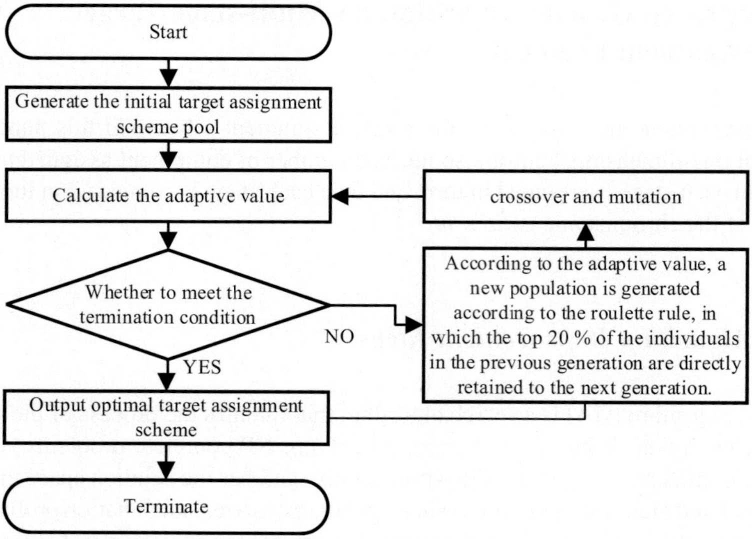

Fig. 2 Algorithm execution process diagram

3.2 Algorithm Execution Steps

The main steps of the algorithm are shown in Fig. 2, in which 20% of the excellent individuals are directly retained when a new population is generated, which improves the average quality of the gene fragments of the parent generation during the crossover and mutation operation, so that the excellent solution fragments obtained under high crossover and mutation probability are more likely to be retained, preventing the instability of the evolution process caused by the additional mutation probability of the chromosome front segment, and reducing the risk of the average quality decline of the new population.

4 Simulation Example

4.1 Parameter Settings

In this example, the general genetic algorithm and the improved genetic algorithm are used to solve the target assignment scheme in the same scene, and the results are compared to verify the improved effect. The evolution generation is set to 1000 generations, and the mutation probability is 20% of the crossover probability. The crossover probability of the two algorithms is set as shown in Table 1.

Table 1 Crossover probability

Generation	0–400 (%)	400–600 (%)	600–700 (%)	700–800 (%)	800–1000 (%)
General genetic algorithm (GGA)	50	50	50	50	50
Improved genetic algorithm (IGA) (preliminary stage)	70	70	50	30	10
Improved genetic algorithm (later stage)	40	30	20	10	5

Table 2 Target information

	A	B	C	D	E	F	G	H	I	J	K	L	M	N	O
Life value	200	200	200	200	200	100	100	100	100	100	100	100	100	100	100
Interception probability	/	/	/	/	/	0.7	0.7	0.7	0.7	0.7	0.7	0.7	0.7	0.7	0.7

Table 3 Interception matrix

	A	B	C	D	E	F	G	H	I	J	K	L	M	N	O
F	1	0	0	0	1	1	0	0	1	1	0	0	1	0	0
G	0	1	0	1	0	1	1	1	1	0	0	0	0	0	1
H	1	0	0	0	1	0	1	1	0	0	0	0	0	1	0
I	0	0	1	0	0	1	0	1	1	1	0	0	0	0	0
J	1	0	0	1	1	0	1	0	0	1	0	0	1	0	0
K	0	0	0	0	1	0	1	0	0	0	1	0	0	0	0
L	1	1	0	1	0	0	0	0	1	1	0	1	0	0	0
M	0	0	1	0	0	0	0	0	0	0	0	0	1	0	0
N	0	1	0	1	0	0	1	0	0	1	0	0	0	1	0
O	0	1	0	0	0	1	0	0	1	0	0	0	0	0	1

The attack wave of the attacker is set to three waves, each wave has 15 equipment, each average lethality is 100, and there are 15 targets. Among them, the F–O target is the interception unit. The specific parameter settings are shown in Tables 2 and 3.

4.2 Simulation Result

The running result diagram, running data table and target assignment scheme of the algorithm are shown in Fig. 3, Tables 4 and 5, respectively. The parameters in Table 4 are the average values obtained from 10 runs.

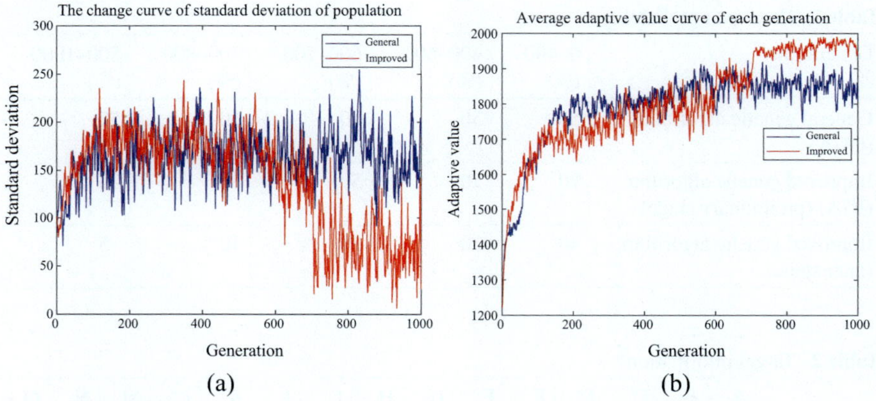

Fig. 3 Running result diagram of the algorithm

Table 4 Running data table of the algorithm

	Running time of algorithm (s)	Adaptive value of optimal scheme	The standard deviation of the adaptive value of the optimal scheme in 10 runs
General GA	0.171	1902	78.44
Improved GA	0.189	1978	20.91

Table 5 Optimal target assignment scheme

Stage	A	B	C	D	E	F	G	H	I	J	K	L	M	N	O
1IGA	0	0	1	1	0	4	0	0	0	0	2	2	1	3	1
2IGA	0	0	2	1	0	1	1	3	0	3	0	0	2	0	2
3IGA	2	3	1	2	2	0	1	0	3	0	0	0	1	0	0
1GGA	0	0	0	1	1	0	0	3	0	3	1	0	2	1	3
2GGA	0	0	4	0	5	3	0	0	1	0	0	2	0	0	0
3GGA	2	3	0	3	2	1	2	0	1	1	0	0	0	0	0

It can be seen from Fig. 3a that the improved genetic algorithm can maintain the same standard deviation of population adaptive value as the general genetic algorithm in the case of increasing the probability of crossover and mutation in the early stage. After convergence, the standard deviation of the population adaptive value is less than 50% than that of the general genetic algorithm. It shows that the adaptive value of the whole population of the improved genetic algorithm is more stable. From the Fig. 3b, the average adaptive value of the population after the final convergence of the improved genetic algorithm is 8.2% higher than that of the general genetic algorithm, and the oscillation after convergence is smaller. It can be seen from the

Table 4 that the adaptive value of the optimal target assignment scheme obtained by the improved genetic algorithm is 4% higher than that of the general genetic algorithm, and the standard deviation of the optimal value obtained by multiple runs is smaller, indicating that the improved genetic algorithm can be more stable to get a better solution.

5 Conclusion

In this paper, a mathematical model of multi-stage target assignment is established for the multi-stage target assignment problem of the attacker. According to the characteristics of the problem, the general genetic algorithm is improved and the example simulation is carried out. The simulation results show that the improved genetic algorithm can get a better solution than the general genetic algorithm, and the stability of the optimal solution and the stability of the population in the evolution process are greatly improved. However, the improvement of genetic algorithm in this paper is not deep enough, and there is still room for optimization.

References

1. Chang X, et al (2023) Adaptive large neighborhood search algorithm for multi-stage weapon target assignment problem. Comput Ind Eng 181
2. Zhang K, et al (2020) A dynamic weapon target assignment based on receding horizon strategy by heuristic algorithm. J Phys Conf Ser 1651.1
3. Gao C, et al (2019) A heuristic algorithm for weapon target assignment and scheduling. Mily Oper Res 24.4
4. Ma Y, et al (2021) Two-stage hybrid heuristic search algorithm for novel weapon target assignment problems. Comput Ind Eng 162
5. Kong L, et al (2021) Solving the dynamic weapon target assignment problem by an improved multiobjective particle swarm optimization algorithm. Appl Sci 11.19
6. Wang T, et al (2023) Unmanned ground weapon target assignment based on deep Q-learning network with an improved multi-objective artificial bee colony algorithm. Eng Appl Artif Intell 117.PB
7. Sonuc E, et al (2017) A parallel simulated annealing algorithm for weapon-target assignment problem. Int J Adv Comput Sci Appl 8(4)
8. Kline AG et al (2019) Real-time heuristic algorithms for the static weapon target assignment problem. J Heuristics 25:377–397
9. Lai CM, et al (2019) Simplified swarm optimization with initialization scheme for dynamic weapon–target assignment problem. Appl Soft Comput 82:105542
10. Wu X, et al (2021) A modified MOEA/D algorithm for solving Bi-objective multi-stage weapon-target assignment problem. IEEE Access 9:71832–71848
11. Liu P, et al (2023) Multi-missile dynamic weapon target assignment algorithm based on particle swarm optimization. J Nanjing Univ Aeronaut Astronaut,\ 55(01):108–115. (in Chinese)
12. Wu W, et al (2021) Improved differential evolution algorithm for solving weapon-target assignment problem. Syst Eng Electron 43(04):1012–1021. (in Chinese)
13. Li M, et al (2023) Developments of weapon target assignment: models, algorithms, and applications. Syst Eng Electron 45(04):1049–1071. (in Chinese)

Design and Analysis of UAV Arresting Device Based on Electromagnetic Damper

Jiayu Hao, Xiaohui Wei, Yiming Peng, Hui Ma, and Guoguo Fu

Abstract An UAV arresting device based on electromagnetic damper is designed. The dynamics mathematical model of UAV arresting process is established. MAXWELL is used to analyze the performance of electromagnetic damper. Then the performance of generator is obtained by experimental. Finally, a multidisciplinary simulation model of the arresting device is established by AMESim. With the increase of UAV mass, the peak acceleration decreases gradually, the arresting distance and the arresting time increase equidistant. The acceleration in the middle and late stages of the arresting process is obviously controlled, and the energy recovery efficiency is the same, which is 9.7% of the UAV kinetic energy.

Keywords UAV arresting device · Electromagnetic damper · Energy recovery · Dynamics · Multidisciplinary simulation

J. Hao · X. Wei (✉) · Y. Peng · G. Fu
Nanjing University of Aeronautics and Astronautics, Nanjing, China
e-mail: wei_xiaohui@nuaa.edu.cn

J. Hao
e-mail: 2249982979@qq.com

Y. Peng
e-mail: yimingpeng@nuaa.edu.cn

G. Fu
e-mail: 1797710957@qq.com

H. Ma
AVIC Jincheng Nanjing Engineering Institute of Aircraft System, Nanjing, China
e-mail: valenmania@163.com

Z. Liu et al. (eds.), *Advances and Challenges in Advanced Unmanned Aerial Systems*, Springer Aerospace Technology, https://doi.org/10.1007/978-981-99-8045-1_11

CCS Concepts: \ begin{CCSXML}

<ccs2012>

<concept>

<concept_id>10010405.10010432.10010433.10010989</concept_id>

<concept_desc>Applied computing~Avionics</concept_desc>

<concept_significance>500</concept_significance>

</concept>

</ccs2012>

\end{CCSXML}

\ccsdesc[500]{Applied computing~Avionics}Introduction

1 Introduction

In the future form of war, network, information, system, speed, etc., have gradually become the new winning factors in war. Swarm, wingman and collaborative combat technologies based on UAV technology meet the requirements of future wars and they will become the key to victory in war [1]. UAV collaborative operations need to decentralize functions. Then arrange multiple-type UAV nodes in the airspace, form an information network and expand information advantages [2]. Maintaining UAV nodes and supplementing losses are the basic conditions for maintaining information networks. To achieve the above effect, it is necessary to quickly recover UAV nodes. However, the decentralization of UAV functions leads to various types of UAV nodes with different mass [3]. How to effectively and safely recover multi-type and multi-quality drones has become an important research direction.

UAV recovery methods include runway recovery, rope-hook recovery [4], net suspended recovery [5], parachute recovery [6], deep stall recovery [7], vertical landing and arresting recovery. Among them, runway recovery is applicable to almost all fixed-wing UAVs with landing gear, but UAV of different mass has different sliding distance, which is not suitable for landing platform with compact structure and limited landing space. Rope-hook recovery, net suspended recovery, parachute recovery and deep stall recovery are suitable for small UAVs, but the recovery efficiency is low, the performance is unstable, and the adaptability to multiple types of UAVs is poor [8]. Vertical landing has high requirements for UAV design and it causes small payload of UAVs [9]. The arresting recovery is relatively mature, the maintenance cost is low, the landing distance can be effectively shortened, and the reset is simple. It is suitable for recovering UAV nodes.

As for the arresting device, it includes side-arm arresting recovery [10], telescopic arm recovery [11] and cable arresting recovery [12]. Side-arm arresting recovery and telescopic arm recovery are complex and difficult to catch the UAVs. There are few researches on cable arresting devices of UAVs.

In order to meet the requirements of recovering multi-type and multi-mass UAVs, an arresting device for multi-type small UAVs based on electromagnetic damper and energy recovery technology is designed. The dynamics mathematical model of UAV arresting process is established. Electromagnetic simulation software-MAXWELL is used to established the electromagnetic damper simulation model. Then the performance of generator is obtained by experimental. The results are input into AMESim to build the whole simulation model of the arresting device. A series of conclusions are drawn from the simulation analysis of the different mass UAV arresting process.

Compared to traditional friction arresting device, new arresting device improved the landing performance of UAVs. With help of electromagnetic damper, it is suitable for arresting multi-type small UAVs. By adding control system, the damping force applied to UAV can be dynamically adjusted. Magneto-resistive effects cannot be used to generate electricity. Therefore, electromagnetic damper only works as a damper. Kinetic energy recovery system is designed for saving energy.

2 Dynamics Mathematical Model of UAV Arresting Process

The arresting device is composed of generator, winch, electromagnetic damper, winch support and arresting cable reversing pulley (see Fig. 1). The arresting device is symmetrically arranged on both sides of the runway, and the two winches are connected by same cable. When the UAV is caught by cable, the arresting cable is pulled out, driving the generator and the electromagnetic damper to rotate. Electromagnetic damper provides the damping moment and the generator recovers the kinetic energy of the UAV. In order to establish the simulation model of the arresting device, it is necessary to conduct the dynamics numerical modeling of the arresting process of UAV (see Fig. 2).

According to the force analysis, the UAV dynamics equation can be obtained as shown in formula (1) to formula (5):

$$F_{\mathrm{ar}} + 2f_2 + f_1 + D_{\mathrm{p}} - T = ma \tag{1}$$

$$L + F_{\mathrm{N1}} + 2F_{\mathrm{N2}} - G = 0 \tag{2}$$

$$f_1 h_1 - F_{\mathrm{N1}} l_1 + 2f_2 h_2 + 2F_{\mathrm{N2}} l_2 + F_{\mathrm{ar}} h_3 = 0 \tag{3}$$

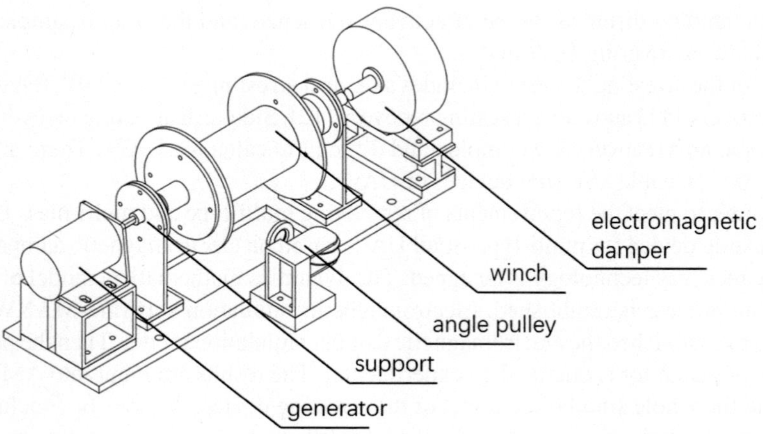

Fig. 1 Structure of arresting device

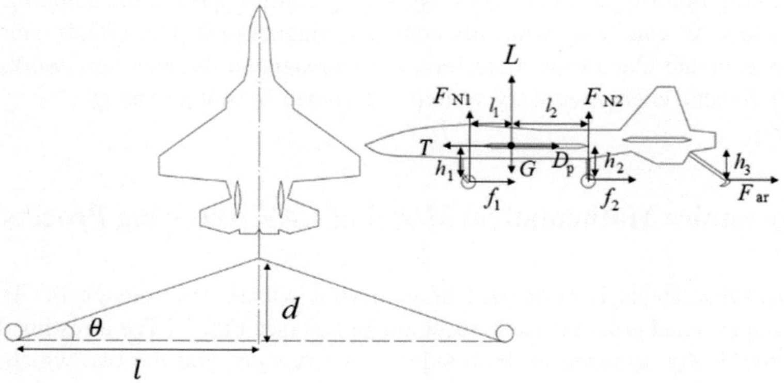

Fig. 2 UAV force diagram

$$L + F_{\mathrm{N1}} + 2F_{\mathrm{N2}} - G = 0 \tag{4}$$

$$\sin\theta = \frac{d}{\sqrt{l^2 - d^2}} \tag{5}$$

where Far is arresting force; a is acceleration of UAV; f_1 is rolling friction of front wheel; f_2 is rolling friction of main wheels; D_{p} is resistance; T is engine thrust; m is the mass of UAV; L is lift of UAV; F_{N1} is the support force of front landing gear; F_{N2} is the support force of main landing gears; G is the gravity; F_{S} is strain inside cable; h_1 and h_2 are the height of the front landing gear and main landing gear respectively; l_1 and l_2 are the distance from landing gear to barycenter; h_3 is vertical distance between the point where the arresting cable contacts the arresting hook and the horizontal line where the center of gravity of the UAV is located; d is the landing distance; l is the half span of arresting cable; θ is the angle between the current position and the initial position.

3 Electromagnetic Damper Simulation Based on MAXWELL

Electromagnetic damper is the main source of damping torque of UAV arresting device. A simulation model of electromagnetic damper is established based on MAXWELL-electromagnetic simulation software (see Fig. 3). The simulation model consists of coil, inner magnetic field conductor, outer magnetic field conductor and rotor. The damping torque is provided by magneto-resistance effect (see Fig. 4).

By changing rotor speed, the relationship between torque and rotational speed is studied. By changing excitation coil current, the relationship between torque and current is studied (see Fig. 5).

It can be seen from the simulation results that as the rotational speed increases from 200 to 1500 rpm, the damping torque slightly increases, and the rising speed slows down in the later stage. The damping torque does not change significantly after the speed rises from 2500 to 3000 rpm, thus it can be seen that the speed has little influence on the damping torque. When the rotation speed is maintained at 3000 rpm and the coil excitation current is changed, the damping torque changes obviously.

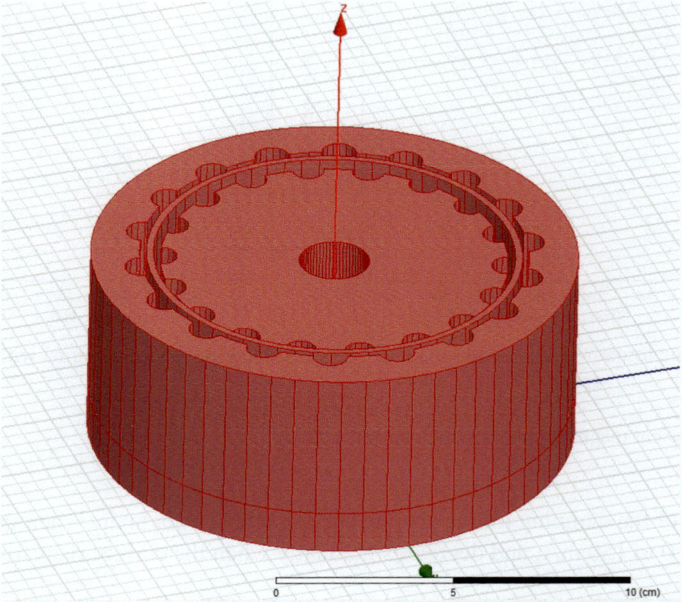

Fig. 3 Simulation model of damper

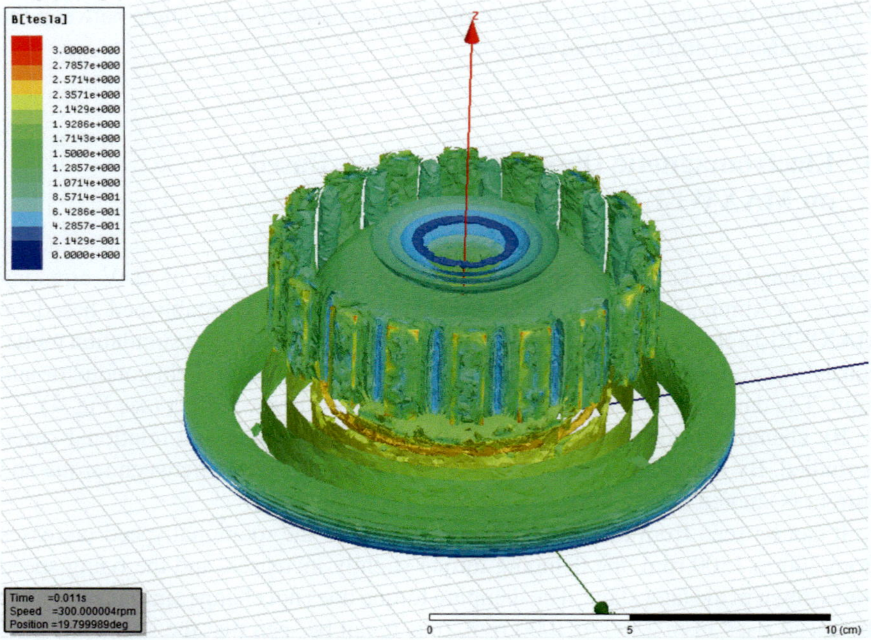

Fig. 4 Magnetic field distribution

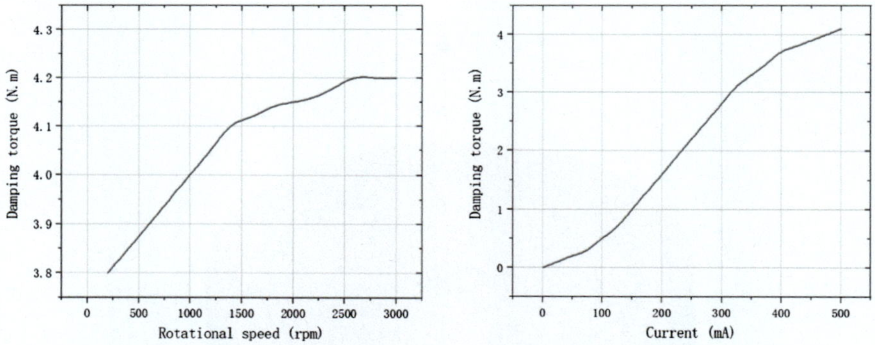

(a) Damping torque variation with rotational speed.

(b) Damping torque variation with current.

Fig. 5 Transient magnetic field simulation results curve

The slope of the damping torque change curve increases gradually when the current rises from 0 to 150 mA, and decreases gradually when the current rises from 350 to 500 mA.

The rotational speed is increased by 2800 rpm, the damping torque is only increased form 3.8N.m to 4.2N.m, an increase of 9.5%. Excitation coil current, in contrast, is more obvious impact. The current is increased by 500 mA and the damping torque is increased from 0N.m to 4.1 N.m. It can be seen that the main factor affecting the performance of the electromagnetic damper is the current in the excitation coil. The rotational speed insignificantly affects the damping torque.

4 Performance Parameter Experiment of Energy Recovery Motor

When the arresting device is working, the energy recovery generator is responsible for recovering part of the UAV kinetic energy. Therefore, it is necessary to measure the generator performance parameters by experiments and compare and verify with the simulation model to increase the reliability of the simulation (see Fig. 6).

By changing the rotational speed of the drive motor, the breaking voltage of the generator under different rotational speed is measured. By connecting the charging circuit, the relationship between the generator rotational speed and output current can be measured under the condition of the initial voltage of the supercapacitor 14.2 V, and the relationship between the generator rotational speed and power can be obtained by measuring the output voltage of the generator, as shown in Fig. 7.

The output voltage of the generator presents a linear relationship with the rotational speed under the open circuit condition (see Fig. 7). In the charging circuit, when the voltage of the energy storage supercapacitor is 14.2 V, the current and power line are similar, showing an approximate linear relationship with the rotational speed. The test result can be used to improve the energy recovery generator model of the arresting device and increase the reliability of the model.

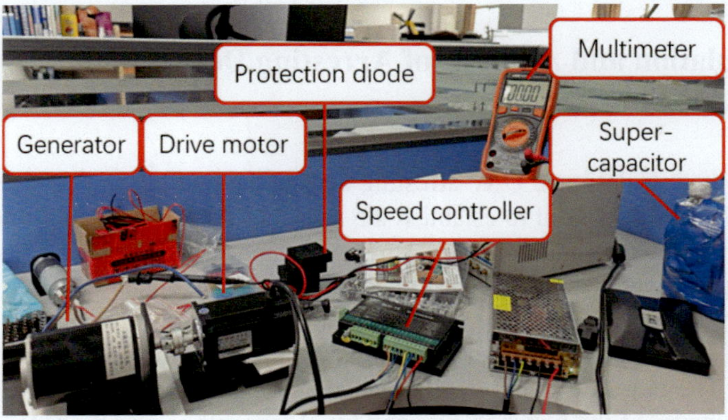

Fig. 6 Energy recovery generator test platform

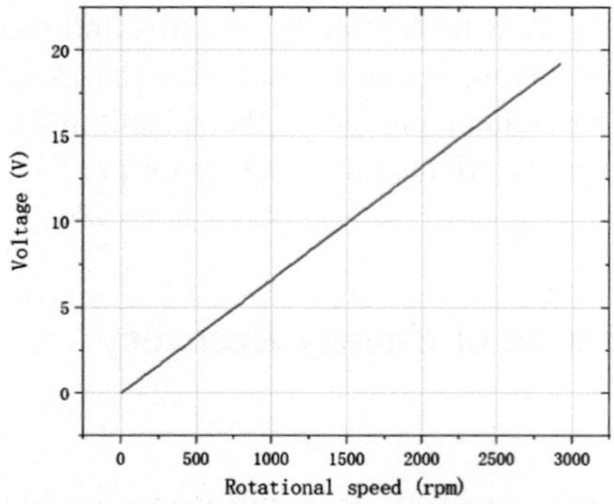

(a) Output voltage variation with rotational speed.

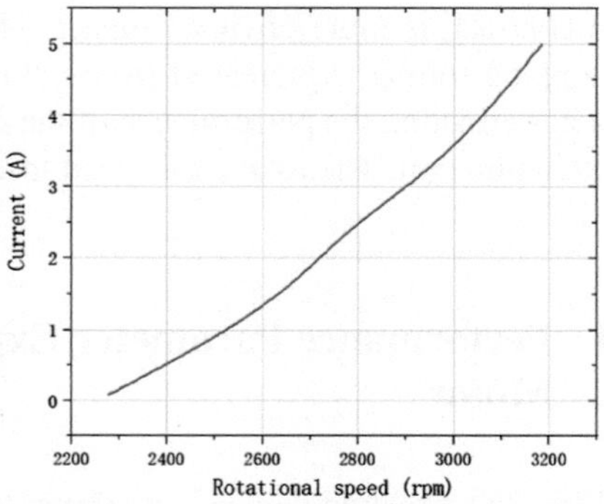

(b) Output current variation with rotational speed.

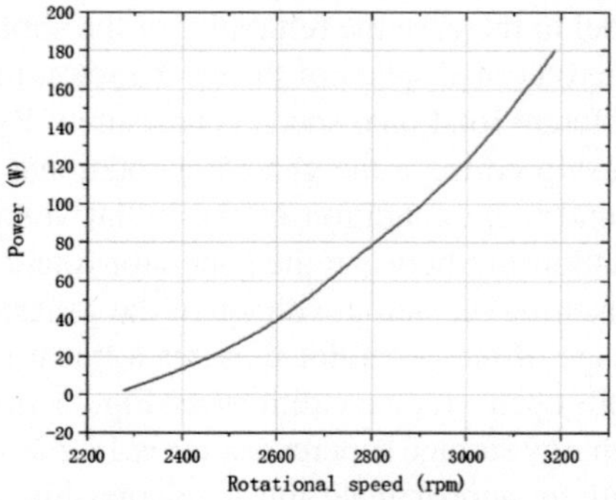

(c) Output power variation with rotational speed.

Fig. 7 Performance test result of energy recovery generator

5 Simulation and Analysis of Arresting Device Based on AMESim

Based on the self-designed UAV arresting device, the multidisciplinary simulation software-AMESim is used to establish the arresting device simulation model (see Fig. 8). Then, run the simulation. The simulation model is composed of electromagnetic damper, generator and energy storage supercapacitor, winch and UAV arresting simulation model. By referring to the measured generator performance parameters and electromagnetic damper simulation results, the arresting device model is improved. The authenticity and credibility of the simulation model are increased. The rotational inertia of the winch is obtained by 3D modelling and the UAV arresting simulation model is built according to the mathematical model.

The mass of the UAV is set as 15, 20, 25 and 30 kg respectively, in order to simulate the arresting process of multiple types of UAV. The course speed of the UAV is 20 m/s, and the simulation results are received (see Fig. 9).

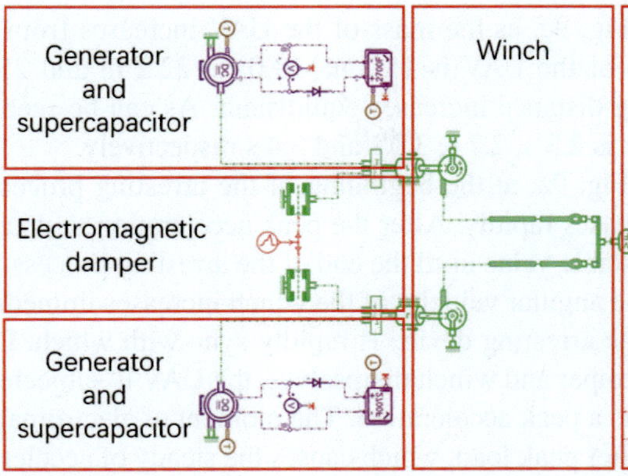

Fig. 8 Simulation model based on AMESim

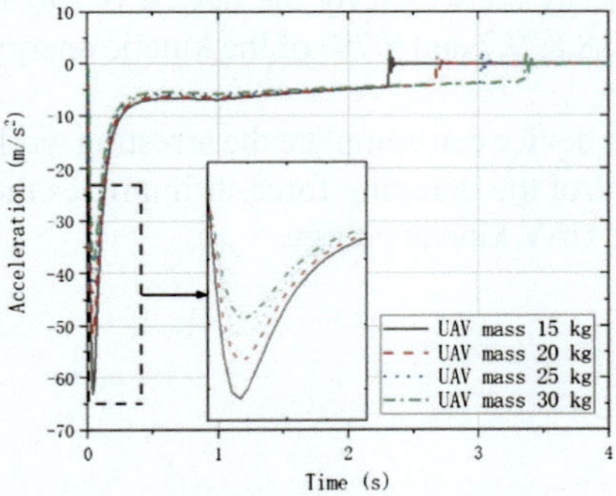

(a) UAV acceleration variation with time.

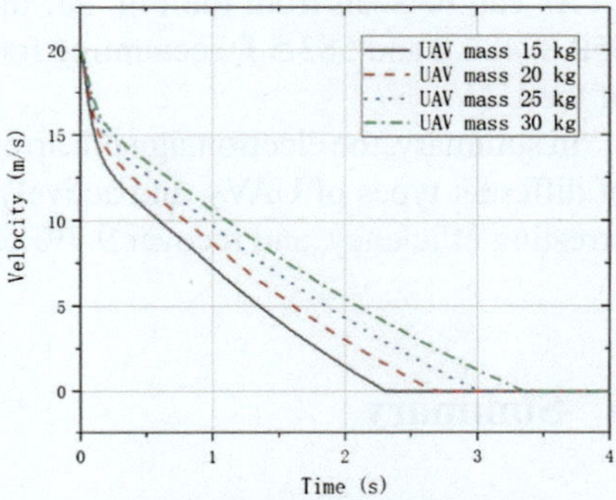

(b) UAV velocity variation with time.

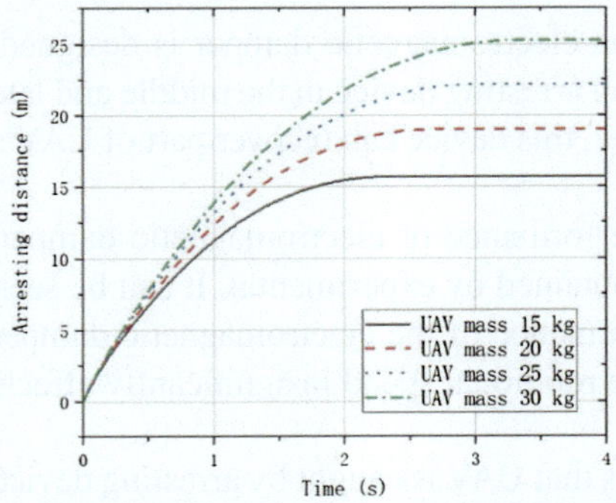

(c) Arresting distance variation with the time.

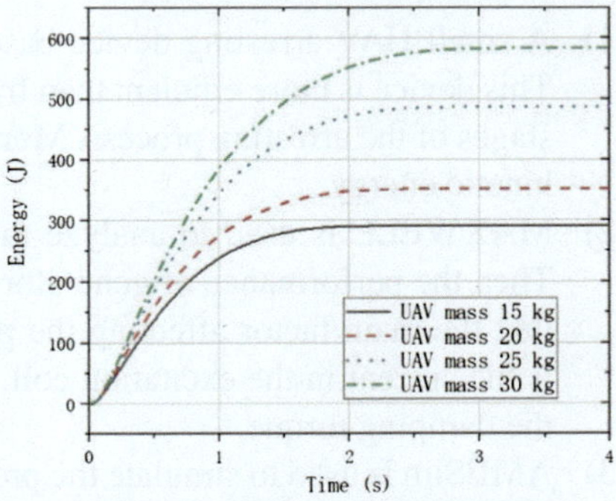

(d) Recovery energy variation with time.

Fig. 9 Simulation result based on AMESim

As can be seen from the Fig. 9c, as the mass of the UAV increases from 15 to 30 kg, the arresting distance of the UAV is 15.7 m, 19.0 m, 22.2 m and 25.3 m, respectively, and the arresting distance increases equidistant. As can be seen from the Fig. 9b, the arresting time is 2.3 s, 2.7 s, 3.0 s and 3.4 s respectively.

As can be seen from the Fig. 9a, at the beginning of the arresting process, the acceleration of the UAV increases rapidly. After the peak acceleration, it decreases rapidly and then maintains a stable value until the end of the arresting process. After the UAV is caught by rope, the angular velocity of the winch increases immediately, and the angular velocity of the arresting device is rapidly sync with winch. Due to the rotational inertia of the damper and winch themselves, the UAV is subjected to a large reaction force, so there is a peak acceleration. The moment of electromagnetic damper increases gradually after peak load, which causes the steady of acceleration. The peak acceleration is -63.2 m/s^2, -51.6 m/s^2, -43.6 m/s^2 and -37.7 m/s^2. With the increase of UAV mass, the interval of peak acceleration decreases gradually.

As can be seen from the Fig. 9d, the energy recovered by the device is 290.9, 350.8, 486.8 and 583.5 J, accounting for 9.7, 8.8, 9.7 and 9.7% of the kinetic energy of the UAV.

In summary, the electromagnetic arresting device can complete the arresting work of different types of UAVs, and actively control the damping force. It improves the arresting efficiency, and recover 9.7% of the UAV kinetic energy.

6 Summary

In summary, this paper draws the following conclusions:

(1) A small UAV arresting device based on electromagnetic damper is designed. This device is more efficient than friction arresting device in the middle and late stages of the arresting process. Moreover, this device can recover part of UAV's kinetic energy.
(2) MAXWELL is used to analyze the performance of electromagnetic damper. Then the performance of generator is obtained by experimental. It can be seen that the main factor affecting the performance of the electromagnetic damper is the current in the excitation coil. The rotational speed insignificantly affects the damping torque.
(3) AMESim is used to simulate the process that UAV is caught by arresting device and stopped by electromagnetic damper. The performance of arresting device is analysed. Result shows that the acceleration curve will quickly reach the peak and quickly fall back after the peak and remain stable until the end of the arresting process. As the mass of the UAV increases from 15 to 30 kg, the peak acceleration reduces from -63.2 m/s^2 to -37.7 m/s^2. The energy recovery efficiency does not change with the increase of the mass of the UAV, and it remains at 9.7%.

Acknowledgements This paper is supported by Aviation Science Foundation of China (No.20200028052010).

References

1. Wang CL, Fan HT (2022) Research on new air combat concepts of US military[J]. Aero Weaponry 29(03):8–13 (in Chinese)
2. Guo JG, Zhou M, Guo ZY et al (2021) Cooperative combat technology under mosaic warfare[J]. Aero Weaponry 28(01):1–5 (in Chinese)
3. Liu L, Liu DW, Guo ZY et al (2022) Analysis of the development status and outlook of UAV clusters and anti-UAV clusters[J]. Acta Aeronautica et Astronautica Sinica 43(S1):4–20 (in Chinese)
4. Lu W, Ma XP, Zhou M et al (2015) Simulation analysis of dynamic characteristic of UAV rope-hook recovery system[J]. Acta Aeronautica et Astronautica Sinica 36(10):3295–3304 (in Chinese)
5. Kristian K, Fossen TI, Johansen TA (2018) Autonomous recovery of a fixed-wing UAV using a net suspended by two multirotor UAVs[J]. J Field Robot 35(5):717–731
6. Wu CF, Shao PY, Ma S et al (2012) Study on assigned point parachute recovery of UAV[J]. Comput Simul 29(06):104–107 (in Chinese)
7. Qu YH, Zhang F, Gu RN, et al Deep stall recovery method for small fixed-wing UAV based on range information: China. CN107991884A[P]. 2018-05-04 (in Chinese)
8. Tan LG, Yang XY, Song SM et al (2019) An overview of marine recovery methods of WAV for small ships[J]. J Harbin Inst Technol 51(10):1–10 (in Chinese)
9. Wang KL, Zhou Z, Ma YW et al (2022) Development and trend analysis of vertical takeoff and landing fixed wing UAV[J]. Adv Aeronaut Sci Eng 13(05):1–13 (in Chinese)
10. Hong D, Zheng ZS, Zhou L (2022) Analysis of US military sidearm system characteristics and its application prospect[J]. Sci Technol Rev 40(05):132–137 (in Chinese)
11. Su ZK, Xu ZN, Li CT et al (2023) Modeling and docking control of aerial recovery in the form of telescopic boom[J]. Acta Aeronautica et Astronautica Sinica 44(01):227–244 (in Chinese)
12. Yang QF (2020) Design and development of rapid prototype for UAV arresting and landing simulation system[D]. Nanjing University of Aeronautics and Astronautics, Nanjing, pp 9–28 (in Chinese)

Research on Obstacle-Crossing Characteristic for Air Cushioned Transport Aircraft

Lincheng Yang, Qiaozhi Yin, Xiaohui Wei, Songyang Zhang, and Hong Nie

Abstract In order to improve the transport aircraft ability to cross obstacles on complex runway, a study was conducted on the characteristics of a military personnel transport aircraft modified with an air cushion landing system (ACLS) for ground obstacle-crossing. Based on the ideal gas equation of state and finite element division, the cross-section of the trunk is simplified and its shape is predicted, based on which the pressure distribution in the cushion is calculated. The effect of the boss, ditch and one-sided ramp on the distance of the centre of gravity from the ground, vertical acceleration, pitch angle and cushion pressure when the air cushioned transport aircraft (ACTA) crosses the obstacle is investigated by adding the obstacle runway function to the calculation of the distance from the ground of the air cushion micro-element, and the mechanism of its influence is analysed. The simulation results show that when the height of the obstacle reaches 70–80% of the height of the airbag, the ACTA can still pass the obstacle runway safely, which reflects the effectiveness of the modified ACLS in improving the transport aircraft ability to cross the obstacle.

L. Yang · Q. Yin · X. Wei (✉) · S. Zhang · H. Nie
Key Laboratory of Fundamental Science for National Defense-Advanced Design Technology of Flight Vehicle, Nanjing University of Aeronautics and Astronautics, Nanjing 210016, China
e-mail: wei_xiaohui@nuaa.edu.cn

L. Yang
e-mail: yanglc@nuaa.edu.cn

Q. Yin
e-mail: yinqiaozhi@nuaa.edu.cn

S. Zhang
e-mail: zhangsongyang@nuaa.edu.cn

H. Nie
e-mail: hnie@nuaa.edu.cn

Q. Yin · H. Nie
National Key Laboratory of Rotorcraft Dynamics, Nanjing University of Aeronautics and Astronautics, Nanjing 210016, China

X. Wei
State Key Laboratory of Mechanics and Control for Aerospace Structures, Nanjing University of Aeronautics and Astronautics, Nanjing 210016, China

Z. Liu et al. (eds.), *Advances and Challenges in Advanced Unmanned Aerial Systems*, Springer Aerospace Technology, https://doi.org/10.1007/978-981-99-8045-1_12

Keyword ACLS · Transport aircraft · Shape prediction · Dynamics modelling · Obstacle-crossing characteristics

CCS Concepts Takeoff and Landing · Launch and Recovery

1 Introduction

With the development of aircraft related technologies, people continue to realize the high standards of their use environment and the singleness of use. Especially when some areas with excellent environments have been fully explored, people have a strong desire to explore the ocean, deserts, polar regions, wetlands, and relatively complex dirt roads and other areas. Moreover, various complex environments cannot be avoided in the material support and personnel transportation for military purposes, and such areas also contain rich resources. The air cushioned transport aircraft (ACTA) is a good fit for this task. Its unique air cushion landing system enables it to take off and land in various complex environments while having the high-speed cruising characteristics of ordinary transport aircraft.

Compared with the ordinary transport aircraft, the ACTA uses an air cushion landing system instead of the wheeled landing gear, which increases the contact area between the transport aircraft and the runway, so that its requirements for the runway are greatly reduced, and it also has the function of taking off and landing on the water. When the ACTA aircraft is traveling at high speed on the complex runway, the strength of its obstacle crossing ability determines whether it can take off and land safely. American scholars were the first to carry out theoretical analysis and prototype testing of ACTA aircraft. Stauffer et al. [1] conducted a feasibility study on the LA-4 aircraft using the air cushion landing system, conducted take-off and landing tests respectively, and studied its rolling motion characteristics. Vaughan et al. [2] conducted a feasibility study on Buffalo aircraft modified with air cushion landing system, and conducted lift test, drop test, skidding test and wind tunnel test on this air cushioned aircraft by using a scale prototype. Perez et al. [3] conducted sliding, take-off and landing tests on different surfaces such as grass and snow for a full-size XC-8A aircraft, and gave a complete test report. Rogers [4] proposed the two-dimensional jet theory to calculate the discharge flow and fan power of the air cushion landing system under a stable cushion lift state, and obtained the optimal jet injection Angle through experiments. Earl et al. [5] studied negative pressure braking of LA-4 aircraft using air cushion landing system in order to improve the braking performance of air cushioned aircraft. Robert et al. [6] tested and analyzed various cornering and braking modes of air cushioned aircraft respectively, proving the reliability of center wheel and negative pressure braking. Digges [7] conducted an in-depth study of the air cushion landing system, gave the design method of the cross section of the air cushion and the deformation rule, and gave the computer simulation program, and established a complete buffering dynamics model. Boghani [8] established a four degree of freedom motion model, analyzed the pitch motion and roll motion of an air cushioned aircraft, and verified it through experiments. Captain et al. [9] analyzed the heave and pitch characteristics of an air cushioned aircraft when it landed and moved on an uneven runway by means of dynamic deduction. Ahmed Sowayan [10] introduced the theoretical model of dynamic characteristics of the incompressible air cushion landing system, and made a simulation analysis. Fevralskikh [11] numerically simulated the splashing phenomenon of the bottom liquid affected by the air flow during the operation of the air cushion landing system on the sea surface. Based on LS-DYNA, Li [12–14] conducted a visual simulation of the drop performance of the air cushion landing system, established a six degree of freedom model of the air cushioned aircraft, and designed the self-correcting control technology of the air cushioned aircraft in the course of landing and taxiing. The development of air cushion landing system in the last century was limited by the technical level at that time and stalled. With the development of simulation software and the increase of environmental complexity, countries began to study the air cushion landing system again, including the buffering performance, liquid splash and control problems, but there were few studies on the obstacle crossing performance of complex runway.

In this paper, the obstacle-crossing performance of a small military personnel transport aircraft modified with an air cushion landing system is studied. Firstly, the cushion is divided into four area by the cross-segmentation method, and the pressure expression of each area in the cavity is solved according to the ideal gas state equation, and the nonlinear cross-section prediction model related to the altitude of the aircraft is established. Secondly, a six degree of freedom dynamic model of the ground taxiing obstacle-crossing of the ACTA is established in MATLAB/Simulink. Finally, the ACTA passes through the boss, ditch and one-sided ramp at different initial speeds, and the changes of the distance from the ground at the bottom of the ACTA, vertical acceleration at the center of gravity, pitch angle and pressure in each area of the cavity in the process are investigated. On this basis, the obstacle-crossing performance of the ACTA passes through different obstacle runway is analyzed.

2 Digital Model of the ACLS

As shown in Fig. 1a, the air cushion landing system consists of an annular airbag mounted under the fuselage and a fan to provide inflatable power, where Q is total flow of the fan, P_t, P_j, P_c are the pressure of the plenum, the trunk and the cushion, respectively, A_{tj}, A_{ja}, A_{jc}, A_{ca} are the discharge areas from the plenum to the trunk, the trunk to the atmosphere, the trunk to the cushion, and the cushion to the atmosphere, respectively. The high-pressure air flow from the fan maintains the trunk inflation and forms a distribution jet at the bottom of the airbag through a nozzle on the surface of the airbag. As the jet approaches the take-off or landing surface, it forms a high-pressure gas at the bottom of the fuselage and in the cushion enclosed by the airbag, providing the aircraft with the support needed for lifting [15]. The establishment of the digital model of the air cushion landing system requires the calculation

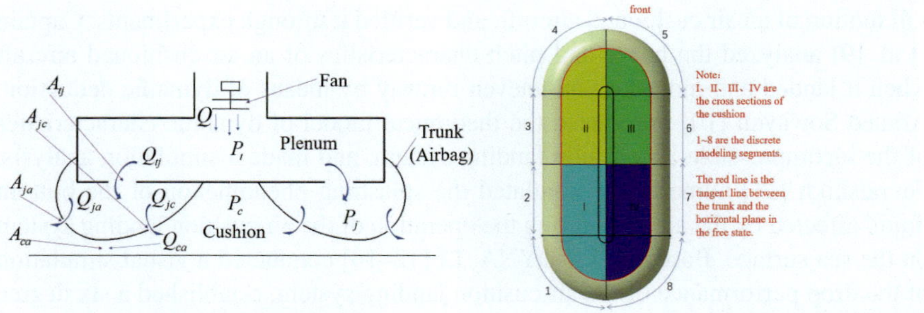

a) Air cushion landing system. b) Segmentation diagram of the ACLS.

Fig. 1 Air cushion landing system

of the pressure, mass, cross-sectional shape and volume in the system, so the cushion is divided into four independent area by the cross-segmentation method (as shown in Fig. 1b) and the finite element division is carried out along the circumference, and each segment is divided into 100 parts.

2.1 Pressure Change Rate

According to the ideal gas equation of state and the first law of thermodynamics, the expression of the pressure change rate in each gas chamber [7]:

$$\dot{P} = kP\left(\frac{\dot{W}}{W} - \frac{\dot{V}}{V}\right) \tag{1}$$

where k is the adiabatic index of the gas, and P, W and V are the pressure, mass and volume of the gas in a gas chamber, respectively.

2.2 Mass Change Rate

Mass change rate of gas in plenum:

$$\dot{W}_t = \rho Q - \sqrt{2\frac{W_t}{V_t}(P_t - P_j)}A_{tj}C_k \tag{2}$$

where ρ is the gas density, V_t is the volume of the plenum, and C_k is the orifice discharge coefficient.

Mass change rate of gas in trunk:

$$\dot{W}_j = \sqrt{2\frac{W_t}{V_t}|P_t - P_j|sign(P_t - P_j)A_{tj}C_k} - \sqrt{2\frac{W_j}{V_j}(P_j - P_a)A_{ja}C_m}$$
$$- \sum_{i=1}^{4}\sqrt{2\frac{W_j}{V_j}|P_j - P_{ci}|sign(P_j - P_{ci})A_{jci}C_k} \quad (3)$$

where V_j is the trunk volume, P_a is the atmospheric pressure, and C_m is the discharge coefficient. P_{ci} is the ith cushion pressure, and A_{jci} is the ith discharge hole areas from the trunk to the cushion.

Mass change rate of gas in cushion:

$$\dot{W}_{ci} = \sqrt{2\frac{W_j}{V_j}|P_j - P_{ci}|sign(P_j - P_{ci})A_{jci}C_k}$$
$$- \sqrt{2\frac{W_{ci}}{V_{ci}}(P_{ci} - P_a)A_{cai}C_m}, i = 1, 2, 3, 4 \quad (4)$$

where V_{ci} is the volume of the ith cushion, and A_{cai} is the ith discharge area from the cushion to the atmosphere.

2.3 Volume Change Rate

The volume of the plenum remains unchanged, and the volume of the trunk and the cushion are related to the compressible height h of the airbag. The trunk is divided into finite elements along the circumference, and Fig. 2 is a schematic diagram of the cross-sectional load state and the free state of a certain section, where the origin is located at the junction between the airbag and the bottom of the fuselage, (x_1, y_1), (x_2, y_2), R_1, R_2, ϕ_1, ϕ_2 respectively, are the center coordinates, radius, and center angle of the two arcs in the figure, and l is the compression length of the compression section. The trunk parameters prediction method [7] is used to calculate curves of the parameters with the compressible height in MATLAB. The parameter curves with compressible height, as shown in Fig. 3, where the right scale is used for the central angle and the left scale is used for the rest. Finally, combined with the ground clearance distance of each micro-segment, the volume of the trunk and each cushion can be calculated, and the volume change rate can be obtained after the time is derived.

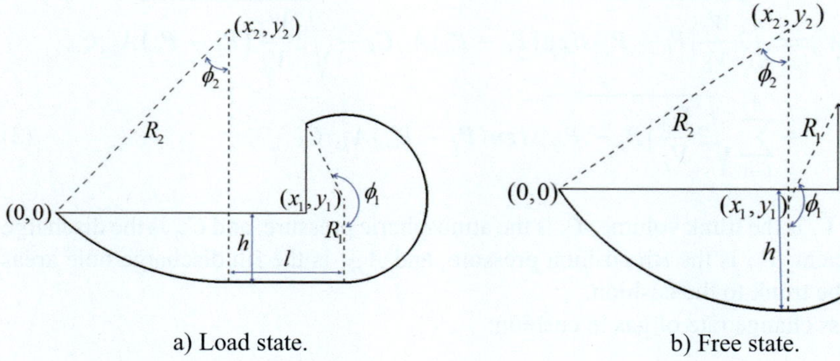

Fig. 2 Shape of cross section

Fig. 3 Calculation results

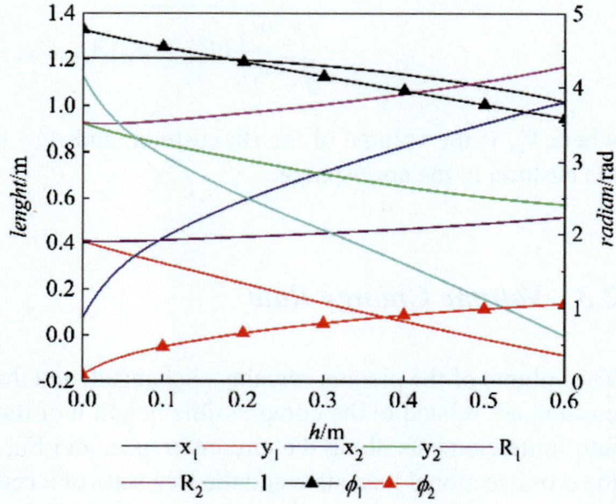

2.3.1 Ground Distance

Establish the ground frame $O_g X_g Y_g Z_g$ and the body frame $O_b X_b Y_b Z_b$. The rotation matrix L_{bg} from the ground frame system to the body frame is:

$$L_{bg} = \begin{bmatrix} b_{11} & b_{12} & b_{13} \\ b_{21} & b_{22} & b_{23} \\ b_{31} & b_{32} & b_{33} \end{bmatrix} \tag{5}$$

The rotation matrix L_{gb} of the body frame to the ground frame is:

Fig. 4 Ground distance diagram

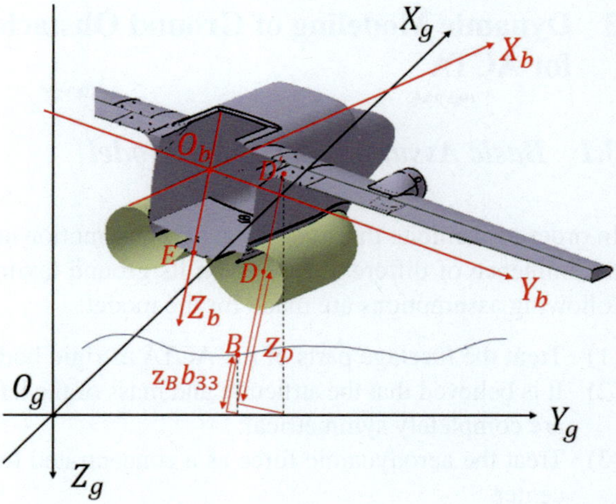

$$L_{gb} = \begin{bmatrix} g_{11} & g_{12} & g_{13} \\ g_{21} & g_{22} & g_{23} \\ g_{31} & g_{32} & g_{33} \end{bmatrix} = L_{bg}^{-1} \tag{6}$$

In Fig. 4, point D is the corresponding point of a section segment center on the bottom plane of the fuselage. The projection of the point in the $X_b O_b Y_b$ plane isD', and the coordinate of the geometric center E of the air cushion landing system in the body coordinate system is $E_b = \left(l_x, l_y, l_z\right)$. The coordinate of point D is $D_b = \left(l_x + x_{cxi}, l_y + y_{cxi}, l_z\right)$. Then the distance of point D to the plane $X_g O_g Y_g$ in the body frame is [9]:

$$Z_D = -\left(z + (x_{cxi} + l_x)g_{31} + \left(y_{cxi} + l_y\right)g_{32}\right) \times b_{33} - l_z \tag{7}$$

where (x_{cxi}, y_{cxi}) is the projection coordinate on the $X_b O_b Y_b$ plane from the center of the ith segment (located in the Fig. 1b in the red line) to the geometric center E.

The coordinate of the aircraft center of mass O_b in the ground frame is $O_b = (x, y, z)$. The corresponding point on the ground at the center segment D is point B, whose coordinate in the ground frame is as follows:

$$B = [x_B y_B z_B]^T = \begin{bmatrix} x + (x_{cxi} + l_x)g_{11} + \left(y_{cxi} + l_y\right)g_{12} + (Z_D + l_z)g_{13} \\ y + (x_{cxi} + l_x)g_{21} + (x_{cxi} + l_x)g_{22} + (Z_D + l_z)g_{23} \\ f(x_B, y_B) \end{bmatrix} \tag{8}$$

where $f(x, y)$ is the runway function, given later. The ground distance in any segment is:

$$z_{bi} = Z_D - z_B b_{33} \tag{9}$$

3 Dynamic Modeling of Ground Obstacle Crossing for ACTA

3.1 Basic Assumptions of the Model

In order to facilitate the study of the taxiing motion state of the ACTA and explore the influence of different runway on its ground taxiing motion characteristics, the following assumptions are made for the model:

(1) Treat the fuselage parts of the ACTA as rigid body models;
(2) It is believed that the structure and mass of the left and right sides of the ACTA are completely symmetrical;
(3) Treat the aerodynamic force as a concentrated force and act on the pneumatic center.

3.2 Force Analysis

In order to establish the ground obstacle crossing dynamic model of the ACTA, the wind frame $O_aX_aY_aZ_a$ and the track frame $O_kX_kY_kZ_k$ are established. L_{ba} is the rotation matrix from the wind frame to the body frame, and L_{gk} is the rotation matrix from the track frame to the ground frame.

In the ground taxiing stage, the ACTA is affected by gravity, aerodynamics, cushion force, ground contact force, engine thrust, air momentum force and their moments. The schematic diagram of the force is shown in Figs. 5 and 6.

3.2.1 Gravity

The expression of gravity in the body frame is as follows:

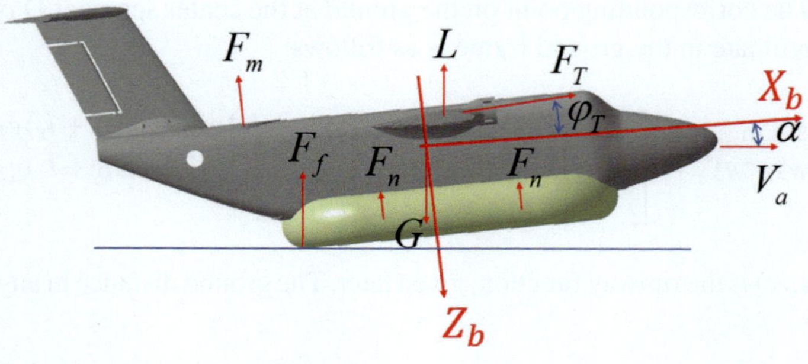

Fig. 5 Side view force diagram

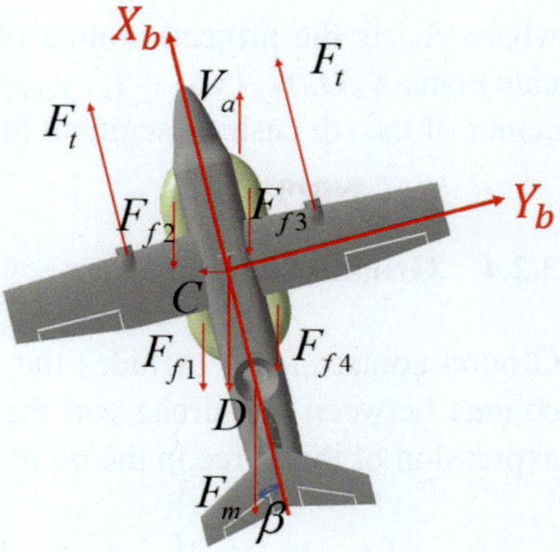

Fig. 6 Top view force diagram

$$G = L_{bg}\left[0\ 0\ mg\right]^T \tag{10}$$

where m is the mass of the ACTA, g is the acceleration of gravity. Gravity acts on the center of mass and has no moment on the center of mass.

3.2.2 Aerodynamic Force

The expression of aerodynamic force in the body frame is as follows:

$$R = L_{ba}[-DC - L]^T \tag{11}$$

The expression of the aerodynamic moment in the body frame is as follows:

$$M_R = [L_a M_a N_a]^T \tag{12}$$

where D, C, and L are drag, side force and lift. L_a, M_a and N_a are roll, pitch and yaw moment, respectively.

3.2.3 Air Cushion Force

The air cushion force is provided by the cushion, and its expression in the body frame is as follows:

$$F_n = \left[0\ 0 - \sum_{i=1}^{800} \sum_{k=1}^{4} P_{ck} A_{ci}\right]^T \tag{13}$$

The expression of air cushion force in the body coordinate system is as follows:

$$M_n = \left[-\sum_{i=1}^{800} \sum_{k=1}^{4} P_{ck} A_{ci}\left(y_{chi} + l_y\right) - \sum_{i=1}^{800} \sum_{k=1}^{4} P_{ck} A_{ci}(x_{chi} + l_x)\ 0\right]^T \tag{14}$$

where A_{ci} is the projection area of the ith cushion segment in the body coordinate plane $X_bO_bY_b$. $\left(x_{chi} + l_x, y_{chi} + l_y\right)$ is the projection coordinate of the pressure center of the ith cushion segment in the body coordinate plane $X_bO_bY_b$.

3.2.4 Ground Contact Force

Ground contact force includes the supporting force and friction generated by the contact between the airbag and the ground. According to Newton's third law, the expression of this force in the body frame is as follows:

$$F_f = \left[F_{fx} F_{fy} F_{fz}\right]^T = L_{bg} L_{gk}\left[-\sum_{i=1}^{800} \mu P_j A_{cni}\ 0 - \sum_{i=1}^{800} P_j A_{cni}\right]^T \tag{15}$$

The expression of the ground moment in the body coordinate system is as follows:

$$M_f = \left[-F_{fy} l_z + F_{fz}\left(y_{jki} + l_y\right)\ F_{fx} l_z - F_{fz}\left(x_{jki} + l_x\right)\ -F_{fx}\left(y_{jki} + l_y\right) + F_{fy}\left(x_{jki} + l_x\right)\right]^T \tag{16}$$

where A_{cni} is the contact area between the ith trunk segment and the ground. $\left(x_{jki} + l_x, y_{jki} + l_y\right)$ is the projection coordinate of the pressure center of the ith trunk segment in the body coordinate plane $X_bO_bY_b$.

3.2.5 Engine Thrust

The expression of the thrust and its moment in the body frame is as follows:

$$F_T = \left[F_{Tx} F_{Ty} F_{Tz}\right]^T \tag{17}$$

$$M_T = \left[M_{Tx}\ M_{Ty}\ M_{Tz}\right]^T \tag{18}$$

3.2.6 Air Momentum Force

According to the law of conservation of momentum, the fan accelerates the compression of the outside air to the plenum, thereby generating the air momentum force acting on the center of the fan, which is expressed in the body frame [16]:

$$F_m = \begin{bmatrix} F_{mx} \\ F_{my} \\ F_{mz} \end{bmatrix} = \begin{bmatrix} -\rho Q V_a \cos\beta \cos a \\ -\rho Q V_a \sin\beta \\ -\rho Q^2 / A_m - \rho Q V_a \cos\beta \sin a \end{bmatrix} \tag{19}$$

$$M_m = \left[-F_{my} l_{mz} \; F_{mx} l_{mz} - F_{mz} l_{mx} \; F_{my} l_{mx} \right]^T \tag{20}$$

where V_a is airspeed; α, β are the angle of attack and side slip angle, respectively. A_m is the area of the fan inlet; (l_{mx}, l_{my}, l_{mz}) is the coordinate of the center of gravity of the fan in the body frame.

3.2.7 Resultant Force and Resultant Moment

Under the body frame, the external forces and moments of the ACTA are expressed as follows:

$$\sum F = \left[F_x F_y F_z \right]^T = G + R + F_n + F_f + F_T + F_m \tag{21}$$

$$\sum M = \left[M_x M_y M_z \right]^T = M_R + M_n + M_f + M_T + M_m \tag{22}$$

3.3 *Modeling of Ground Obstacle Crossing Motion of ACTA*

The ground obstacle crossing model of the ACTA has six degrees of freedom. The kinematic model and dynamic model of the ACTA are established in the body frame according to the force analysis of Sect. 3.2.

3.3.1 Kinematic Model

In the ground frame, the translational and rotational kinematics model expression of the ACTA is:

$$\begin{cases} \dot{x} = u\cos\theta\cos\psi + v(\sin\phi\sin\theta\cos\psi - \cos\phi\sin\psi) + w(\cos\phi\sin\theta\cos\psi + \sin\phi\sin\psi) \\ \dot{y} = u\cos\theta\sin\psi + v(\sin\phi\sin\theta\sin\psi + \cos\phi\cos\psi) + w(\cos\phi\sin\theta\sin\psi - \sin\phi\cos\psi) \\ \dot{z} = -u\sin\theta + v\sin\phi\cos\theta + w\cos\phi\cos\theta \\ \dot{\phi} = p + q\sin\phi\tan\theta + r\cos\phi\tan\theta \\ \dot{\theta} = q\cos\phi - r\sin\phi \\ \dot{\psi} = q\sin\phi\sec\theta + r\cos\phi\sec\theta \end{cases} \tag{23}$$

where x, y, and z are the displacement of the body under the ground frame, and u, v, w are the ground velocity of the body under the body frame. p, q, and r are the angular rotational velocities of the body.

3.3.2 Dynamic Model

In the body frame, the translational and rotational dynamic model expression of the ACTA is:

$$\begin{cases} F_x = m(\dot{u} - rv + qw) \\ F_y = m(\dot{v} - pw + ru) \\ F_z = m(\dot{w} - qu + pv) \\ M_x = I_x\dot{p} - I_{xz}(\dot{r} + pq) + (I_z - I_y)qr \\ M_y = I_y\dot{q} - I_{xz}(r^2 - p^2) + (I_x - I_z)pr \\ M_z = I_z\dot{r} - I_{xz}(\dot{p} - qr) + (I_y - I_x)pq \end{cases} \tag{24}$$

where I_x, I_y, I_z are the moment of inertia of ACTA around O_bX_b, O_bY_b and O_bZ_b. I_{xz} is the cross product of inertia around the O_bX_b and O_bZ_b axes.

3.4 Model Reliability Verification

In order to verify the reliability of the model, an 1/8 scale prototype was produced and a drop test was conducted in Fig. 7a. The prototype dropped to a height of 8 cm in the test, the model dropped to a height of 64 cm in the simulation, and the initial angle was 0. The changes in acceleration at the center of mass of the prototype during this process were recorded and compared with the simulation results. In Fig. 7b, the results show that the simulation model can fit the test well, verifying the reliability of the model.

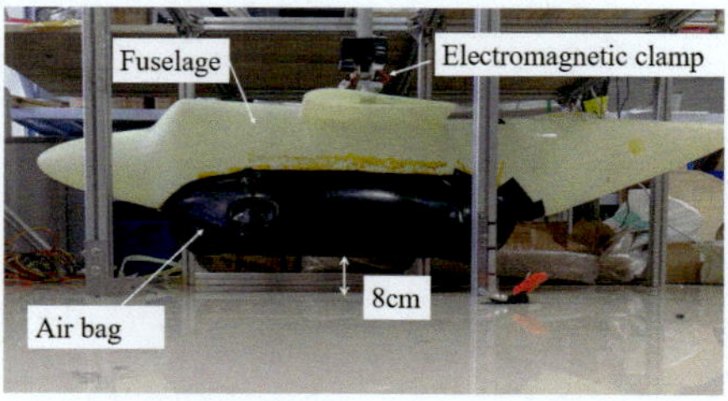

(a) 1/8-scale model.

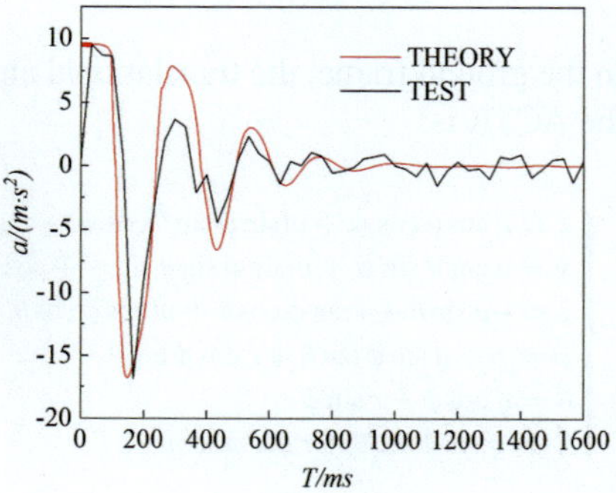

(b) Drop test.

Fig. 7 Model reliability verification

Table 1 Model simulation parameters

Parameter name	Symbol (unit)	Numeric value
Quality	m (kg)	4980
Airbag length	L_c (m)	6.2
Airbag width	B_c (m)	2.4
Airbag height	H_c (m)	0.6
Ground clearance	Z_c (m)	0.05

4 Analysis of the Characteristics of Ground Taxiing Obstacle Crossing

4.1 Simulation Parameters

In this section, we will establish different runway functions in MATLAB/Simulink, explore the obstacle crossing ability of air cushion transporters, and analyze the reasons for the changes of various parameters in this process.

The model simulation parameters are shown in Table 1.

4.2 Ground Obstacle Crossing Simulation Analysis

In order to explore the ability of the ACTA to taxiing over obstacles on the ground, three different runway models were built, namely boss, ditch, and one-sided ramp. The ACTA passed through different runway at the initial speed of ug = 7 m/s, 20 m/s and 40 m/s, respectively. The changes of the height of the center of mass from the ground Z, the vertical acceleration a, the pitch angle θ, and the cushion pressure Pc with time T were recorded. The engine thrust maintains only the speed of the ACTA.

4.2.1 Ditch-Boss Runway

The ditch-boss runway is composed of a ditch and a long boss, the depth of the ditch is about 77% of the height of the airbag, and the length is about 33% of the length of the airbag. The height of the boss is about 70% of the height of the airbag, and the length is about 500% of the length of the airbag. In this paper, the influence of ditch and boss obstacles on the motion state of ACTA is analyzed.

The function expression for the ditch-boss runway is as follows:

$$f(x,y)=\begin{cases}0 & (x<50\cup x\geq 130.84)\\ -x+50 & (50\leq x<50.5)\\ -0.5 & (50.5\leq x<51.5)\\ x-51.5 & (51.5\leq x<52)\\ 0 & (52\leq x<100)\\ x-100 & (100\leq x<100.42)\\ 0.42 & (100.42\leq x<130.42)\\ -x+130.84 & (130.42\leq x<130.84)\end{cases} \tag{25}$$

In Fig. 8a, when passing through the ditch at a speed of 7 m/s, the distance from the ground has the greatest effect, and the body drops by 0.3 m. When passing the boss at a speed of 40 m/s, the distance from the ground has the greatest effect, and the body jumps by 0.5 m. In Fig. 8b, when passing through the ditch at a speed of 20 m/s, the vertical acceleration is maximum due to the impact of the front airbag contacts the runway. At 40 m/s through the ditch, the aircraft navigates the obstacles with a minimal change in distance from the ground, but its acceleration is maximum at -26.2 m/s^2 over the boss. As can be seen in Fig. 8c, the influence of ditch or boss obstacles on the pitch angle is related to the speed of the aircraft, and the faster the speed, the smaller the impact.

Figure 8d–f shows the variation curves of cushion pressure in different area during obstacle crossing at three speeds, and the trends are roughly the same. Taking Fig. 8d as an example, when the front end of the airbag contacts the ditch, the discharge area of the second and third cushion increases, resulting in a decrease in cushion pressure and a positive pitching moment for the aircraft. As a result, the total pressure of the aircraft decreases and the discharge area of the first and fourth cushion decreases, while the pressure in the area increases briefly. Then the front end of the airbag leaves the ditch, the pressure increases, and the back end goes over the ditch, the pressure decreases. After a certain period of shock, it gradually stabilizes. When the ACTA passes over the boss, the pressure changes opposite to the ditch during the initial collision, after which the aircraft reaches the top. The higher the body speed, the higher the jump height, when the speed reaches 40 m/s, the air cushion aircraft leaps over the boss. In this process, the maximum pressure is 12.3 kPa.

4.2.2 One-Sided Ramp

The height of the one-sided ramp is about 36% of the height of the airbag, which is used to simulate the influence of the asymmetrical obstacle type on the taxiing characteristics of ACTA.

The function expression for a unilateral ramp is as follows:

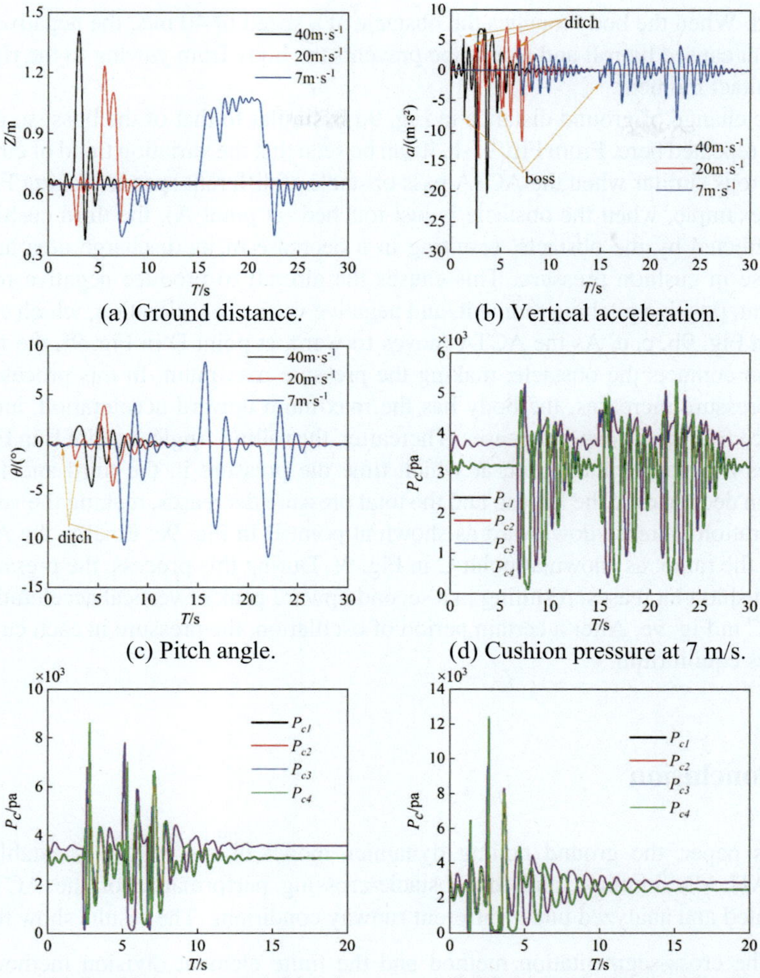

(a) Ground distance. (b) Vertical acceleration.

(c) Pitch angle. (d) Cushion pressure at 7 m/s.

(e) Cushion pressure at 20 m/s. (f) Cushion pressure at 40 m/s.

Fig. 8 Influence of the ditch-boss runway on each parameter

$$f(x, y)=\begin{cases} y & (0 \le y < 0.232 \cap 50 \le x \cap x - y - 50 \ge 0) \\ x - 50 & (50 \le x < 50.232 \cap y > 0 \cap x - y - 50 < 0) \\ 0.232 & (80.432 \le x \cap 0.232 \le y) \\ 0 & (80.432 \le x \cap 0.232 \le y) \end{cases} \quad (26)$$

Figure 9a–c respectively show the change curve of the attitude Angle of the body over time. The change trend shows that the faster the aircraft speed, the less the attitude Angle is affected. In Fig. 9c, a positive yaw angular velocity appears both when touching the obstacle and when leaving the ramp, which is caused by the ground contact force generated by the collision between the right-side airbag and the ground. When the body touches the obstacle at a speed of 40 m/s, the negative yaw moment caused by roll and sideswipe prevents the body from yawing to the right at the contact moment.

The change of ground distance in Fig. 9d is similar to that of the boss, so it will not be repeated here. From Fig. 9f–h, it can be seen that the variation trend of cushion pressure is similar when the ACTA pass obstacle at different speeds. Taking Fig. 9f as an example, when the obstacle is just touched (at point A), the third cushion is first affected by the obstacle, resulting in a decrease in its discharge area and an increase in cushion pressure. This causes the aircraft to produce negative rolling moment, positive pitching moment, and negative vertical acceleration, which can be seen in Fig. 9b, c, e. As the ACTA moves forward, at point D in Fig. 9f, the fourth cushion contacts the obstacle, making the pressure maximum. In this process, the total pressure increases, the body has the maximum upward acceleration, and the distance from the ground increases. Thereafter, the rolling Angle at point B in Fig. 9f reaches a negative maximum, at which time the pressure in the third and fourth cushion decreases to the lowest, and the total pressure decreases, making the vertical acceleration increase downward as shown at point B in Fig. 9e. Finally, the ACTA leaves the ramp, as shown at point C in Fig. 9f. During this process, the pressure in each cushion increases, resulting in a second upward peak of vertical acceleration at point C in Fig. 9e. After a certain period of oscillation, the pressure in each cushion reaches equilibrium.

5 Conclusion

In this paper, the ground taxiing dynamics model of the ACTA is established in MATLAB/Simulink, and the obstacle-crossing performance of the ACTA is simulated and analyzed under different runway conditions. The results show that:

(1) The cross-segmentation method and the finite element division method can calculate the pressure of each cushion well, which makes the ACTA have the recovery moment, enhances the stability of the model, and improves the calculation accuracy of the model.
(2) When the ACTA passes through the boss obstacle, the faster its speed, the greater the change in distance from the ground, the smaller the change in attitude angle. When passing through the ditch obstacle, the faster the speed, the smaller the change in distance from the ground and the smaller the change in attitude angle.
(3) When the ACTA passes through the obstacle with a height of 70% of the airbag height and a length of 500% of the airbag length at a speed of 40 m/s, its maximum pressure is 1.23×10^4 pa and the maximum vertical acceleration is -26.2 m/s^2, which indicates that the ACTA can safely pass most obstacles, reflecting its excellent obstacle performance.
(4) When passing through the one-sided ramp, there is obvious sideslip and yaw, which requires yaw control.

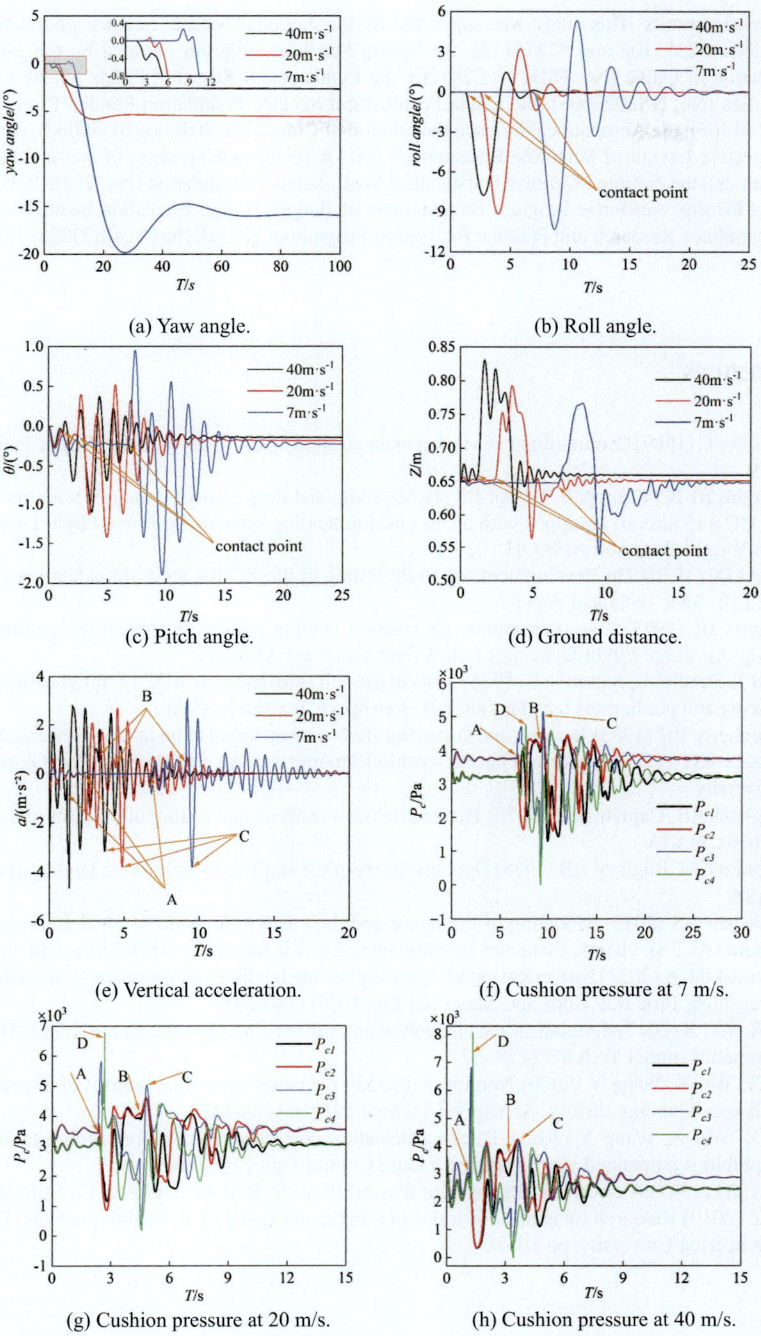

(a) Yaw angle.

(b) Roll angle.

(c) Pitch angle.

(d) Ground distance.

(e) Vertical acceleration.

(f) Cushion pressure at 7 m/s.

(g) Cushion pressure at 20 m/s.

(h) Cushion pressure at 40 m/s.

Fig. 9 Influence of one-sided ramp on each parameter

Acknowledgements This study was supported by the National Natural Science Foundation of China (Nos. 52375102 and 52275114), the Spring Sunshine Program initiated by the Ministry of Education of China (No. HZKY20220126), the Fundamental Research Funds for the Central Universities (No. NS2023005), the China Postdoctoral Science Foundation Funded Project (No. 2021M691565), the Aeronautical Science Foundation of China (No. 20200041052002), the Fund of Prospective Layout of Scientific Research for NUAA (Nanjing University of Aeronautics and Astronautics), the National Defense Outstanding Youth Science Foundation (No. 2018-JCJQ-ZQ-053), the Priority Academic Program Development of Jiangsu Higher Education Institutions, and the Postgraduate Research and Practice Innovation Program of NUAA (No. xcxjh20220112).

References

1. Stauffer C (1969) Ground/flight test report of air cushion landing gear (LA-4). Wright-Patterson AFB
2. Vaughn III JC, Campbell S, Pool DJ (1972) Static and drop tests of a quarter scale model of the CC-115 aircraft equipped with an air cushion landing system. Air Force Flight Dynamics Lab Wright-Patterson AFB OH
3. Perez DJ (1976) The development and flight testing of the XC-8A air cushion landing system (ACLS). SAE Technical Paper
4. Rogers JR (1973) Two-dimensional air cushion landing system peripheral jet confirmation study. Air Force Flight Dynamics Lab Wright-Patterson AFB OH
5. Earl T, Stauffer C, Satterlee C (1975) Tests of the bell aerospace LA-4 ACLS fitted with suction braking and predictions for other aircraft. Aerospace Textron Buffalo
6. Daugherty RH (1983) Braking and cornering studies on an air cushion landing system. NASA
7. Digges KH (1970) Theory of an air cushion landing system for aircraft. The Ohio State University
8. Boghani AB, Captain KM (1978) Heave-pitch-roll analysis and testing of air cushion landing systems. NASA
9. Captain KM, Boghani AB (1975) Dynamic heave-pitch analysis of air cushion landing systems. NASA
10. Sowayan AS (2012) Modeling of the heave and the cushion pressure of air cushion landing system (ACLS): chaotic dynamics investigation. Am Soc Mech Eng 45202:619–628
11. Fevralskikh A (2021) Numerical simulation of splashing landing gear on an amphibious aircraft air cushion. Proc IOP Conf Ser: Mater Sci Eng 1027(1):012005
12. Li X, Wei X (2017) Simulation and analysis of air cushion landing system based on LS-DYNA. Aeronaut Comput Tech 47(2):59–62
13. Li X, Wei X, Wang Y (2020) Nonlinear tracking deviation correction control of airplanes in high-speed landing taxiing. Aerosp Sci Technol 43(12):2413–2424
14. Li X, Wei X, Wang Y (2021) Tracking deviation correction control for landing taxiing of amphibious unmanned aerial vehicles. J Guid Control Dyn 116:106863
15. Earl TD (1967) Air cushion landing gear feasibility study. Bell Aerosystems Co Buffalo
16. Li Z (2019) Research on control of trajectory angle and speed of air cushion vehicle. Harbin Engineering University, pp 16–21

Analysis and Optimization Design of the Locking Performance of the Self-folding Strut Lock Mechanism for Landing Gear Based on Singularity Theory

Zhipeng Zhang, Yin Yin, Shengxiao Wu, Yangyang Li, Xiaohui Wei, and Hong Nie

Abstract The locking mechanism of the landing gear is crucial for ensuring stable load-bearing capability in the deployed position. However, the locking mechanism's performance is subject to more stringent requirements due to the high landing loads of modern aircraft. Currently, the locking mechanism's operational principles are not well understood which makes the design of a high-performance locking mechanism a challenging problem. To address these issues, an analysis and optimization design method is proposed to study the locking performance of the locking mechanism using singularity theory combined with numerical continuation method. Firstly, taking the self-folding strut locking mechanism of the main landing gear of unmanned aerial vehicles as the research object, the influence of different singular positions of the landing gear on the locking performance is analyzed. The critical unlocking condition is obtained. Furthermore, the load-carrying ratio is defined as the indicator of locking performance. Finally, a comprehensive optimization model with locking performance as the optimization objective is established. The optimization results demonstrate a significant improvement in the critical unlocking load. The study

Z. Zhang · Y. Yin (✉) · S. Wu · Y. Li · X. Wei · H. Nie
State Key Laboratory of Mechanics and Control for Aerospace Structures, Nanjing University of Aeronautics and Astronautics, Nanjing 210016, China
e-mail: yinyin@nuaa.edu.cn

Z. Zhang
e-mail: zhangzhipeng@nuaa.edu.cn

S. Wu
e-mail: shengxiao_WU@nuaa.edu.cn

Y. Li
e-mail: liyangyang@nuaa.edu.cn

X. Wei
e-mail: wei_xiaohui@nuaa.edu.cn

H. Nie
e-mail: hnie@nuaa.edu.cn

Z. Liu et al. (eds.), *Advances and Challenges in Advanced Unmanned Aerial Systems*, Springer Aerospace Technology, https://doi.org/10.1007/978-981-99-8045-1_13

demonstrates that singularity analysis combined with numerical continuation method can effectively capture the variation trends of locking performance with different structural parameters, and the proposed optimization method can provide effective support for the design of landing gear locking mechanisms.

Keywords Landing gear · Locking mechanism · Singularity · Numerical continuation

CCS Concepts Computing methodologies · Modeling and simulation

1 Introduction

As the sole interface between an aircraft and the ground, the landing gear mechanism is responsible for absorbing the impact loads during landing. According to Part 25 of the US Federal Aviation Regulations [1], the locking mechanism must ensure that the landing gear remains locked in the down position, taking into account various possible ground loads. The design of landing gear locking mechanisms often utilizes mechanism singularities to lock them in singular positions. However, due to factors such as machining errors and structural deformations, it is often impossible to guarantee that the mechanism is in a singular position, making it possible for the locking mechanism to unintentionally unlock under ground loads. Currently, there is no research analyzing the locking performance around the singular points of the locking mechanism.

Methods for determining singularities include the Jacobian matrix rank deficiency method [2], screw theory, and Grassmann line geometry. For many mechanical systems, their structural forms can be described by a system of polynomial equations. However, due to the difficulty of directly obtaining algebraic solutions for these multivariable nonlinear equations, numerical continuation methods have been extensively applied in the synthesis and analysis of mechanisms [3]. Therefore, this paper will use the Jacobian matrix rank deficiency method combined with numerical continuation methods to solve for singular configurations.

The numerical continuation method has been widely applied in the analysis and optimization of locking mechanisms in landing gear. Knowles utilized the numerical continuation method to analyze the bifurcation characteristics of the unlocking action in the nose landing gear under different flight conditions [4]. Xukui, based on singular value bifurcation theory, analyzed the reasons why the dual-sidestay landing gears cannot achieve proper locking [5]. Additionally, Xukui proposes a bifurcation analysis method for synchronous locking of dual-sidestay landing gears, which takes into consideration the joint clearance [6]. Yinyin used numerical continuation and bifurcation theory to analyze the unlocking process of the hook lock and presented an optimization method based on the critical unlocking point [7].

This paper focuses on the research of a self-folding strut locking mechanism. Firstly, the singularity analysis of the landing gear locking mechanism will be conducted using singularity discriminants and numerical continuation methods, explaining the situation of the landing gear under ground loads from a kinematic perspective. Furthermore, an optimization route will be developed to achieve the optimized design of the strut locking structure. Finally, the optimized design conclusions will be validated using a multibody dynamics simulation platform. The research shows that this singularity-based optimization method for locking struts can effectively support the design of landing gear locking mechanisms.

2 Singularity Analysis of the Self-folding Strut Locking Mechanism

2.1 *Kinematic Modeling and Configuration Equations*

2.1.1 Composition of the Retraction Mechanism and Ground Loads

The simplified schematic diagram of the retraction mechanism of the landing gear is shown in Fig. 1a. It can be divided into five components: upper lock link, lower lock link, upper sidestay link, main strut, and lower sidestay link. Although the landing gear retracts in space, the motion planes of its components are parallel. Therefore, the components are projected onto the same retraction plane $O - XZ$ and a coordinate system is established as shown in Fig. 1b. The ground loads are divided into the take-off and landing loads (in the same direction as the horizontal load in Fig. 1a) and the rebound loads (in the opposite direction of the horizontal load in Fig. 1a).

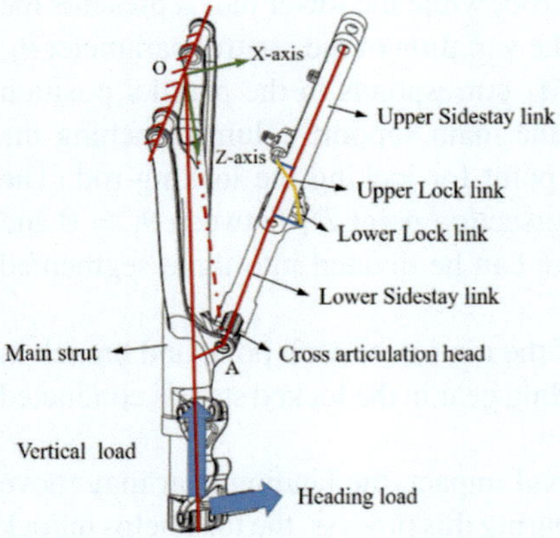

(a) Schematic diagram of landing gear retraction mechanism

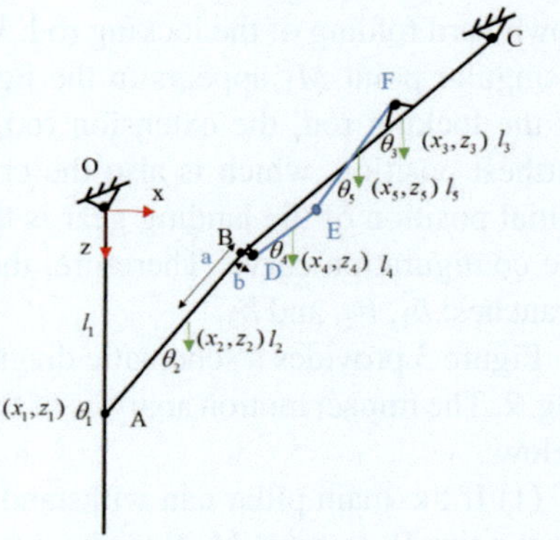

(b) Simplified diagram of landing gear retraction mechanism

Fig. 1 Schematic diagram of landing gear retraction mechanism modeling

2.1.2 Configuration Equations

The structural parameters of each component are shown in Fig. 1b. The main state parameters of the components are denoted as (x_i, z_i, θ_i), where i represents the ith component, (x_i, z_i) represents the position coordinates of the ith component in the coordinate system $O - XZ$, and θ_i represents the angle between the ith component and the Z-axis of the coordinate system, where the angle of the main strut θ_1 represents the retraction angle Θ. The constraint equation is shown in formula (1).

$$\begin{cases} x_1 = 0 & x_4 - x_D - 0.5 \cdot l_4 \cdot \sin\theta_4 = 0 & x_F - x_5 - 0.5 \cdot l_5 \cdot \sin\theta_5 = 0 \\ z_1 - \cos\theta_1 \cdot l_1 = 0 & z_4 - z_D + 0.5 \cdot l_4 \cdot \cos\theta_4 = 0 & z_F - z_5 + 0.5 \cdot l_5 \cdot \cos\theta_5 = 0 \\ x_2 - x_1 - 0.5 \cdot l_2 \cdot \sin\theta_2 = 0 & x_5 - x_4 - 0.5 \cdot l_4 \cdot \sin\theta_4 - 0.5 \cdot l_5 \cdot \sin\theta_5 = 0 & x_2 + a \cdot \sin\theta_2 + b \cdot \cos\theta_2 - x_D = 0 \\ z_2 - z_1 + 0.5 \cdot l_2 \cdot \cos\theta_2 = 0 & z_5 - z_4 + 0.5 \cdot l_4 \cdot \cos\theta_4 + 0.5 \cdot l_5 \cdot \cos\theta_5 = 0 & z_2 - a \cdot \cos\theta_2 + b \cdot \sin\theta_2 - z_D = 0 \\ x_3 - x_2 - 0.5 \cdot l_2 \cdot \sin\theta_2 - 0.5 \cdot l_3 \cdot \sin\theta_3 = 0 & x_C - x_3 - 0.5 \cdot l_3 \cdot \sin\theta_3 = 0 & x_3 - c \cdot \sin\theta_3 - d \cdot \cos\theta_3 - x_F = 0 \\ z_3 - z_2 + 0.5 \cdot l_2 \cdot \cos\theta_2 + 0.5 \cdot l_3 \cdot \cos\theta_3 = 0 & z_C - z_3 + 0.5 \cdot l_3 \cdot \cos\theta_3 = 0 & z_3 + c \cdot \cos\theta_3 - d \cdot \sin\theta_3 - z_F = 0 \end{cases} \quad (1)$$

A kinematic mathematical model is established on the MATLAB platform, and the trajectory of the mechanism's state variables with respect to the control parameter is solved using the numerical continuation algorithm in the COCO toolbox. The presence of motion singularities can be determined by checking whether the Jacobian matrix is singular.

2.2 *Single Singular Point Configuration Shock Locking Motion Analysis*

The configuration motion curve in Fig. 2 is symmetric about $\theta_{ov} = 0$. The upper half represents the upward folding of the locking rod, while the lower half represents the downward folding of the locking rod. With the variation of the control parameter θ_1, a singular point M_1 appears in the figure. M_1 corresponds to the parallel position of the locking rod, the extension rod, and the main support column reaching the furthest position, which is also the critical point for locking the locking rod. The initial position of the landing gear is the intersection point D_1 between $\theta_1 = 0$ and the configuration curve. Therefore, the curve can be divided into three segmented branches: b_1, b_2, and b_3.

Figure 3 provides a schematic diagram of the motion of each point and branch in Fig. 2. The impact motion analysis of the landing gear in the locked state is conducted below.

(1) If the main pillar can withstand the load impact, the landing gear may move from point D_1 to point M_1 along branch b_2. During this process, the load helps unlock the movement, while the spring force and pillar gravity hinder the movement. After reaching point M_1, the landing gear may continue to move along branch b_3, where the load of the load, spring force, and pillar gravity all hinder the movement.

Therefore, under the load impact condition, the lock rod may oscillate and even unlock. The reason for this is that when the landing gear is in the down position, the

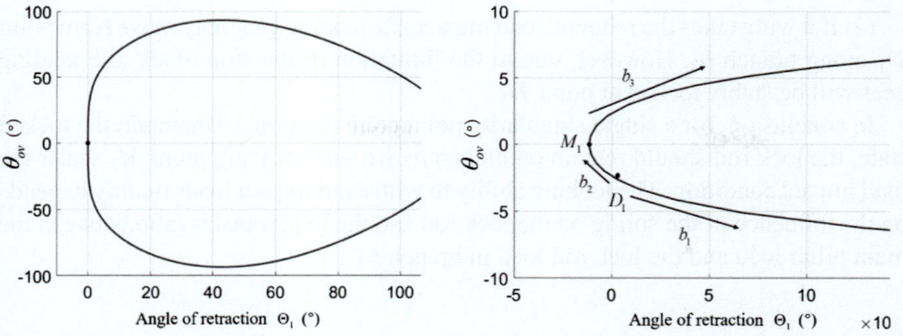

(a) Retraction angle-Locking rod angle (overall)
(b) Retraction angle-Locking rod angle (local)

Fig. 2 Motion curve of retraction mechanism

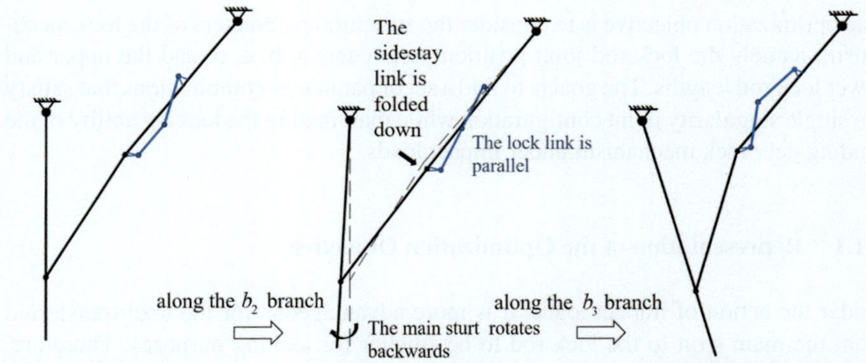

(a) Release the lock state D_1
(b) Release the lock state M_1
(c) retract the process state

Fig. 3 Landing gear motion diagram

strut is not in a parallel state but is inclined upward by 0.2°. Compared to the parallel state of the strut, more load is transferred to the lock rod under the load impact condition, causing the lock rod to move upward along branch b_2. At this time, the lock rod may have the following two motion trajectories.

One is that the lock rod cannot cross point M_1 and can maintain the locked state. This indicates that in branch b_2, under the obstruction of spring force, pillar gravity, and other loads, the impact load fails to make the lock rod cross point M_1. However, due to the characteristic of the spring force changing with the compression amount, the lock rod will experience repeated oscillations.

The other is that the lock rod crosses point M_1 and cannot maintain the locked state continuously. This means that in branch b_2, under the loads of spring force, pillar gravity, and other loads, the lock rod fails to successfully hinder the impact load from crossing point M_1. In branch b_3, the loads of the load, spring force, and pillar gravity hinder the movement, causing the lock rod to move back to the downward position.

(2) If it withstands the rebound load impact, the landing gear may move from point D_1 along branch b_3. However, due to the limitation of the stop block, the landing gear will be stably locked at point D_1.

In conclusion, for a single singularity point configuration, to maintain the locked state, the lock rod should remain on branch b_2 without crossing point M_1 under the load impact condition. The locking ability to withstand impact loads mainly depends on the influence of the spring on the lock rod and the load transfer ratio between the main pillar load and the lock rod load in branch b_2.

3 Main Landing Gear Lock Structure Parameter Design

3.1 *Optimization Approach*

The optimization objective is to consider the structural parameters of the lock mechanism, namely the lock rod joint position parameters a, b, c, d, and the upper and lower lock rod lengths. The goal is to find a set of parameter combinations that satisfy the single singularity point configuration while maximizing the locking ability of the landing gear lock mechanism under impact loads.

3.1.1 Representation of the Optimization Objective

Under the action of impact loads, it is more advantageous for the load transferred from the main strut to the lock rod to be smaller for locking purposes. Therefore, the moment of the strut/lock rod should be maximized. Due to the conservation of energy, the above relationship can be transformed into the fact that the smaller the motion angle of the main strut/lock rod, the more advantageous for locking. Here, the load transfer ratio of the lock rod can be represented by the range of motion of the main strut during the motion of the branch b_2, given the locked angle of the lock rod. The locked angle of the lock rod $\theta_{ov}^{D_1}$ in the lowered position of the landing gear is generally around 2°. Since the rotation angle of the main strut is very small during the unlocking process, and the optimization design does not change the motion relationship between the brace and the main strut, the maximum angle ($\theta_{of}^{M_1}$) of the brace at position M_1 is used as the optimization objective in the final simulation optimization.

3.1.2 Definition of Design Variables

The design variables are the joint position parameters a, b, c, d of the upper and lower lock rods on the strut (as shown in Fig. 4) and the length of the lower lock rod. Since the lock rod angle in the locked state is determined, the length of the upper lock rod can be solved from the constraint equation.

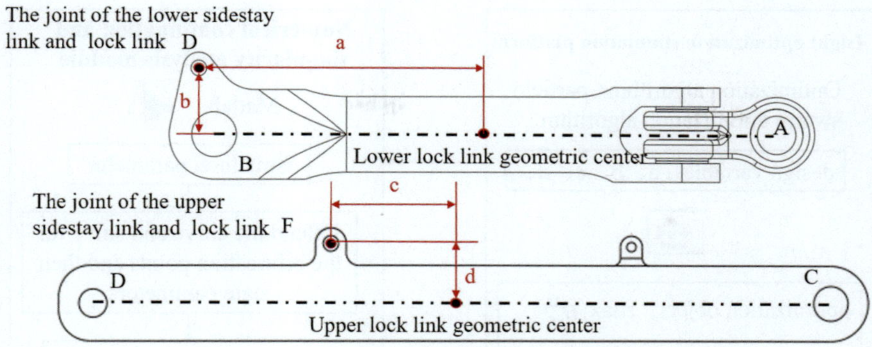

Fig. 4 Schematic diagram of lock rod joint position parameters

3.1.3 Constraint Definition

The constraints are defined as follows:

(1) Since the lock mechanism is a single singular point configuration, the number of singular points, n, should be limited to 1.
(2) If the motion angle range of the lock rod during the retraction process is too large, it may cause interference between the upper and lower lock rod structures. Therefore, the maximum lock rod angle ($\theta_{ov}^{\max}$) during the retraction process is set to an upper limit of 135° in the optimization.
(3) During the retraction process, the lock rod motion has a return path. If the retraction angle is large enough, the lock rod may fold from the upper position to the lower position. To avoid this situation, the lower limit of the lock rod angle (θ_{ov}^{U}) when the landing gear is fully retracted is set to 20° in the optimization.
(4) To avoid excessive or insufficient length of one side of the lock rod. In addition, the upper lock rod is directly connected to the spring, and the larger the ratio between the lengths of the lower and upper lock rods, the larger the dynamic range of the upper lock rod in the b_2 branch, which is beneficial for compressing the spring and applying a greater spring force. In the optimization, the range of this ratio, p, is set to (1, 1.9).

3.1.4 Optimization Design Technical Route

This optimization is based on the Isight software calling the MATLAB software platform for simulation implementation. For the optimization problem of the lock mechanism's structural parameters, combined with the above analysis, the technical route shown in Fig. 5 is outlined.

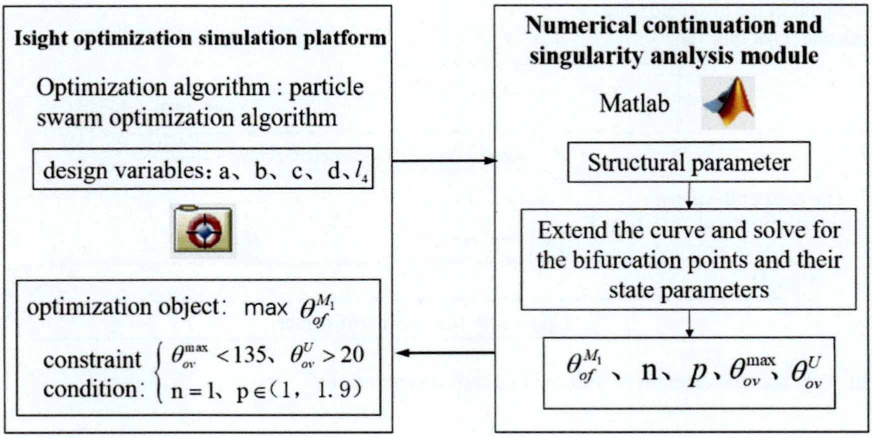

Fig. 5 Optimization design technical route

Table 1 Optimization results and corresponding constraint parameters

	Design variables					Objective		Constraint conditions		
	a (mm)	b (mm)	c (mm)	d (mm)	l_4 (mm)	$\theta_{of}^{M_1}$ (°)	n	$\theta_{ov}^{\max}$ (°)	θ_{ov}^{U} (°)	p
After optimization	516.21	180	240.51	113.47	221.86	0.1726	1	129.52	36.17	1.2
Before optimization	469.79	105.23	210.79	95	169.4	0.1538	1	93.43	43.91	0.71

3.2 Optimization Results

The optimization results are shown in the table, and the optimized length of the upper lock rod, l5, is determined to be 184.112 mm based on the solution of the constraint equations (Table 1).

3.3 Verification of Mechanism Performance Through Post-optimization Simulation

A multibody dynamics simulation model is established using LMS Virtual Lab Motion software to verify the optimization results. Under the same locking spring torque, the starting yaw load is iteratively changed until the lock rod undergoes oscillation and unlocks. The starting yaw load at this moment is recorded to characterize the locking performance of the lock rod. After simulation, the critical unlocking starting yaw loads before and after optimization are shown in Table 2.

Table 2 Comparison of critical unlocking starting yaw loads before optimization

	Critical unlocking starting yaw load	Optimization design objective
Before optimization	158782N	0.1538
After optimization	329800N	0.1726

A larger critical unlocking starting yaw load in the table indicates a stronger locking performance of the lock rod. It can be seen from the table that the locking performance of the lock rod in terms of bearing the starting yaw load has significantly improved after optimization. The dynamic responses of the locking mechanism under three operating conditions, namely, loading the starting yaw load separately, loading the rebound load separately, and loading the starting yaw load first and then the rebound load, are shown in Fig. 6.

From Fig. 6a, b, it can be seen that, when loaded with ground load alone, the optimized lock rod jumping situation is significantly improved, better maintaining the locked state. The maximum load between lock rods during the unlocking process decreased from 10250 to 5800N. From Fig. 6c, it can be seen that, when loaded with rebound load alone, the optimized effect of reducing the load transmitted to the lock rod during rebound load is also significant. From Fig. 6d, it can be seen that,

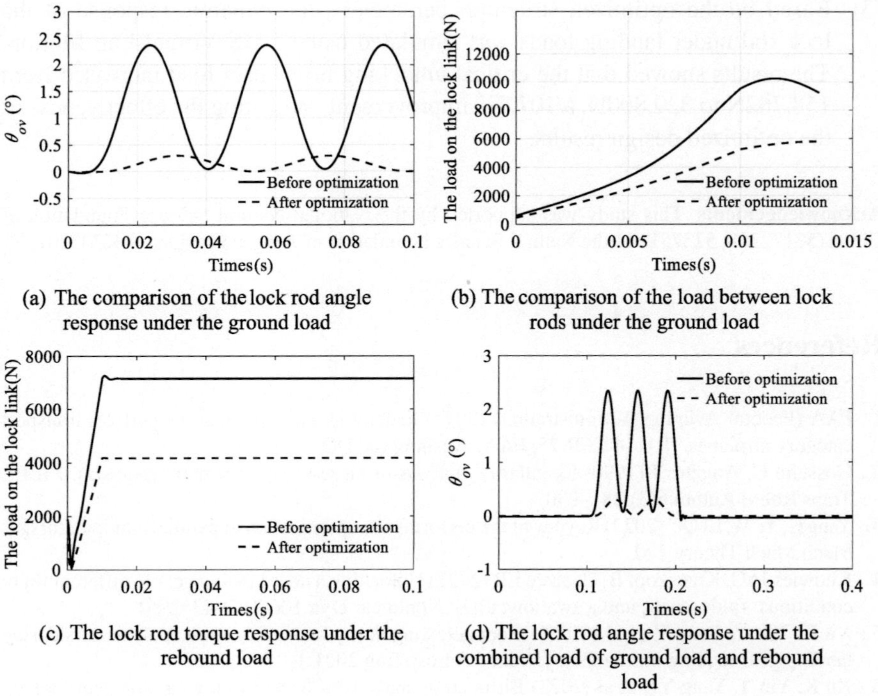

(a) The comparison of the lock rod angle response under the ground load

(b) The comparison of the load between lock rods under the ground load

(c) The lock rod torque response under the rebound load

(d) The lock rod angle response under the combined load of ground load and rebound load

Fig. 6 Shows the dynamic response of the lock mechanism under different ground load conditions under the condition of first loading the ground load and then the rebound load, the optimized lock rod motion during the ground load phase is smaller. In conclusion, the optimized landing gear has a better advantage in terms of locking performance under ground load.

4 Summary

In this section, the parameters of the main landing gear strut lock structure were optimized, and the content and conclusions are as follows:

(1) Singular and bifurcation theories were used to analyze the bifurcation of the landing gear lock mechanism kinematic configuration. It was pointed out that the oscillation of the lock rod is caused by favorable and obstructive loads on both sides of the singular point, and the critical unlocking condition is to cross the next singular point beyond the locked position.
(2) An optimization scheme was constructed with the goal of reducing the main landing gear lock rod load ratio. The optimization problem was solved using the isight and coco numerical extrapolation tools, resulting in the optimal lock rod installation position and length.
(3) Based on the optimized structural parameters, the dynamic response of the lock rod under landing loads was simulated using LMS Virtual Lab Motion. The results showed that the critical unlocking nosewheel load increased from 158,782N to 329,800N, a 107.7% improvement, validating the effectiveness of the optimized design results.

Acknowledgements This study was supported by the National Natural Science Foundation of China (52172368, 52275114), the Natural Science Foundation of Jiangsu Province (BK20220135).

References

1. FAA (Federal Aviation Administration) (2011) Federal aviation regulations part 25: transport category airplanes. [R]. 14 CFR 25. FAA, Washington, DC
2. Gosselin C, Angeles J (1990) Singularity analysis of closed-loop kinematic chains[J]. J IEEE Trans Robot Autom 6(3):281–290
3. Yang C, Ye W, Li QC (2022) Review of the performance optimization of parallel manipulators[J]. Mech Mach Theory 170
4. Knowles JAC, Krauskopf B, Coetzee EB (2021) Unlocking a nose landing gear in different flight conditions: folds, cusps and a swallowtail[J]. Nonlinear Dyn 106(4):2943–2961
5. Xu K, Yin Y, Yang Y et al (2021) Kinematic singularity and bifurcation analysis of sidestay landing gear locking mechanisms[J]. Int J Aerosp Eng 2021:1–15
6. Xu K, Yin Y, Yang YX et al (2022) Bifurcation analysis of dual-sidestay landing gear locking performance considering joint clearance[J]. Chin J Aeronaut 35(7):209–226
7. Yin Y, Xu K, Nie H et al (2019) Research on the performance of an aircraft landing gear hook lock based on bifurcation analysis[J]. Appl Sci Basel 9(24):15

Design Method for Landing Gear Mechanisms Based on Characterization Matrix

Boliang Wang, Yin Yin, Haifeng Wei, Xiaohui Wei, and Hong Nie

Abstract With the advancement of aerospace technology, the development cycles of unmanned aerial vehicle have become increasingly shorter, and the performance requirements have become higher. The space to store the landing gear has become more constrained, necessitating the design of more complex landing gear mechanisms and posing greater design challenges. Traditional design methods rely on engineers' design experience and iterative processes based on CAD technology. This approach is time-consuming, labor-intensive, and only yields feasible solutions rather than optimal ones. Moreover, it lacks guidance in design and hampers innovative mechanism design. To address this issue, this paper innovatively establishes a rapid and universal method for mechanism design and optimization. Firstly, a comprehensive synthesis of mechanisms with different numbers of components is conducted, and a mechanism characterization method is proposed to facilitate innovative mechanism configuration design using characterization matrices. Next, the principles of multibody systems are employed to realize generalized kinematics simulation of complex mechanisms. Finally, particle swarm optimization algorithm is applied to optimize the mechanism parameters. Taking into account the constraints of a certain aircraft's compartment and the design requirements for the landing gear retraction position, this method is utilized to design a retraction mechanism that meets the specified requirements, demonstrating the superiority of this approach and providing a new perspective for landing gear mechanism design.

B. Wang · Y. Yin (✉) · H. Wei · X. Wei · H. Nie
State Key Laboratory of Mechanics and Control for Aerospace Structures, Nanjing University of Aeronautics and Astronautics, Nanjing 210016, China
e-mail: yinyin@nuaa.edu.cn

B. Wang
e-mail: 2581750917@nuaa.edu.cn

X. Wei
e-mail: Wei_xiaohui@nuaa.edu.cn

H. Nie
e-mail: hnie@nuaa.edu.cn

Z. Liu et al. (eds.), *Advances and Challenges in Advanced Unmanned Aerial Systems*, Springer Aerospace Technology, https://doi.org/10.1007/978-981-99-8045-1_14

Keywords Mechanism synthesis · Kinematics simulation · Multibody systems · Characterization matrix · Optimization-driven

CCS Concepts: Computing Methodologies ~ Modeling and Simulation

1 Introduction

Mechanism design is widely applied in the aerospace field, including the design of retraction mechanisms for landing gears and doors, as well as motion mechanisms for aircraft flaps, elevators, and other components.

Conventional methods of mechanism design involve relating the design requirements to the inherent principles of matured and applied mechanisms, and then conducting theoretical calculations and analyses based on this foundation. With the advancement of aerospace technology, retraction mechanisms design meats greater design challenges [1, 2]. Zhu et al. proposed an innovative method in solving the design of landing gear retraction trajectory in narrow cabin space, the 3D models of the landing gear and the cabin segment are converted to mesh models, the landing gear retraction trajectory is optimized by combining collision detection algorithm [3]. Ye established an intelligent optimization model of landing gear retraction mechanisms based on physical simulation and optimize the minimum clearance between landing gear and cabin segment, which can effectively guide the intelligent design of key parameters [4]. Chen simplified the optimization problem of actuating force within the context of inverse dynamic analysis of a three-dimensional landing gear mechanism, and use sequential quadratic programming to solve the simplified optimization problem with only two design variables [5]. However, for complex mechanism design problems, it is difficult to grasp the inherent principles of the mechanisms, and the theoretical calculation process can be cumbersome. Eric presented a graph-theory based approach for representing planar mechanisms, generating an extended matrix containing the complete characterization of a planar mechanism, including all the information about both topology and geometry, which can be used for the optimal design of mechanisms [6]. Jonathan developed simultaneous structural and kinematic optimization methodology to design lightweight retractable landing gears, and used the proposed optimization methodology for 3 different commercial aircraft applications, demonstrating that multidisciplinary design optimization methodology can accelerate the preliminary design phase of retractable nose landing gears with locking links, out-of-plane load cases and actuator performance also considered [7].

In this research, the characterization matrix is established and modified by computer-driven processes to achieve configuration synthesis of the mechanism. Different configurations of mechanisms are simulated and modeled by directly reading the data from the characterization matrix based on the principles of multi-body systems. The simulation models are embedded into an optimization algorithm framework, enabling an integrated process of mechanism design, including task-based design, configuration synthesis, and scale synthesis. The design and optimization of the retraction mechanism are illustrated using the design task of the main landing gear of a specific aircraft.

2 Innovative Mechanism Design

Retraction mechanisms can be classified into spatial retraction mechanisms and planar retraction mechanisms. For spatial retraction mechanisms that rotate around a single axis can be transformed into planar mechanisms in a specific coordinate system. From an engineering application perspective, planar retraction mechanism design is more general. In the analysis of retraction mechanisms, a single degree-of-freedom retraction mechanism satisfies the equation:

$$F = 3n - 2p = 4 \tag{1}$$

It can be concluded that planar retraction mechanisms must consist of an even number of components. By employing graph theory methods for mechanism synthesis, it is found that the six-bar kinematics chain has two topological configurations, while the eight-bar kinematics chain has 16 topological configurations. After applying specific constraint conditions to the topological configurations of the mechanism, all feasible configuration solutions can be obtained. This process can be accomplished using a characterization matrix. For a mechanism composed of N rigid bodies, the size of the characterization matrix M is $N \times (N + 1)$. The characterization matrix M forms a square matrix with the first N columns. Matrix elements $M_{i,j}(i < j)$ represent the adjacency relationship between body i and body j.

$$\begin{cases} M_{i,j} = 0, body\, i\, and\, body\, j\, are\, not\, adjacent \\ M_{i,j} = 1, body\, i\, and\, \text{body}\, j\, are\, adjacent \end{cases} \tag{2}$$

$M_{i,j}(i > j)$ represents the positions of the kinematics pairs corresponding to the adjacency relationships. The main diagonal of the characterization matrix represents the pose information of rigid bodies. The last column of the characterization matrix is used to represent specific information about the rigid body. 0, 1, and 2 are used to represent ordinary members, the base, and the end effector, respectively. For the landing gear system, the fuselage is the base, the landing gear is the end effector, and the retraction mechanism is the ordinary member. In the initial design of the landing gear mechanism, the ordinary members are unknown (as shown in Fig. 1). When applying specific constraint conditions to the components during configuration synthesis, only the base and the end effector can be specified.

Therefore, for each topological configuration of the mechanism with N components, by traversing and modifying the specific information in the column N + 1 of

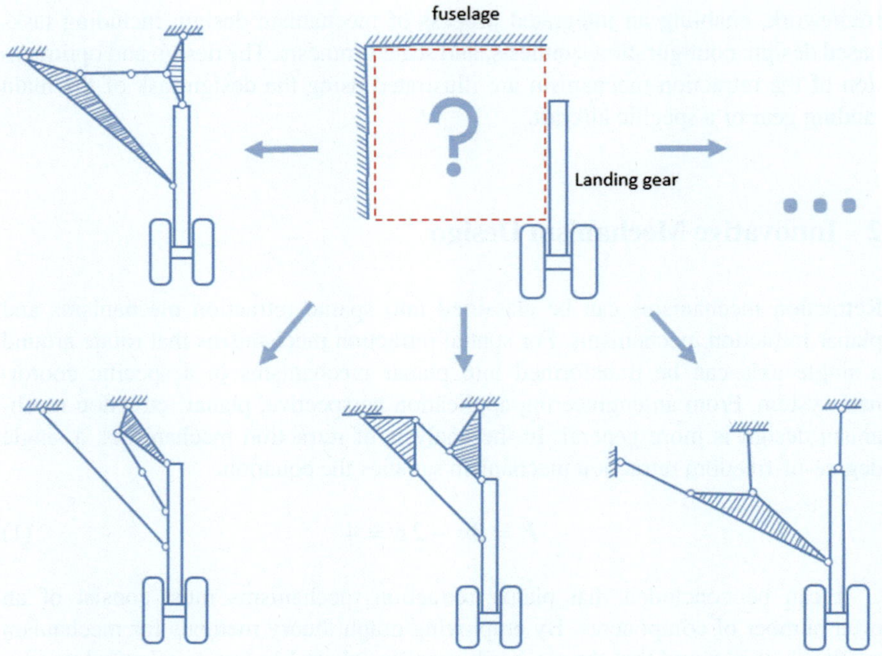

Fig. 1 The process of configuration synthesis

the characterization matrix, a total of M configurations can be generated and stored in the computer in matrix form.

3 Multibody Kinematics Modeling, Simulation, and Optimization

A planar rigid body has three degrees of freedom, and its pose is represented using a coordinate $[x_i, y_i, \phi_i]^T$, which fixed to the body. The kinematics constraints between rigid bodies are established by constructing constraint equations based on the connecting vectors between the bodies. Based on the principles of multibody systems, a kinematics simulation model is established. Firstly, the hinge constraint information and the corresponding indices of the rows and columns in the characterization matrix are traversed. Then, the corresponding module group in the hinge constraint library is called to assemble the constraint equations $\underline{\Phi}(\underline{q}, t_0)$ and the Jacobian matrix $\underline{\Phi}_q(\underline{q}, t_0)$ for the rigid bodies.

$$\underline{\Phi}(\underline{q}, t_0) = \begin{pmatrix} \underline{\Phi}^K(\underline{q}) \\ \underline{\Phi}^D(\underline{q}, t_0) \end{pmatrix} = \underline{0} \tag{3}$$

$$\underline{\Phi}_q(\underline{q}, t_0) = \begin{pmatrix} \frac{\partial \Phi_1}{\partial q_1} & \cdots & \frac{\partial \Phi_1}{\partial q_{3n}} \\ \vdots & \ddots & \vdots \\ \frac{\partial \Phi_{3n}}{\partial q_1} & \cdots & \frac{\partial \Phi_{3n}}{\partial q_{3n}} \end{pmatrix}, \quad \underline{\Phi}_q(\underline{q}, t_0) \in R^{3n \times 3n} \tag{4}$$

In the case of driving constraints $\underline{\Phi}^D(\underline{q}, t_0)$, they vary with time according to a given motion law. By solving the constraint equations $\Phi(q, t) = 0$, the equilibrium solution at time t is obtained, which provides the pose information of each rigid body at time t. The equilibrium solution is obtained using the Newton's method.

$$q_t^{i+1} = q_t^i - \frac{\Phi(q, t)}{\Phi_q(q, t)} \tag{5}$$

Both configuration synthesis and kinematics simulation of the mechanism are implemented through the characterization matrix. The characterization matrix tightly integrates configuration synthesis and position synthesis.

The optimization design of retraction mechanisms is a multi-constraint non-convex optimization problem, which can be efficiently solved using swarm intelligence algorithms [8]. This study employs the particle swarm optimization algorithm (PSO) to solve the optimization problem. PSO algorithm offers advantages such as direct solution in the problem domain using real number encoding, parallel computation handling multiple particles in the swarm, few adjustable parameters, and strong optimization capability. An optimization design model for landing gear and retraction mechanisms is established in the following model.

$$\begin{aligned} &\max f = -||r - r_e||_2 - k|\phi - \phi_e| \\ &s.t.\, X \subset A \end{aligned} \tag{6}$$

r, ϕ represents the pose of the rigid body. r_e, ϕ_e represents the desired pose of the rigid body.

k is the proportional coefficient that balances the impact of position and attitude differences on the objective function. $X = (X_1, X_2, \cdots, X_n)$, $X_i = (x_i, y_i)$ represents the initial position of the N kinematics pairs in the mechanism. Closed curve A is a polygon approximated by N1 points, which represents the space constraints of the cabin section. Closed curve A is constructed by using N1 points to fit a polygon as shown in Fig. 2. And the number of edges in this polygon is also N1.

During the optimization iterations, the ray method is used to determine if the kinematics pair is located inside the closed curve A.

Fig. 2 Definition of cabin section boundaries

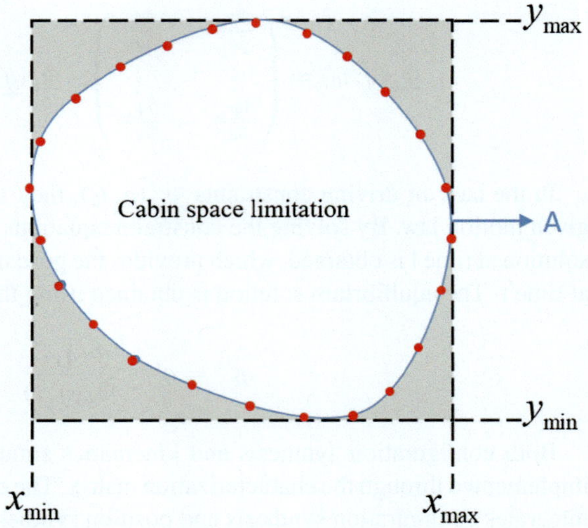

4 Example

The cabin section constraints and design requirements for a certain unmanned aerial vehicle are shown in Fig. 3. The retracted position of the landing gear is (−651, −364, 0°), and the extended position is (−960, −1040, −20.87°). Based on these specifications, the design and optimization of the retraction mechanism are conducted.

The retraction mechanism is designed as an eight-bar mechanism, with additional constraints of the landing gear not rotating around a fixed axis and the main strut as a binary bar. Configuration synthesis is performed based on the specific constraints of the base and the end effector.

$$\begin{cases} M_{\mathrm{i,j}} = 0 \\ \sum_{k=1}^{8} M_{k,\mathrm{j}} = 2 \end{cases} \quad (7)$$

There are multiple results of mechanism kinematics simulations that satisfy the requirements, but due to limited space, only one configuration and its corresponding schematic diagram are presented as shown in Figs. 4 and 5.

The optimized solution of the kinematics pairs for the given configuration of the mechanism is shown in Table 1.

The change in the fitness function values with respect to the iteration number is depicted in the Fig. 6. Based on the optimization results, the specific retraction mechanism was modeled in CATIA, as shown in Fig. 7.

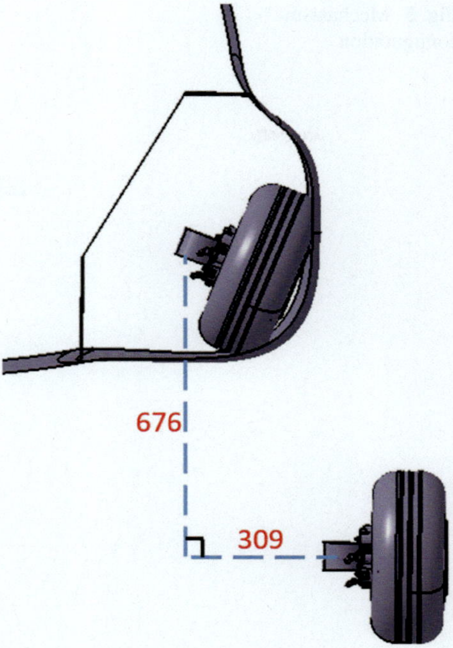

Fig. 3 Design task diagram

Fig. 4 Mechanism characterization matrix

$$\begin{pmatrix} B_1 & 1 & 0 & 1 & 0 & 1 & 0 & 1 & 0 \\ J_1 & B_2 & 1 & 0 & 0 & 0 & 0 & 0 & 1 \\ 0 & J_2 & B_3 & 1 & 0 & 0 & 0 & 0 & 0 \\ J_8 & 0 & J_3 & B_4 & 1 & 0 & 0 & 0 & 0 \\ 0 & 0 & 0 & J_4 & B_5 & 1 & 0 & 0 & 0 \\ J_9 & 0 & 0 & 0 & J_5 & B_6 & 1 & 0 & 0 \\ 0 & 0 & 0 & 0 & 0 & J_6 & B_7 & 1 & 0 \\ J_{10} & 0 & 0 & 0 & 0 & 0 & J_7 & B_8 & 2 \end{pmatrix}$$

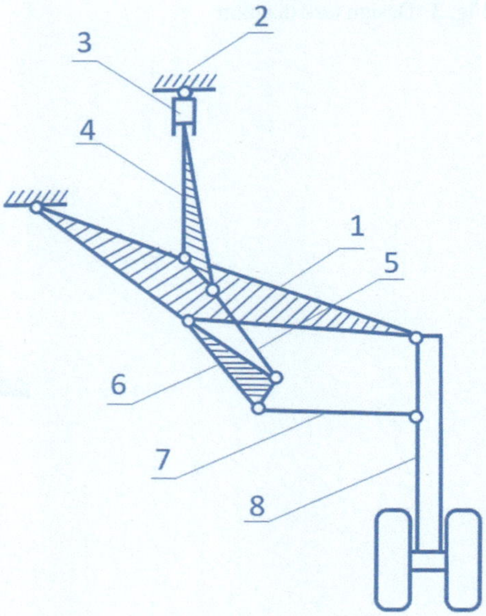

Fig. 5 Mechanism configuration

Table 1 The optimized positions of the kinematics pairs

Joints	J_1	J_2	J_3	J_4	J_5	J_6	J_7	J_8	J_9	J_{10}
x	−474.28	−38.11	−436.49	−447.30	−412.05	−165.90	−142.68	−212.68	−231.82	−364.35
y	−496.95	−720.74	−550.57	−592.55	−639.19	−647.35	−850.31	−805.44	−623.86	−651.65

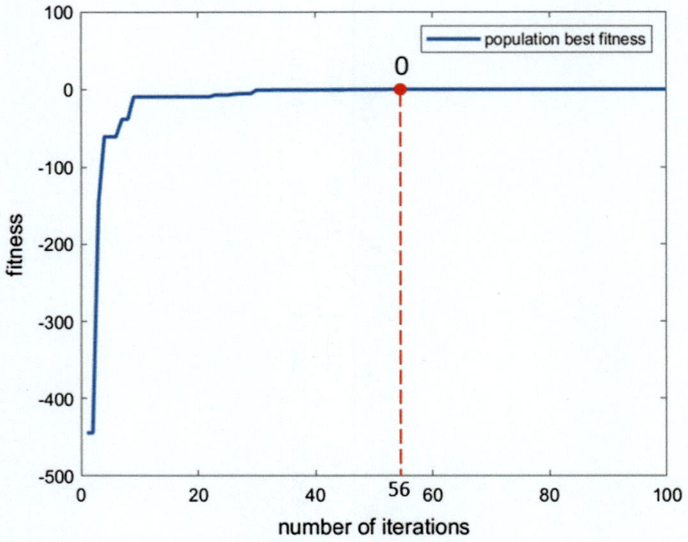

Fig. 6 Fitness curve off particle swarm algorithm

Fig. 7 Physical model of mechanism

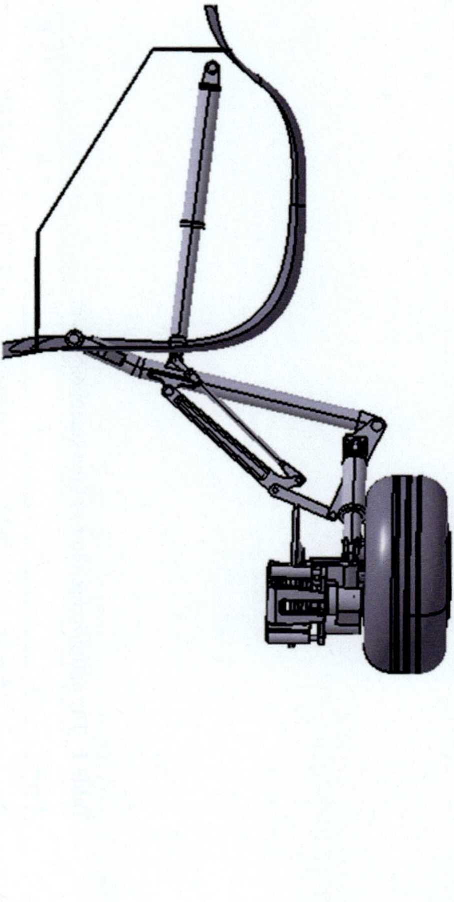

5 Conclusion

This paper presents a novel approach to mechanism design. Firstly, a mechanism characterization method is proposed, utilizing a characterization matrix for innovative configuration design. Based on the principles of multibody systems, a kinematics simulation model is established for the mechanism. Finally, the model is embedded into a particle swarm optimization algorithm framework for optimization. By employing this approach, the design of the landing gear retraction mechanism for a specific UAV was completed in just one day, yielding a variety of outstanding results. In contrast, employing conventional methods would have taken several weeks. Compared to conventional design methods, this approach demonstrates higher design efficiency, greater diversity in design results, and certain advantages.

Acknowledgements This study was supported by the National Natural Science Foundation of China (52172368, 52275114), the Natural Science Foundation of Jiangsu Province (BK20220135).

References

1. Udayakumar R, Ibrahim M (2019) Design analysis of landing gear system of an aircraft. In: 2019 3rd International conference on electronics, communication and aerospace technology. India, pp 1326–1330
2. Douglas C, et al (2021) Multidisciplinary design optimization of aircraft landing gear. AIAA AVIATION 2021 FORUM
3. Zhu L, Zou Z, Yin Y et al (2021) Intelligent design method of landing gear retraction and extension trajectory for narrow space. J Beijing Univ Aeronaut Astronaut 47(12):2560–2570
4. Ye Y, et al (2023) Intelligent optimization of landing gear retraction based on physical simulation. J Phys Conf Ser 2469(1) (IOP Publishing)
5. Chen H, Yuhong J (2018) Optimal design of aircraft landing gear mechanism actuating force. In: 2018 3rd Joint international information technology, mechanical and electronic engineering conference (JIMEC 2018). Atlantis Press
6. Santiago-Valentin E, Portilla-Flores EA, et al (2018). A graph-theory based method for topological and dimensional characterization of planar mechanisms as a computational tool for engineering design. IEEE Access 1–1. https://doi.org/10.1109/ACCESS.2018.2885563
7. Ryu J, et al (2023) Gradient based simultaneous structural and kinematics optimization of landing gear members based on the modified input-output equation for multibody kinematics. AIAA SCITECH 2023 Forum
8. Wang D, Tan D, Liu L (2018) Particle swarm optimization algorithm: an overview. Soft Comput 22:387–408

A Novel Method for Camera Focal Lengths Calibration Based on Active Vision

Zhanfei Chen, Renfu Li, Fengnian Tian, Jiqiu Liang, Xuelong Si, and Bin Zhang

Abstract It is often difficult to align the camera principal point with the rotation center of the motion mechanism during active vision calibration, and this disalignment could lead to inaccuracy of the constraint equations and calibration results. To circumvent such issues, this paper proposes a novel method for active vision camera calibration and facilitates its real-world implementation. In this method, the rotational motion axis is assumed not necessarily to pass through the camera's optical center, and the constraint equations are established by correlating image matched points before and after camera motions. The displacements caused by rotation are incorporated into the constraint equations to improve the accuracy of calibration. Numerical simulations and experiments are conducted. The feasibility of this proposed method is justified by using both synthetic and experimental data, and the results show that this method could estimate camera focal lengths with high accuracy. This study suggests that this novel method has great potential for visual and image applications.

Keywords Active vision · Camera calibration · Intrinsic parameters · Machine vision

---

Z. Chen · R. Li (✉) · F. Tian (✉)
School of Aerospace Engineering, Huazhong University of Science and Technology, Wuhan, China
e-mail: renfu.li@hust.edu.cn

F. Tian
e-mail: tianfengnian@hust.edu.cn

J. Liang · X. Si · B. Zhang
System Design Institute of Hubei Aerospace Technology Academy, Wuhan, China

Z. Liu et al. (eds.), *Advances and Challenges in Advanced Unmanned Aerial Systems*, Springer Aerospace Technology, https://doi.org/10.1007/978-981-99-8045-1_15

1 Introduction

Precise calibration of the camera's intrinsic parameters is a fundamental issue for vision-related technology [1–3]. Generally, existing camera calibration techniques can be grouped into three major categories: traditional calibration [4–6], self-calibration [7–9] and active vision calibration.

Traditional camera calibration methods rely on a calibration object with known structure to establish the correspondence between 3 and 2D coordinates. To achieve accurate calibration, this approach necessitates a highly precise calibration object. And the object should match the camera's field of view and be positioned at various locations during the calibration process.

Self-calibration methods utilize geometric invariance of image features to calibrate a camera. However, these methods are challenging to implement due to the complexity of geometry, such as determining the absolute conic. Furthermore, the accuracy and robustness of the results cannot be guaranteed.

In this paper, we focus our attentions on active vision calibration. For active vision calibration, it is essential to have the knowledge of specific motion parameters of the camera, which can be used to establish constraint equations that allow for the determination of the camera's intrinsic parameters. This approach enables high accuracy calibration without relying on highly precise calibration objects. Moreover, it is a simple and straightforward technique to implement and can be easily automated. In other words, active vision calibration method maintains high calibration accuracy and improves efficiency.

Some studies have explored active vision calibration techniques. Du and Brady proposed a pure rotation method around the optical center, which could solve the camera's intrinsic parameters by utilizing the camera's motion and optical flow [10]. Stein considered that when the rotation axis doesn't pass through the optical center, the camera produces translations during rotation, which introduces depth information into the constraint equation [11]. However, this method didn't provide a solution to eliminate the depth information. Basu proposed a method based on small-angle rotation around the optical center to calibrate the camera's intrinsic parameters, which approximates the small angle using $\sin\theta = \theta$ and $\cos\theta = 1$ [12]. Later, in order to overcome the shortcoming in Ref. [12] which is sensitive to noise, Basu and Ravi added rolling motion(rotation about the optic axis)for optimization [13], which improves the robustness. However, a roll rotation is difficult to implement and has high requirements on the motion equipment. Similarly, several studies, including Refs. [14–18], control the camera to perform pure rotation motion to calculate the camera intrinsic parameters. Nonetheless, achieving pure rotation around the optical center in reality is difficult, so efforts have been made to address this problem [19–21]. Davis and Chen proposed a camera model that rotates around arbitrary axes [19]. But for proper calibration results, the rotation angle needs to be adjusted so that the camera provides full coverage of space. In Ref. [20], Sanchez-Riera et al. considered that the camera generates displacements when performing pan and tilt movements. Then they added the displacements to the constraint equations for solving the camera intrinsic parameters. However, to obtain better results, a reasonable estimation of the initial value of the intrinsic parameters is needed. The calibration method for freely moving cameras was discussed in Ref. [21]. The results show that in order to ensure the unique results, it is necessary to constrain the camera intrinsic parameters. For example, the solution for the focal length, aspect ratio is unique if two image pairs are used, and the principal point's coordinates are known.

To circumvent this hard situation, we propose a novel method for the camera focal lengths calibration in this paper. The main idea of this method is to establish the correspondence between 3 and 2D coordinates using rotational motion, and to eliminate the depth information using translational motion, then calibrate the camera. The proposed approach eliminates the need for rotation around the optical center and complicated motion. It only requires the camera to perform two rotations and two translations according to the set procedure. Besides, The proposed approach does not require high-precision calibration objects and the calculation process is relatively simple.

The rest of this paper is structured as follows. Section 2 presents the camera calibration method proposed in this work. Section 3 includes an analysis of our method using both simulated data and real experiments. Finally, we conclude the whole paper in Sect. 4.

2 Formulation of an Active Visual Calibration

In visual technique applications, the camera is usually mounted on a platform of six-degree-of-freedom, or a robotic arm so that the camera can move rotationally and translationally. The camera's intrinsic parameters can be retrieved by the information of camera motions. Next, we will introduce the specific implementation of the method proposed in this paper.

2.1 Coordinate Systems

In this section, we involve three coordinate systems, namely, the platform coordinate system fixed on the motion mechanism, the camera coordinate system fixed on the camera, and the image coordinate system after the camera imaging. Figure 1 illustrates the relationship between these coordinate systems. The definitions of the coordinate systems are as follows:

- Platform coordinate system: $O_p - X_pY_pZ_p$. The origin O_p is defined at the center of rotation, with the axes determined by the right-hand rule.
- Camera coordinate system: $O - XYZ$. Origin O is defined at the camera's optical center position, with the axis parallel to the corresponding platform coordinate

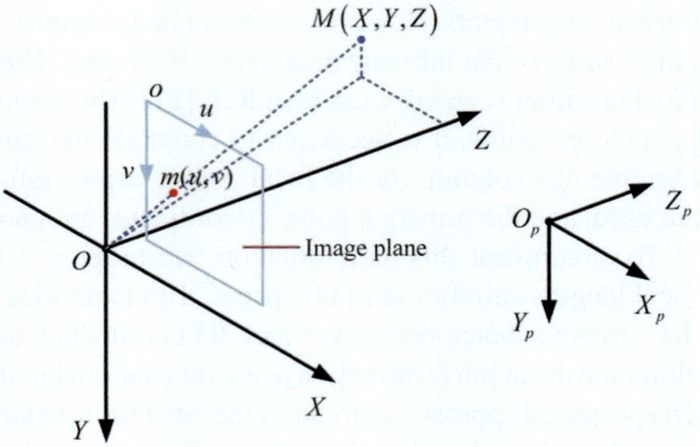

Fig. 1 Relationships between coordinate systems

system axis (e.g., X_p axis and X axis parallel). The camera is fixed to the platform and the optical center in the platform coordinate system is $\begin{bmatrix} x_1 & y_1 & z_1 \end{bmatrix}^T$.

- Pixel coordinate system: $o - uv$. The pixel coordinate represents the position of the pixel in the image, and the origin o is defined in the upper left corner of the image.

2.2 Rotation

In this paper, we adopt Euler angles to represent the camera's rotation. Our proposed calibration method focuses on the rotation around the X_p axis and Y_p axis, and does not consider the rotation around the Z_p axis. We define pan rotation and tilt rotation similar to the PTZ camera's rotation. The corresponding rotation matrices are defined as follows:

(1) Tilt: We define the tilt rotation of the camera as a rotation about the X_p axis, which can be determined by the following matrix:

$$R_x = \begin{bmatrix} 1 & 0 & 0 \\ 0 & \cos\psi & -\sin\psi \\ 0 & \sin\psi & \cos\psi \end{bmatrix} \tag{1}$$

where ψ represents the angle of rotation.

(2) 2) Pan: For the pan rotation, we define it as a rotation about the Y_p axis, which can be obtained by the following matrix:

$$R_y = \begin{bmatrix} \cos\theta & 0 & \sin\theta \\ 0 & 1 & 0 \\ -\sin\theta & 0 & \cos\theta \end{bmatrix} \tag{2}$$

where θ represents the angle of rotation.

With the rotation matrix defined around the different axes, the final rotation matrix. can be represented as $R = R_y R_x = [r_{ij}]_{3\times3}$, where r_{ij} represents the elements of the $i - th$ row and $j - th$ column of the rotation matrix.

2.3 Camera Model

The camera imaging process is to map the real-world object to the image plane, and the imaging principle can be simplified to a hole imaging model. The imaging process can be illustrated in Fig. 1. The mapping relationship between the two can be expressed by Eq. (3).

$$Z\begin{bmatrix} u \\ v \\ 1 \end{bmatrix} = \begin{bmatrix} f_x & 0 & u_0 \\ 0 & f_y & v_0 \\ 0 & 0 & 1 \end{bmatrix}\begin{bmatrix} X \\ Y \\ Z \end{bmatrix} = K\begin{bmatrix} X \\ Y \\ Z \end{bmatrix} \tag{3}$$

The matrix K is the camera intrinsic matrix, f_x and f_y are the focal lengths of the camera in the u direction and v direction, respectively. (u_0, v_0) represents the coordinates of the principal point.

2.4 Image Acquisition Process

In this part, we design a picture acquisition process that conforms to the method of this paper. In order to control the camera motion, we need to mount the camera on a specific mechanism that can rotate and translate according to a set program, as well as acquire the parameters of the motion. Once these requirements are satisfied, we can acquire the required images by the following steps:

(1) Fix the camera on the six-degree-of-freedom platform, so that the corresponding axes between the camera coordinate system and the platform coordinate system are parallel. The platform is in the initial position (0, 0, 0, 0, 0, 0)(the first three items of the position vector represent the distance along the X_p axis, Y_p axis, Z_p axis, and the last three represent the rotation angle along the X_p axis, Y_p axis, Z_p axis, respectively). Then the camera takes image 1;
(2) The platform is moved to the position $(\Delta x, 0, 0, 0, 0, 0)$, and the camera takes image 2;
(3) The platform is moved to the position $(0, \Delta y, 0, 0, 0, 0)$, and the camera takes image 3;
(4) The platform is rotated into position $(0, 0, 0, 0, \theta, 0)$ and the camera takes image 4;
(5) The platform rotates to the position $(0, 0, 0, \psi, 0, 0)$, the camera takes image 5.

2.5 Focal Lengths Calibration

We consider the initial coordinates of a point in the camera coordinate system as $\begin{bmatrix} X & Y & Z \end{bmatrix}^T$, and the pixel coordinates after imaging as $\begin{bmatrix} u & v \end{bmatrix}^T$. The coordinates of the point in the camera coordinate system after a certain movement of the camera are $\begin{bmatrix} X' & Y' & Z' \end{bmatrix}^T$, and the pixel coordinates after imaging as $\begin{bmatrix} u'_n & v'_n \end{bmatrix}^T$, where n represents the $n - th$ image, $n = 2, 3, 4, 5$.

We rotate the platform around the Y_p axis by a set angle θ. Using the correspondence of matched point before and after the motion. We can obtain:

$$u'_4 = u_0 + \frac{r_{11}(u - u_0) - r_{11}\frac{\Delta X}{Z} f_x + r_{31}\left(1 - \frac{\Delta Z}{Z}\right) f_x}{r_{13}\left(\frac{u-u_0}{f_x} - \frac{\Delta X}{Z}\right) + r_{33}\left(1 - \frac{\Delta Z}{Z}\right)} \tag{4}$$

$$v'_4 = v_0 + \frac{v - v_0}{r_{13}\left(\frac{u-u_0}{f_x} - \frac{\Delta X}{Z}\right) + r_{33}\left(1 - \frac{\Delta Z}{Z}\right)} \tag{5}$$

where ΔX and ΔY are the displacement of the optical center after rotation.

We note that Eqs. (4) and (5) contains the depth information Z. To eliminate the effect of depth information, we move the platform along the X_p axis by a set distance Δx, we can obtain:

$$\frac{f_x}{Z} = -\frac{u'_2 - u}{\Delta x} \tag{6}$$

Relating Eqs. (4), (5) and (6), we have:

$$f_x = \frac{\frac{u'_4 - u_0 - k_1 \cos\theta (u - u_0)}{k_1} - [x_1(1 - \cos\theta) + z_1 \sin\theta]\frac{u'_2 - u}{\Delta x}}{-\sin\theta} \tag{7}$$

where $k_1 = \frac{v'_4 - v_0}{v - v_0}$. Since the coordinates of the principal point are quite close to the center of the image, we can make $u_0 = c_u = \frac{w}{2}$, $v_0 = c_v = \frac{h}{2}$((c_u, c_v) represents the coordinates of the center point of the image, w and h represent the width and height of the image respectively). Then Eq. (7) can be written as follows:

$$f_x = \frac{\frac{u'_4 - c_u - k_1 \cos\theta (u - c_u)}{k_1} - [x_1(1 - \cos\theta) + z_1 \sin\theta]\frac{u'_2 - u}{\Delta x}}{-\sin\theta} \tag{8}$$

We can notice that Eq. (8) is linear and with only a single unknown, so only one matched point corresponding to before and after the motion needs to calculation f_x. If for more robust results, multiple sets of matched points can be selected, and the average of the results obtained from these points can be used to improve the accuracy of the solution.

In the same way as solving for f_x, We rotate the platform around the X_p axis by a set angle ψ, and move the platform along the Y_p axis by a set distance Δy. We make $k_2 = \frac{u'_5 - u_0}{u - u_0}$, then the function of f_y can be written as:

$$f_y = \frac{\frac{v'_5 - c_v - k_2 \cos\theta(v - c_v)}{k_2} - [y_1(1 - \cos\psi) - z_1 \sin\psi]\frac{v'_3 - v}{\Delta y}}{\sin\psi} \tag{9}$$

3 Experiments and Analyses

In order to validate the proposed method, we conducted several sets of simulation experiments with different rotation angles and translation distances. Synthetic data was used in the simulations, where a 3D apple was observed by a camera. Figure 2 illustrates the apple image obtained by the camera. In addition, we mounted the camera on a six-degree-of-freedom platform to collect real data, and the correctness of the simulation results was confirmed.

3.1 Simulation Results Analysis

We perform the simulation based on MATLAB, setting the intrinsic matrix of the camera as Eq. (10):

$$K = \begin{bmatrix} 4137.8 & 0 & 2128 \\ 0 & 4147.3 & 1416 \\ 0 & 0 & 1 \end{bmatrix} \tag{10}$$

It is shown that the theoretical value of the focal length in the u direction is $4137.8 pixel$, and the focal length in the v direction is $4147.3 pixel$. In the following, we will discuss the simulation results in detail.

Based on the proposed method in Sect. 2, the focal lengths of the camera in the u direction and in the v direction can be solved by Eqs. (8) and (9), respectively. Despite the fact that only one matched point is needed to calculate the results, we use more matched points in sake of making the results more accurate. In the simulation, the pan and tilt angles are set from $-7°$ to $7°$ (excluding $0°$), and the translation distances are set to 10mm, 30mm and 50mm in turn. The errors between the obtained results and the theoretical values are shown in Fig. 3.

In the ideal case, Fig. 3 illustrates that the proposed method has a high accuracy in solving the focal lengths. The calculated mean square error is $0.468 pixel$ for f_x and $0.145 pixel$ for f_y.

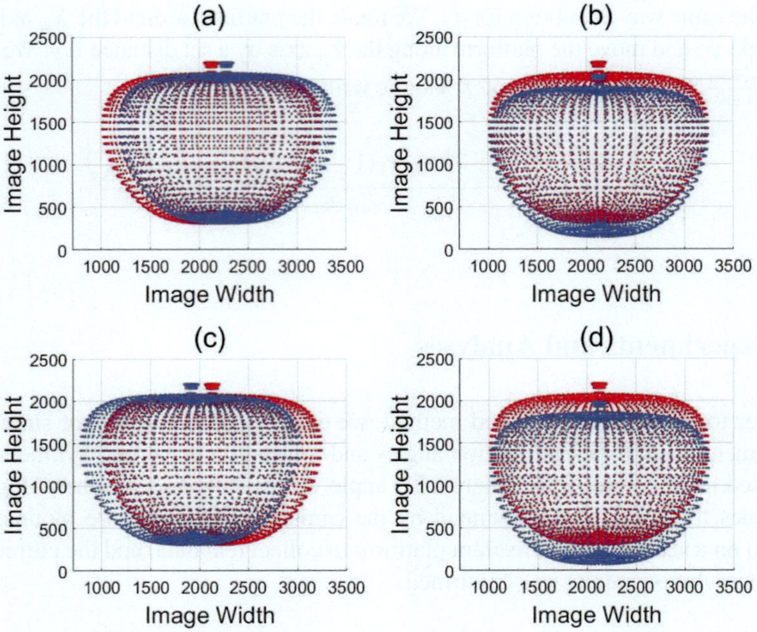

Fig. 2 Synthesis data used in the simulation. **a** The projected images of the apple on the camera before (red apple) and after (blue apple) moving along the X_p axis. **b** The projected images of the apple on the camera before (red apple) and after (blue apple) moving along the Y_p axis. **c** The projected images of the apple on the camera before (red apple) and after (blue apple) a pan rotation. **d** The projected images of the apple on the camera before (red apple) and after (blue apple) a tilt rotation

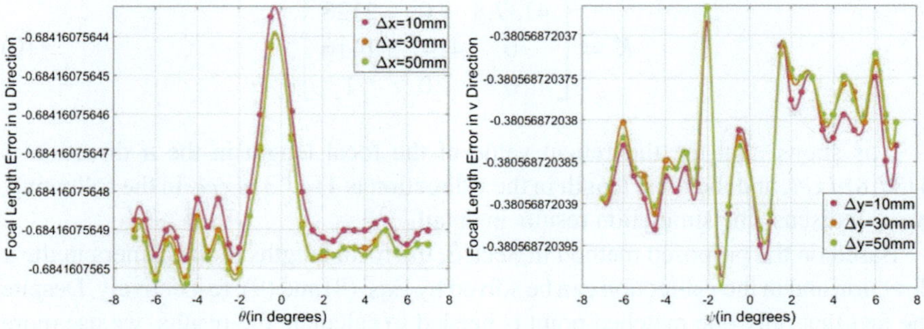

Fig. 3 Focal length error in u direction and v direction

Table 1 Results of the proposed method for various motions of the camera. $|\delta_{f_x}|$ and $|\delta_{f_y}|$ are the pixel errors with the reference value (calculated by [4])

Zhang's calibration results [4]:
$f_x = 10683$, $f_y = 10687$

#	Motion parameters	f_x	f_y	$\lvert\delta_{f_x}\rvert$	$\lvert\delta_{f_y}\rvert$
1	$\theta = \psi = 1.2°$, $\Delta x = \Delta y = 10\text{mm}$	10,901	11,040	218	353
2	$\theta = \psi = 1.2°$, $\Delta x = \Delta y = 30\text{mm}$	10,902	11,035	219	348
3	$\theta = \psi = 1.2°$, $\Delta x = \Delta y = 50\text{mm}$	10,904	11,042	221	355
4	$\theta = \psi = 2.0°$, $\Delta x = \Delta y = 10\text{mm}$	10,894	11,046	211	359
5	$\theta = \psi = 2.8°$, $\Delta x = \Delta y = 10\text{mm}$	10,882	11,065	199	378

3.2 Real Data

Several experiments are performed on real scene. We used a HIKVISION MV-CE200-10UM camera with 5472*3648 resolution mounted on a six-degree-of-freedom platform. Based on the proposed method in Sect. 2, since the distortion of the camera is ignored in this paper, in order to decrease the error as well as to improve the computational efficiency, we use only one corresponding point near the center of the image to calculate f_x. The same is true in solving for f_y.

In order to evaluate the calculation results of real images, we use the method proposed in Ref. [4] to calibrate the camera and implementation is available as a MATLAB Toolbox [22]. We take the MATLAB Toolbox calibration results as a reference, and analyze the calculation results of the method proposed in this paper. Table 1 lists the results of camera intrinsic parameters in real images. As we can see, the calibration results are close to Zhang's results, which indicates that the proposed approach is reliable.

4 Conclusion

In this paper, we introduce a novel method for calibrating focal lengths of a camera, which avoids fixing the camera's optical center to the rotation center. The proposed method requires controlling the camera to make two rotational motions and two translational motions. Based on the matched points in the images, the focal lengths can be solved correctly. In addition, only one matched point is required for solving the focal length, which can improve the computational efficiency.

We conducted several sets of simulation experiments using synthetic data, and achieved a low error rate in solving for the focal lengths. Furthermore, we verified that our proposed method is accurate and workable through experiments in the real world.

Overall, the proposed method enables to accurately solve for the focal lengths. At the same time, it is simple to operate, easy to automate, and has a significant potential to be applied in engineering practice.

References

1. Jiang T et al (2019) Dual-camera-based method for identification and location of scattered self-plugging rivets for robot grasping. Measurement 134:688–697
2. Liu W, Hu J, Wang W (2020) A novel camera fusion method based on switching scheme and occlusion-aware object detection for real-time robotic grasping. J Intell Robot Syst 100(2020):791–808
3. Chen M et al (2019) High-accuracy multi-camera reconstruction enhanced by adaptive point cloud correction algorithm. Opt Lasers Eng 122:170–183
4. Zhang Z (2000) A flexible new technique for camera calibration. IEEE Trans Pattern Anal Mach Intell 22(11):1330–1334
5. He D et al (2013) Eccentricity error identification and compensation for high-accuracy 3D optical measurement. Meas Sci Technol 24.7:075402
6. He H et al (2020) A novel efficient camera calibration approach based on K-SVD sparse dictionary learning. Measurement 159:107798
7. Faugeras O, Quan L, Strum P (2000) Self-calibration of a 1d projective camera and its application to the self-calibration of a 2d projective camera. IEEE Trans Pattern Anal Mach Intell 22(10):1179–1185
8. Jin J, Li X (2013) Efficient camera self-calibration method based on the absolute dual quadric. JOSA A 30(3):287–292
9. Tang Z et al (2019) ESTHER: joint camera self-calibration and automatic radial distortion correction from tracking of walking humans. IEEE Access 7:10754–10766
10. Du F, Brady M (1993) Self-calibration of the intrinsic parameters of cameras for active vision systems. In: Proceedings of IEEE conference on computer vision and pattern recognition. IEEE
11. Stein GP (1995) Accurate internal camera calibration using rotation, with analysis of sources of error. In: Proceedings of IEEE international conference on computer vision. IEEE
12. Basu A (1995) Active calibration of cameras: theory and implementation. IEEE Trans Syst Man Cybern 25(2):256–265
13. Basu A, Ravi K (1997) Active camera calibration using pan, tilt and roll. IEEE Trans Syst Man Cybern Part B (Cybernetics) 27.3:559–566
14. Hartley RI (1997) Self-calibration of stationary cameras. Int J Comput Vision 22:5–23
15. Collins RT, Tsin Y (1999) Calibration of an outdoor active camera system. In: Proceedings. 1999 IEEE computer society conference on computer vision and pattern recognition (Cat. No PR00149), vol 1. IEEE
16. de Agapito L, Hartley RI, Hayman E (1999) Linear self-calibration of a rotating and zooming camera. In: Proceedings. 1999 IEEE computer society conference on computer vision and pattern recognition (Cat. No PR00149), vol 1. IEEE
17. Trajković M (2002) Interactive calibration of a PTZ camera for surveillance applications. Asian Conference Computer Vision
18. Faraji M, Basu A (2019) Simplified active calibration. Image Vis Comput 91:103799
19. Chen (2003) Calibrating pan-tilt cameras in wide-area surveillance networks. In: Proceedings ninth IEEE international conference on computer vision. IEEE
20. Sanchez-Riera J, Salvador J, Casas JR (2009) Indoor PTZ camera calibration with concurrent PT axes. In: 4th international conference on computer vision theory and applications (VISAPP'09), vol 2. INSTICC Press

21. Frahm and Koch (2003) Camera calibration with known rotation. In: Proceedings ninth IEEE international conference on computer vision. IEEE
22. Fetić A, Jurić D, Osmanković D (2012) The procedure of a camera calibration using camera calibration toolbox for MATLAB. In: 2012 Proceedings of the 35th international convention MIPRO. IEEE

Influence on Vertical Landing Stability of Multirotor UAVs with Adaptive Landing Gears

Qiaozhi Yin, Xinlei Ni, Jiayi Song, Xiaohui Wei, and Rujie Sun

Abstract The vertical taking-off and landing stability of multirotor UAVs directly affects the fight safety. To improve the adaptability of a multirotor UAV when landing on complex terrains, a new type of an adaptive landing gear which can be adjusted according to various terrains is designed. The dynamic model of adaptive taking-off and landing UAV considering the friction between the landing gear and the ground is established in Simcenter3D. And the adaptive landing gear shock absorber dynamic model is established in AMESim. The influences of the landing gear frictional force and the buffering force on the landing stability are studied using a co-simulation method of Simcenter3D and AMESim. The results show that a small frictional coefficient of the landing leg leads to the fuselage more vulnerable to lose its landing stability in the rebound phase. Increasing the spring stiffness of the shock absorber and the throttle hole diameter can effectively reduce the maximum inclination angle of the fuselage, so that the vertical landing stability of a UAV can be improved remarkably.

Keywords Adaptive landing gear · Landing shock absorber · Friction · Landing stability · Co-simulation method

Q. Yin · X. Ni · X. Wei (✉)
National Key Laboratory of Helicopter Aeromechanics, State Key Laboratory of Mechanics and Control for Aerospace Structures, College of Aerospace Engineering, Nanjing University of Aeronautics and Astronautics, Nanjing, China
e-mail: wei_xiaohui@nuaa.edu.cn

Q. Yin
e-mail: yinqiaozhi@nuaa.edu.cn

X. Ni
e-mail: nxl_99@nuaa.edu.cn

J. Song
AVIC Chengdu Aircraft Design and Research Institute, Chengdu, China
e-mail: SJY98@nuaa.edu.cn

R. Sun
Department of Materials, Imperial College London, London, UK
e-mail: r.sun@imperial.ac.uk

Z. Liu et al. (eds.), *Advances and Challenges in Advanced Unmanned Aerial Systems*, Springer Aerospace Technology, https://doi.org/10.1007/978-981-99-8045-1_16

CCS Concepts Takeoff and Landing · Launch and Recovery

1 Introduction

Due to the strong crypticity and flexibility, multirotor unmanned aerial vehicles (UAVs) can fulfill a variety of important tasks, including island protection, border patrol, rescue in disaster districts, covert detection and reconnaissance, but it is difficult for UAVs to find a suitable taking-off and landing position in harsh terrain conditions. The development of a new adaptive landing gear can overcome the limitations of landing conditions and the autonomous adjustment function of the landing gear when taking-off and landing can greatly expand the application range of UAVs. However, the load and compression stroke of each shock absorber is different due to the different adjustment height of each landing gear when landing on complex terrains, which will lead to a drastic change in the attitude angle of the fuselage. Therefore, it is necessary to find out the influences of key factors on landing stability of terrain adaptive UAVs, which will provide a guidance for adaptive landing gear design and landing control.

In terms of adaptive landing structure design and landing dynamics, Boix [1] proposed an articulated landing leg and established a ground interaction model based on thrust controller and level controller. Leo [2] designed a new collision-resistant cable-driven robotic landing gear which can adapt to ground with 20º slope angle. Grigoriy [3] established an adaptive landing platform for multirotor aircraft and designed a feed-forward controller to enable the aircraft to complete the adaptive landing on uneven ground. Yuri [4] designed a multi-rotor adaptive landing platform called DroneGear, which is equipped with torque sensors and inertial measurement units to enable it to adapt to rough terrains. Boris [5] proposed a leg-based terrain adaptive landing gear and designed motor/brake drive devices to reduce the impact load during hard landing. Ding [6] established a parametric dynamic analysis model of adaptive legged landing gear, analyzed the landing performance under different landing conditions with the optimized Latin hypercube experimental design method, and optimized the buffer parameters and other factors by using the layered optimization method. Sun [7] designed a new landing gear integrated for buffering actuation and walking with two stage buffering system, which adjusts the landing attitude of landing gear through the hydraulic actuator. Li [8] designed an adaptive landing gear suitable for shipborne helicopters, which adjusts the horizontal attitude of the fuselage through the feedback control method to ensure the helicopter lands smoothly on the inclined ship deck. In summary, most of the existing studies adjust the landing attitude of a rotorcraft by actuating the landing legs actively to achieve the adaptive landing function on complex terrains.

In order to improve the stability of the aircraft during landing, NASA Langley Research Center carried out research on active control technology of landing gear [9], and realized the ideal landing of aircrafts by adjusting the strut displacement by the servo system at a given landing speed, which could effectively reduce the load transmitted from the landing gear to the airframe. Yakia [10] proposed an adaptive landing control strategy for landing gear, which could control the length of landing gear pillars actively according to the feedback of UAV and achieve a safe landing with all landing gear touching the ground at the same time. Lukasz [11] proposed an active/semi-active control strategy to design an adaptive control method, which actively adapted the landing gear to specific landing conditions by adjusting the landing gear through controlled hydraulic pressure. Sancak [12] used magnetorheological buffer landing gear instead of passive buffer landing gear, and designed a Fuzzy logic controller and PID controller to numerically simulate the landing gear landing dynamics, which improved the buffering characteristics during the landing process. Jia [13] combined the optimal control theory and self-correcting control theory to propose an adaptive control strategy and semi-active controller for semi-active landing gear. Wang [14] studied the application of active control technology in buffer-control system of landing gear, and analyzed the influence of the buffer parameters on the system performance. Hu [15] designed an adaptive taking-off and landing system based on the feedback of sensors and verified the robustness and instant response of the control system. Ni [16] designed a four-legged landing gear with passive buffer structure of multirotor UAV and searched the landing stability boundary under different combinations of crosswind speeds and slope angle. It can be seen that the adaptive landing buffer performance and the landing stability had been thoroughly studied, but the influence factors and rules of the mechanical characteristics on the landing stability of an adaptive landing gear need to be conducted further.

In this paper, the influences on landing stability of a multirotor UAV with a new type of adaptive landing gear are studied. In Sect. 2, a dynamic model of an adaptive taking-off and landing UAV is established in Simcenter3D, and the adaptive landing gear shock absorber model is established in AMESim. The influences of the landing gear frictional force and the buffering force on the landing stability are studied by a co-simulation method in Sect. 3. Conclusions are drawn in Sect. 4.

2 Dynamic Modeling of UAV Adaptive Landing

The adaptive landing process of a UAV can be described as: the UAV adjusts the landing leg according to the feedback of distance sensor and then starts to drop; the landing legs touch the ground and the shock absorbers are compressed for the energy dissipation; the fuselage decelerates and stabilizes in the landing buffering process.

2.1 UAV Adaptive Landing Dynamic Modeling

The dynamic model of the UAV adaptive landing is established in the multibody dynamic software Simcenter3D and the complex terrain is set as shown in Fig. 1.

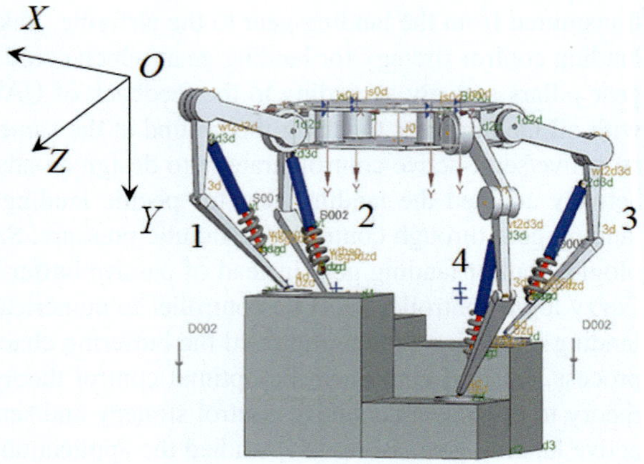

Fig. 1 Adaptive landing gear kinematic model in Simcenter3D

The initial distances of the four adaptive landing gears from the ground to the landing gear foots are 300 mm, 100 mm, 200 mm and 250 mm respectively. Then the four landing gears should be adjusted to touch the ground at the same time in the drop simulation. The kinematic relationship between the landing leg and the ground are considered and the friction model between the rubber on the landing gear foot and the cement pavement are established. The initial state of the UAV after touching the ground is shown in Fig. 1. The landing legs No. 1 and No. 3 are a pair of symmetrical legs, the shock absorber buffering coefficients of those are the same. While the landing legs No. 2 and No. 4 are a pair of symmetrical legs with different buffering coefficients. (The coefficient of leg No. 4 is greater than that of leg No. 2.)

The aerodynamic load of UAV during landing is mainly provided by the rotors, which can be calculated by the throttle instructions given by the operator and the parameters of the blade and the motor used. To simulate the common landing process of multirotor UAVs, we did not apply rotor aerodynamic forces on the fuselage in the dynamic model and let the UAV free fall from a height of 200 mm. (When the UAV approaches the ground, the operator always disconnects the throttle and makes the UAV withstand a hard landing.)

2.2 Shock Absorber Dynamic Modeling

An adaptive landing dynamic model of a shock absorber is established in AMESim. The initial compressing speed of the shock absorber is zero. The gravity and the inertial force of the fuselage are transmitted to the shock absorber through the arm of the adaptive landing gear, where the shock absorber is compressed and the drop energy is dissipated. A spring-damper system between the mass block and the shock absorber is established to simulate the contact between the adaptive landing legs and the piston rod of the shock absorber, as illustrated in Fig. 2. The AMESim simulation model is imported into Simcenter3D for co-simulation. During the co-simulation, the

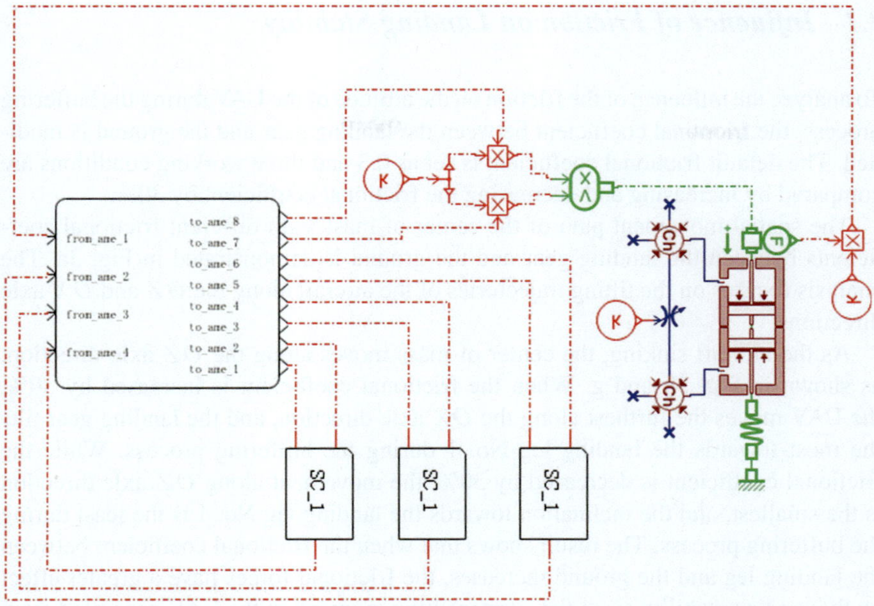

Fig. 2 Shock absorber dynamic model in AMESim

compression stroke and the corresponding speed of the shock absorber are calculated in the landing dynamic model in Simcenter3D, while the real-time feedback of the buffering force is obtained from the shock absorber model in AMESim.

3 Analysis of Influence on Landing Stability

During the vertical landing phase, various structural postures of the four landing legs will result in the differences in the frictional forces and buffering forces after contacting the ground. This will lead to the incline of the UAV fuselage in the OX and OZ axle directions during the fuselage sinking process. Therefore, the landing stability of the adaptive landing gear in current configuration is influenced by the frictional coefficient and the buffering force between the landing gear and the ground. Thus it is necessary to carry out specific analysis of the influences on the inclination of the UAV fuselage and the adaptive landing gears.

3.1 Influence of Friction on Landing Stability

To analyze the influence of the friction on the attitude of the UAV during the buffering process, the frictional coefficient between the landing gear and the ground is modified. The default frictional coefficient is set as 0.3 and three working conditions are compared by increasing and decreasing the frictional coefficient by 30%.

The spatial movement path of the center of mass with different frictional coefficients between the landing gear and the ground is demonstrated in Fig. 3a. The analysis focuses on the tilting trajectories of the aircraft along the OZ and OX axle directions.

As the aircraft sinking, the center of mass moves along the OZ axle direction, as shown in Fig. 3b and g. When the frictional coefficient is increased by 30%, the UAV moves the furthest along the OZ axle direction, and the landing gear tilts the most towards the landing leg No. 1 during the buffering process. While the frictional coefficient is decreased by 30%, the movement along OZ axle direction is the smallest, and the inclination towards the landing leg No. 1 is the least during the buffering process. The result shows that when the frictional coefficient between the landing leg and the ground increases, the frictional forces have a greater effect on the landing stability since the greater tilting moment on the UAV generated from the frictional forces will result in a larger inclination angle on the UAV fuselage.

Towards the end of the buffering process, the center of mass moves along the negative direction of OZ axle and the landing gear tilts towards leg No. 3. In the working condition of -30% frictional coefficient, the center of mass moves significantly further along the negative direction of OZ axle. In the condition of $+30\%$ frictional coefficient, the movement along the negative direction of OZ axle is the smallest. Besides, the larger the frictional coefficient, the smaller the distance of the center of mass movement along the negative direction of OZ axle.

As the sinking distance of the fuselage rises, the sinking velocity drops, resulting in a reduction of the damping force produced by the shock absorber so that a balance is finally obtained between the spring force of the shock absorber and the gravity. However, once the shock absorber reaches its maximum compression stroke, the spring force exceeds the gravity load on landing legs, which can lead to a significant rebound of the fuselage. At this point, landing leg No. 1, which has a greater compression stroke along the OZ axle direction (see Fig. 4a), experiences a larger buffering force. Then the UAV tilts towards leg No. 3, leading to an increase in the buffering force on leg No. 3 and a decrease in the buffering force on leg No. 1.

From the perspective of energy, when the fuselage reaches the maximum sinking distance, the initial kinetic energy and gravitational potential energy of the UAV are converted into dissipated energy between the landing leg and the ground, damping dissipated energy and elastic potential energy of the shock absorbers. Since the compression of the shock absorber in the final state is smaller than that at the moment of the maximum sinking state, the spring force stored in the shock absorbers is greater than the gravity it bears, resulting in a rebound. And a greater frictional coefficient can lead to a smaller the travel of the rebound. If the frictional force is sufficiently

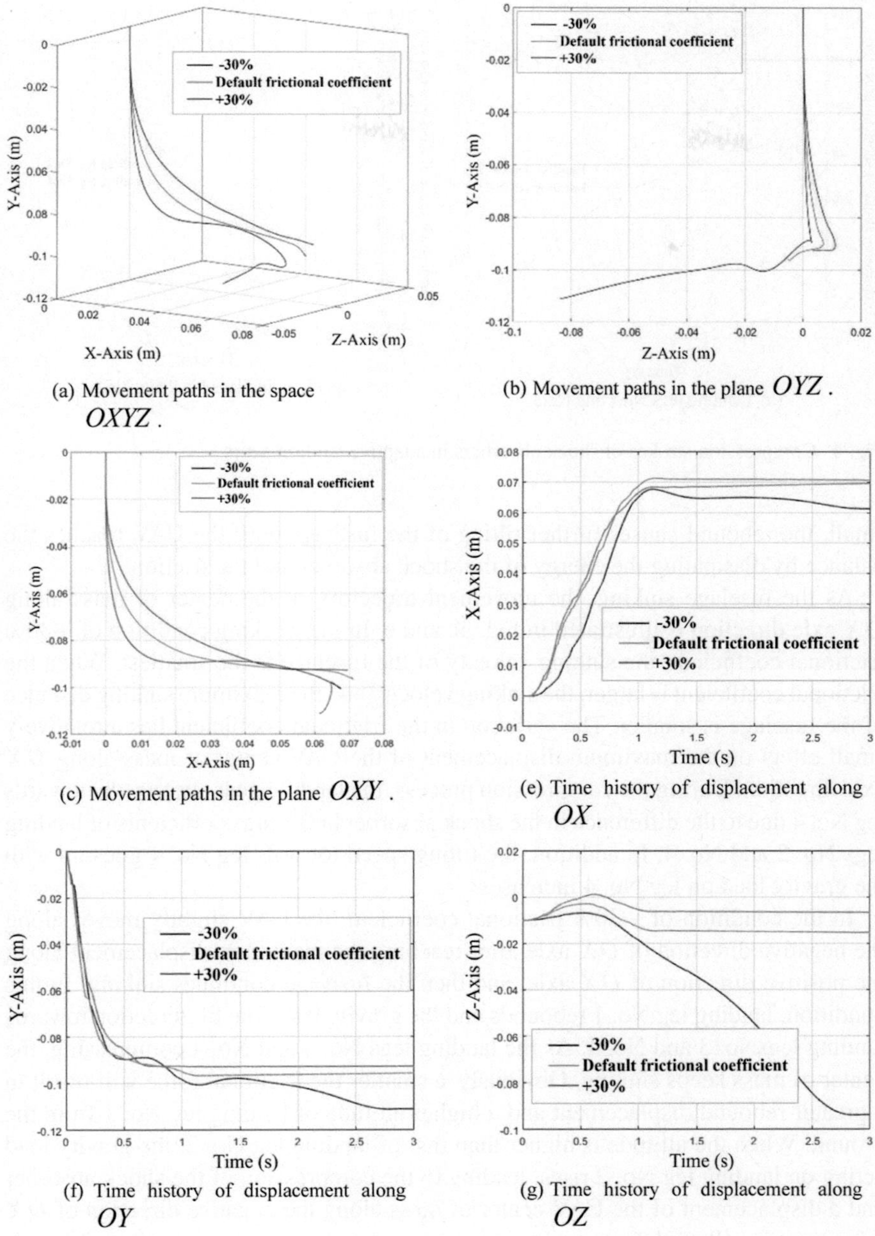

(a) Movement paths in the space $OXYZ$.

(b) Movement paths in the plane OYZ.

(c) Movement paths in the plane OXY.

(e) Time history of displacement along OX.

(f) Time history of displacement along OY

(g) Time history of displacement along OZ

Fig. 3 The movement paths of the UAV center of mass with various frictional coefficients

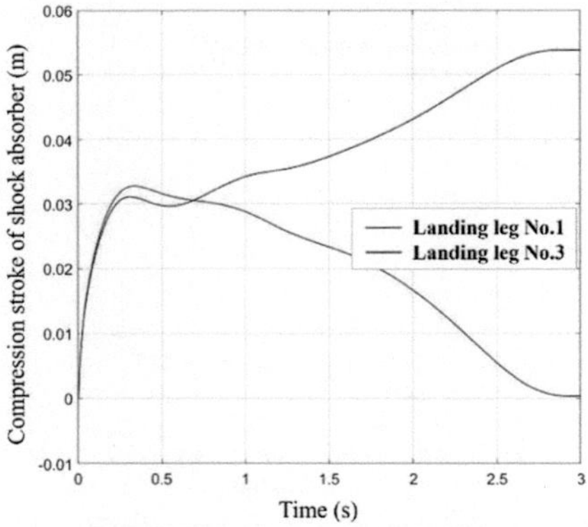

(a) Landing legs No.1 and No.3.

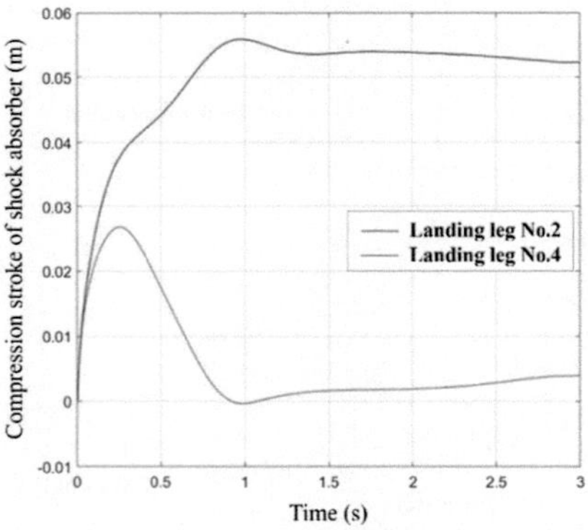

(b) Landing legs No.2 and No.4.

Fig. 4 Compression strokes of shock absorbers in adaptive landing gears

small, the rebound causes further tilting of the fuselage until the UAV reaches the balance by dissipating the energy of the shock absorber and the friction.

As the fuselage sinking, the movement trajectory of the center of mass along OX axle direction is illustrated in Fig. 3c and e. In the working condition of $+30\%$ frictional coefficient, the sinking velocity of the fuselage is the smallest. When the frictional coefficient is larger, the sinking velocity and the maximum sinking distance of the fuselage is smaller. The variation in the frictional coefficient has a relatively small effect on the maximum displacement of the UAV center of mass along OX axle direction. During the compression process, the landing gear always tilts towards leg No. 4 due to the difference in the shock absorber buffering coefficients of landing legs No. 2 and No. 4. In addition, the tilting speed towards leg No. 4 goes up with the gravity load on leg No. 4 increases.

In the condition of -30% frictional coefficient, the UAV slightly moves along the negative direction of OX axle after reaching the maximum displacement along the positive direction of OX axle, and then the fuselage continues sinking. In this condition, landing leg No. 1 rebounds and the gravity leads the tilt direction towards landing legsNo. 3 and No. 4. As the landing legs No. 3 and No. 4 compressing, the center of mass keeps sinking. Obviously, a smaller the frictional force will result in a greater rebound displacement and a higher altitude of landing leg No. 1 from the ground. When the altitude is higher than that of landing leg No. 2, the gravity load acting on landing leg No. 2 rises, leading to the compression of the shock absorber and a displacement of the UAV center of mass along the negative direction of OX axle (see Figs. 4b and 3c).

Various frictional coefficients between the landing gear and the ground have a significant effect on the displacement of the UAV along the OZ axle direction, which indicates that frictional force has a significant influence on the landing stability when the shock absorber buffering coefficients on the symmetrical landing legs (No. 1 and No. 3) are similar. On the other hand, the frictional coefficient has a minor effect on the displacement of the UAV along the OZ axle direction, indicating that the buffering forces have a greater influence on the landing stability than the frictional force when a remarkable difference exists in the buffering coefficients on the symmetrical landing legs (No. 2 and No. 4).

3.2 Influence of Buffering Force on Landing Stability

Spring Stiffness. The default spring stiffness of the shock absorber is set at 7812.5 N/m. Three sets of working conditions are analyzed by increasing and decreasing the spring stiffness by 30%。

As demonstrated in Fig. 5a and b, the UAV center of mass moves along the negative direction of OX axle in almost a same displacement in all the conditions of various spring stiffnesses during the initial sinking stage. As the sinking distance of the fuselage increases and the spring stiffness decreases, the maximum displacement of center of mass along the OX axle direction and the sinking distance of fuselage increase.

The displacement of the UAV center of mass drops quickly after reaching the maximum along the OX axle direction in that the fuselage reaches the maximum sinking displacement and the maximum displacement along the OX axle direction at the same time. From the perspective of energy, it can be seen that the landing gear on the lower side has a greater compression stroke and a larger buffering force during the buffering process. As a result, the landing leg on the lower side (No. 4) rebounds at a higher speed than that of leg No. 2 so that the center of mass moves along the negative direction of OX axle. As the tilt angle of the fuselage during the rebound is larger, the displacement of the center of mass along the negative direction of OX axle is greater. In the condition of +30% spring stiffness, the shock absorber stroke is smaller in the final state. Thus the displacement of the UAV center of mass along the negative direction of OX axle is the smallest, indicating that the landing stability is better.

The final parking states after adaptive landing with various spring stiffnesses are illustrated in Fig. 6.

Throttle Hole Diameter. The damping force of the shock absorber is mainly related to the piston area and the throttle hole diameter. To study the influence of damping force on landing stability, the influence of the throttle hole diameter is studied, whose default value is set as 1.59 mm. Then three sets of working conditions (±30% and the default value) are analyzed and compared.

As demonstrated in Fig. 7, the displacement of the UAV center of mass along the negative direction of OX axle is essentially the same at the initial sinking stage in all the conditions of various throttle hole diameters. As the sinking distance of the fuselage increases and the throttle hole diameter decreases, the buffering force on the landing legs at the beginning of the buffering process rises. As the difference in the buffering forces between landing legs No. 2 and No. 4 goes up, the moment

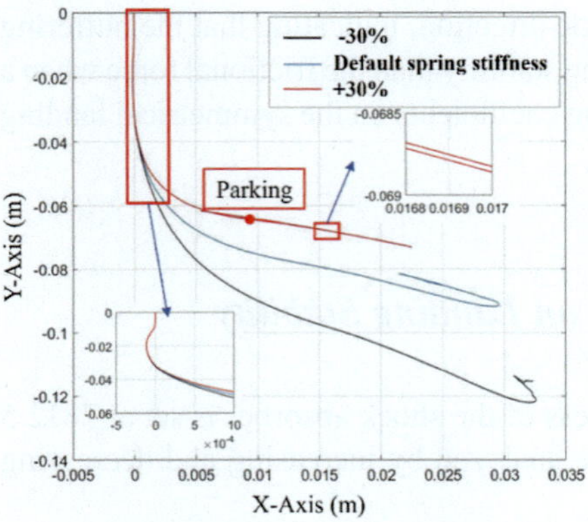

(a) Movement paths in the space $OXYZ$.

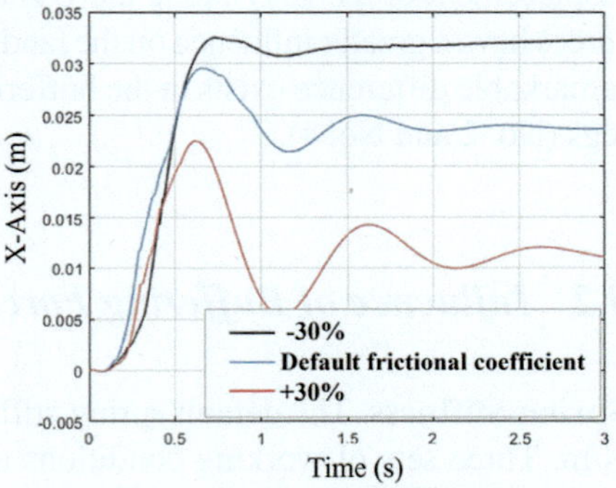

(b) Time history of displacement along OX.

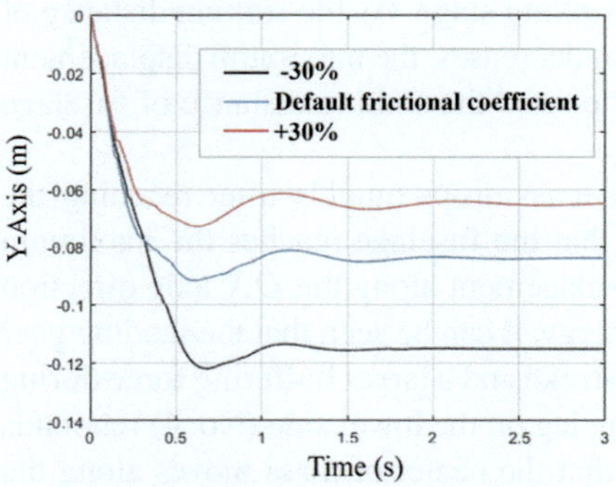

(c) Time history of displacement along OY.

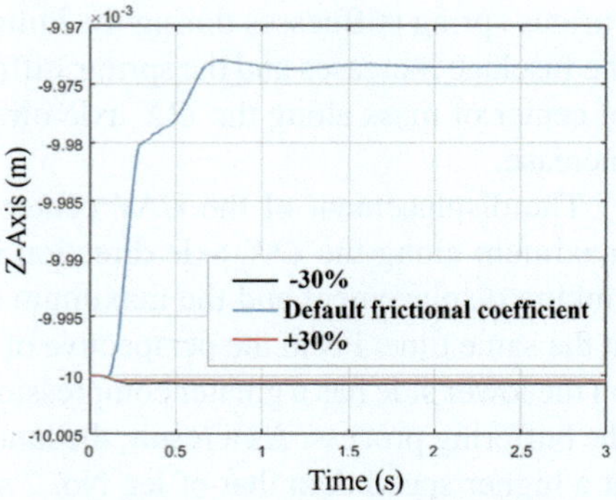

(d) Time history of displacement along OZ.

Fig. 5 The movement paths of UAV center of mass with various spring stiffnesses

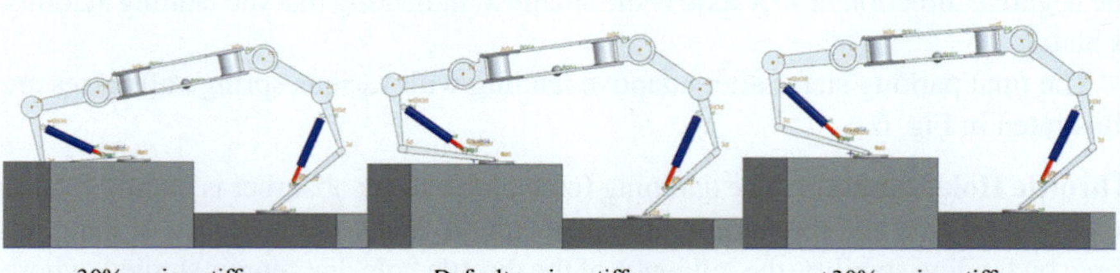

-30% spring stiffness   Default spring stiffness   +30% spring stiffness

Fig. 6 The landing stability states of landing gear with various spring stiffnesses

that tilts the fuselage towards landing leg No. 4 becomes larger and the movement of the UAV center of mass along the OX axle direction becomes faster. Therefore, a smaller throttle hole diameter leads to a larger maximum displacement of fuselage along the OX axle direction.

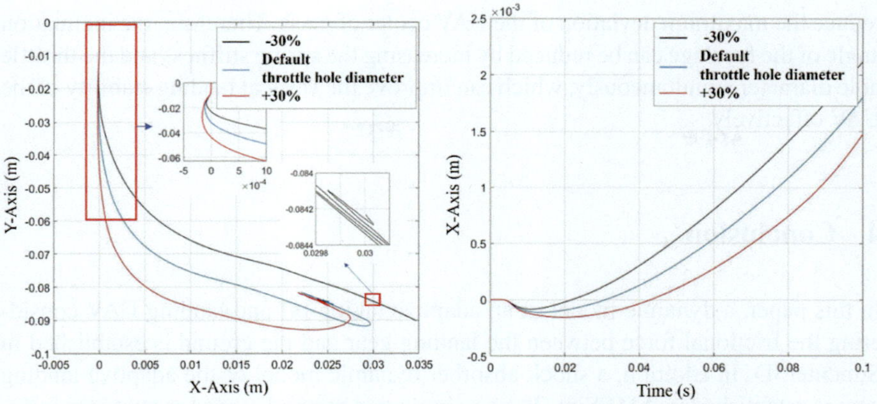

(a) Movement trajectory in OXY plane. (b) Displacement along OX direction.

Fig. 7 The movement trajectories of UAV center of mass with various throttle hole diameters

During the sinking process, landing legs No. 2 and No. 4 compress simultaneously. However, the duration of landing leg No. 2 reaching the maximum compression stroke is always shorter than that of landing leg No. 4 due to the continuous movement of the UAV center of mass towards landing leg No. 4. Landing leg No. 2 starts to rebound after reaching the maximum compression stroke, causing the center of mass to rise. However, before landing leg No. 4 reaching the maximum compression stroke, the UAV center of mass continues moving along the OX axle direction. A larger throttle hole diameter leads to a smaller damping force of the shock absorber and a greater rebound displacement of landing leg No. 2. When the damping force is relatively large, the point of the fuselage maximum sinking distance and the point of the maximum displacement of the UAV center of mass along the OX axle direction overlap.

The final parking states of the UAV adaptive landing gear with various throttle hole diameters are shown in Fig. 8. It can be seen that the damping force of the landing gear has a relatively minor influence on the tilt angle of the landing gear.

In conclusion, increasing the spring stiffness of the shock absorber can reduce the maximum deviation of the UAV center of mass and result in a smaller tilt angle of the landing gear in the final parking state. Increasing the throttle hole diameter can also

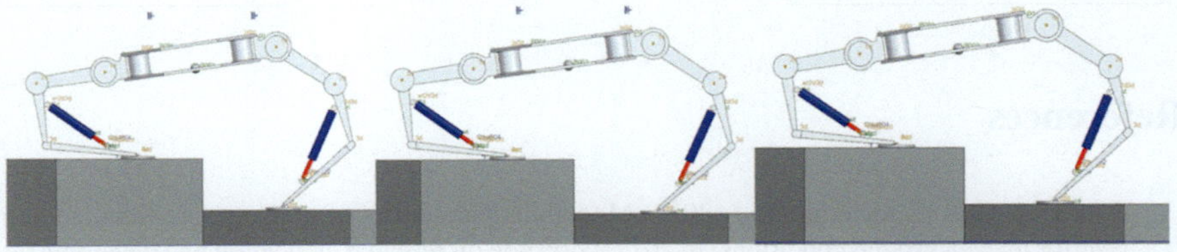

-30% throttle hole diameter Default throttle hole diameter +30% throttle hole diameter

Fig. 8 The landing stability states of landing gear with various throttle hole diameters reduce the maximum deviation of the UAV center of mass. Therefore, the inclination angle of the fuselage can be reduced by increasing the spring stiffness and the throttle hole diameter simultaneously, which can improve the vertical landing stability of the UAV effectively.

4 Conclusion

In this paper, a dynamic model of an adaptive taking-off and landing UAV considering the frictional force between the landing gear and the ground is established in Simenter3D. In addition, a shock absorber dynamic model of the adaptive landing gear is established in AMESim. The co-simulation method is used to study the influences of frictional force and buffering force on the adaptive vertical landing stability. The following conclusions are drawn below:

(1) During the initial phase of buffering, the frictional force at the landing gear foot has an influence on the attitude angles of the landing gears. When the shock absorber buffering coefficients of symmetrical legs are similar, the frictional force has a significant influence on the landing stability. A smaller frictional coefficient can lead to the fuselage more vulnerable to lose its landing stability in the rebound phase.
(2) Compared with the throttle hole diameter of shock absorber, the spring stiffness has a greater influence on the maximum inclination angle of the fuselage and the tilt angle of the landing gear in the final parking state. Increasing the spring stiffness and the throttle hole diameter can effectively reduce the maximum deviation of the UAV, indicating that the adaptive vertical landing stability can be improved remarkably.

Acknowledgements This study was supported by the Spring Sunshine Program initiated by the Ministry of Education of China (No. HZKY20220126), the National Natural Science Foundation of China (No. 52375102), the Fundamental Research Funds for the Central Universities (No. NS2023005), the China Postdoctoral Science Foundation Funded Project (No. 2021M691565), the Aeronautical Science Foundation of China (No. 202000410520002), the Fund of Prospective Layout of Scientific Research for NUAA (Nanjing University of Aeronautics and Astronautics), the National Defense Outstanding Youth Science Foundation (No. 2018-JCJQ-ZQ-053), and the Priority Academic Program Development of Jiangsu Higher Education Institutions.

References

1. Boix DM, Goh K, McWhinnie J (2017) Modelling and control of helicopter robotic landing gear for uneven ground conditions. In 2017 workshop on research, education and development of unmanned aerial systems (RED-UAS). IEEE, pp 60–65
2. Di Leo CV, León B, Wachlin J, Kurien M, Krishnan A, Krishnan A, Costello M (2020) Design of a crashworthy cable-driven four-bar link robotic landing gear system. J Aircr 57(2):224–244

3. Yashin G, Egorov A, Darush Z, Zherdev N, Tsetserukou D (2020) LocoGear: locomotion analysis of robotic landing gear for multicopters. IEEE J Miniat Air Space Syst. 1(2):138–147
4. Sarkisov YS, Yashin GA, Tsykunov EV, Tsetserukou D (2018) Dronegear: a novel robotic landing gear with embedded optical torque sensors for safe multicopter landing on an uneven surface. IEEE Robot Autom Lett 3(3):1912–1917
5. Stolz B, Brödermann T, Castiello E, Englberger G, Erne D, Gasser J, Hutter M (2018) An adaptive landing gear for extending the operational range of helicopters. In 2018 IEEE/RSJ international conference on intelligent robots and systems (IROS). IEEE, pp 1757–1763
6. Ding Z, Wu H, Wang C, Ding J (2019) Hierarchical optimization of landing performance for lander with adaptive landing gear. Chin J Mech Eng 32:1–12
7. Sun H, Yin Q, Wei X, Chen S, Nie H (2021) Research of single leg drop performance of new adaptive landing gear. J Beijing Univ Aeronaut Astronaut 49(4):990–998 (in chinese)
8. Li G (2019) Research on key technologies of adaptive landing gear of shipborne helicopters. Harbin Institute of Technology. (in chinese)
9. Howell WE, Mcgehee JR, Daugherty RH, Vogler WA (1990) F-106B airplane active control landing gear drop test performance (No. NASA-TM-102741)
10. Ikura M, Miyashita L, Ishikawa M (2020) Real-time landing gear control system based on adaptive 3D sensing for safe landing of UAV. In 2020 IEEE/SICE international symposium on system integration (SII). IEEE, pp 759–764
11. Mikułowski G, Jankowski Ł (2009) Adaptive landing gear: optimum control strategy and potential for improvement. Shock Vib 16(2):175–194
12. Sancak ÖF, Güçlü R (2018) Semi active control of fighter jet landing gear comparing with PID and fuzzy logic controller equipped magnetorheological damper. Sigma J Eng Nat Sci 36(4):1215–1234
13. Jia Y, Wang J (2006) Adaptive control for semi-active landing gear. J Vib Shock 25(2):55–58 (in chinese)
14. Wang T (2003) Analysis and research on active control technology of shock absorber of landing gear system. Northwestern Polytechnical University. (in chinese)
15. Hu D, Li Y, Xu M, Tang Z (2018) Research on UAV adaptive landing gear control system. In Journal of physics: conference series, vol 1061, no 1. IOP Publishing, p 012019
16. Ni X, Yin Q, Wei X, Zhong P, Nie H (2022) Research on landing stability of four-legged adaptive landing gear for multirotor UAVs. Aerospace 9(12):776

Six-Wheel Trolley Type Landing Gear Ground Slip-Run Load Analysis Considering Fuselage Flexibility

Xiazheng Shi, Hong Nie, Ming Zhang, Shuang Ruan, and Zimin Yang

Abstract This paper investigates the effect of fuselage flexibility on the landing gear ground skidding load in the landing gear load design of a large unmanned aerial vehicle (UAV). Since the traditional load design method often assumes that the aircraft fuselage is a rigid body, the influence of fuselage flexibility is neglected. In this paper, on the basis of applying the basic principles of dynamics, a six-wheeled trolley landing gear and the whole aircraft dynamics model are established; the fuselage is flexibly processed, and the influence of the fuselage flexibility on the landing gear ground loads is analysed. The results show that after considering the fuselage flexibility, the acceleration at the centre of gravity of the fuselage decreased by 35%, and the vertical load of the main landing gear decreased by 7.4%. This indicates that the flexible fuselage deformation absorbs energy and helps to attenuate the fuselage vibration. Therefore, in the design process of UAV landing gear, the maximum value of work absorbed by the buffer system composed of flexible fuselage and landing gear system should be the design goal.

Keywords Fuselage flexibility · Aircraft taxiing dynamics · Landing gear · Ground load analysis

X. Shi · H. Nie (✉) · M. Zhang · S. Ruan · Z. Yang
College of Aerospace Engineering, Nanjing University of Aeronautics and Astronautics, Nanjing, China
e-mail: hnie@nuaa.edu.cn

X. Shi
e-mail: sxzemail@nuaa.edu.cn

M. Zhang
e-mail: zhm6196@nuaa.edu.cn

S. Ruan
e-mail: ruanshuang@nuaa.edu.cn

Z. Liu et al. (eds.), *Advances and Challenges in Advanced Unmanned Aerial Systems*, Springer Aerospace Technology, https://doi.org/10.1007/978-981-99-8045-1_17

1 Introduction

In the design and development process of large aircraft, the design of landing gear systems is one of the most critical research elements and research challenges. The main function of the landing gear is to support the aircraft body during take-off and landing and taxiing phases, to mitigate impact loads and to complete aircraft turning maneuvers [1]. With the development of the times, the size of the aircraft is gradually developing from the micro size in the past to the super size. On the one hand, the increasing ground load during landing, and on the other hand, the increasing diameter of the wheels make it very difficult to retract the landing gear. The trolley type landing gear is a good solution to both problems, so the trolley type landing gear is more often used in the landing gear of large aircraft [2].

Yin [3] established a nonlinear dynamic landing gear model considering the influence of the coupling of the shock absorber stroke variation and the landing gear longitudinal motion with an anti-skid PID braking control system. Cao [4] carried out the drop tests of a twin tandem landing gear with different filling parameters in two different landing attitudes. Zhang [5] obtaining its first six natural frequencies and mode shapes of six-wheel bogie landing gear. Liao [6] investigates vibration on six-wheel main landing gear induced by anti-skid brake based on rigid-flexible dynamic model and co-simulation. Wang [7] studied the force and transmission processes of the tires, buffers, braking system, front wheel turning and other components that make up a single landing gear for a multi-wheel, multi-pillar landing gear system of a certain type of aircraft, established a six-degree-of-freedom model of the aircraft, and carried out experimental verification on a dynamic base simulator of a certain type of aircraft. Fan [8] established a set of three-dimensional mathematical models of the multi-pillar trolley type landing gear and carried out the simulation of the whole aircraft dynamics in Adams. Kuang [9] established a trolley-type landing gear dynamics balance equation by drawing on previous experience, and studied in depth the problems related to tire load distribution of multi-wheel landing gear. Xuan [10] controlled the front and rear wheel load ratios by appropriately adjusting the dimensional arrangement of the landing gear, and designed the corresponding load distribution algorithm by taking the unevenness of the airport as an influencing factor. Zhang [11] established a virtual prototype of a multi-wheel and multi-pillar landing gear, which can simulate the process of asymmetric landing and asymmetric anti-skid braking of trolley-type landing gear.

In the above analysis of trolley-type landing gear dynamics, a two-mass simplified model is usually used to study the dynamic performance of the landing gear, with the fuselage replaced by a rigid mass point, failing to consider the effect of fuselage flexibility on aircraft ground loads. In this paper, the effect of flexible fuselage nature on the aircraft glide ground load is mainly considered.

2 Six-Wheel Trolley Type Landing Gear Dynamics Modelling

2.1 Trolley Type Landing Gear Movement Geometry Relationship

A coordinate system is established by selecting the point of connection between the fuselage and the landing gear as the origin, with the Z-axis vertical downwards as positive, the X-axis vertical Z-axis and the heading as positive. A local coordinate system is established at the point of connection between the frame and the piston rod. The differential equations of motion, the geometric equations of motion and the force relations of the landing gear are established separately. The force analysis diagram is shown in Fig. 1. The motion of the entire landing gear cushioning system should include the motion of the body and the motion of the inelastic support mass of the landing gear. The motion of the body includes vertical rigid body translation and horizontal forward direction uniform translation, which is obtained from Newton's second law as follows:

$$/M\ddot{Z} = (1 - a)Mg - 2F_z \tag{1}$$

In the above equation, M is the mass of the fuselage, a is the lift-to-weight ratio, F_z is the ratio of lift to gravity, $\ddot{Z}$ is the vertical acceleration of the fuselage.

The landing gear motion consists of horizontal and vertical translation of the masses of the moving parts, front, middle and rear wheel rotation and frame rotation, with the equations of motion shown below:

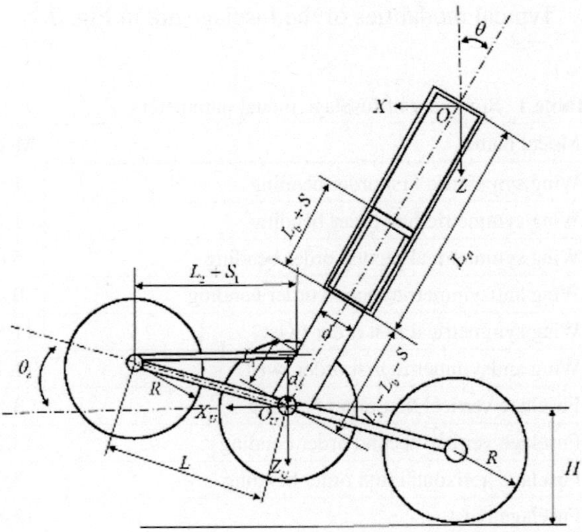

Fig. 1 Diagram of landing gear geometry and forces on various parts

$$
\begin{aligned}
& M_U \ddot{X}_U = F_X - 2D_1 - 2D_2 - 2D_3 \\
& M_U \ddot{X}_U = F_Z - 2V_1 - 2V_2 - 2V_3 \\
& I_C \ddot{\theta}_C = -2V_1 L \cos\theta_c - 2D_1 L \sin\theta_c + 2V_3 \sin\theta_c \\
& +2V_3 L \cos\theta_c + 2D_3 L \sin\theta_c + F_e d_s \\
& I_1 \dot{\omega} = D_1 (R - \delta_1) \\
& I_2 \dot{\omega} = D_2 (R - \delta_2) \\
& I_3 \dot{\omega} = D_3 (R - \delta_3)
\end{aligned}
\tag{2}
$$

In the above equation, M_U is the inelastic support mass of the landing gear, V_1, V_2, V_3 is the vertical load of the front, middle and rear tires respectively, D_1, D_2, D_3 is the heading load of the front, middle and rear tires respectively, I_C is the rotational inertia of the frame, I_1, I_2, I_3 is the rotational inertia of the front, middle and rear tires, $\dot{\omega}$ is the corresponding rotational angular acceleration and δ is the corresponding tire compression.

2.2 Full Aircraft Modelling with Flexible Fuselage

The fuselage flexibility begins with the creation of a beam unit fuselage with wing planes based on the shape of the aircraft, with the fuselage section connected to the landing gear and the tail section connected to the main landing gear, and the aircraft is segmented by the aircraft mass share. The fuselage strength is adjusted by adjusting the cross-sectional shape of the beam unit. By changing the distribution of the fuselage mass points, the modalities can be changed and the overall modalities are calculated after adjustment as shown in Table 1.

Typical modalities of the fuselage are in Fig. 2.

Table 1 Summary of fuselage modal parameters

Modal names	Modal frequency (HZ)
Wing symmetric first order bending	1.5
Wing symmetric first order bending	1.7
Wing symmetrical second order bending	5.6
Wing antisymmetric second order bending	9.4
Wing symmetrical first order twist	11.5
Wing antisymmetric first order twist	11.7
Fuselage vertical first order bending	3.5
Fuselage vertical second order bending	13.2
Fuselage horizontal first order bending	3.9
Fuselage twist	13.1

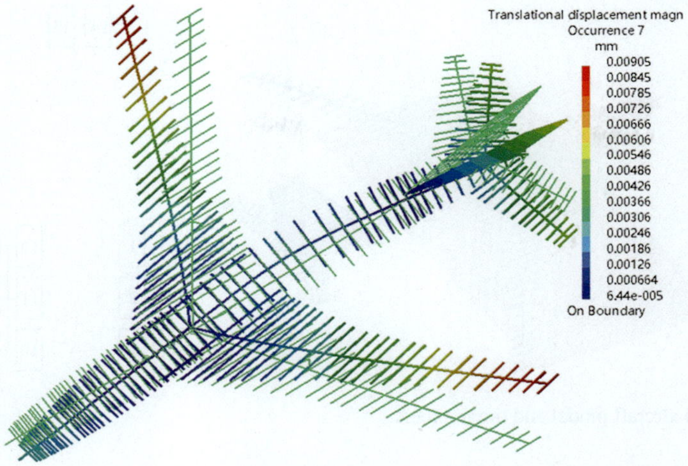

Fig. 2 Symmetric first order bending mode of wing

In this paper, the overall aircraft parameters and the main simulation parameters are shown in Table 2.

Full aircraft models with flexible fuselage with tire numbers as follows in Fig. 3.

Table 2 List of parameters

Simulation parameters	Values
Longitudinal distance from centre of gravity to front landing gear	2610 mm
Longitudinal distance from centre of gravity to main landing gear	13390 mm
Vertical height from centre of gravity to ground	3645 mm
Distance between left and right main landing gear	11600 mm
Tyre radius	602 mm
Tyre radial stiffness	5.5e + 007 N/m
Tyre vertical stiffness	3.35e + 006 N/m
Tyre damping values	3500 kg/s
Sliding mass	160000 kg
Sliding speed	100 km/h
Gravitational acceleration	100 km/h

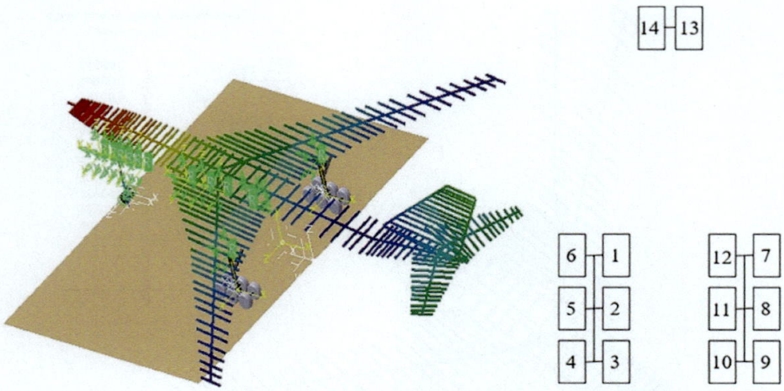

Fig. 3 Full aircraft model and tire numbers

3 Full Aircraft Slip Run Results Analysis

As shown Fig. 4, the acceleration at the aircraft's center of gravity oscillates between 0.3 and 0.3 m/s^2 with a rigid body and between 0.2 and 0.2 m/s^2 with a flexible body during the full glide. It can be seen that the acceleration response at the aircraft's center of gravity is significantly reduced when considering the elasticity of the fuselage.

During the full taxi run, the displacement value at the center of gravity oscillates around 2430 mm under the rigid body fuselage and 2320 mm under the flexible body fuselage. It can be seen in Figs. 5 and 6, after considering the elasticity of the fuselage, the elastic deformation of the fuselage causes the center of gravity of the aircraft to decrease. During the full aircraft taxi run, the left main take-off vertical load is less with the flexible body fuselage than with the rigid body fuselage, due to the deformation generated by considering the flexible body fuselage, which absorbs part of the energy.

As shown in the Figs. 7 and 8, Under the rigid body at t = 30 s, the vertical load of tire No. 1 is 110496N, the vertical load of tire No. 2 is 120860N and the vertical

Fig. 4 Centroid acceleration of aircraft

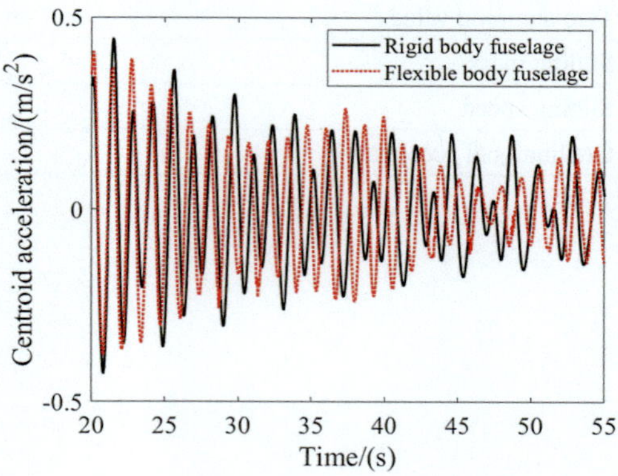

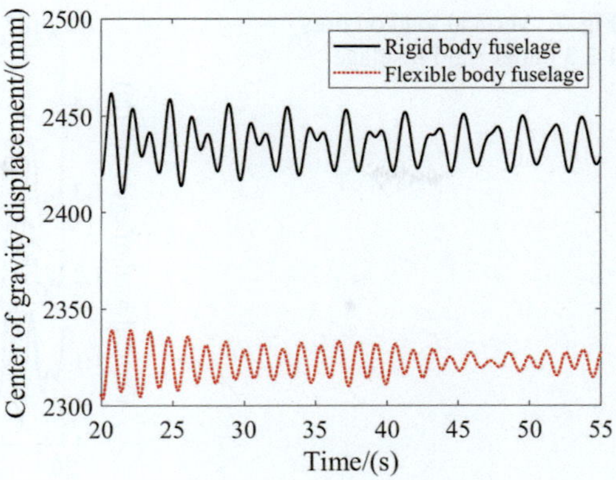

Fig. 5 Displacement at the center of gravity of the aircraft

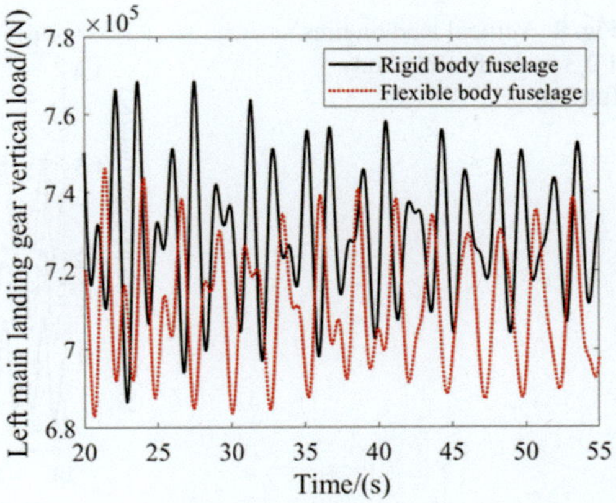

Fig. 6 Left main landing gear vertical load

load of tire No. 3 is 131225N. During the full skid run, the frame is kept in a raised position due to the frame stabilizing buffer, the vertical load of tire No. 3 is the largest and the vertical load of tire No.1 is the smallest. The sliding load under the flexible body, at t = 30 s, is 102,736 N for tire No. 1, 111,389 N for tire No. 2 and 12,042 N for tire No. 3. Compared with the sliding load under the rigid body, the load on each tire is significantly reduced.

The key data results under rigid and flexible fuselage at t = 30 s are included and compared in the Table 3, which shows that the fuselage is flexible compared to the flexible fuselage; the magnitude of the acceleration at the fuselage center of gravity, the main take-off vertical load and the tire vertical load values are reduced.

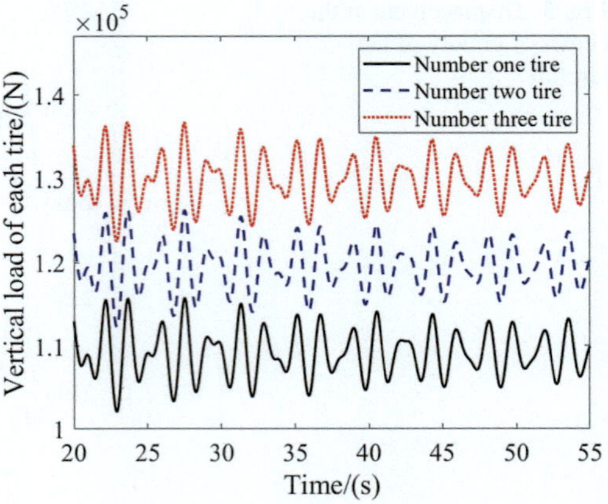

Fig. 7 Vertical load on tires 1 2 3 under rigid fuselage

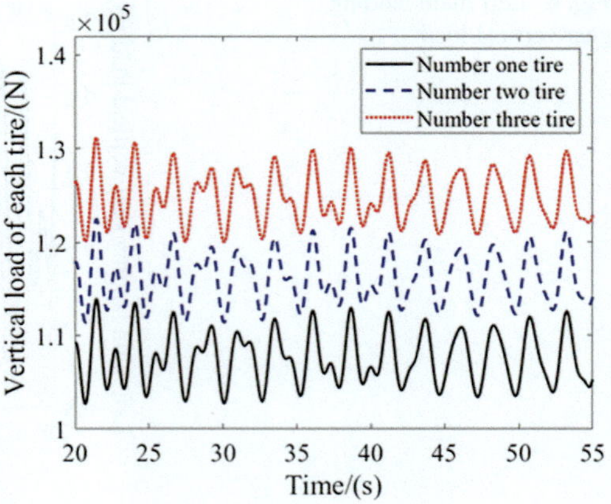

Fig. 8 Vertical load on tires 1 2 3 under flexible body fuselage

Table 3 Analysis of key data results

Key parameters	Rigid fuselage	Flexible body	Rate of change (%)
Acceleration at the centre of gravity of an aircraft	0.253 m/s^2	0.164 m/s^2	−35.2
Displacement at the aircraft's centre of gravity	2430 mm	2320 mm	−4.5
Left main landing gear vertical load	738327N	683741N	−7.4
No. 1 tyre vertical load	110496N	102736N	−7.0
No. 2 tyre vertical load	120860N	111389N	−7.8
No. 3 tyre vertical load	131225N	120042N	−8.5

4 Conclusion

Comparing the slip run loads under a rigid fuselage with those under a flexible fuselage, the acceleration at the centre of gravity is reduced by 35% after considering the fuselage flexibility, and the position of the centre of gravity is reduced by 4.5%, indicating that the flexible fuselage helps to attenuate fuselage vibration; the left main landing gear vertical load is reduced by 7.4%, and for the tyre vertical load, the tyre vertical load is reduced by 7% for tyre 1, 7.8% for tyre 2 and This indicates that the deformation of the elastic fuselage absorbs some of the energy to reduce the vertical load on the landing gear and tyres.

The elastic body can absorb and dissipate energy through vibration, and the energy stored in the elastic body should be considered to reduce the pressure on the landing gear buffer system. In this way, the traditional design target of only the energy absorbed by the buffer appears to be conservative, resulting in a large and bulky landing gear. Therefore, the design objective should be the maximum amount of work absorbed by the buffer system consisting of the elastic fuselage and landing gear system. With this design concept, the load on the tyres and buffers can be reduced and the life of the landing gear extended. Therefore, according to this method, the load design of large UAV landing gear can be guided.

References

1. Nie H (2008) Key technologies for landing gear of large civil aircrafts. Nanjing University of Aeronautics and Astronautics. J Nanjing Univ Aeronaut Astronaut (04):427–432
2. Cao X (2015) Research on landing impact characteristic of multi-wheel bogie landing gear's truck. Int J Aerosp Syst Eng 2(2):83–86
3. Yin Q (2019) Investigation of gear walk suppression while maintaining braking performance in a main landing gear. Aerosp Sci Technol (91):122–135
4. Fang W (2022) Landing performance study for four wheels twin tandem landing gear based on drop test. Aerospace 9:334
5. Zhang Z (2020) Modal analysis of bogie landing gear based on hypermesh and nastron. ICMCCE 2020. IEEE, Harbin, China, pp 316–320
6. Liao D (2021) Vibration analysis on six-wheel landing gear induced by anti-skid brake. ICEEMT 2021. IEEE, Qingdao, China, pp 232–236
7. Wang B (2014) Simulation research on multiple wheels and multiple strut aircraft landing gear's kinetic characteristic. Adv Aeronaut Sci Eng 5(4):503–507
8. Fan H (2006) Research on ground loads of aircraft with multi-wheels and multi-landing gear. Nanjing University of Aeronautics and Astronautics
9. Kuang A (2009) Method to calculate load distribution among wheels for the six-wheeled landing gear. J Mech Strength. 31(2):256–259
10. Xuan J (2009) Impact of roughness of airport on tricycle-type landing gear wheel load distribution. Acta Aeronaut et Astronaut Sinica (1):52–55
11. Zhang M (2009) Research on some key technologies of aircraft ground dynamics. Nanjing University of Aeronautics and Astronautics

Research on Landing and Rollout Dynamic Characteristics of Skid-Equipped Aircraft

Taotao Liang, Yin Yin, Qiaozhi Yin, Xiaohui Wei, and Hong Nie

Abstract Skid landing gear is a specialized landing device that satisfies design requirements such as compact space and light weight. However, current research on the dynamic characteristics of skid-equipped aircraft during the landing phase is limited, making the feasibility and optimization direction of skid landing gear unclear. To address this gap, a general aircraft landing and rollout dynamics model based on modular concept is first constructed, considering the coupled effects of aerodynamic loads, buffer loads, and ground loads. On this basis, the motion characteristic differences between skid-equipped and wheeled aircraft during the landing and rollout phase are compared in detail. Simulation results show that the lack of cushioning capability provided by rubber tires results in a 10.57% increase in overload of skid-equipped aircraft and a 17.1% decrease in the efficiency of the main landing gear buffer. Due to the differences in the characteristics of lateral forces on the nose and main landing gear, skid-equipped aircraft during the rollout phase is a statically unstable system. The intervention of asymmetric disturbances would pose a serious threat to rollout safety. The relevant conclusions provide a clear direction for optimizing the design of skid landing gear and help enhance its feasibility for the application in special aircrafts.

Keywords Landing gear · Aircraft ground motion · Dynamic characteristic · Course instability

T. Liang
College of General Aviation and Flight, Nanjing University of Aeronautics and Astronautics, Liyang 213300, China

Y. Yin · Q. Yin · X. Wei (✉) · H. Nie
College of Aerospace Engineering, Nanjing University of Aeronautics and Astronautics, Nanjing 210016, China
e-mail: Wei_xiaohui@nuaa.edu.cn

Z. Liu et al. (eds.), *Advances and Challenges in Advanced Unmanned Aerial Systems*, Springer Aerospace Technology, https://doi.org/10.1007/978-981-99-8045-1_18

1 Introduction

Skid landing gear is a kind of important aircraft landing device. It has compact structure, light weight and excellent pavement adaptability, which can effectively make up for the shortcomings of wheeled landing gear under the special design requirements of aircraft [1]. With the rapid development of hypersonic vehicles in recent years [2], the landing gear design problems are becoming more and more prominent. The flat aerodynamic shape and ultra-high landing velocity seriously restrict the design of the wheel landing gear. The current solution is usually to directly adjust the aerodynamic shape to provide sufficient storage space for the wheeled landing gear, but it will seriously affect the overall performance of the aircraft in the whole speed region. In contrast, the skid landing gear of lightweight and compact is more suitable for the design requirements of hypersonic vehicles. However, due to the lack of systematic research on the dynamic characteristics of the skid landing gear, the feasibility and shortcomings of its application are always vague. Therefore, it is of great engineering value to carry out in-depth research on the motion characteristics of the skid-equipped aircraft in the landing and rollout stage.

An accurate and effective dynamic model is a prerequisite for studying the characteristics of aircraft landing and rollout motion. Considering the influence of the error caused by the kinematic assumption in the modeling process, Gu [3] established a dynamic model of aircraft taxiing and turning including the dynamic response of the front wheel based on the actual control signal. Jiao [4] added the landing gear walking equivalent model to the traditional aircraft taxiing model and explored the ground motion response of the aircraft under the independent action of nose wheel steering and the braking system in detail. In addition, in the research field of aircraft landing characteristics, Yuan [5] established the load transfer process of the landing gear buffer by using the mass-spring-damper system, and evaluated the dynamic response of the aircraft during the landing stage. Tang [6] established a more perfect aircraft landing dynamics model considering the landing gear structure size and assembly form. In addition to the research on the motion characteristics of aircraft by deriving the mechanism model, the construction of virtual prototype based on multi-body dynamics software is also an effective research method. Liu [7] constructed an effective aircraft flexible body model based on the experimental data and carried out the whole aircraft drop simulation. They pointed out that the flexibility of the fuselage reduced the peak load of the nose landing gear and the main landing gear by 4.11% and 2.46%, respectively.

Combined with the above literature, the current research on aircraft dynamics modeling and motion characteristics mainly focuses on the wheeled aircraft, and there is a lack of systematic research on the special configuration of skid-equipped aircraft. To clarify the application feasibility and shortcomings of the skid landing gear, this paper constructs a general nonlinear dynamic model of the aircraft based on the modular concept to effectively characterize the dynamic response of the aircraft during the landing and rollout phase. On this basis, the landing and rollout motion characteristics of the skid-equipped aircraft are systematically analyzed, and the application defects and potential optimization directions of the skid-equipped aircraft are clarified.

2 Architecture of the Aircraft Dynamics Model

In order to effectively analyze the trajectory, attitude change and force condition of the aircraft during the landing and rollout phase, five coordinate systems shown in Fig. 1 are introduced, and the transformation relationship between them is described.

The aircraft is continuously affected by a variety of nonlinear external forces and external moments during the landing and rollout phase. The force analysis is shown in Fig. 2.

It mainly includes aerodynamic force, buffer force, contact force between tire (or skid) and the ground. Therefore, it is a prerequisite for accurate modeling to carry out the force analysis of the aircraft in the coordinate system on which the external forces depend. The specific derivation of each load will be detailed in the subsequent modeling process. The aircraft in the landing and rollout phase is a complex nonlinear dynamic system. If all factors are considered at the same time, the calculation equation will be extremely complicated. In order to ensure the accuracy of the model and improve the efficiency of the solution, this paper conducts modeling research based on several reasonable assumptions [8].

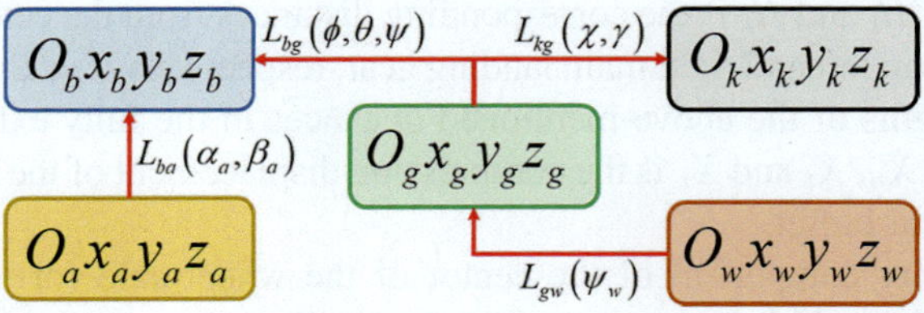

Fig. 1 Transformation relation of coordinate systems

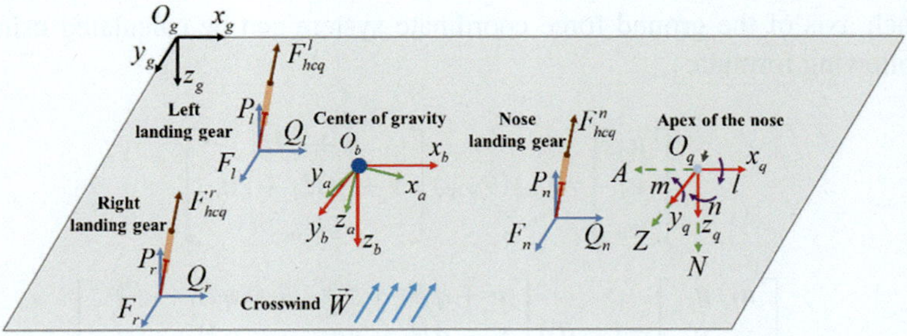

Fig. 2 Force analysis of aircraft during landing and rollout phase

3 General Dynamics Model of the Aircraft

The detailed derivation of the aerodynamic force model and buffer force model can be found in relevant references, so they are not repeated here. The full derivation of the ground force is listed as follows [9].

3.1 Ground Force Model

There is a significant difference in the ground mechanical characteristics between skid landing gear and wheeled landing gear, including vertical, longitudinal and lateral force. Therefore, it is necessary to carry out targeted construction of ground loads. The detailed derivation of the required parameters for solving ground force of the landing gear is as follows:

(1) The distance from the center of the wheel axle (or the installation position of the skid) along the axis of the buffer to the center of gravity H_i:

$$\begin{bmatrix} H_n \\ H_l \\ H_r \end{bmatrix} = \begin{bmatrix} L_n - X_n \\ L_m - X_l \\ L_m - X_r \end{bmatrix} \tag{1}$$

where H_n, H_l and H_r is the corresponding distances from the center of the nose landing gear, left and right main landing gear, respectively. L_n, L_l and L_r are the initial lengths of the above-mentioned distances in the fully extended state of the buffer. X_n, X_l and X_r is the compression displacement of the corresponding landing gear buffer.

(2) The velocity component of the center of the wheel axle (or the installation position of the skid) in the ground force coordinate system u_i, v_i, w_i:

Based on the distance obtained above and the real-time motion status of the aircraft, the velocity components of the landing gear touchdown point along each axis of the ground force coordinate system can be calculated using the following formula:

$$\begin{bmatrix} u_n \\ v_n \\ w_n \end{bmatrix} = L_{wb}(\theta_{NW}) \begin{bmatrix} u + qH_n \\ v - pH_n + ra_n \\ V_{hn} \end{bmatrix} \tag{2}$$

$$\begin{bmatrix} u_l & u_r \\ v_l & v_r \\ w_l & w_r \end{bmatrix} = L_{wb}(0) \begin{bmatrix} u + qH_l + \frac{rb_m}{2} & u + qH_r - \frac{rb_m}{2} \\ v - pH_l - ra_m & v - pH_r - ra_m \\ V_{hl} & V_{hr} \end{bmatrix} \tag{3}$$

where θ_{NW} is the deflection control angle of the nose landing gear. u, v are the components of the ground velocity vector in the aircraft coordinate system $O_b x_b y_b z_b$. p, q and r represent roll, pitch, and yaw rate in sequence. a_n and a_m are the longitudinal distance between the nose landing gear and the main landing gear to the center of mass of the aircraft, respectively.b_m is the lateral distance between the main landing gears.

(3) Sideslip angle of the landing gear during the rollout phase β_i:

The sideslip angle refers to the angle between the direction of landing gear motion speed and the longitudinal symmetry plane of the landing gear during aircraft motion. When the motion direction of the landing gear is on the right side of the plane of symmetry, the sideslip angle is positive. The calculation formula is as follows:

$$\beta_i = \begin{cases} \arctan \frac{v_i}{u_i} & u_i \geq 0 \\ \pi \operatorname{sgn}(v_i) + \arctan \frac{v_i}{u_i} & u_i < 0 \end{cases} \tag{4}$$

where β_i is the sideslip angle of a specific landing gear during the rollout phase.

(4) Reaction force of the landing gear P_i:

The reaction force of the landing gear based on a linear spring damping system has unreasonable phenomena. Therefore, this paper introduces the Gunther contact force model to characterize the ground reaction force of different types of landing gears, and the expression is [10]:

$$P_i = k_i \delta_i^{1.5} + \frac{1 - c_{ri}^2}{c_{ri}} \cdot \frac{k_i}{w_{i0}} \delta_i^{1.5} w_i \tag{5}$$

where $P_i (i = n, l, r)$ represents the reaction force of the nose landing gear, left and right main landing gears. k_i and c_{ri} represent the contact stiffness coefficient and restitution coefficient of different types of landing gear. δ_i, w_i and w_{i0} represent the compression amount, compression speed, and compression speed at the time of landing gear touchdown.

In this paper, the skid-equipped aircraft with nose-wheel and rear-skids is used as the specific modeling object.

The Lateral and Longitudinal Ground Forces of Nose Wheel. The lateral force of a wheeled landing gear is a nonlinear model related to its sideslip angle and vertical load. The force analysis is shown in Fig. 3.

where β_n is the sideslip angle of tire. The longitudinal force of the tire is related to the real-time slip rate of the wheels during the aircraft motion, and the equation is as follows:

$$\kappa = \frac{u_n - \omega_n r_n}{u_n} \tag{6}$$

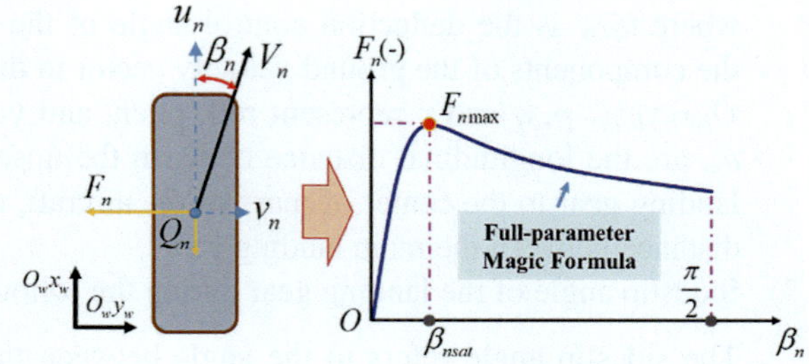

Fig. 3 Force analysis of wheeled landing gear

where ω_n and r_n are the rotational angular velocity of the nose wheel and the real-time radius of the nose wheel under load.

In subsequent research, a systematic comparative analysis will be conducted on the landing and rollout dynamic characteristics of wheeled aircraft and skid-equipped aircraft. Therefore, it is necessary to take into account the frictional elliptical effect of tire ground forces to avoid the independent lateral and longitudinal loads on wheeled aircraft during the braking phase [11]. Therefore, the full-parameter magic formula is introduced to build the contact force between the tire and the pavement.

The Lateral and Longitudinal Ground Forces of Rear Skid. The loading condition of metal skid on the ground is similar to that of plate friction, so the longitudinal force and lateral force are the components of skid sliding friction force along the axes of the ground force coordinate system $O_w x_w y_w z_w$, which are expressed as follows:

$$\begin{cases} F_l = -P_l \mu_l \sin \beta_l \\ Q_l = -P_l \mu_l \cos \beta_l \\ F_r = -P_r \mu_r \sin \beta_r \\ Q_r = -P_r \mu_r \cos \beta_r \end{cases} \tag{7}$$

where Q_i and $F_i (i = l, r)$ denote the longitudinal and lateral forces of the skid landing gear, respectively. μ_l and μ_r are the friction coefficients of the left and right skid landing gears, respectively. β_l and β_r are the sideslip angles of the skid during the rollout phase.

According to the lateral force variation curve of the skid in Fig. 4, it can be found that the lateral force provided by the skid is much smaller than that of the rubber tire on the condition with small sideslip angles and equal positive pressure conditions. The existence of the above differences makes it impossible to use the existing research conclusions of traditional wheeled aircraft as a reference for the motion characteristics of skid-equipped aircraft.

The ground forces of the nose and main landing gears and their moments on the center of mass of the aircraft in the aircraft coordinate system are defined as $\boldsymbol{F}_g$ and $\boldsymbol{M}_g$.

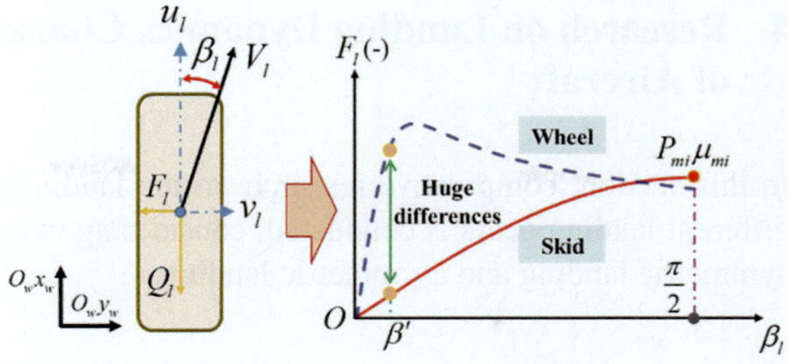

Fig. 4 Force analysis of left skid landing gear

3.2 Differential Equation of Aircraft

The gravity of the aircraft is expressed in the ground coordinate system as follows:

$$\boldsymbol{G} = \begin{bmatrix} 0\ 0\ mg \end{bmatrix}^T \tag{8}$$

where m is mass of the aircraft, g is gravitational acceleration. The vector dynamics equation of the aircraft in the dynamic coordinate system is as follows [12]:

$$\left(\frac{\delta \boldsymbol{V}}{\delta t} + \boldsymbol{\omega} \times \boldsymbol{V}\right) = \frac{\boldsymbol{F}}{m} \tag{9}$$

where $\boldsymbol{V} = \begin{bmatrix} u\ v\ w \end{bmatrix}$ is the ground velocity vector of the aircraft, $\boldsymbol{\omega} = \begin{bmatrix} p\ q\ r \end{bmatrix}$ is the angular velocity vector of the body coordinate system relative to the ground coordinate system. The resultant force vector acting on the center of mass of the aircraft $\boldsymbol{F}$ is:

$$\boldsymbol{F}^T = \boldsymbol{F}_a + \boldsymbol{F}_d + \boldsymbol{F}_g + L_{bg}\boldsymbol{G} \tag{10}$$

The dynamic equation expression for the aircraft to rotate around the center of mass in the body coordinate system is [12]:

$$\left(\frac{\delta \boldsymbol{h}}{\delta t} + \boldsymbol{\omega} \times \boldsymbol{h}\right) = \boldsymbol{M} \tag{11}$$

where $\boldsymbol{h}$ is the moment of momentum of the aircraft towards the origin of the aircraft coordinate system. $\boldsymbol{M}$ is the vector of the resultant moment acting on the center of mass, expressed as:

$$\boldsymbol{M}^T = \boldsymbol{M}_a + \boldsymbol{M}_d + \boldsymbol{M}_g \tag{12}$$

4 Research on Landing Dynamics Characteristics of Aircraft

In this section, comparative research on the landing performance of aircraft with different landing gears is conducted, considering two types of working conditions: symmetric landing and asymmetric landing.

4.1 Symmetric Landing Condition

During the landing phase of the aircraft, the initial pitching angle is 0 degree, and the initial longitudinal velocity and sinking velocity in the ground coordinate system are 70 m/s and 2.5 m/s, respectively. The coefficient of friction between the metal skid of the skid-equipped aircraft and the runway surface is 0.4. When the speed of the wheeled aircraft is less than 65 m/s, the braking function is activated, and the slip ratio of the main landing gear wheels κ is maintained at 0.1.

The state changes of the aircraft under symmetric condition are shown in Fig. 5. From Fig. 5a, it can be seen that due to the compressibility of rubber tires, the minimum height and stable height of CG of the wheeled aircraft during the landing phase are lower than those of the skid-equipped aircraft.

The initial variation curve of descent velocity in Fig. 5b indicates that the skid-equipped aircraft experiences a larger vertical load at the moment of touchdown, resulting in a more rapid decrease in descent velocity. Figure 5c and d depict the trends of pitching angle and pitching angular velocity for the two types of aircraft during the landing phase. It can be observed that the skid-equipped aircraft exhibits a more significant nose-down phenomenon, with a minimum pitching angle of −0.252 degrees.

Figure 6 shows the variation trend of ground loads on the two types of aircraft during the landing phase. From Fig. 6a, it can be observed that the landing overload of the skid-equipped aircraft is 2.679, which is 10.57% higher than that of the wheeled aircraft under the same hydraulic shock absorber parameters. The core reason for this difference is that the metal skid lacks cushioning capability. The pitching moment curve in Fig. 6b effectively illustrates the variation difference in pitching angular velocity during the initial landing stage of the two aircraft.

Figure 6d indicates that the skid-equipped aircraft has a significantly higher rate of increase and peak value in the reaction force of the main landing gear compared to the wheeled aircraft. Meanwhile, considering the phenomenon of wheel spin during the moment of touchdown, the longitudinal friction force on the tire rapidly increases and gradually decreases with the increase of wheel speed. Therefore, the longitudinal force on the main landing gear of the wheeled aircraft only exists for about 0.2 s and then quickly approaching zero. When the braking function is activated after 2.1 s, the wheeled main landing gear can provide similar longitudinal force as the skid landing gear.

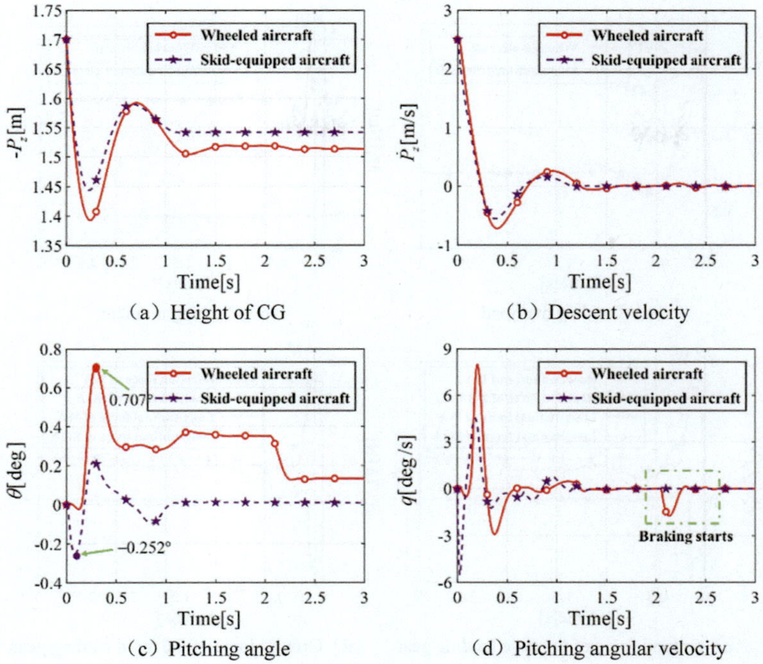

(a) Height of CG (b) Descent velocity

(c) Pitching angle (d) Pitching angular velocity

Fig. 5 Change trend of aircraft states under symmetric landing condition

The variation curves of the shock absorber states during the landing phase of the aircraft are shown in Fig. 7. From Fig. 7a, it can be seen that the compression velocity of the nose landing gear of the skid-equipped aircraft rapidly increases from 0 to 2.279 m/s, with a peak value of compression velocity increased by 21.03% compared to the wheeled aircraft. The reason for this is that the skid-equipped aircraft experiences a more severe pitch-down process due to the influence of friction forces generated by the main landing gears. Figure 7b indicates that the compression velocity of the main landing gear of the skid-equipped aircraft rapidly increases from 0 to 2.243 m/s, which is close to the initial sinking velocity of the aircraft.

Figure 7c and d depict the power diagrams of the nose landing gear and the main landing gear, respectively. The buffer efficiency of the nose and main landing gear applied in the wheeled aircraft is 84.03% and 90.12% respectively. Under the same buffer parameters, the efficiency of the nose and main landing gear used in skid-equipped aircraft is 84.34% and 74.73% respectively. Due to the larger initial compression velocity of the skid-equipped aircraft's main landing gear shock absorber, the peak load dominated by hydraulic damping force is significantly higher than the peak load corresponding to the gas spring force, which adversely affects the overall efficiency of the shock absorber.

Compared to the main landing gear applied in the wheeled aircraft, the buffer efficiency of landing gear used in the skid-equipped aircraft decreases by 17.1%. Although increasing the diameter of the oil hole or reducing the pressure oil area of

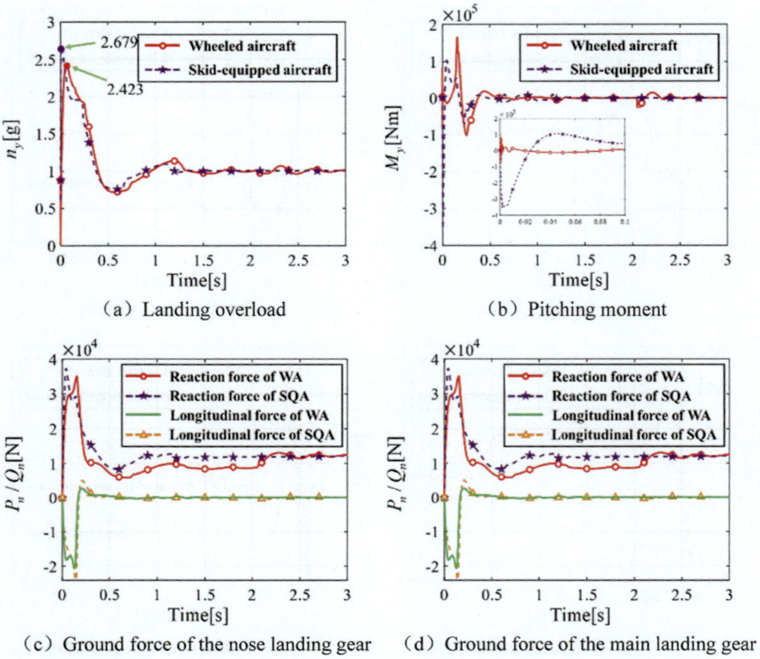

Fig. 6 Change trend of ground force under symmetric landing condition

the buffer can reduce the hydraulic damping force in the early stage of landing, this will be detrimental to the rapid dissipation of landing energy by the buffer.

4.2 Asymmetrical Landing Conditions

This section primarily analyzes the landing dynamic responses and the associated changes in ground forces of two types of aircraft under asymmetrical landing conditions. The initial rolling angle during the aircraft landing phase is 5 degrees. The initial longitudinal velocity and sinking velocity in the ground coordinate system are 70 m/s and 2.5 m/s, respectively. The initial height of the aircraft CG ensures the preliminary contact between the main landing gear and the ground. When the velocity is less than 65 m/s, the wheeled aircraft activates the main wheel braking function and maintains a slip ratio of 0.1.

Figure 8 illustrates the changes in the motion state of the two types of aircraft during asymmetrical landing conditions. From Fig. 8a, it can be observed that the lateral displacement of the skid-equipped aircraft increases to 1.82 m after 6 s. At the same time, its lateral velocity does not show a trend of converging to zero. In contrast, the wheeled aircraft exhibits better heading stability, as its lateral velocity converges to zero within 1.18 s. Similarly, Fig. 8c and d reflect the tendency of the

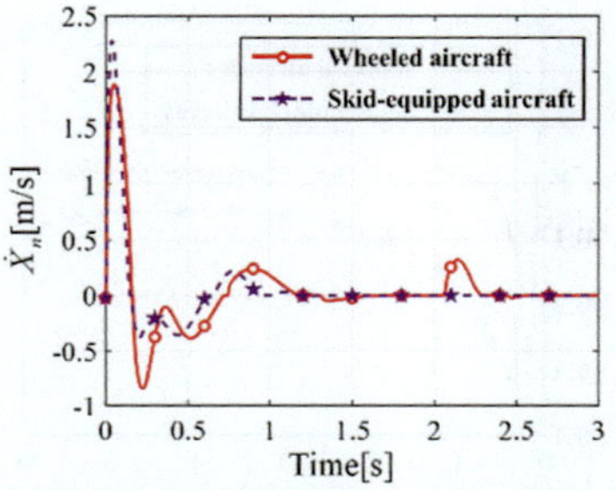

(a) Compression velocity of the nose landing gear buffer

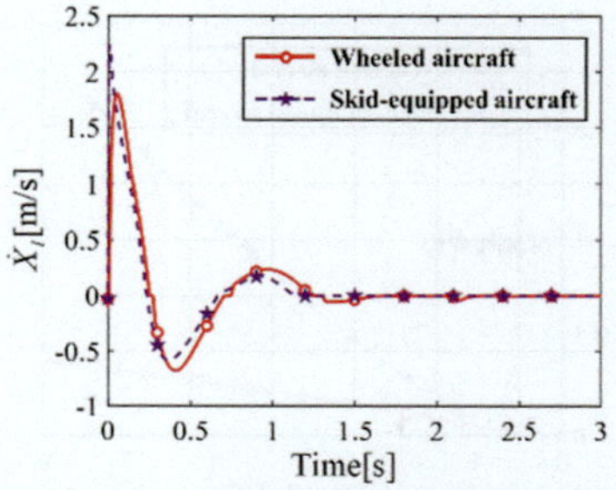

(b) Compression velocity of the main landing gear buffer

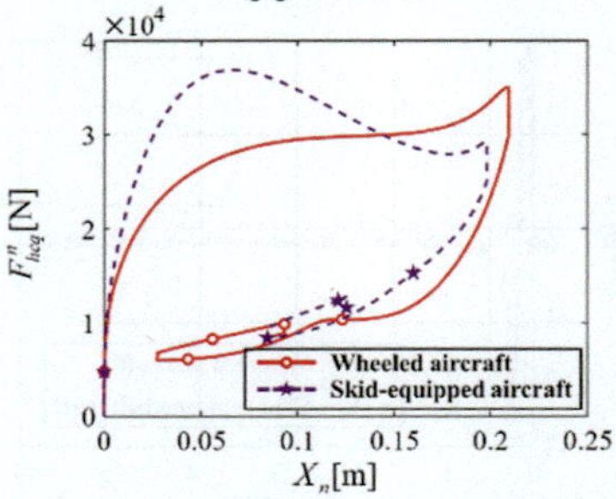

(c) Power diagram of the nose landing gear buffer

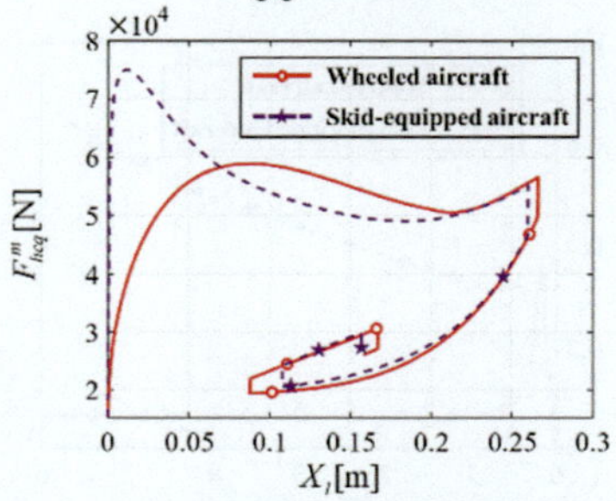

(d) Power diagram of the main landing gear buffer

Fig. 7 Change trend of shock absorber states under symmetric landing condition

skid-equipped aircraft to experience heading instability after asymmetrical landing, with the yawing angle continuously increasing and the yawing rate gradually rising.

Compared to the trend of landing gear reaction forces in Fig. 9a for the wheeled aircraft, the curve in Fig. 9c indicates that the main landing gear load of the skid-equipped aircraft is significantly affected by asymmetrical landing. The rolling rate caused by the initial touchdown of the right main landing gear substantially increases the vertical velocity of the left main landing gear, leading to a significant increase in the peak load of the reaction force.

Comparing Fig. 9b and d, it can be observed that both the nose and main landing gears of the wheeled aircraft can provide significant lateral forces in the early stages of landing to quickly eliminate the lateral velocity component of the aircraft. However, the nose and main landing gears of the skid-equipped aircraft only provide smaller lateral forces after landing, and the lateral force generated by the nose wheel is significantly greater than that generated by two main landing gears. Considering that the longitudinal distance from the nose landing gear to the center of gravity is significantly greater than the longitudinal distance from the main landing gears to the center of gravity, the difference in lateral forces will exert a significant yawing moment on the aircraft, causing the aircraft to accelerate its yawing motion. As depicted by the variation in lateral forces of tire and skids in Fig. 4, the difference in lateral forces between the nose and main landing gears will be further amplified by

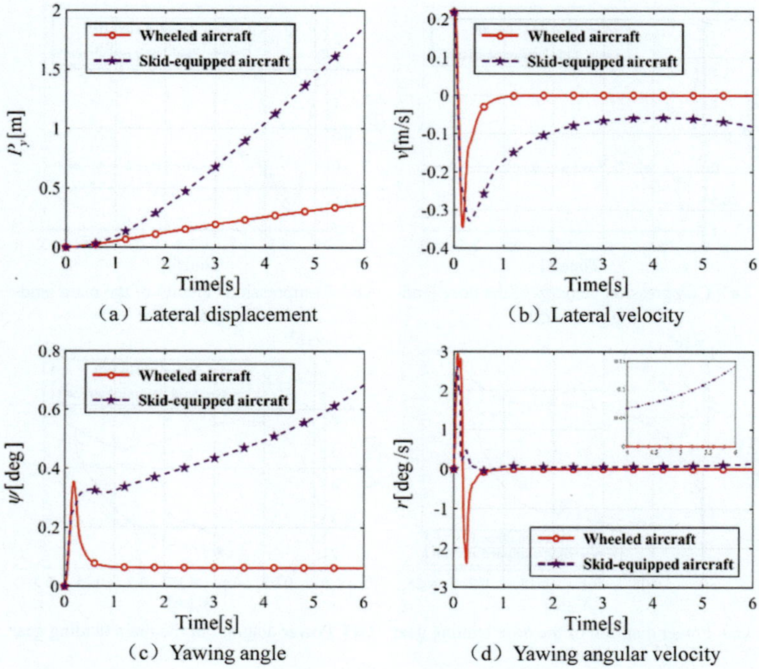

Fig. 8 Change trend of aircraft states under asymmetric landing condition

the presence of the yawing moment. This may result in uncontrollable yawing that poses a risk to the rollout safety.

5 Conclusions

The main conclusions are summarized as follows:

(1) The established generic dynamics model of the aircraft effectively reflects the dynamic response of the aircraft in different landing phases. It can serve as a benchmark model for comparative studies on the differences in landing and rollout characteristics of skid-equipped and wheeled aircraft.

(2) Due to the lack of cushioning capability provided by tires, the landing overload and the peak compression velocity of the main landing gear buffer of the skid-equipped aircraft increase by 10.57 and 21.03%, compared to the wheeled aircraft, when the buffer filling parameters are consistent. At the same time, the buffer efficiency of the main landing gear applied in the skid-equipped aircraft is significantly lower than that of the wheeled aircraft.

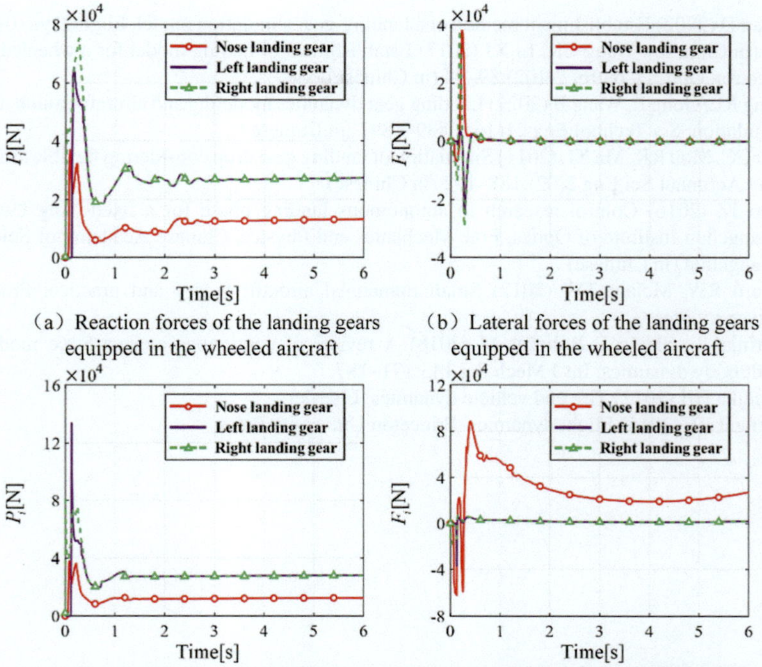

(a) Reaction forces of the landing gears equipped in the wheeled aircraft

(b) Lateral forces of the landing gears equipped in the wheeled aircraft

(c) Reaction forces of the landing gears equipped in the skid-equipped aircraft

(d) Lateral forces of the landing gears equipped in the skid-equipped aircraft

Fig. 9 Change trend of ground force under asymmetric landing condition

(3) The skid-equipped aircraft exhibits weak yawing stability according to the significant differences in the mechanical properties of the nose and main landing gears. The motion direction of the aircraft will experience continuous deviation under asymmetric disturbances, and failure to intervene with effective control can lead to serious runway accidents. Therefore, to improve the feasibility of skid-equipped aircraft, emphasis should be placed on enhancing its directional control capability during the ground rollout phase.

References

1. Airoldi A, Lanzi L (2006) Design of skid landing gears by means of multibody optimization. J Aircr 43(2):555–563
2. Li LG (2013) Research on the location and tracking of near space aircraft and the guidance of interceptor. Harbin Institute of Technology, Harbin, 3–5 (in Chinese)
3. Gu HB (2001) Dynamic model of aircraft ground handling. Acta Aeronautica et Astronautica Sinica 02:163–167 (in Chinese)
4. Li B, Jiao ZX (2007) Aircraft landing gear system dynamic modeling and simulation. J Beijing Univ Aeronaut Astronaut 01:46–49 (in Chinese)

5. Yuan D (2002) Establishment method of a landing-gear simulation model. Flight Dyn (04):44–47 (in Chinese). Wang CK, Li XJ (2017) Establishment of taxiing model for a wheeled UAV. Electron Optics Control 24(02):89–94 (in Chinese)
6. Tang RL, Gong L, Wang B (2021) Landing gear dynamics modeling and aircraft ground motion simulation. Sci Technol Eng 21(16):6889–6897 (in Chinese)
7. Liu CC, Mou RK, Ma XL (2011) Simulation of landing gear drop considering flexible airframe. Adv Aeronaut Sci Eng 2(02):188–192 (in Chinese)
8. Gao JZ (2016) Control research in autonomous landing phase for a fixed-wing UAV[D]. Changchun Institute of Optics, Fine Mechanics and Physics, Chinese Academy of Sciences, Changchun (in Chinese)
9. Beard RW, Mclain TW (2012) Small unmanned aircraft: theory and practice. Princeton university press
10. Skrinjar L, Slavič J, Boltežar M (2018) A review of continuous contact-force models in multibody dynamics. Int J Mech Sci 145:171–187
11. Pacejka HB (2012) Tire and vehicle dynamics. Elsevier
12. Stengel RF (2015) Flight dynamics. Princeton University Press

A Multi-legged Terrain Adaptive Take-Off and Landing Ducted Aircraft

Peilin Zhong, Qiaozhi Yin, Hao Sun, Xiaohui Wei, and Le Zhou

Abstract An unmanned rotorcraft usually faces the risk of airframe capsizing when taking off or landing on uneven terrains in the wild. A new multi-legged adaptive take-off and landing ducted aircraft is put forward which can adapt to various terrains. The configuration design of the aircraft is proposed based on the duct and the bionic multi-legged robot. Then, the kinematics of the adaptive take-off and landing configuration is analyzed. The adaptive adjustment scheme of the aircraft which focuses on the semi-slope and the full-slope terrain models is designed. The multibody dynamic model of the aircraft is established in Simcenter3D, and the attitude adjustment and drop simulation under the two terrains are carried out. The results show that the torque of each joint is within a reasonable range and the capability of adaptation to terrains is proved. The adaptive aircraft prototype is verified that it can land safely on a 15° full-slope terrain by the attitude adjustment experiments.

Keywords Adaptive landing · Ducted-fan vehicle · Configuration design · Kinematic analysis · Dynamic simulation

1 Introduction

In recent years, adaptive vertical take-off and landing UAVs have gradually become a research hotspot. It not only has vertical take-off and landing function, but also has good adaptability to complex terrain environments, which can expand the take-off

P. Zhong · Q. Yin · H. Sun · X. Wei (✉) · L. Zhou
Key Laboratory of Fundamental Science for National Defense-Advanced Design Technology of Flight Vehicle, Nanjing University of Aeronautics and Astronautics, Nanjing 210016, China
e-mail: wei_xiaohui@nuaa.edu.cn

Q. Yin
National Key Laboratory of Helicopter Aeromechanics, Nanjing University of Aeronautics and Astronautics, Nanjing 210016, China

X. Wei
State Key Laboratory of Mechanics and Control for Aerospace Structures, Nanjing University of Aeronautics and Astronautics, Nanjing 210016, China

Z. Liu et al. (eds.), *Advances and Challenges in Advanced Unmanned Aerial Systems*, Springer Aerospace Technology, https://doi.org/10.1007/978-981-99-8045-1_19

and landing places of traditional UAVs. It plays a crucial role in outdoor exploration, border patrols, and other similar scenarios [1]. Therefore, higher design requirements are put forward for adaptive take-off and landing UAVs.

There have been a large number of research results on adaptive take-off and landing UAV design at home and abroad. Yuri et al. [1] designed the multi-rotor adaptive landing platform DroneGear, which combines a multi-rotor drone with adaptive legs to adapt the DroneGear to rough terrain by installing torque sensors at the knee joints of the legs. Boris et al. [2] proposed an outrigger type terrain adaptive landing gear. Liu et al. [3] designed a bionic multi-fingered UAV based on eagle claws, combined with and verified that the take-off and landing system can effectively land on slopes, steps and unstructured terrain. Cabuk et al. [4] designed an adaptive take-off and landing octocopter UAV. Ren et al. [5] designed a multi-link bionic landing gear multi-rotor UAV, which adopts the configuration design method of multi-rotor and linkage landing gear and adapts to the terrain through the multi-link hybrid mechanism. Liu et al. [6] designed a hexapod landing gear for unmanned helicopters and applied adaptive landing gear to unmanned helicopters. Sun et al. [7] designed a new integrated adaptive landing gear with buffer actuation for walking based on a helicopter platform.

Most of the existing adaptive take-off and landing UAVs are based on multi-rotor platforms for research. When the external rotor volume of multi-rotor UAVs become too large, there is often a significant risk associated with performing tasks in confined spaces. Additionally, it presents a certain level of danger [8]. Among various types of vertical take-off and landing aircraft, the ducted fan aircraft is shielded by its annular fuselage, which protects the aircraft itself and the operators, enabling it to fly in narrow, cluttered, and hazardous environments [9]. In 2003, DARPA designed a ducted fan unmanned aerial vehicle that could be carried by a single soldier for ground combat squads [10]. Rones et al. [11] proposed a conceptual design for a multi-ducted fan aircraft. Ackerman et al. [12] proposed a ducted fan UAV called Cleo, which possesses safety, user-friendliness, and portability. Its compact size also allows for operations in enclosed spaces [13].

A multi-legged terrain adaptive take-off and landing ducted aircraft is proposed in this paper, and the research of the adaptive take-off and landing UAV based on ducted flight platforms are carried out. The adaptive take-off and landing capability of the aircraft is effectively validated by integrating the adaptive landing gear with the ducted aircraft through a compact structure.

2 UAV Overall Configuration Design

In the application scenario of UAVs for take-off and landing in outdoor environments, it is necessary for the UAV to have the ability to take-off and landing on different terrains, withstand certain impact loads, and possess sufficient structural strength to withstand unexpected crash accidents. The overall structure of the UAV should be designed to be as streamlined and compact as possible, without any redundant components.

2.1 Overall Structural Design of Aircraft

The design of a multi-legged adaptive take-off and landing ducted aircraft is presented in this paper. A three-dimensional geometric model is established in Solidworks, and the structural design and leg landing gear scheme design of the whole aircraft are carried out. The entire aircraft model is displayed in Fig. 1. The aircraft takes a single bypass as the main body, combined with adaptive leg landing gear to form two systems, the main components of which are ducted fan, servos, anti-twist spoilers, rudder pads, ring connector and leg landing gear.

Dai et al. [14] provided a detailed introduction to the structural design of the ducted fan aircraft. Taking inspiration from this concept, the ducted UAV is divided into two parts, the upper part is the ducted fan and motor, and the lower part is the rudder pads that control the airflow. Its structural design is shown in Fig. 2. A 90 mm ducted fan is chosen as the full power source to provide flight lift, and the reverse torque is balanced by the anti-twist spoilers. In order to ensure that the ducted fan has a suitable deflection torque for manipulation, the rudder adopts symmetrical airfoil, and its width is determined according to the size of the ducted fan, just within the tail blowing area of the duct, and the spoilers should not be too long to avoid touching the ground. Drone navigation is performed by adjusting the rudder pads to change the attitude of the airframe, while placing the spoilers longitudinally away from the center of gravity as much as possible. The deflection of the spoilers is controlled by the servos on the connecting arm through the screw connection. The duct is connected to the ring connector of which the diameter is 170 mm through the ears, and the connection space between the ducted fan and the connector is used to place the control board, batteries, etc. The cylindrical structure arrangement makes the structure of the ducted aircraft relatively compact, which can reduce the volume and weight well compared with other forms of rotary wing UAV.

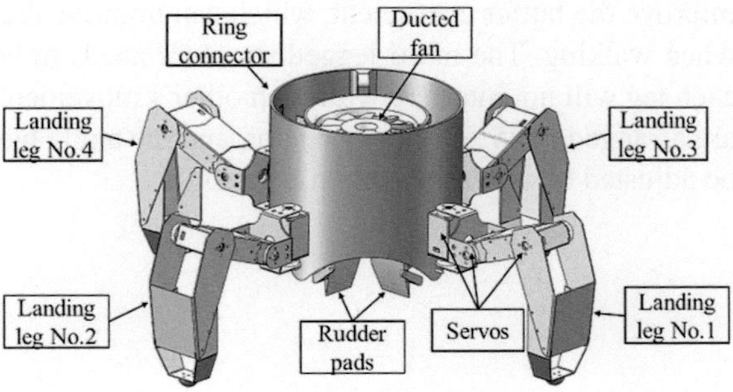

Fig. 1 Structural design of adaptive take-off and landing ducted aircraft

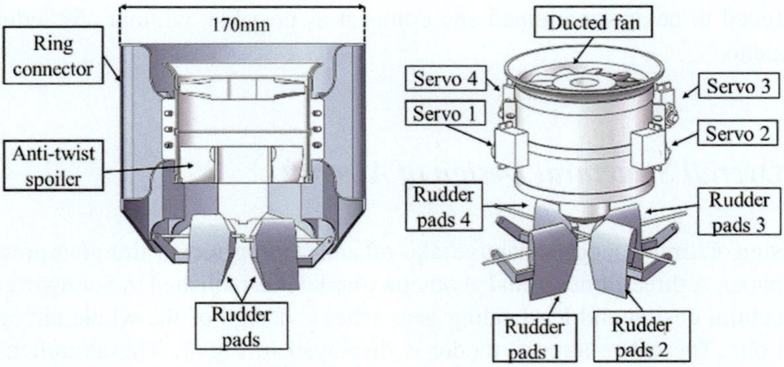

Fig. 2 Ducted aircraft structural design

The landing gear adopts a four-legged three-joint type, and the four supporting legs are distributed symmetrically in a central symmetry relative to the center of mass of the UAV. The three joints are divided into hip, knee, and foot joints, as shown in Fig. 3a. When the aircraft stands on the ground normally, the maximum distance from the hip joint to the ankle joint is 163 mm, and the length of the hip joint to the ground is 123 mm. Because the landing gear does not have a large impact load, so the three joints are driven by the servos, in the face of different terrain, mainly by adjusting the angle of the foot joints and knee joints for terrain adaptation. If the drone encounters more complex ground situations, it can better adapt to the ground by adjusting the angle of the hip joints, but the hip joints mostly work when walking. In the structural design of a single leg, the leg is mainly divided into three parts: upper arm, middle arm, lower arm. The movement direction of upper arm and middle arm is perpendicular, so it is necessary to design a cross plate connecting the hip joint and knee joint of the servo. The middle arm needs to connect the two servos together. The lower arm and the middle arm of the motion pair is a rotary pair. In the design of the lower arm, in order to reduce the position of the center of gravity and enhance stability, the lower arm is increased in the fold. Additionally, the lower arm's foot end has a hemispherical design, and rubber with a relatively high elastic coefficient is chosen to increase friction between the leg landing gear and the ground and to improve the buffer coefficient, which can improve the stability of the entire aircraft when walking. The multi-legged structure needs to be reasonably arranged so that each leg will not interfere with each other's movement when walking. The four legs are arranged at 45° of the connector, and the angle between each leg is 90° and can be adjusted at will, as shown in Fig. 3b.

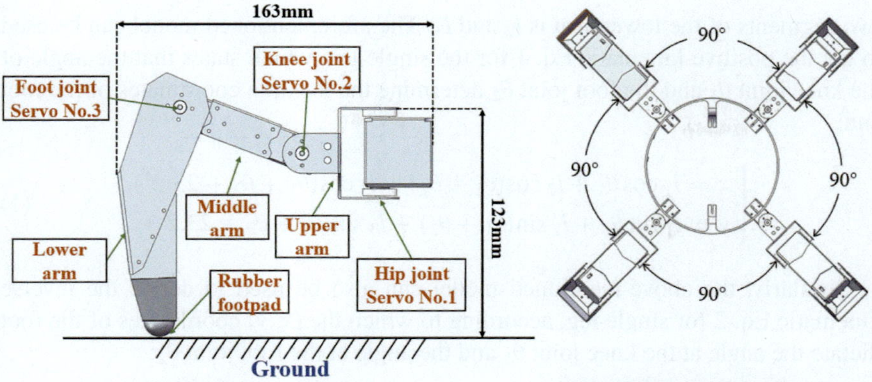

(a) Single-leg structure (b) Multi-leg structure layout

Fig. 3 Landing gear structure diagram

2.2 *Kinematic Analysis of Adaptive Take-Off and Landing Single-Leg*

In order to study the single-leg motion of a leg-type landing gear and the coordination of multiple legs to achieve adaptive functionality on different terrains, it is necessary to conduct kinematic analysis of the single leg. Since the ground walking function of the landing gear is not involved and the hip joint of the landing gear only participates in ground walking, without considering ground adaptation functionality, the landing gear structure can be simplified as shown in Fig. 4. The motion of the hip joint is ignored, and only the motion of the knee joint and ankle joint is studied.

From the above simplified model, the single-legged kinematic model is a two-degree-of-freedom model, and the coordinate system is established in the initial state as shown in Fig. 4. It is stipulated that the counterclockwise rotation direction around the joint is positive, the length of the middle arm is l_1, and the length of the

Fig. 4 Establishment of a single-leg coordinate system

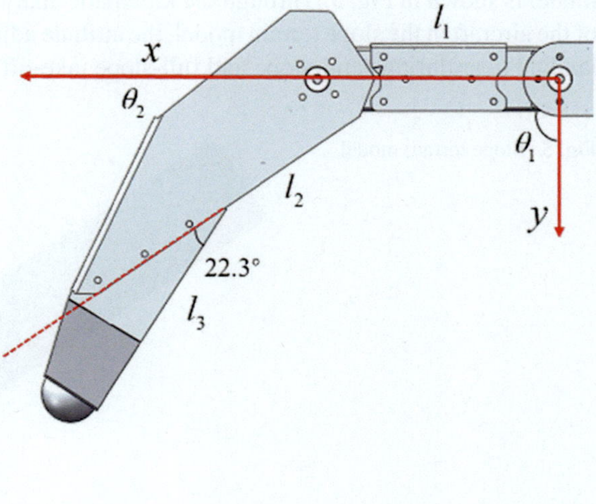

two segments of the lower arm is l_2 and l_3. The aforementioned model can be used to get the positive kinematic Eq. 1 for the single leg, which states that the angle of the knee joint θ_1 and the foot joint θ_2 determine the location coordinates of the foot end:

$$\begin{cases} x = l_1 \cos\theta_1 + l_2 \cos(\theta_1 + \theta_2) + l_3 \cos(\theta_1 + \theta_2 + 23.2^\circ) \\ y = l_1 \sin\theta_1 + l_2 \sin(\theta_1 + \theta_2) + l_3 \sin(\theta_1 + \theta_2 + 23.2^\circ) \end{cases} \tag{1}$$

Similarly, the above-mentioned model can also be used to derive the inverse kinematic Eq. 2 for single leg, according to which the (x, y) coordinates of the foot dictate the angle at the knee joint θ_1 and the angle at the foot joint θ_2:

$$\begin{cases} \theta_1 = \frac{\pi}{2} - \sin^{-1}\frac{x}{\sqrt{x^2+y^2}} - \cos^{-1}\frac{l_1^2 + x^2 + y^2 - l_2^2 - l_3^2 + 2l_2 l_3 \cos 156.8^\circ}{2l_1\sqrt{x^2+y^2}} \\ \theta_2 = \tan^{-1}\frac{y - l_1 \sin\theta_1}{x - l_1\cos\theta_1} - \theta_1 - \sin^{-1}\frac{l_3 \sin 156.8^\circ}{l_2} \end{cases} \tag{2}$$

3 Adaptive Take-Off and Landing Dynamic Simulation

3.1 Adaptive Landing Gear Take-Off and Landing Mathematical Model

3.1.1 Terrain Model

The adaptive capability of the landing gear to complex terrains can be investigated by studying landing of the aircraft on a sloped terrain model, and the slope terrain model is shown in Fig. 5. Through the kinematic analysis of the take-off and landing of the aircraft in the slope terrain model, the attitude adjustment ability of the adaptive landing gear during semi-slope and full-slope take-off and landing is studied.

Fig. 5 Slope terrain model

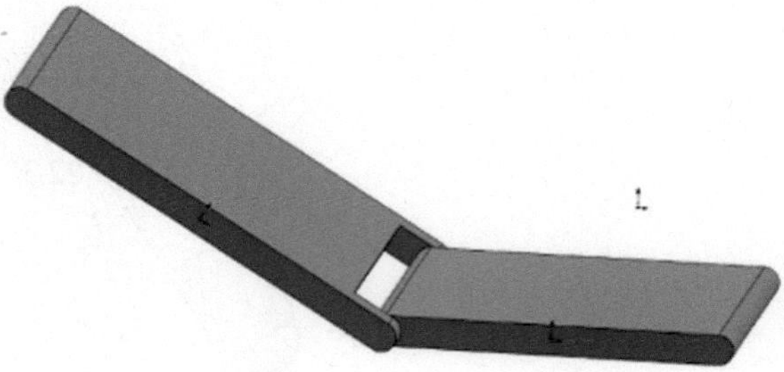

3.1.2 Semi-Slope Take-Off and Landing Model

Considering the possibility of the aircraft landing with one side on flat ground and the other side on the slope, a semi-slope take-off and landing model is established to solve the actual situation of this landing, as shown in Fig. 6.

Based on this model, the relationship between the deflection angles of the joints is established. The knee angles θ_1 and foot joint angles θ_2 of leg 1 and leg 3 are expressed as follows:

$$\theta_1 = \sin^{-1}\left(\frac{H - l_3 - l_2 \cos 23.2^\circ}{l_1}\right) \tag{3}$$

$$\theta_2 = \frac{\pi}{2} - \theta_1 - 23.2^\circ \tag{4}$$

The knee angles θ_1 and foot joint angles θ_2 of leg 2 and leg 4 are expressed as follows:

$$\theta_1 = \sin^{-1}\left(\frac{H - h - l_3 \cos\theta - l_2 \cos(\theta + 23.2^\circ)}{l_1}\right) \tag{5}$$

$$\theta_2 = \frac{\pi}{2} - \theta_1 - \theta - 23.2^\circ \tag{6}$$

where, θ is the inclination angle of the slope, H is the dropped height of the center of mass of the aircraft, and h is the height difference between leg 1 and 2. The semi-slope tilt angle and the joint servo deflection angle curve is established as shown in Fig. 7. From the change trend of the four curves, it can be seen that the joint angles of the No. 1 leg and the No. 3 leg located on the plane do not change, and the two curves of the No. 2 and No. 4 leg are approximately linear changes. The angle change trend of the knee joint and the foot joint of the single leg is opposite, demonstrating that the leg adapts to the semi-slope at different angles is smooth.

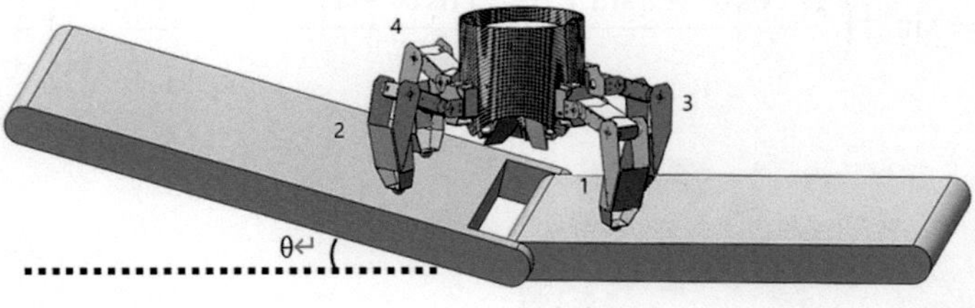

Fig. 6 Semi-slope ground take-off and landing model

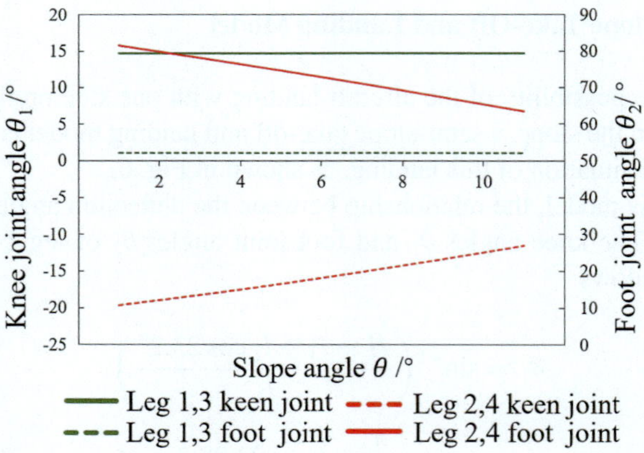

Fig. 7 The relationship between the angle of the semi-slope and the deflection angle of the servo

3.1.3 Full-Slope Take-Off and Landing Model

Considering that the landing gear of the aircraft is all on the slope, a full-slope take-off and landing model is established to solve the actual situation of this landing, as shown in Fig. 8.

In order to maintain the stability of the aircraft on the inclined plane, that is, the main plane of the aircraft remains horizontal, the model keeps the ends of each leg on the left side perpendicular to the bevel. Based on this model, the relationship between the joint angles of each leg and the slope angle is established as follows:

$$\theta' = \sin^{-1}\frac{\sin\theta}{\sqrt{1+(\cos\theta)^2}} \tag{7}$$

The knee angles θ_1 and foot joint angles θ_2 of leg 1 and leg 3 are expressed as follows:

$$\theta_1 = \sin^{-1}\left(\frac{H\cos\theta' + l\sin\theta' - l_3\cos\theta' - l_2\cos(\theta' + 23.2°)}{l_1}\right) + \theta' \tag{8}$$

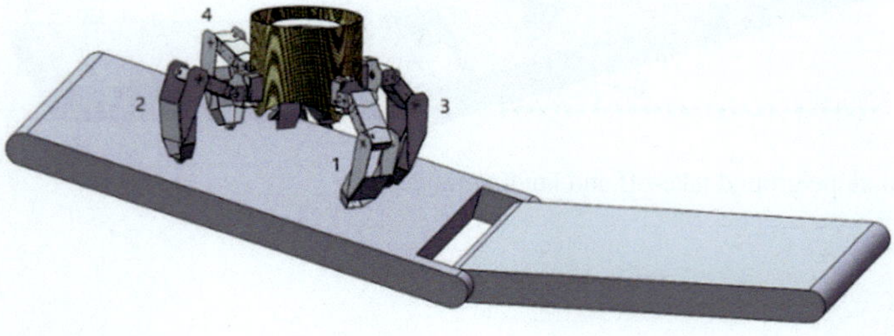

Fig. 8 Full-slope ground take-off and landing model

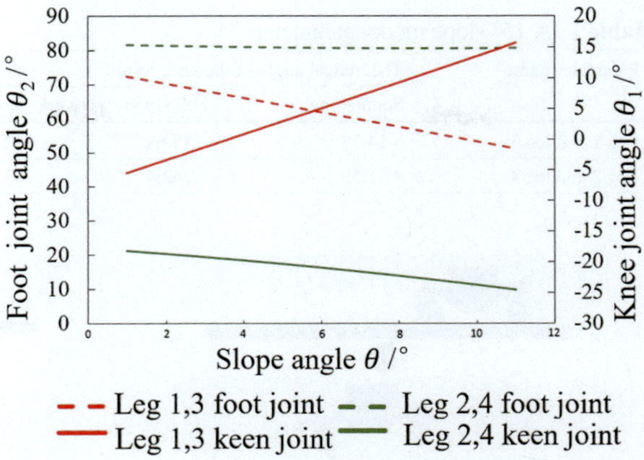

Fig. 9 The relationship between the angle of the full slope and the deflection angle of the servo

$$\theta_2 = \frac{\pi}{2} - \theta_1 - 23.2^\circ \tag{9}$$

The knee angles θ_1 and foot joint angles θ_2 of leg 2 and leg 4 are expressed as follows:

$$\theta_1 = \sin^{-1}\left(\frac{H\cos\theta' - l\sin\theta' - l_3 - l_2\cos 23.2^\circ}{l_1}\right) + \theta' \tag{10}$$

$$\theta_2 = \frac{\pi}{2} - \theta_1 - 23.2^\circ - \theta' \tag{11}$$

where, θ is the slope inclination angle, θ' is the projection angle of the corresponding slope inclination angle in the direction of 45° of plane, H is the dropped height of the center of mass of the aircraft, and l is half the length of the landing gear main body. The influence curves of slope inclination angle on the angle of the joint servos are established, as shown in Fig. 9.

3.2 *Take-Off and Landing Model Simulation*

Set the angle of the slope terrain to 15°, and calculate the theoretical angles of each joint of the legs under the semi-slope model and the full slope model by using Eq. 3 to Eq. 11, respectively, as shown in Table 1. The simulation is carried out in SolidWorks/motion, the corresponding results of which are shown in Fig. 10. The results indicate that the stable horizontal attitude angles on the slope are 1.41° and 1.27°, respectively. This closely aligns with the theoretical data and ensures a nearly level attitude angle for the aircraft, which shows that the mathematical model established above can provide a theoretical basis for the attitude planning of the prototype landing test.

Table 1 A 15° slope model parameters

Parameter name	Theoretical angle of the knee joint/°		Theoretical angle of foot joint/°	
	Semi-slope	Full slope	Semi-slope	Full slope
Leg 1 and leg 3	14.73	15.68	52.07	51.12
Leg 2 and leg 4	−11.59	−24.64	63.39	80.72

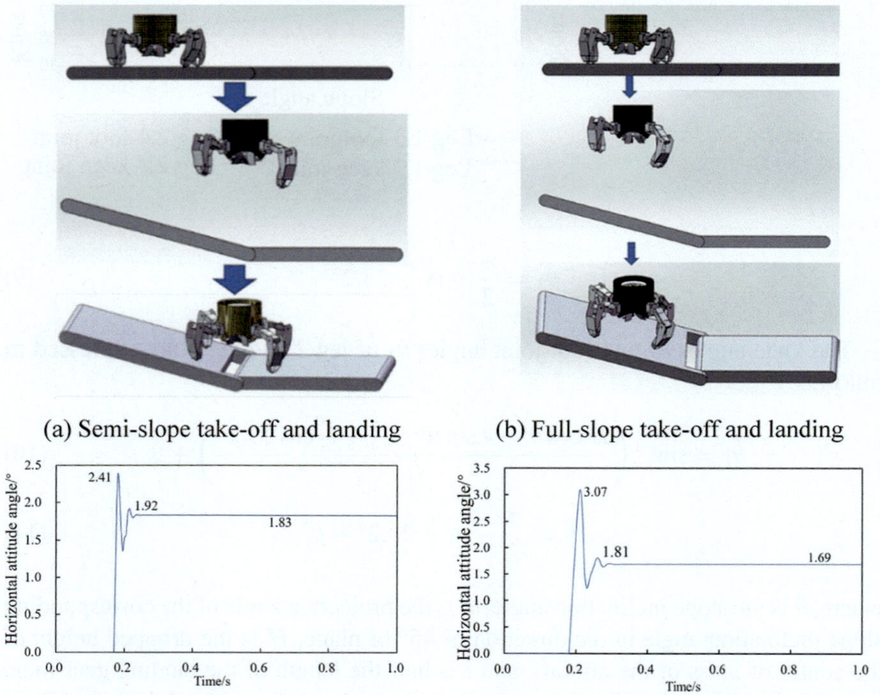

(a) Semi-slope take-off and landing (b) Full-slope take-off and landing

(c) Semi-slope aircraft horizontal attitude angle curve (d) Full-slope aircraft horizontal attitude angle curve

Fig. 10 Slope take-off and landing model simulation diagrams

3.3 Landing Dynamics Simulation

In view of the landing of the aircraft in the semi-slope and full-slope terrain, in order to ensure that the adaptive landing gear has certain structural strength requirements when the aircraft lands, it is necessary to analyze the torque of each joint servo to ensure that the selection of the servos meet the torsion resistance requirements, so the aircraft drop simulation analysis is carried out.

A full-aircraft rigid body dynamic model is established in Simcenter3D, and drop simulation is analyzed for semi-slope ground landing and full-slope ground landing.

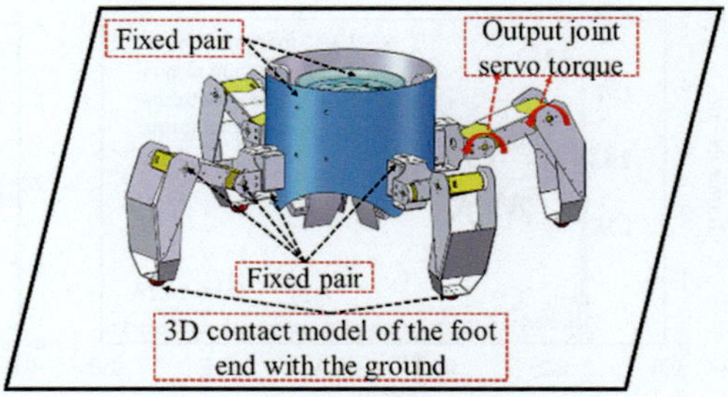

Fig. 11 Schematic diagram of establishing the kinematic joints for the aircraft

The aircraft is mainly divided into three parts: duct, connector and adaptive landing gear. There is no relative displacement between the ducted fan and the rudder, which can be regarded as a whole. In the existing model, the control board, batteries, etc. are converted into particle. Since only the joint torques need to be analyzed and the hip joints of the leg landing gear do not work, so the upper arm and the connector can be integrated as a whole. The simplified establishment of the aircraft motion pair is shown in Fig. 11. The connecting parts are connected to the duct and landing gear parts by a fixed pair, respectively. The foot joint and knee joint of each leg are fixed pairs, and the foot end makes contact with the ground through a 3D contact model.

The entire aircraft is set to undergo free-fall motion, landing at a descent rate of 1.4 m/s on both a 15° semi-slope and 15° full-slope terrain. The simulation diagram is shown in Fig. 12. According to the mathematical model of ground take-off and landing in the previous section, it can be seen that leg 1 and leg 3, leg 2 and leg 4 have symmetrical structures, so leg 1 and leg 4 are taken for analysis and the foot joint torque and knee joint torque change diagrams are obtained, respectively, as shown in Fig. 13.

From Fig. 13, it can be observed that under both operating conditions, the knee joint and ankle joint experience the highest torque load during the instant of the aircraft landing. Due to the force exerted by the slope, the normal force on the inclined surface causes a rebound phenomenon in the body, resulting in the appearance of a second peak. The overall torque variation curve shows a decaying trend, eventually reaching a steady-state value. Specifically, in the two operating conditions of landing on the semi-slope and the full-slope, the maximum torque values of the joint servos are 4.45 N·m and 3.37 N·m, respectively. These values are generally lower than the 60 kg·cm servos commonly used for robots in the market. This ensures the usability

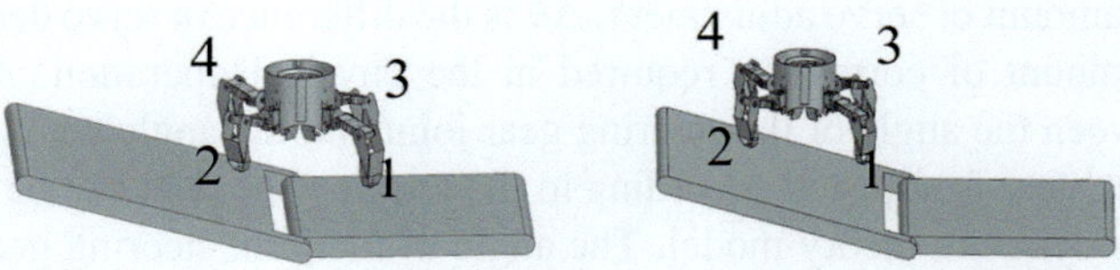

(a) Semi-slope landing simulation (b) Full-slope landing simulation

Fig. 12 Simcenter3D simulation diagrams

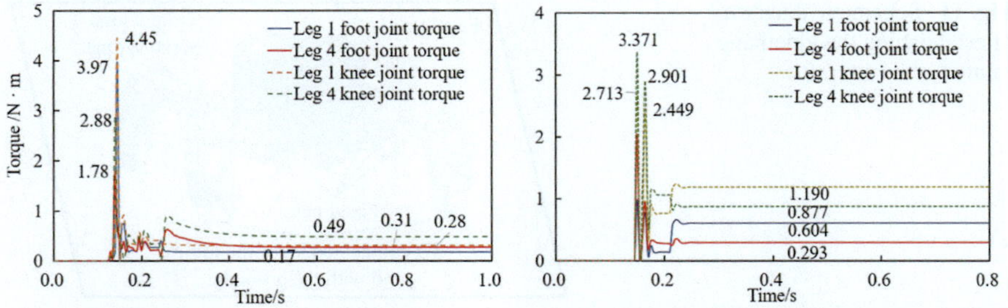

(a) Semi-slope terrain landing joint torque curve

(b) Full-slope terrain landing joint torque curve

Fig. 13 Landing gear joint torque curves

of the load-bearing joint structure of the adaptive take-off and landing aircraft's landing gear and provides a theoretical simulation basis for the servo selection in the subsequent physical prototype.

4 Adaptive Aircraft Prototype Attitude Adjustment Experiment

Based on the scheme configuration design of the aircraft, the entire aircraft parts are processed and assembled to form a multi-legged adaptive take-off and landing ducted aircraft prototype. The total weight of the entire aircraft is approximately 2.5 kg.

In order to verify the accuracy of the adaptive landing gear ground dynamic model, the attitude adjustment experiment of the adaptive landing prototype for 15° full slope terrain is carried out. Based on the Raspberry Pi motherboard through the servo directly control the leg attitude. The adjustment ranges of servo in upper computer 2000 degrees corresponds to the angle deflection of the actual servo of 180 degrees. Thus, formula conversion is performed by adjusting the formula, as shown in Eq. 12.

$$\delta = \frac{\Delta\theta}{0.09} + \varepsilon \tag{12}$$

where, θ is the amount of servo adjustment, $\Delta\theta$ is the difference in servo declination, and ε is the amount of correction required in the physical operation. $\Delta\theta$ is the difference between the angle of the steering gear joint and the angle when standing after the landing gear is adjusted according to the terrain. The data can be obtained by the previous dynamic theory model. The angle of the joint steering gear for the full-slope landing is shown in Table 2.

The experiment was carried out based on the adaptation angle, which is shown in Fig. 14. Firstly, the initial angle of each joint is obtained by standing flat on the

Table 2 A 15° ramp landing joint servo adaptation angle

Leg number	1, 3		2, 4	
Joint name	Knee joint	Foot joint	Knee joint	Foot joint
Practical angle/°	−22	80	18	76
Initial angle/°	−16	42	−16	42
$\Delta\theta$/°	6	38	34	34

(a) The aircraft standing on a flat ground (b) Leg-type landing gear before changing its configuration (c) Leg-type landing gear before changing its configuration (d) Attaining a balanced posture on the slope.

Fig. 14 Experimental procedure for slope adaptation test

ground of the aircraft. Then, the joint angle is settled according to the dynamic model in the air and the adaptive leg posture is adjusted accordingly. The terrain adaptation is carried out on the slope ground after changing the attitude. Results show that the aircraft can make corresponding attitude adjustment for 15° full slope terrain to ensure that the body attitude angle is horizontal to achieve balance on the slope, verifying that the adaptive take-off and landing ducted aircraft can adapt to sloping terrains.

5 Conclusion

A multi-legged terrain adaptive take-off and landing ducted aircraft is designed with the advantages of a compact structure and certain terrain adaptability. The configuration of the aircraft model is established in Solidworks. The kinematic and dynamic model are established in SolidWorks/motion and Simcenter3D. Finally, the expriment is carrid out on the prototype aircraft. The conclusions are summarized as follows:

(1) Through the drop simulation of aircraft, the knee joints and foot joints bear torsional loads within a reasonable range when the aircraft landing at a descent rate of 1.4 m/s on both a 15° semi-slope and 15° full-slope terrain, which verifies that the aircraft possesses a certain load-bearing capacity.

(2) Through kinematic analysis simulations and attitude adjustment experiments of adaptive landing gear, the designed multi-legged terrain adaptive take-off and landing ducted aircraft is verified to be capable of performing multi-legged posture adjustments and possesses the ability to adapt to sloping terrains.

Acknowledgements This study was supported by the Fundamental Research Funds for the Central Universities (No. NS2023005), the Fund of Prospective Layout of Scientific Research for NUAA (Nanjing University of Aeronautics and Astronautics), the Spring Sunshine Program initiated by the Ministry of Education of China (No. HZKY20220126), the China Postdoctoral Science Foundation Funded Project (No. 2021M691565), the National Natural Science Foundation of China (No. 52275114), the Aeronautical Science Foundation of China (No. 202000410520002), the National Defense Outstanding Youth Science Foundation (No. 2018-JCJQ-ZQ-053), and the Priority Academic Program Development of Jiangsu Higher Education Institutions.

References

1. Sarkisov YS , Yashin GA, Tsykunov EV et al (2018) A novel robotic landing gear with embedded optical torque sensors for safe multicopter landing on an uneven surface. IEEE Robot Autom Lett 3(3):1912–1917
2. Stolz B, Brodermann T, Castiello E (2018) An adaptive landing gear for extending the operational range of helicopters. In: 2018 IEEE/RSJ international conference on intelligent robots and systems (IROS), pp 1757–1763
3. Liu J, Zhang D, Wu C, et al (2021) A multi-finger robot system for adaptive landing gear and aerial manipulation. Robot Auton Syst 146, 103878
4. Çabuk N (2022) Design and kinematic analysis of proposed adaptive landing gear for multirotor UAV. El-Cezeri 9(1):159–170
5. Ren J et al (2023) Design of multi-link hybrid bionic landing gear and landing test. Sci Technol Eng 23(11):4881–4893. (in chinese)
6. Haolin L et al (2023) Design and control algorithm for hexapod landing gear of unmanned helicopter. China Mech Eng 34(4). (in chinese)
7. Sun H, Yin Q, Wei X et al (2021) Research of single leg drop performance of new adaptive landing gear. J Beijing Univ Aeronaut Astronaut 49(4):990–998. (in chinese)
8. Chou T, Ying Q, Qian Y, Zhuge W, Zhang Y (2021) Study on overall design of a vertical take-off and landing unmanned aerial vehicle powered by electric ducted fans. In Fluids engineering division summer meeting, vol 85291, p V002T03A008. American Society of Mechanical Engineers
9. Tayyab M, Xia Y, Yasir A et al (2022) Flight control techniques and classification of ducted fan aerial vehicles. Control Theory Technol 39(2):201–221
10. Reavis B, Hem B (2011) Honeywell T-Hawk aids Fukushima Daiichi disaster recovery: unmanned micro air vehicle provides video feed to remote monitors. Honeywell Aerospace Media Center
11. Rones A, Hoff SO (2017) Modular design of integrated duct fan quadrotor structure. UiT Norges arktiske universitet
12. Ackerman E (2018) Cleo robotics demonstrates uniquely clever ducted fan drone. https://spectrum.ieee.org/automaton/robotics/drones/cleo-robotics-demonstrates-uniquely-clever-ducted-fan-drone
13. Pireva K, Dragusha K, Karahoda B (2013) Cleaner robot in the cloud–" Cleo"
14. Dai M, Ren Z, Huang S et al (2018) Structure design of ducted fan unmanned aerial vehicle. Science and Technology Innovation Herald. (in chinese)

Perception-Constrained Vision-Based Quadrotor Control

Chao Qin and Hugh H.-T. Liu

Abstract One of the most critical issues in vision-based quadrotor control is to keep the target within the camera field of view. This is challenging in aggressive flights since the quadrotor is an under-actuated system that cannot translate without rotation. We present a perception constrained vision-based controller for quadrotors that can ensure target visibility. Our framework leverages nonlinear model predictive control to formulate a constrained visual servo control problem, where the actuation limit and target visibility are explicitly considered. We utilize image error instead of position error to accomplish the target tracking tasks which make our system robust to camera calibration error. Simulation results show that our method can work without the need for positional information, and it successfully completes the target tracking tasks from multiple initial configurations while keeping the target visible. It verifies the effectiveness of the defined perception constraint in aggressive quadrotor maneuvers.

Keywords Visual servoing · Quadrotors · Nonlinear model predictive control

1 Introduction

With the advancement of vision and navigation technology, unmanned aerial vehicles (UAVs) have been widely used in videography, delivery, surveillance, etc. In particular, the quadrotor UAV has attracted significant research attention due to its high maneuverability, agile mobility, and vertical take-off and landing (VTOL) capability [1]. The design of the flight controller typically assumes the availability of an onboard position estimator, e.g., the global positioning system (GPS) or the simultaneous localization and mapping (SLAM) system. However, this assumption does

C. Qin · H. H.-T. Liu (✉)
University of Toronto, Toronto, Canada
e-mail: hugh.liu@utoronto.ca

C. Qin
e-mail: chao.qin@mail.utoronto.ca

Z. Liu et al. (eds.), *Advances and Challenges in Advanced Unmanned Aerial Systems*, Springer Aerospace Technology, https://doi.org/10.1007/978-981-99-8045-1_20

not hold in some cases: GPS and motion capture systems cannot work in unknown indoor environments; in dynamic or feature-less scenes, VIO and SLAM will degrade seriously; in high-speed flights, the high latency of an onboard localization system is unacceptable, and one needs to trade off accuracy for speed.

We seek to answer the following question. Can we remove the dependence on the positioning system but preserve the agility of the quadrotor? There are two benefits if this can be achieved. First, it immunizes the controller from the deficiencies of the localization system. Second, we can free the computational resources preserved for onboard localization. We note that although image-based visual servo (IBVS) does not depend on any positioning system, it can only produce low-speed control with underactuated robots like quadrotors [2]. The reason is that that target can easily escape the image frame due to large rotations, which immediately fails the mission. For example, when tracking a quarter-circle trajectory at a high speed as shown in Fig. 1, the quadrotor needs to change its orientation very frequently, and it is very likely for traditional IBVS methods to lose the target.

In our previous work [3], we have proposed a novel framework to handle target visibility and enable an aggressive racing drone to reach a maximum speed of 6.87 m/s in simulation without losing the target visibility. It takes into account both the quadrotor and feature's kinematic functions and enforces visibility constraints for all predicted features in a time horizon. However, it is susceptible to camera calibration error as it requires reconstruction of relative positional information from image information. As an extension of this prior work, we propose a new objective function expressed in the image space which leads to reduced sensitivity to the calibration

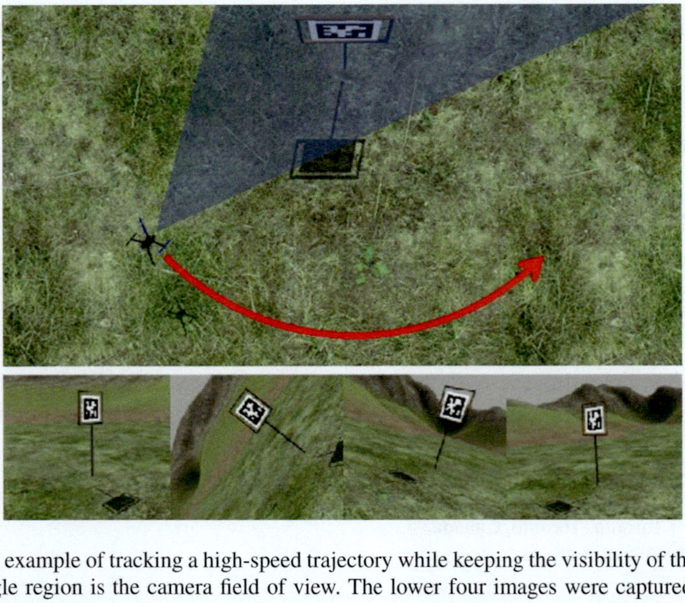

Fig. 1 An example of tracking a high-speed trajectory while keeping the visibility of the gate. The blue triangle region is the camera field of view. The lower four images were captured while the proposed method was tracking this trajectory at a maximum reference speed of 9 m/s

error. Extensive simulations show that our method can perform robust visual servoing in high-speed and aggressive quadrotor flight.

2 Problem Formulation

In this section, we detail the system modeling including dynamic, constraints, and objective functions, and assemble these modules using an optimal control framework.

2.1 Notation Definition

We define the world frame as $\mathcal{F}_w$, the body frame as $\mathcal{F}_b$, the camera frame as $\mathcal{F}_c$, and the landmark frame as $\mathcal{F}_l$. A clear overview of the reference frames is depicted in Fig. 2. The position of the landmark frame with respect to the camera frame, expressed in the camera frame is denoted by $\mathbf{p}_w^{lc} = (\mathbf{q}_{wb} \otimes \mathbf{q}_{bc}) \odot \mathbf{p}_c^{lc}$. We use quaternions to represent the orientation between frames. For example, $\mathbf{p}_c^{lc} = (x_c, y_c, z_c)^T$ represents the rotation from the landmark frame to the camera frame. An operator $\otimes$ is used to denote the Hamilton product of two quaternions, and an operator $\odot$ is used to express the multiplication between a quaternion and a vector. For example, rotating the landmark position from the camera frame to the world frame can be written as $\mathbf{p}_w^{lc} = (\mathbf{q}_{wb} \otimes \mathbf{q}_{bc}) \odot \mathbf{p}_c^{lc}$.

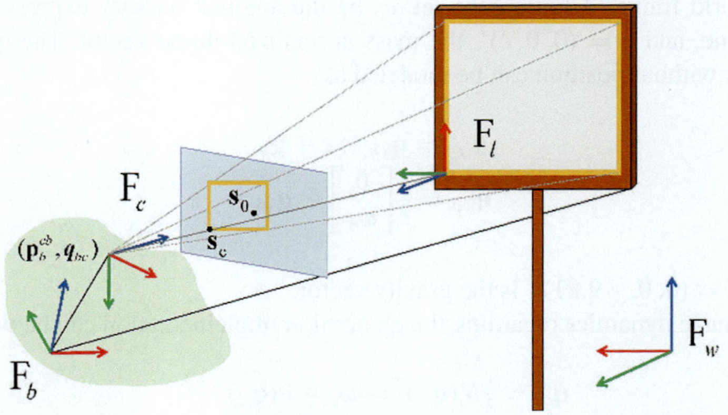

Fig. 2 A schematics representing the world $\mathcal{F}_w$, body frame $\mathcal{F}_b$ camera frame $\mathcal{F}_c$ and landmark frame $\mathcal{F}_l$. The relative pose from the camera frame to the body frame is represented by $(\mathrm{p}_b^{cb}, \mathrm{q}_{bc})$. A landmark centered at the origin of $\mathcal{F}_l$ is projected onto image coordinate s_c. The image center is denoted by s_0.

2.2 Feature Representation

A bearing vector is a 3D unit vector on 2-sphere manifold. Following our previous work, we employ it to represent an image feature and use a quaternion $\mathbf{q}_{cl}$ as its underlying implementation. The benefits of using it for feature representation are illustrated in [4]. Two functions involved in the bearing vector transformation are defined below:

$$\begin{aligned} n(\mathbf{q}_{cl}) &= \mathbf{q}_{cl} \odot \mathbf{e}_z \subset \mathbb{R}^3, \\ N(\mathbf{q}_{cl}) &= [\mathbf{q}_{cl} \odot \mathbf{e}_x, \mathbf{q}_{cl} \odot \mathbf{e}_y] \in \mathbb{R}^{3\times 2}, \end{aligned} \tag{1}$$

where $n(\mathbf{q}_{cl})$ returns the original 3D unit vector and $N(\mathbf{q}_{cl})$ linearly projects a 3D vector onto the 2D tangent space around the bearing vector; $(\mathbf{e}_x, \mathbf{e}_y, \mathbf{e}_z)$ is the standard basis for the 3D space. We use an operator $[\cdot]z$ to transform a 3D vector to homogeneous coordinates. For example, the image coordinate of a feature can be represented by $\mathbf{s}_c = [\mathbf{p}_c^{lc}]_z = (x_c/z_c, y_c/z_c, 1)^T$. Note that to simplify notations, in this paper, the image coordinate refers to homogeneous coordinates instead of pixel coordinates. The following relationship exists among different representation: $\mathbf{p}_c^{lc} = \mathbf{s}_c \cdot z_c = n(\mathbf{q}_{cl}) \cdot d$. where z_c denotes the depth and d the distance to the feature.

2.3 System Dynamics

Let $\mathbf{v}_w$ and $\mathbf{q}_{wb}$ be the velocity and the orientation of the body frame with respect to the world frame. Additionally, let $\boldsymbol{\omega}_b$ be the angular velocity expressed in the body frame, and $\mathbf{c} = (0, 0, c)^T$ the mass-normalized thrust vector. The quadrotor dynamics without position can be modeled as:

$$\begin{aligned} \dot{\mathbf{v}}_w &= \mathbf{q}_{wb} \odot \mathbf{c} + \mathbf{g}_w, \\ \dot{\mathbf{q}}_{wb} &= \frac{1}{2}\begin{bmatrix} 0 \\ \boldsymbol{\omega}_b \end{bmatrix} \otimes \mathbf{q}_{wb}, \end{aligned} \tag{2}$$

where $\mathbf{g}_w = (0, 0, -9.81)^T$ is the gravity vector.

The image dynamics regarding the quaternion implementation can be written as

$$\begin{aligned} \dot{\mathbf{q}}_{cl} &= \tfrac{1}{2} N(\mathbf{q}_{cl})^T (-\boldsymbol{\omega}_c - n(\mathbf{q}_{cl})^{\times} \tfrac{\mathbf{v}_c}{d}), \\ \dot{d} &= -n(\mathbf{q}_{cl})^T \mathbf{v}_c, \end{aligned} \tag{3}$$

where the skew operator $(\cdot)^{\times}$ produces the cross-product matrix of a 3D vector. $\mathbf{v}_c$ and $\boldsymbol{\omega}_c$ denote the linear velocity and angular velocity represented in the camera frame, respectively. The transformations from $\mathbf{v}_w$ to $\mathbf{v}_c$ and $\boldsymbol{\omega}_b$ to $\boldsymbol{\omega}_c$ are formulated as:

$$\begin{aligned} \mathbf{v}_c &= \mathbf{q}_{bc}^{-1} \odot (\mathbf{q}_{wb}^{-1} \odot \mathbf{v}_w + \boldsymbol{\omega}_b^{\times} \mathbf{p}_b^{cb}), \\ \boldsymbol{\omega}_c &= \mathbf{q}_{bc}^{-1} \odot \boldsymbol{\omega}_b. \end{aligned} \tag{4}$$

The detailed derivation of the above image dynamics is provided in [3]. Finally, the state and the input vectors of the system are defined as:

$$\begin{aligned} \mathbf{x} &= [\mathbf{v}_w^T, \mathbf{q}_{wb}^T, \mathbf{q}_{cl}^T, d]^T \in \mathbb{R}^{12}, \\ \mathbf{u} &= [c, \boldsymbol{\omega}_b^T]^T \in \mathbb{R}^4. \end{aligned} \tag{5}$$

2.4 Image Error Minimization

The visual servoing objective aims at minimizing the errors in terms of image coordinates and distances, between reference states and current states. We define the visual servoing objective function at each time step as

$$\mathcal{L}_{vs}(\mathbf{x}, \mathbf{u}) = \|\mathbf{s}_{c^*} - \mathbf{s}^*\|_{\mathbf{Q}_s}^2 + \|d - d^*\|_{Q_d}^2, \tag{6}$$

where $\|\cdot\|_{\mathbf{Q}}^2$ denotes the weighted squared Euclidean norm; $\mathbf{Q}_s$ is the diagonal weight matrix for image errors; Q_d is the scalar weight for distance errors; d^* and d are reference distance and current distance, respectively; $\mathbf{s}^*$ denotes the reference image coordinate while $\mathbf{s}_{c^*}$ denotes the current image coordinate after rotational compensation:

$$\mathbf{s}_{c^*} = [n((\mathbf{q}_{bc}^{-1} \otimes \mathbf{q}_{wb} \otimes \mathbf{q}_{bc} \otimes \mathbf{q}_{cl}))]_z. \tag{7}$$

Different from [3], this error term is defined in the image space instead of the Euclidean space, leading to increased robustness to calibration error.

2.5 Optimal Control Problem Formulation

Using the system dynamics and the objective function, we can formulate a nonlinear model predictive control problem as

$$\begin{aligned} \min_{\mathbf{u}} \; & \int_{t_0}^{t_f} \mathcal{L}_{vs}(\mathbf{x}, \mathbf{u}) + \mathcal{L}_a(\mathbf{x}, \mathbf{u}) dt \\ \text{s.t.} \; & \dot{\mathbf{x}} = f(\mathbf{x}, \mathbf{u}), \\ & \mathbf{s}_{min} \le \mathbf{s}_c \le \mathbf{s}_{max}, \\ & c_{min} \le c \le c_{max}, \\ & \boldsymbol{\omega}_{min} \le \boldsymbol{\omega}_b \le \boldsymbol{\omega}_{max}, \end{aligned} \tag{8}$$

where $\mathcal{L}_a(\mathbf{x}, \mathbf{u})$ is the action objective aiming to stabilize a quadrotor, such as tracking a specific orientation or velocity. The first equality constraint is the full dynamic constraint containing the quadrotor and image dynamics, the second one is the target visibility constraint which bounds the feature into a pre-determined image area, and the last two constraints correspond to thrust and body-rate limits, respectively. t_0 is the start time and t_f is the end time.

We apply multiple shooting with N samples as a transcription method and the Runge–Kutta-4 method for integration. To account for the drifts in visual prediction, our framework works in a receding-horizon fashion by iteratively solving the above optimization problem, with only the first control input being executed.

3 Simulation

We present realistic simulation results using Gazebo and the Pixhawk4 controller [5] with software-in-the-loop (SITL) simulation. Our goal is to validate the proposed method in varied scenarios and show its potential in high-speed applications with agile maneuvers.

3.1 Setup

The system is implemented in C++ and runs in ROS environment. Our NMPC framework is based on ACADO [6] with qpOASES as the solver. We set the time step as $dt = 0.5$ s with a horizon of $N = 20$. After each iteration, we put all solved control inputs in a queue and publish them at 20 Hz in another loop. A dynamic $\mathbf{Q}_{vs}$ is applied for the visual servoing objective function to compensate the magnitude inconsistency between distance error and image error.

We simulated a 3DR Iris quadrotor with a forward-looking depth camera. The controller was configured into the offboard mode to enable thrust and body rate control. As shown in Fig. 3, the landmark is a rectangular gate with a fiducial marker [7] inside. The target feature is selected as the center of the fiducial marker, whose position is $\mathbf{p}_w^{lw} = (6, 0, 3)^T$ for all simulations. Like [8], the decision of using fiducial markers comes from the emphasis of this work on designing a generic perception-aware control framework instead of image processing, and thus the feature extraction part can be suitably abstracted. Users can replace the marker with their own landmarks to accomplish different tasks.

In terms of state measurements, we assume that the velocity and orientation (i.e., roll, pitch, and heading) of the quadrotor can be accurately measured. For a real quadrotor that is equipped with an optical flow sensor and IMUs, this is a reasonable assumption. Furthermore, we assume that the size of the fiducial marker is given as a prior and we can estimate the distance from the robot to the marker center using the perspective-n-point (PnP) method [9]. Additionally, the camera intrinsic/extrinsic

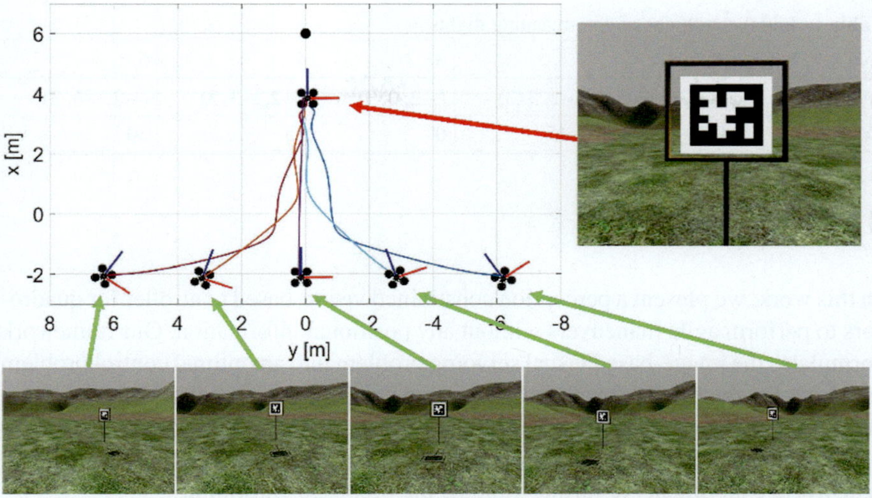

Fig. 3 Five quadrotor paths from five different initial poses to the same goal. Five initial images and the reference image are captured

parameters [10] are assumed to be well-calibrated. For simplicity, we take the image center and a small distance as the reference state.

3.2 Result

To demonstrate the effectiveness of our method, we let the controller reach the gate from varied positions. This simulation aims to show that our method can converge to the reference image from different initial conditions. Table 1. lists the initial poses and Fig. 3 showcases the initial images and the goal image, as well as the tracking trajectories. The results show that the controller can track the reference image from all initial configurations. The average velocity in these five tasks is 2.4 m/s. However, the velocity along each axis is quite different. From Fig. 3, we observe that the position errors in the y-axis converge much faster than those in the x-axis. That means the quadrotor flew much more aggressively in rolling than in pitching. One reason is that when the gate is around the image center, conducting large rolling will not lose the gate, whereas conducting pitching to accelerate the drone forward can easily move the gate to the image upper border and then lose it. As a result, the controller will prevent large forward acceleration to maintain visibility, which also limits the forward speed.

Table 1 Initial conditions of gate-reaching flights

No	1	2	3	4	5
Pos. (m)	(−2, 6, 3)	(−2, 3, 3)	(−2, 0, 3)	(−2, −3, 3)	(−2, −6, 3)
Yaw (deg.)	−30	−15	0	15	30

4 Conclusion

In this work, we present a perception constrained vision-based controller for quadrotors to perform agile maneuvers without any positional information. Our framework formulates the image-based visual servoing problem into an optimal control problem in which the quadrotor dynamics, image dynamics, actuation constraints, and visibility constraints are considered. Simulation results demonstrate that our method can ensure target visibility during aggressive flights. In future work, we will conduct real-world experiments to further validate the proposed algorithm.

References

1. Mellinger D, Michael N, Kumar V (2012) Trajectory generation and control for precise aggressive maneuvers with quadrotors. Int J Robot Res 31(5):664–674
2. Zhang K, Shi Y, Sheng H (2021) Robust nonlinear model predictive control based visual servoing of quadrotor UAVs. IEEE/ASME Trans Mechatron 26(2):700–708
3. Qin C, Yu Q, Go HH, Liu HHT (2023) Perception-aware image-based visual servoing of aggressive quadrotor UAVs. IEEE/ASME Transactions on Mechatronics
4. Bloesch M, Omari S, Hutter M, Siegwart R (2015) Robust visual inertial odometry using a direct EKF-based approach. In: 2015 IEEE/RSJ international conference on intelligent robots and systems (IROS), 298–304
5. Px4 User Guide. https://docs.px4.io/master/en/. Accessed 13 Aug 2023
6. Houska B, Ferreau HJ, Diehl M (2011) ACADO toolkit—an open-source framework for automatic control and dynamic optimization. Optimal Control Appl Methods 32(3):298–312
7. Wang J, Olson E (2016) AprilTag 2: efficient and robust fiducial detection. In: 2016 IEEE/RSJ international conference on intelligent robots and systems (IROS), 4193–4198
8. Jacquet M, Franchi A (2020) Motor and perception constrained NMPC for torque-controlled generic aerial vehicles. IEEE Robot Autom Lett 6(2):518–525
9. Lepetit V, Moreno-Noguer F, Fua P (2009) EP n P: An accurate O (n) solution to the P n P problem. Int J Comput Vision 81:155–166
10. Hartley R, Zisserman A (2003) Multiple view geometry in computer vision. Cambridge university press

Study on the Aerodynamics Interaction Between Propeller and Stratospheric Airship with Multi-propeller Layout

Xiangyan Chen, Hui Chen, Kaiyan Chen, and Jiwei Tang

Abstract The aerodynamics interaction between propeller and Euler-shaped stratospheric airship with muti-propeller layout is studied through computational fluid dynamics method. It is found that due to the airship body, the thrust of the front propeller increases and the torque decreases. The thrust and torque of the rear propeller are obviously increased. As for the side propeller, its thrust is greatly influenced by the airship, while the torque is almost unchanged comparing with that of single propeller. Therefore, the efficiency of the front propeller is the highest which is about 7% higher than that of the single propeller, and the efficiency of the side propeller is the lowest. Compared with the resistance of the airship body alone, the pressure drag and viscous resistance of the airship with propeller interference are larger. Under the effects of the interference of propeller, the pressure drag of airship is increased by about 60%, which leads to the increase of total resistance of airship. On the windward side of the airship, due to the influence of the front propeller, the pressure surged in a small range. However, the rear of the airship was affected by the rear propeller, and the pressure dropped obviously. At the front of the side propeller, there is a local pressure drop on the airship surface.

Keywords Propeller · Stratospheric airship · Aerodynamics interaction · Euler-shaped

---

X. Chen
Dlian University of Technology, Dlian, China

X. Chen · H. Chen · K. Chen
Beijing Institute of Aerospace Systems Engineering, Beijing, China

H. Chen · K. Chen
Northwestern Polytechnical University, Xi'an, China

J. Tang (✉)
School of Aeronautics and Astronautics, Shanghai Jiao Tong University, Shanghai, China
e-mail: tangjw@sjtu.edu.cn

Z. Liu et al. (eds.), *Advances and Challenges in Advanced Unmanned Aerial Systems*, Springer Aerospace Technology, https://doi.org/10.1007/978-981-99-8045-1_21

1 Introduction

Stratospheric airship is a lighter-than-air aircraft, and its working height is generally about 20 km above sea level. Stratospheric airship is very suitable as a new information platform because of its long-term stay ability in the air and reusability. Stratospheric airship can play the role of communication relay, earth monitoring, weather forecast and emergency rescue, and has great military and civilian strategic value [1–5].

Propeller is a kind of propulsion device of aircraft, and scholars have done a lot of research on the interference between propeller and aircraft. Yang et al. [6] carried out a numerical analysis of the interference effect of propeller slipstream on aircraft flow field by panel method. The numerical example results show that slipstream has a significant impact on aircraft lift characteristics, which in turn leads to a great change in aircraft torque performance. Hawkswell et al. [7] discussed how to choose the position and diameter of propeller to maximize the lift-drag ratio of wing in cruising state by using the method of actuating disk. The results show that when only the propeller and the wing are considered, the radius of the propeller is 0.2 of the chord length, and it is the best choice to install the propeller below the wing at 0.21 times the chord length. Wang et al. [8] used MRF method to carry out aerodynamic simulation on a new four-propeller-wing structure. Sun et al. [9] used the rotating coordinate system method to study the interference of the rear propeller to the double-tailed UAV. Taniguchi et al. [10] studied the influence of propeller installation position on the aerodynamic characteristics of fixed wing by solving URANS using overlapping grid technology. The results show that the pressure drag coefficient of wingtip layout is small and the pressure drag coefficient of midwing layout is large, which leads to the drag coefficient of wing nacelle mechanism with wingtip layout being smaller than that with midwing layout.

In view of the interference between the propeller and the airship, Lin and Guo [5] used the Realizable k-ε turbulence model and MRF technology to simulate the smoothness around the stratospheric airship with propeller, which showed that the aerodynamic coefficient of the airship increased with the increase of propeller thrust coefficient, and the closer the propeller was to the airship, the more obvious the influence on the airship was. However, only the flow field between airship and single propeller is simulated, and the research on the interference between airship and multiple propellers is lacking. Xie and Ye [11] calculated the flow field around the airship with propeller blowing by using the method of actuating disk. They found that the blowing action of the propeller can drive away the separation vortex on the hull surface, thus reducing the drag coefficient of the airship. The propeller is arranged behind the suction peak, which is the best blowing position, and the larger the diameter of the propeller and the denser the arrangement, the more obvious the drag reduction effect, but the airship studied is conventional in shape. Witkowski et al. [12] studied the effect of propeller slipstream on fixed wings through wind tunnel experiments and numerical simulations. Giorgi et al. [13] used OpenFOAM to numerically predict the performance of the counter rotating propeller system of remote-controlled aircraft.

In this paper, the stratospheric airship with Euler body shape and propeller are taken as the research object, and the flow field of their interaction is analyzed by CFD method. Firstly, the research model is introduced. Secondly, the numerical methodology is illustrated. Then, the CFD simulation results are present and discussed. Finally, the main conclusions are drawn.

2 Research Model

As shown in Fig. 1, the stratospheric airship with Euler body shape has a maximum diameter of 36 m, and the propeller is arranged on the horizontal symmetry plane of the airship. A three-blade high-altitude propeller with a radius of 1.35 m is adopted, as shown in Fig. 2. Because the bracket od propulsion system has little influence on its overall aerodynamic performance, the effects of these parts are ignored in aerodynamic simulations.

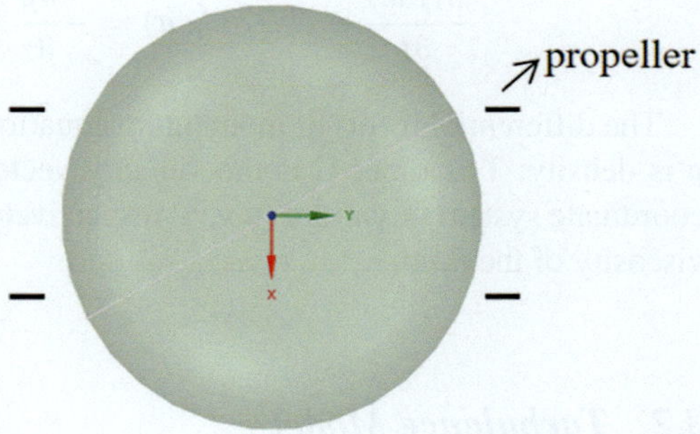

Fig. 1 The layout of propellers and airship

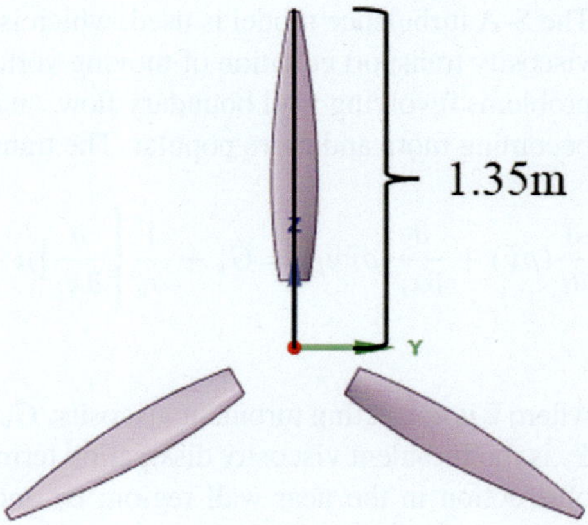

Fig. 2 Schematic diagram of high altitude propeller

3 Numerical Method

3.1 Control Equation

The interaction between airship and propeller is simulated by solving RANS equations. When the source term is not considered, the differential form of the continuity equation in the control equation is as follows:

$$\frac{\partial \rho}{\partial t} + \frac{\partial (\rho u)}{\partial x} + \frac{\partial (\rho v)}{\partial y} + \frac{\partial (\rho w)}{\partial z} = 0 \tag{1}$$

The differential forms of momentum equations in three directions are as follows:

$$\frac{\partial (\rho u)}{\partial t} + \mathrm{div}(\rho u \cdot \boldsymbol{u}) = -\frac{\partial p}{\partial x} + \mathrm{div}(\mu \cdot grad(u)) \tag{2}$$

$$\frac{\partial (\rho v)}{\partial t} + \mathrm{div}(\rho v \cdot \boldsymbol{u}) = -\frac{\partial p}{\partial y} + \mathrm{div}(\mu \cdot grad(v)) \tag{3}$$

$$\frac{\partial (\rho w)}{\partial t} + \mathrm{div}(\rho w \cdot \boldsymbol{u}) = -\frac{\partial p}{\partial z} + \mathrm{div}(\mu \cdot grad(w)) \tag{4}$$

The differential forms of momentum equations in three directions are as follows: ρ is density; T is time; U is the velocity vector, and its components in Cartesian coordinate system (x,y,z) are (u,v,w) respectively; P is the pressure; μ is the dynamic viscosity of the fluid.

3.2 Turbulence Model

The S-A turbulence model is used, which is a single equation model for solving the viscosity transport equation of moving vortex. It is widely used to simulate aviation problems involving wall boundary flow, and its application in rotating machinery is becoming more and more popular. The transport equation of is:

$$\frac{\partial}{\partial t}(\rho \tilde{\nu}) + \frac{\partial}{\partial x_i}(\rho \tilde{\nu} u_i) = G_\nu + \frac{1}{\sigma_{\tilde{\nu}}}\left\{\frac{\partial}{\partial x_j}[\mu + \rho \tilde{\nu}]\frac{\partial \tilde{\nu}}{\partial x_j} + C_{b2}\rho\left(\frac{\partial \tilde{\nu}}{\partial x_j}\right)^2\right\} - Y_\nu + S_{\tilde{\nu}} \tag{5}$$

where $\tilde{\nu}$ is correcting turbulent viscosity, G_ν is a turbulent viscosity generation term; Y_ν is the turbulent viscosity dissipation term caused by viscosity inhibition and wall obstruction in the near-wall region; $\sigma_{\tilde{\nu}}$ and C_{b2} is a constant; ν is the molecular kinematic viscosity; $S_{\tilde{\nu}}$ is a custom source item;

Turbulent viscosity is calculated by the following formula:

$$\mu_t = \rho \tilde{\nu} f_{\nu 1} \tag{6}$$

$f_{\nu 1}$ is the viscosity inhibition function, which is specifically expressed as:

$$f_{\nu 1} = \frac{\chi^3}{\chi^3 + C_{\nu 1}^3} \tag{7}$$

χ defined as:

$$\chi \equiv \frac{\tilde{\nu}}{\nu} \tag{8}$$

The turbulent viscosity generation term is expressed as:

$$G_\nu = C_{b1} \rho \tilde{S} \tilde{\nu} \tag{9}$$

where

$$\tilde{S} \equiv S + \frac{\tilde{\nu}}{\kappa^2 d^2} f_{\nu 2} \tag{10}$$

$$f_{\nu 2} = 1 - \frac{\chi}{1 + \chi f_{\nu 1}} \tag{11}$$

C_{b1} and κ is a constant; D is the distance from the grid to the wall; S is the scalar value of the deformation tensor.

The turbulent viscosity dissipation term is expressed as:

$$Y_\nu = C_{w1} \rho f_w \left(\frac{\tilde{\nu}}{d} \right)^2 \tag{12}$$

among

$$f_w = g \left(\frac{1 + C_{w3}^6}{g^6 + C_{w3}^6} \right)^{1/6} \tag{13}$$

$$\mathrm{g} = \mathrm{r} + C_{w2}\left(r^6 - r\right) \tag{14}$$

$$\mathrm{r} \equiv \frac{\tilde{\nu}}{\tilde{S} \kappa^2 d^2} \tag{15}$$

The values of the above correlation constants are as follows: $C_{b1} = 0.1355$, $C_{b2} = 0.622$, $\sigma_{\tilde{\upsilon}} = 2/3$, $C_{\upsilon 1} = 7.1$, $C_{w1} = \frac{C_{b1}}{\kappa^2} + \frac{(1+C_{b2})}{\sigma_{\tilde{\upsilon}}}$, $C_{w2} = 0.3$, $C_{w3} = 2.0$, $\kappa = 0.4187$.

3.3 Calculation Method and Boundary Conditions

The rotation motion of propeller is simulated by MRF method. The governing equation is discretized by finite volume method, the second-order upwind scheme is used for the convection term, the least square cell-based method is used for the gradient term, and the second-order scheme is used for the pressure term. In order to speed up the convergence of calculation, this paper first uses Simple algorithm to simulate, obtains the initial flow field and then changes to Coupled algorithm to continue the calculation until the final result is obtained.

Stratospheric airship generally runs at an altitude of 20 km, when the air density is 0.0889 kg/m^3, the dynamic viscosity is 1.42161×10^{-5} Ns/m^2 and the air pressure is 5529 Pa. The boundary conditions used in this paper are mainly velocity inlet, pressure outlet and wall surface. The inflow speed at the speed inlet is 5 m/s, the pressure outlet is a pressure gradient-free outlet, and the operating pressure is 5529 Pa, so the gauge pressure is 0 Pa. The outer boundary of the calculation domain is set as the wall boundary condition except the velocity inlet and the pressure outlet. The propeller surface is set as a non-slip wall boundary condition and is stationary relative to the rotation domain.

3.4 Mesh Setup

In order to give consideration to both grid quality and computational efficiency, this paper selects the unstructured grid composed of polyhedron and hexahedron to divide the computational domain, with polyhedron grid near the boundary and hexahedron grid in the rest. In order to accurately capture the viscous effect, mesh encryption is carried out on the propeller surface and its nearby area. The grid height of the first layer on the wall is 4×10^{-5} m, which makes y + around 1. Details of the grid are shown in Fig. 3.

4 Results and Discussions

Table 1 shows the aerodynamic performance of propellers arranged around the airship. It can be seen from the table that due to the existence of the airship, the wake generated by the front propeller will impact the surface of the airship, making the wake flow back, thus increasing the thrust of the front propeller and reducing the

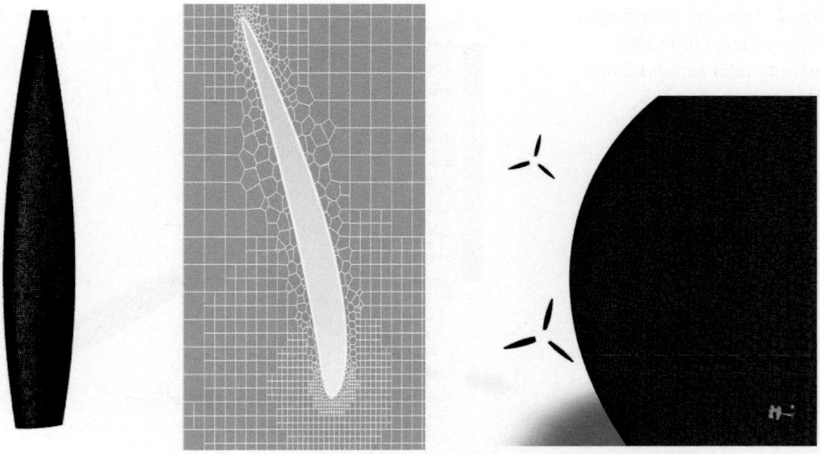

(a) blade surface mesh (b) blade section mesh at 90% radius (c) grid with propeller arranged in series

Fig. 3 The sketch of computational grid

torque. Due to the shielding effect of airship, the thrust and torque of rear propeller will increase obviously. As for the side propeller, its thrust is greatly influenced by the airship, while the torque and the ratio of single propeller are almost unchanged. Therefore, according to the previous analysis, the efficiency of the front propeller is the largest, which is about 7% higher than that of the single propeller. The thrust torque of the rear propeller is increasing, but the thrust is increasing more obviously, and its efficiency is slightly higher than that of the single propeller. The efficiency of the side propeller is the lowest because of the smaller thrust. Figure 4 shows the pressure distribution of propellers arranged around the airship. As can be seen from the figure, the absolute value of negative pressure of the rear propeller is the largest, followed by the front propeller and the smallest is the side propeller. And the negative pressure on the windward surface is the smallest in the rear propeller and the largest in the front propeller. The above analysis of propeller pressure distribution just reflects the previous analysis of aerodynamic performance of front propeller, rear propeller and side propeller.

Figure 5 shows the pressure drag, viscous drag and total drag of the airship under the interference of propellers arranged around. Compared with the resistance of the airship alone, the pressure drag and viscous resistance of the airship with propeller

Table 1 Aerodynamic performance of propeller with surrounding arrangement under airship interference

	Thrust (N)	Torque (Nm)	Efficiency (%)
Front propeller	74.9	14.7	39.9
Rear propeller	81.5	18.6	34.4
Side propeller	66.3	16.9	30.8
Single propeller	69.2	16.4	33.1

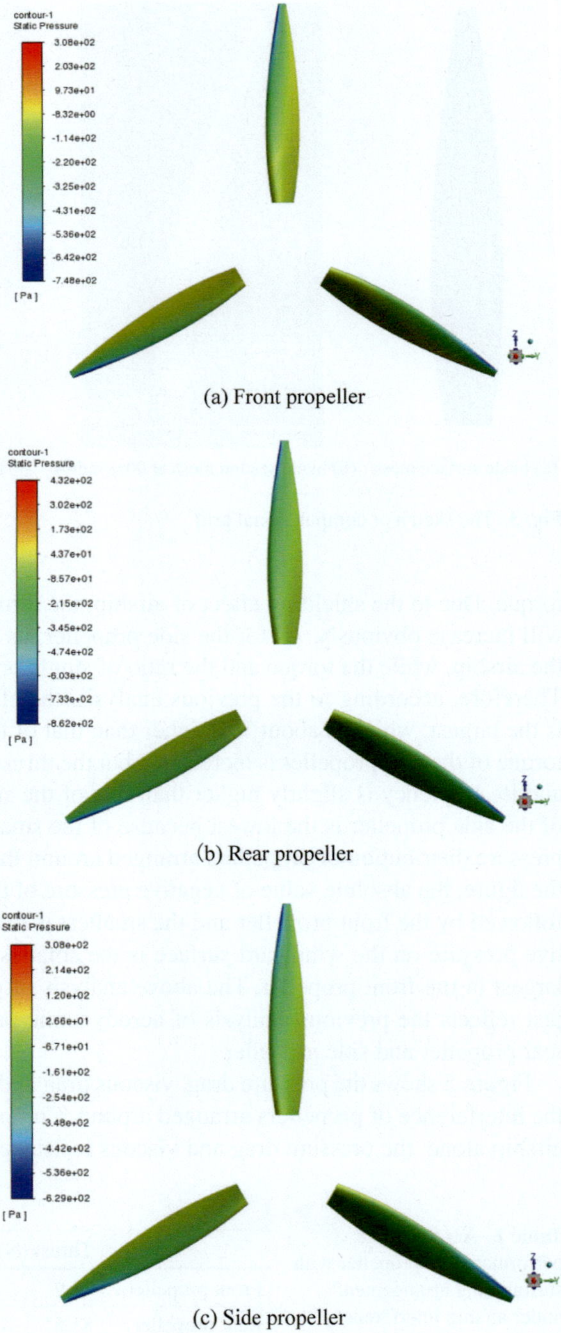

Fig. 4 Pressure distribution of propellers around the airship under aerodynamics interference

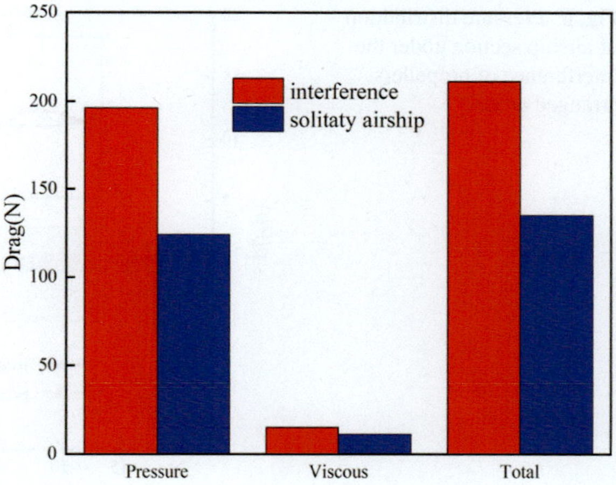

Fig. 5 Resistance of airship with or without propeller around

interference are larger than those of the airship alone. The interference of propeller has little influence on the viscous resistance of airship, and the situation with interference is only slightly larger than that without interference, while there is obvious difference in pressure drag. Under the interference of propeller, the pressure drag of airship is increased by about 60% compared with that without interference, which leads to the total resistance of airship under the influence of propeller is greater than that without influence.

Figure 6 shows the pressure distribution of z = 0 section and y = 0 section under the interference of propellers arranged around. As can be seen from Fig. 6a, on the windward side of the airship, due to the influence of the front propeller, the pressure surged in a small range, from a small negative pressure to a large positive pressure. After that, the pressure distribution of the airship disturbed by the propeller is similar to that without interference within a certain distance, until the front of the side propeller, the pressure drops locally. Finally, under the influence of the rear rotor, there is an obvious pressure drop at the tail of the airship. Similar to Fig. 6a, the pressure distribution influenced by propeller in Fig. 6b shows a pressure surge at the front end of the windward side of the airship, and a pressure drop at the rear end of the airship. However, the pressure distribution of the airship with or without propeller interference has similar changes in a long range in the middle of the airship. Because there is no influence of the propeller, the pressure distribution of z = 0 profile does not have the peak value of negative pressure in the middle of Fig. 6a. Figure 7 shows the pressure distribution on the airship surface under propeller interference. As previously analyzed, there is a sharp increase in pressure at the front end of the windward side of the airship, a sudden drop in pressure at the rear end of the airship, and a local pressure reduction can be observed at the front end of the side propeller.

Fig. 6 Pressure distribution of airship section under the interference of propellers arranged around

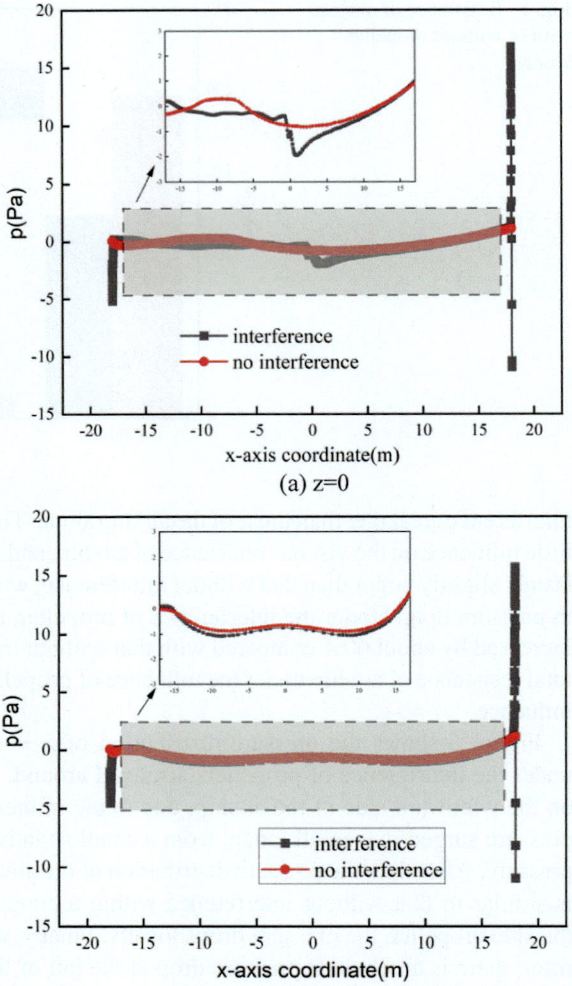

5 Conclusions

In this paper, the interaction between propeller and airship with Euler-shaped body is simulated. It is found that due to the airship, the thrust of the front propeller increases and the torque decreases. The thrust of the rear propeller is obviously increased, and the torque is also obviously increased. As for the side propeller, its thrust is greatly influenced by the airship, while the torque and the ratio of single propeller are almost unchanged. Therefore, the efficiency of the front propeller is the highest and the efficiency of the side propeller is the lowest. Compared with the resistance of the airship alone, the pressure drag and viscous resistance of the airship with propeller

Fig. 7 Surface pressure distribution of airship under the interference of propellers arranged around

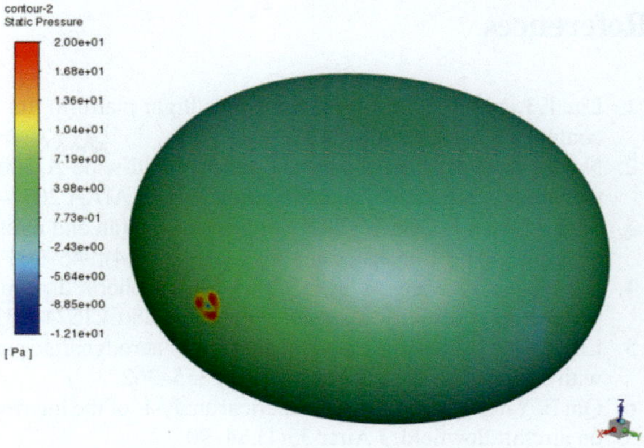

(a) The front of airship

(b) the rear of airship

interference are larger than those of the airship alone. Under the interference of propeller, the pressure drag of airship is increased by about 60% compared with that without interference, which leads to the total resistance of airship under the influence of propeller is greater than that without influence. On the windward side of the airship, due to the influence of the front propeller, the pressure surged in a small range, while the tail of the airship was affected by the rear propeller, and the pressure dropped obviously. At the front of the side propeller, there is a local pressure drop on the airship surface. The conclusions are beneficial to the design of propulsion system layout of stratospheric airship with Euler-shaped body.

References

1. Liu T, Liou W (2006) Aeroship: a new flight platform. In: 24th AIAA applied aerodynamics conference. AIAA 2006–3922
2. Nayler A (2001) Airship development world-wide-A 2001 review. In: 1st AIAA, aircraft, technology integration, and operations forum. AIAA 2001–5263
3. Tang J, Duan D, Xie W (2022) Shape exploration and multidisciplinary optimization method of semirigid nearing space airships. J Aircr 59(4):946–963
4. Yao W, Li Y, Wang W et al (2007) Stratospheric airship optimization method and design parameters sensitivity analysis. J Astronaut 28(6):1524–1528
5. Lin R, Guo X (2010) Numerical analysis of aerodynamic performance for stratospheric airship with a propeller. Chin Q Mech 31(03):355–362
6. Qin E, Yang G, Li F (1998) Numerical analysis of the interference effect of propeller slipstream on aircraft flowfield. J Aircr 35(1):84–90
7. Hawkswell GN, Miller R, Pullan G (2022) Selection of propeller-wing configuration for blown wing aircraft. In: AIAA SCITECH 2022 forum. AIAA 2022–1024
8. Wang K, Zhou Z, Zhu X, Ping X (2019) Aerodynamic design of multi-propeller/wing integration at low Reynolds numbers. Aerosp Sci Technol 84:1–17
9. Sun C, Zhao W (2001) Analysis of aerodynamic interaction between mounted porpeller and twin-tailboom UAV. J Ordnance Equip Eng 42(06):118–122
10. Taniguchi S, Oyama A (2022) Numerical analysis of propeller mounting position effects on aerodynamic propeller/wing interaction. In: AIAA SCITECH 2022 forum. AIAA 2022–0153
11. Xie F, Ye Z (2009) Drag reduction for an airship with proper arrangement of propellers. Chin J Aeronaut 22(6):575–582
12. Witkowski D, Lee AKH., Sullivan JP (1989) Aerodynamic interaction between propellers and wings. J Aircr 26(9):829–836
13. Grazia DGM, Teresa D, Antonio F (2017) Numerical investigation of the performance of contra-rotating propellers for a remotely piloted aerial vehicle. Energy Procdia 126:1011–1018

Rapid Trajectory Optimization of Continuous-Low-Thrust Spacecraft Based on Bezier Shape Function and Inverse Dynamics

Tianchen Wang, Xiaoyu Shang, Jianfeng Wang, Mingying Huo, and Naiming Qi

Abstract Deep space exploration missions are a great fit for continuous-thrust spacecraft because of their benefits of high specific impulse and extended operating period. However, when it comes to tackling the trajectory optimization problem for continuous-low-thrust spacecraft, conventional approaches for trajectory optimization have drawbacks including low computing efficiency and challenging beginning value guessing. In this research, a fast way to produce workable transfer trajectories for small-thrust spacecraft is given using a form optimization approach based on Bezier curves. The dynamics model is first processed, and the inverse dynamics concept is used to parameterize the model. The transfer trajectory is then shaped using various Bezier curves. Finally, we mix several boundary conditions to maximize the shape function's unknown parameters. The Earth-to-Mars transfer simulation example of the continuous-thrust spaceship is chosen for validation. The findings confirm the applicability and effectiveness of this algorithm to the continuous-low-thrust spacecraft trajectory optimization problem and demonstrate a significant computational efficiency advantage for the Bezier shape function method for trajectory optimization over the direct method.

Keywords Bezier curve · Trajectory optimization · Low-thrust spacecraft

T. Wang · X. Shang · M. Huo (✉) · N. Qi
Areospace Engineering, Harbin Institute of Technology, Harbin, China
e-mail: huomingying@hit.edu.cn

T. Wang
e-mail: 18745095295@163.com

N. Qi
e-mail: qinmok@163.com

J. Wang
Aerodynamics, Aerodynamic Research Institute, Harbin, China
e-mail: 13208479940@163.com

Z. Liu et al. (eds.), *Advances and Challenges in Advanced Unmanned Aerial Systems*, Springer Aerospace Technology, https://doi.org/10.1007/978-981-99-8045-1_22

1 Introduction

The trajectory optimization of the spacecraft entails designing and modifying the system control variable parameters for all or a portion of the flight mission period of the spacecraft in order to more effectively accomplish the flight mission objectives while satisfying dynamic constraints, boundary constraints, and other process constraints. This is carried out to guarantee that the process's one or more performance metrics are optimized.

The core of the trajectory optimization problem is the optimal control problem. To solve this kind of issue, both the direct method and the indirect method are routinely used. However, the direct method solves the continuous problem by discretizing it. The excess of optimization variables is the root cause of the high computing expense and low computation efficiency. Fast answers to large-scale, high-precision trajectory optimization problems are difficult to come by.

The indirect method modifies the problem based on the variational methodology and the Pontryagin minimal value concept. Even if the efficiency of the computation and optimality of the solution may both be guaranteed, the issue transformation process may be difficult, especially in complicated dynamic models. Due to difficult transformations, complex constraint conditions, and the indirect approach's co-state variables' unclear physical significance, the actual computation procedure has a low convergence rate.

The two methods discussed above cannot be used to handle large-scale trajectory optimization problems with ease. Using exponential sinusoidal forms to suit spacecraft trajectories, a rapid generating method for airplane trajectories based on shape approaches was first proposed in literature [1]. The shape method works by selecting a certain form as the spacecraft's reference route. The shape can then be designed to be optimal under a certain optimization index while still fulfilling a number of constraints by developing and optimizing the shape's attributes. Accuracy can be guaranteed since the reference trajectory offered by the shape method is continuous. Additionally, none of these variables have any impact on the number of optimization variables because it only depends on the form parameter and not on the trajectory transfer time or distance. When constructing a transfer trajectory optimization, it functioned superbly.

In the literature [2], the shape-based method was used for spacecraft trajectory analysis and optimization. The literature [3] offered a pseudo equinox shaping strategy, employing this method to create small-thrust spacecraft's initial orbits while simultaneously utilizing a variety of gravity-assistance strategies. The shape approach and genetic algorithm are used in literature [4] to solve the continuous low-thrust spacecraft rendezvous issue and produce suboptimal transfer paths both with and without time limitations. Literature [4] developed a trajectory fitting approach employing velocity components based on the spacecraft's velocity curve. The velocity function may readily satisfy the boundary constraints of location and velocity since it is made up of the sum of many basic basis functions. A shape approximation method was put out in the literature [5] for the continuous small-thrust transfer trajectory between two coplanar circular orbits. A broad foundation for the continuous low-thrust trajectory design approach based on the shape method was provided in the literature [6]. A novel elevation and radius shape combination for 3D low-thrust trajectories utilizing the initial orbital plane as the reference plane is presented in literature [7]. The specified modest thrust transfer trajectory can serve as a beginning value for a more precise trajectory optimization approach, according to literature [8] that examined the shape method based on the initial and target trajectory elements. The sequence search of solar sail spacecraft for a variety of mission circumstances is carried out in [9] using the shape method. The continuous small-thrust transfer orbit was created using the shape approach in [10] within the context of the circular constrained three-body issue. In the literature [11], the adjoint estimation approach was used to optimize the continuous small-thrust transfer trajectory and compare the computational effectiveness across several coordinate systems. A logarithmic spiral curve was used in literature [12] to depict the analytical solution of the solar sail spacecraft trajectory. References [13–15] produce on/off thrust profiles, stretch trajectories from two-dimensional planes to three-dimensional space trajectories, and construct approximation space trajectories using finite Fourier series. For the purpose of studying the Earth-Mars transmission job, literature [16] employed the shape method based on a Fourier series approximation. The literature [17] demonstrated that this approach can be employed in the quick design of the electric sail trajectory while taking into account the real properties of the thrust vector by using it to construct the time-optimal transfer trajectory of the electric sail. The aforementioned method has been used in literature [18–20] for the quick production of numerous asteroid detection trajectories, and it is also used for the rapid optimization of satellite formation trajectories [21].

The Bezier curve, which was widely used in computer graphics, was first used to design the main body of the car. The Bezier curve, because to its intrinsic qualities, may readily and rigorously satisfy the start and end points of the boundary limitations of the flight trajectory, particularly those related to position and speed. The Bezier curve is a great reference shape for creating and fitting trajectories as a result. In literature [22], the Bezier shape function was used to anticipate the course of the spacecraft and fit the actual orbital data. In literature [23], the Bezier curve is used to fit the output variables of the system, and the system state variables and control variables are derived using the inverse operator.

Inverse dynamics and the Bezier shape function method will be used in this article to tackle the transfer trajectory optimization problem for continuous low-thrust spacecraft. The material in the article is arranged as follows. The second chapter introduces the form function trajectory optimization approach, the third chapter uses numerical simulation to show the method's value, and the fourth chapter discusses this paper's conclusion and its future directions.

2 Shape Function Trajectory Optimization Algorithm

The Bezier curve is first mathematically described, and the equation for the n-order Bezier curve is as follows:

$$B(\tau) = \sum_{i=0}^{n} \binom{n}{i} (1-\tau)^{n-i} \tau^i P_i \tag{1}$$

Which $\binom{n}{i} = \frac{n!}{i!(n-i)!}$, Eq. (1) can be expanded to

$$\begin{aligned} B(\tau) &= (1-\tau)^n P_0 + \binom{n}{1} (1-\tau)^{n-1} \tau P_1 \\ &+ \cdots + \binom{n}{n-1} (1-\tau)\tau^{n-1} P_{n-1} + \tau^n P_n,\ \tau \in [0, 1] \end{aligned} \tag{2}$$

Which, P_0, P_n is the start point and end point of the Bezier curve, and $P_1, P_2, ..., P_{n-1}$ is the middle control point; τ is the time parameter, and its value range is $\tau \in [0, 1]$ [0, 1]. The above equation can also be expressed as:

$$B(\tau) = \sum_{i=0}^{n} b_{i,n}(\tau) P_i \qquad \tau \in [0, 1] \tag{3}$$

Which

$$b_{i,n}(\tau) = \binom{n}{i} \tau^i (1-\tau)^{n-i}, \quad i = 0, ..., n \tag{4}$$

Which polynomial $b_{i,n}$ is called a Bernstein basis polynomial of order n.

The Bezier curve can be used to solve the challenge of optimizing the trajectory of a spaceship with continuous propulsion since it is an n-order differentiable smooth curve. The Bezier curve's first derivative with respect to the time parameter τ is:

$$\frac{d}{d\tau} B_n(\tau) = \sum_{i=0}^{n-1} \binom{n}{i+1} (i+1)\tau^i (1-\tau)^{n-i-1} P_{i+1} - \binom{n}{i} (n-i)\tau^i (1-\tau)^{n-i-1} P_i \tag{5}$$

When the Bezier curve's order n is known, Eq. (4) may be used to calculate its Bernstein basis polynomial, and the control points primarily dictate the curve's form. Create some indexes for optimization to get the best results.

The continuous low-thrust spacecraft's dynamic model is provided below.

$$\begin{aligned}\dot{\boldsymbol{r}} &= \boldsymbol{v} \\ \dot{\boldsymbol{v}} &= -\frac{\mu}{r^3}\boldsymbol{r} + \frac{\boldsymbol{T}}{m} \\ \dot{m} &= -\frac{T}{I_{sp} g_0}\end{aligned} \tag{6}$$

Which, $\boldsymbol{r}$, $\boldsymbol{v}$ are the position and velocity vector of the spacecraft respectively, $\boldsymbol{T}$ is the thrust vector of the spacecraft, m is the mass of the spacecraft, I_{sp} is the fuel specific impulse of the spacecraft, and g_0 is the gravitational acceleration constant at sea level.

The position and velocity vectors $\boldsymbol{r}$, $\boldsymbol{v}$ are the system state variables in the analysis of the dynamic model mentioned above. The control variable is thrust acceleration $\boldsymbol{u} = \boldsymbol{T}/m$, and the change of spacecraft mass Δm is determined by the change of thrust. Using the Bezier curve to fit the trajectory of the state variable $\boldsymbol{r}$, the parameterized form of the state variable $\boldsymbol{v}$ and the control variable $\boldsymbol{T}$ can be obtained through Eq. (6). The specific process is as follows:

First, because the Bezier curve's temporal parameters are specified in the interval $\tau \in [0, 1]$, there must be a transformation of the real time domain. The linearized time domain transformation approach is used in this study:

$$\tau = \frac{t}{t_f} \cdot t = [0, t_f] \tag{7}$$

Which, t_f is the end time, Then

$$\frac{d(\cdot)}{dt} = \frac{d(\cdot)}{d\tau} \cdot \frac{d\tau}{dt} = \frac{1}{t_f} \cdot \frac{d(\cdot)}{d\tau} \tag{8}$$

In order to simplify the expression, $\frac{d(\cdot)}{d\tau}$ will be represented by $(\cdot)'$.
Fitting the reference trajectory with the Bezier shape function:

$$\boldsymbol{r} = B(\tau) \tag{9}$$

From Eq. (9), we can get:

$$\boldsymbol{v} = \dot{\boldsymbol{r}} = B'(\tau)/t_f \tag{10}$$

$$\boldsymbol{u} = \dot{\boldsymbol{v}} + \frac{\mu}{r^3}\boldsymbol{r} = \frac{B''(\tau)}{t_f^2} + \frac{\mu}{|B(\tau)|^3}B(\tau) \tag{11}$$

The state variables and control variables have thus far been parameterized to allow for their representation by a collection of Bezier curves $B(\tau)$. The following are the restrictions on the trajectory optimization issue for a low-thrust spacecraft:

$$\begin{aligned} r(0) = r_0 \; r(t_f) = r_f \\ v(0) = v_0 \; v(t_f) = v_f \end{aligned} \tag{12}$$

$$\begin{aligned} 0 < \|\boldsymbol{u}\|_2 < \frac{T_{\max}}{m_0} \\ m_n < m < m_0 \end{aligned} \tag{13}$$

Which, Eq. (12) is the boundary constraint condition, Eq. (13) is the thrust process constraint, which $T_{\max}$ represents the maximum thrust of spacecraft and m_n is the mass of the spacecraft body without fuel mass.

Simultaneous Eqs. (2), (5), (9), (10) and (12), when the boundary constraints are determined, part of the curve parameters of the Bezier curve P_0, P_1, P_{n-1}, P_n can be directly solved, namely:

$$\begin{aligned} P_0 &= r_0 \quad\quad P_1 = v_0 t_f / n + r_0 \\ P_n &= r_f \; P_{n-1} = -v_f t_f / n + r_f \end{aligned} \tag{14}$$

The Bezier curve can rigorously satisfy the boundary constraint constraints as long as Eq. (14) is satisfied, and the aforementioned curve parameters P_0, P_1, P_{n-1}, P_n can be taken for granted as known and not computed as optimization variables.

It is thought to set Gauss–Legendre collocation points for the process constraint processing of Eq. (13), and the constraints are employed as nonlinear inequality constraints, with the thrust acceleration being restricted on the collocation points. It is sufficient to restrict the ultimate mass to be more than the dry weight of the spacecraft when treating mass constraints.

The optimization index is listed below, and the fuel optimum index and the time optimal index are the two separate optimization indexes that are taken into consideration. The optimal index is attained when the index function value is the lowest, and the index is as follows:

$$J_1(P_2, \ldots, P_{n-2}, t_f) = m(t_f) \tag{15}$$

$$J_2(P_2, \ldots, P_{n-2}, t_f) = t_f \tag{16}$$

The free Bezier curve parameter points and trajectory transfer time are the optimization variables.

A parameterized representation is needed for spacecraft mass. Combining Eq. (6) and $\boldsymbol{u} = \boldsymbol{T}/m$, we arrive to the following differential equation for spacecraft mass:

$$\dot{m} = -\frac{m\|\boldsymbol{u}\|_2}{I_{sp} g_0} \tag{17}$$

Since the parameterized propulsion acceleration vector $\boldsymbol{u}$ has been obtained, the final mass of the spacecraft $m(t_f)$ can be obtained through integration.

After obtaining the parameterized propulsion acceleration vector $\boldsymbol{u}$, integration can be used to determine the spacecraft's ultimate mass $m(t_f)$.

3 Numerical Simulation

The Bezier shape function method is used to resolve the continuous thrust Earth-Mars transfer problem, and the orbital data is based on the J2000 ephemeris. The following Table 1 are a few parameters:

3.1 *Time Optimal Trajectory Optimization Based on Bezier Shape Function*

Combining the start and end position constraints to solve the matching 3rd-order Bezier curve yields an optimized initial value that is then entered into the optimization solver for optimization. The Bezier curve has a 6th order, and Homan transfer is used to infer the time's beginning value. The index for optimization in solution is transfer time.

The exact same problem is tackled directly, and the results are contrasted. The calculator software is GPOPS2. Table 2 are the outcomes of the simulation:

The results show that even though the direct method's solution result is better than the Bezier shape function's in terms of the optimization index (transfer time), the direct method's solution time is still significantly longer than the Bezier form function method's. The transfer trajectory computed using the two methods is shown in Fig. 1. Figure 2 is respectively (a) Bezier and GPOPS2 thrust amplitude curves, (b) Bezier and GPOPS2 position component curves, (c) Bezier thrust component curves, (d) GPOPS2 thrust component Graph. It is obvious that the Bezier shape function approach's calculation result is not the best solution, but rather a practical but

Table 1 Astronomical constant

Symbol	Parameter name	Value	Unit
μ	Sun's gravitational constant	$1.32712440018 \times 10^{20}$	m^3/s^2
AU	astronomical unit	149597870700	m

Table 2 Simulation results

Method	Calculate time [s]	Transfer Time [year]
Bezier function	**0.0985**	0.3071
hp-adaptive Legendre–Gauss-Radau pseudospectral method	0.6303	**0.2491**

subpar one. The advantages of this strategy are apparent. The solution is sufficiently continuous and smooth and has a high computational efficiency, which makes it easier to design a control system.

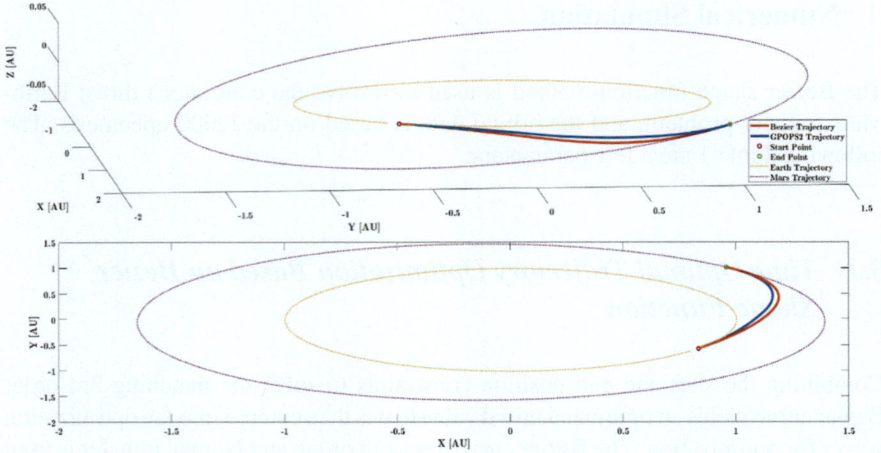

Fig. 1 Earth-Mars transfer trajectory with Bezier and Direct method

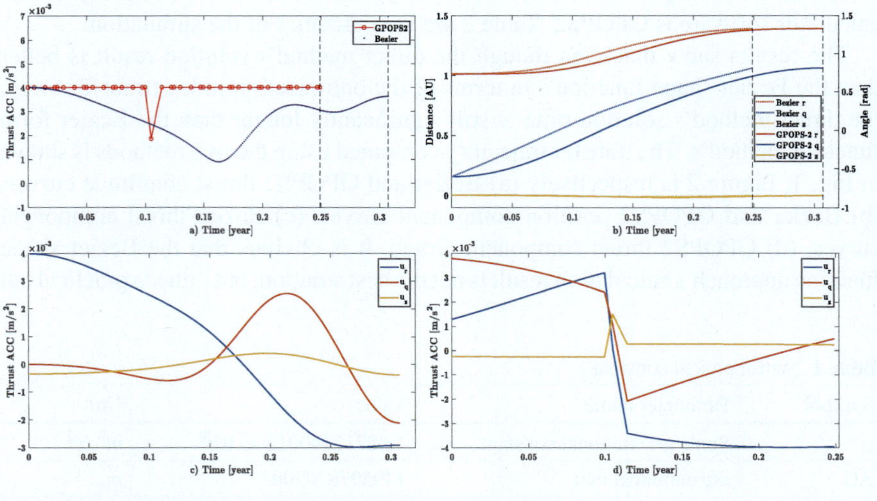

Fig. 2 A figure caption is always placed below the illustration

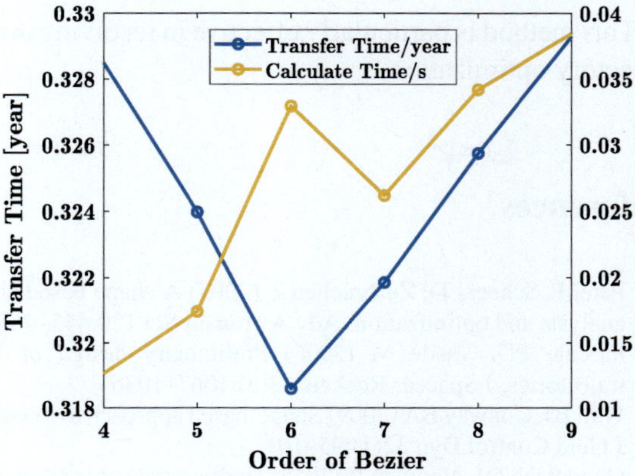

Fig. 3 Influence of different Bezier curve orders on results

3.2 Influence of Different Bezier Curve Orders on Results

Figure 3 depicts the transfer time and calculation time for the continuous thrust Earth-Mars transfer problem that was addressed using the form function strategy of Bezier curves of various orders. In general, it can be demonstrated that as the order of Bezier curves increases, the number of optimization variables also increases, leading to an equivalent increase in calculation time.

However, there is no positive or negative correlation between the order and the value of the optimization indicator function (transition time), most likely because it is easy to guess the beginning value incorrectly as the order of the Bezier curve increases. When there is a poor local optimal solution, the optimizer terminates too soon.

4 Conclusions

The continuous thrust trajectory optimization problem is quickly addressed in this paper utilizing a form method based on Bezier curves. The dynamic equation may be used to parametrize the state variables and control variables by fitting the transfer trajectory through the Bezier curve. The following advantages of the Bezier shape method over a direct solution to the identical problem:

1. The Bezier shape method may be used to rigourously satisfy the start and terminal boundary limitations;
2. Because the control curve produced by the Bezier shape method is sufficiently smooth, the effect on the control system may be reduced;
3. The calculation efficiency of the Bezier shape approach is excellent. The aforementioned simulation's computation time is 15.63% faster than the direct method's.

This method is particularly effective in resolving problems with continuous thrust trajectory optimization.

References

1. Patel P, Scheers D, Zurbruchen T (2005) A shape based approach to spacecraft trajec-tories: analysis and optimization. Adv Astronaut Sci 120:445–463
2. Pascale PD, Vasile M (2006) Preliminary design of low-thrust multiple gravity-assist trajectories. J Spacecr Rockets 43(5):1067–1076
3. Wall BJ, Conway BA (2009) Shape-based approach to low-thrust rendezvous trajectory design. J Guid Control Dyn 32(1):95–101
4. Gondelach DJ, Noomen R (2015) Hodographic-shaping method for low-thrust inter-planetary trajectory design. J Spacecr Rockets 52(3):728–738
5. Xie C, Zhang G, Zhang YC (2015) Simple shaping approximation for low-thrust trajectories between coplanar elliptical orbits. J Guid Control Dyn 38(12):2448–2455
6. Novak DM, Vasile M (2011) Improved shaping approach to the preliminary design of low-thrust trajectories. J Guid Control Dyn 34(1):128–147
7. Xie C, Zhang G, Zhang YC (2016) Shaping approximation for low-thrust trajectories with large out-of-plane motion. J Guid Control Dyn 39(12):2776–2785
8. Zeng K, Geng YH, Wu BL (2017) Shape-based analytic safe trajectory design for spacecraft equipped with low-thrust engines. Aerosp Sci Technol 62:87–97
9. Peloni A, Dachwald B, Ceriotti M (2018) Multiple near-earth asteroid rendezvous mission: solar-sailing options. Adv Space Res 62(8):2084–2098
10. Vellutini E, Avanzini G (2014) Shape-based design of low-thrust trajectories to cislunar Lagrangian point. J Guid Control Dyn 37(4):1329–1335
11. Jiang F, Tang G, Li J (2017) Improving low-thrust trajectory optimization by adjoint estimation with shape-based path. J Guid Control Dyn 40(12):3280–3284
12. Bassetto M, Niccolai L, Quarta AA (2018) Logarithmic spiral trajectories generated by Solar sails. Celest Mech Dyn Astron 130(2):18
13. Abdelkhalik O, Taheri E (2012) Approximate on-off low-thrust space trajectories using Fourier series. J Spacecr Rockets 49(5):962–965
14. Taheri E, Abdelkhalik O (2015) Fast initial trajectory design for low-thrust restricted-three-body problems. J Guid Control Dyn 38(11):2146–2160
15. Taheri E, Abdelkhalik O (2016) Initial three-dimensional low-thrust trajectory design. Adv Space Res 57(3):889–903
16. Taheri E, Kolmanovsky I, Atkins E (2016) Shaping low-thrust trajectories with thrust-handling feature. Adv Space Res 61(3):879–890
17. Huo MY, Zhang G, Qi NM (2019) Initial trajectory design of electric solar wind sail based on finite Fourier series shape-based method. IEEE Trans AerospElectr Syst 55(6):3674–3683
18. Huo M, Mengali G, Quarta AA, Qi N (2019) Electric sail trajectory design with Bezier curve-based shaping approach. Aerosp Sci Technol 88:126–135
19. Fan ZC, Huo MY, Qi NM (2019) Fast preliminary design of low-thrust trajectories for muti-asteroid exploration. Aerosp Sci Technol 93:105295
20. Fan ZC, Huo MY, Xu Y (2020) Continuous transfer trajectory generation of low-thrust spacecraft for multi-asteroid exploration. Basic Clin Pharmacol Toxicol 126:249–250
21. Fan ZC, Huo MY, Xu S (2020) Fast cooperative trajectory optimization for close-range satellite formation using Bezier shape-based method. IEEE Access 8:20918–30927
22. Pan X (2020) Trajectory design for continuous thrust maneuvering based on Bezier curve. Harbin Inst Technol:34–45 (in Chinese)
23. Lee S, Kim Y (2021) Optimal output trajectory shaping using Bezier curves. J Guid Control Dyn 44(5):1027–1035

Research on Tail-Sitter VTOL UAV Landing Technology Based on Pre-trained Bionic Technology

Hao Qi, Yiming Peng, Hong Nie, and Xiaohui Wei

Abstract The Tail-sitter VTOL UAV adopts fuselage erection, VTOL, and overall commutation, resulting in poor stability of the UAV during landing and low adaptation to traditional landing devices. To solve the above problems, this paper proposes a Tail-sitter VTOL UAV landing gear based on Free-tail technology. Through the design of free tail structure and function integration, the landing gear is adapted to the vertical and level flight dual mode of the Tail-sitter VTOL UAV. In addition, with the pre-training learning of the dynamic landing process of the cat, the effective bionic simulation of the biological structure of the Tail-sitter VTOL UAV is realized, followed by the multi-objective optimization combined with the need for parameters in different modes of the UAV itself, which effectively improves the landing adaptability and stability.

Keywords VTOL · Tail-sitter UAV · Landing gear · Adaptive · Free-tail · Bionic technology

H. Qi · X. Wei (✉)
State Key Laboratory of Mechanics and Control for Aerospace Structures, Nanjing University of Aeronautics and Astronautics, Nanjing, China
e-mail: Wei_xiaohui@nuaa.edu.cn

H. Qi
e-mail: Qihao2020@nuaa.edu.cn

Y. Peng · H. Nie
Key Laboratory of Fundamental Science for National Defense Advanced Design Technology of Flight Vehicle, Nanjing University of Aeronautics and Astronautics, Nanjing, China
e-mail: yimingpeng@nuaa.edu.cn

H. Nie
e-mail: hnie@nuaa.edu.cn

Z. Liu et al. (eds.), *Advances and Challenges in Advanced Unmanned Aerial Systems*, Springer Aerospace Technology, https://doi.org/10.1007/978-981-99-8045-1_23

1 Introduction

The Tail-sitter VTOL UAV adopts vertical seated takeoff/landing, followed by turning into level flight during airborne. When landing, the Tail-sitter VTOL UAV first climbs and adjusts the nose upward, subsequently reducing thrust to land vertically. Tail-sitter VTOL UAV fix the thrust direction on the longitudinal axis of the fuselage, making it possible for the thrust and fuselage to rotate in tandem when the thrust changes direction [1–7].

However, the Tail-sitter VTOL UAV has a high center of gravity and a large windward area due to its fixed-wing structure and upright fuselage, resulting in poor stability in the takeoff/landing phase. At present, the test machine or civil aircraft have been launched are generally used in a single function and limited role of fixed landing gear, through the use of fixed landing gear bending deformation of its own structure to achieve the absorption and dissipation of vertical energy during the landing process. In terms of conventional landing gear, only the "Sky Eye [8]", "Tern", "Wisdom Bee" Tail-sitter VTOL UAV and a few other models use wheeled landing gear, and all of them are only effective during skate takeoff/landing, and are supported by the airframe structure during vertical takeoff/landing. The same skid type landing gear is only adopted by the "Spirit Stilts" of Beijing University of Aeronautics and Astronautics, which uses the curved skid of the abdomen to realize the attitude change from flat to vertical takeoff/landing before the fuselage leaves the ground, as well as the vertical landing at any angle. As a result, the typical vertical takeoff/landing phase of the Spirit Stilts consumes only 1/5th of the energy of a tilt-rotor or complex powered VTOL vehicle, or only 2% of the entire mission profile. However, the belly skid curve coupled with the power vector requires a special design, resulting in a more complex skid landing gear design compared to conventional skids. "XFY-1", "XFV-1", "Sky-Tote", "Golden-Eye", "TERN", "X-Plane" and other well known Tail-sitter VTOL UAV have added strut type buffers at the bottom [9–12] to improve the energy release buffering capacity of the landing gear, but the problems of single function fixed landing gear, weak impact resistance, poor takeoff/landing stability, and susceptibility to rebound and rollover still exist. Even the "THU-1600 [13]" of Tsinghua University in 2017 proposed a "controlled forward landing scheme" to try to avoid the above defects. When the UAV lands vertically by giving the body a forward tilt angle to ensure that after the landing gear touches the ground and the strutted landing leg at the bottom absorbs the impact, the body uses the tipping trend to ensure the safe landing of the UAV with the cushioning device on the belly [14–16].

Therefore, although the above solutions have a huge improvement in local performance relative to fixed landing gear, they still have their own types of deficiencies. Takeoff/landing constraints have become one of the key technologies limiting the development of Tail-sitter VTOL UAV.

Therefore, based on the Free-tail bionic technology, this paper refers to the bionic method similar to the Free-tail technology structure, carries out motion capture and data acquisition, and improves the landing stability of the Tail-sitter VTOL UAV through the identification and tracking of multi-data fusion.

2 The Landing Gear Structure Design of Tail-Sitter VTOL UAV Based on Free-Tail Bionic Technology

In order to make the landing gear better adapt to the two typical modes of VTOL / cruise with the tail seat VTOL UAV, this paper combines the structural design of the landing gear itself with the optimization of the overall layout of the Tail-sitter VTOL UAV based on the idea of structural functional integration design. The landing gear consists of a foldable two section structure of tail support and support strut, which are symmetrically installed on both sides of the fuselage. The connection between the fuselage, the tail support and the support strut is equipped with a driving mechanism, which can drive the tail support and the support strut to open and close, enabling the landing gear to be adapted with the mode change of the aircraft and adjusting the attitude of the landing gear for adaptive support during takeoff/landing.

The assembled Tail-sitter VTOL based on Free-tail bionic technology is shown in Fig. 1.

To better solve the problem of aerodynamic drag in the cruise mode of the Tail-sitter VTOL UAV, the embedded structure causes the landing gear to be unable to have an external buffer device like other conventional multi-link schemes, and the tail support driving force arm which is equivalent to the length of the support rod greatly increases the tail support load torque. Therefore, in this paper, the gear rack structure shown in Fig. 2 is used to replace the connecting rod tail support driving structure.

By employing conventional landing techniques which involve utilizing a model resume, alongside performing inverse dynamics calculations based on mission trajectory data to acquire the relevant drive angle, as well as monitoring motion trajectory utilizing positive kinematics and sensors, a compensating control mechanism is implemented to adjust for deviations from the intended trajectory.

3 Data Acquisition and Processing of Bionic Example Based on Free-Tail Bionic Technology

Since the introduction of the ZMP theory by Vukobratovic [17], its associated approaches have become commonly utilized for evaluating robot stability. A significant advantage of this method lies in its utilization of mathematical models to provide a detailed analysis and description of the kinematic constraints and kinetic stability conditions of the robot motion process. However, with the increasing scope of robot applications, the effectiveness of mathematically based designs is often limited by the

Fig. 1 Conceptual diagram of the Free-tail for Tail-sitter VTOL UAV

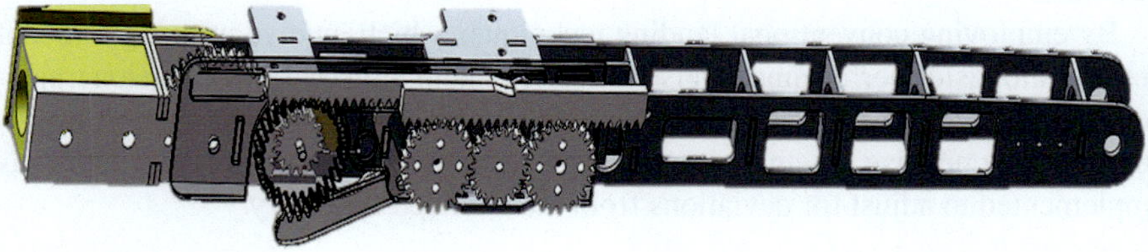

Fig. 2 Tail-sitter VTOL UAV adaptive landing gear structure

form of the motion equations. Moreover, the coordination between joint trajectories in robots differs significantly from that in natural organisms. In particular, for instantaneous and complex dynamic tasks, the establishment and solution of analytical motion equations are challenging, resulting in poor operational feasibility.

The adoption of bionic technology has emerged as a potential solution to enhance the intelligence of robots and accomplish complex and diverse tasks. This approach draws inspiration from the evolutionary traits of biological organisms, which have developed superior control over their limb muscles and skeletal joints, allowing for natural and seamless transitions of movements and optimized energy consumption during motion. By emulating the task trajectories and motion behaviors of living organisms based on their surrounding environment and experiences, bionic robots can fully exploit their morphological advantages, leading to an overall improvement in their intelligence and motion behavior.

This study aimed to identify a biomimetic target for quadrupedal locomotion by comparing various organisms, ultimately selecting the cat for its superior landing performance. Leveraging experimental data previously conducted by other researchers on the biological structure of the cat and its landing states, including flat jump, longitudinal jump, and vertical landing [18–20], we identified five key nodes that encapsulate the entire biological motion of the cat, as illustrated in Fig. 6. Specifically, these key nodes comprise the forelimb scapular joint, forefoot paw, hindlimb hip joint, hindfoot paw, and lumbar torsion center, and we obtained the trajectory coordinates of each key node across different motion processes.

After comparing a variety of quadruped creatures, this paper chooses cats with good landing performance as bionic targets. According to the experimental basis of other researchers on the cat's biological structure and typical landing states such as flat jump, vertical jump and vertical landing [12–14], the cat's motion data is divided into five key nodes according to Fig. 3. The motion of these five key nodes can represent the whole biological motion of the cat, and the trajectory coordinates of the five key nodes of the scapular joint of the forelimb, the forefoot, the hip joint of the hindlimb, the hindfoot and the waist torsion center in different motion processes are obtained in turn.

The bionic strategy is represented by π_E. The bionic data generated by the strategy is stored in the bionic data set D_E, and the data distribution conforms to:

$$D_E \overset{\Delta}{=} \{(s_i, a_i)\}_{i=1}^{N} \overset{i.i.d.}{\sim} \rho_{\pi_E}(s, a) \tag{1}$$

In the formula: $\rho_{\pi_E}(s, a)$ represents the non-normalized distribution of the data generated by the bionic strategy π_E; *i.i.d.* indicates that the data in the bionic data set D_E is independent and identically distributed.

To enable the application of bionic data for guiding reinforcement learning training, we have formulated the teaching objective function of strategy learning

Fig. 3 Feature map transformation

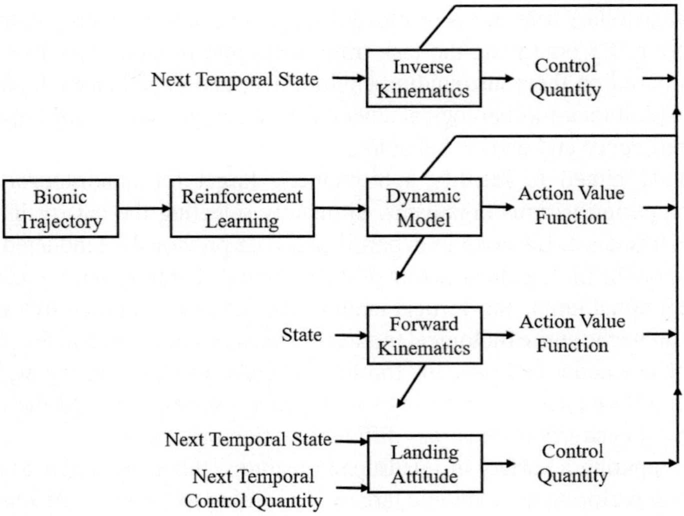

Fig. 4 Bionic learning system workflow

in the following manner:

$$J(\pi) = \mathbb{E}_{s_i \sim D, a_i \sim \pi}\left[\sum_{t=0}^{T} \gamma^t \big(r(s_t, a_t) + \mathbb{I}_{s_t \in D_\varepsilon}\big(M - a_t - \pi_E(s_t)^2\big)\big)\right] \quad (2)$$

In the formula $\mathbb{I}_{s_t \in D_\varepsilon}$ is an indicator function, when satisfying $s_t \in D_\varepsilon$, the function value is 1, when not satisfying $s_t \in D_\varepsilon$, the function value is 0; M is a normal number, and the value is as small as possible on the basis of ensuring $M - a_t - \pi_E(s_t)^2 \geq 0$.

The workflow of the bionic learning system is shown in Fig. 4.

The policy objective function in the original reinforcement learning algorithm is represented by the first term in the aforementioned equation. The second term, on the other hand, is intended to steer the learned policy closer to the bionic policy. If a state s_t is absent from the bionic dataset, the second term is assigned a value of zero, and the policy is updated using the original reinforcement learning algorithm. Conversely, if state s_t is included in the bionic dataset, the second term is assigned a positive value, promoting the current policy to learn the bionic policy. The magnitude of the second term increases as the discrepancy between the actions generated by the current strategy and those generated by the bionic strategy decreases. Nevertheless, the equation's assumption of optimal affine experience restricts further optimization of the strategy network, particularly during the late training phase, when the constraint effect becomes more prominent. To address this issue, a state action value function $Q(s, a)$ is employed to determine the value of the imitation action, and if the Q value of the imitation action exceeds the Q value of the action generated by the strategy, the demonstrative knowledge's guidance is integrated. Consequently, the demonstrative objective function of the strategy undergoes modification:

$$J(\pi) = \mathbb{E}_{s_i \sim D, a_i \sim \pi}\left[\sum_{t=0}^{T} \gamma^t r(s_t, a_t) + \sum_{t=0}^{T} \gamma^t \mathbb{I}\left(M - a_t - \pi_E(s_t)^2\right)\right] \quad (3)$$

In the formula:

$$\mathbb{I} = \mathbb{I}_{s_t \in D_E} \mathbb{I}_{Q(s_t, \pi_E(s_t))} > Q(s_t, a_t) \quad (4)$$

4 Optimization of Free-Tail Parameters of Tail-Sitter VTOL UAV

Since the Tail-sitter VTOL UAV landing gear based on Free-tail bionic technology proposed in this paper functions as both the structural body and the tail spar of the Free-tail wing, and its structural parameters involve both the landing gear itself and the overall layout of the UAV, this paper firstly investigates the aerodynamic interference of the double tail spar layout under cruise conditions by changing the structural length of the landing gear while considering the finite wingspan three dimensional effect The pattern of aerodynamic disturbance in cruise condition is studied by changing the length of the landing gear structure.

In order to study the law of aerodynamic interference between rotor propeller and landing gear and Free-tail, the influence of different lengths of landing gear is calculated in this paper. The landing gear length increased by two thirds and three quarters, respectively, where the size of the rotating area is the same, but in order to minimize the impact of the rotating area circumferential slip interface location on the calculation results, according to the closer the slip interface to the rotor, the smaller the impact of rotor rotation on the surrounding area fluid; the closer the slip interface, the greater the impact of rotor rotation on the surrounding area The closer the slip interface is to the rotor, the smaller the influence of rotor rotation on the surrounding area fluid; the closer the slip interface is to the stationary parts, the greater the influence of rotor rotation on the surrounding area fluid; the location of the slip interface should be taken between the rotor and the stationary parts near the rotor 1/8 ~ 1/4. In the calculation of this paper, considering the characteristics of the model itself, the value is chosen to be around 1/5.

In order to minimize the influence of the grid on the calculation results, the grid independence study is conducted first. In this paper, four sets of calculation grids are established for the model with the grid numbers of 700,000, 2.1 million, 4.2 million, and 8.4 million, respectively, and when the grid numbers are above 4 million, the calculated values of the propeller pull coefficient and the drag coefficient of the wing tend to be smooth and no longer have large variations, indicating that the grid numbers on this rotor propeller The resolution of the aerodynamic interference problem of the propeller/landing gear and Free-tail meets the requirements. Therefore, the grid of 4.2 million is used in the subsequent calculations as shown in Fig. 5.

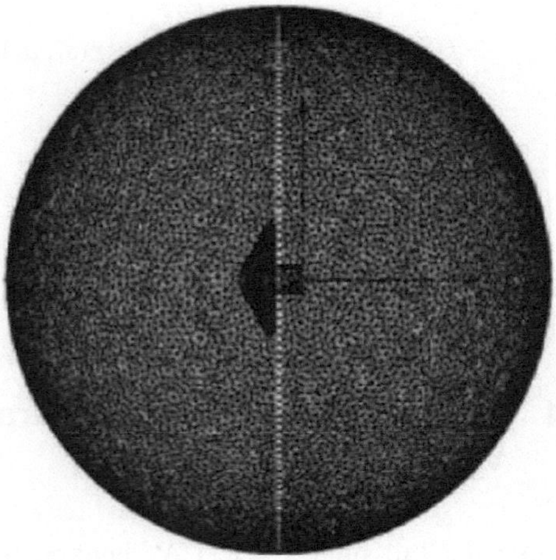

Fig. 5 CFD calculation mesh

According to the simulation results as shown in Fig. 6, in terms of landing gear length, the horizontal tail is not beneficially too close to the wing. When the horizontal tail is too close to the wing, the air will form a vortex after converging at the rear end of the wing due to the different flow directions of the upper and lower surfaces of the wing. The existence of the propeller slipstream makes the air flow vortex volume, that is, the instantaneous acceleration has a significant increase. In addition, the propeller tip vortex still has a strong effect after bypassing the lower surface of the wing, making the whole flow field distribution at the trailing edge of the wing become more irregular, while in the horizontal tail is far away from the wing, not only the above coupling effect is reduced, and can effectively avoid the rotating slipstream, but too far away will affect the Tail-sitter VTOL UAV landing center of gravity height, while reducing the landing gear stiffness and increasing the structural design difficulty.

In addition to aerodynamic effects, landing gear structure length and support attitude are also two key factors that affect the maximum slope of a Tail-sitter VTOL UAV landing. The increase of structural length gives more cushioning margin to the landing gear, while the increase of the spacing between the legs of the landing gear due to the change of the support attitude can improve the stability of the Tail-sitter VTOL UAV, and both of them also have a coupling effect on the height of the center of gravity of the Tail-sitter VTOL UAV. In this paper, the 1/3 times landing gear length parameter obtained from the bionic structure is used as an interval to obtain the influence law of landing gear length and spacing parameter on the maximum landing slope as shown in Fig. 6.

Given that the length of the landing gear impacts both the aerodynamic efficiency and landing stability of the Tail-sitter VTOL UAV, it is necessary to consider this variable as a component of multi-objective optimization.

Objective function:

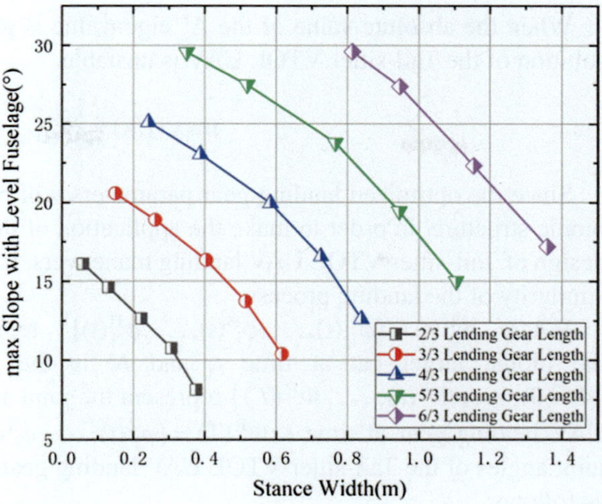

Fig. 6 The influence of landing gear length and spacing parameters on the maximum landing slope

$$\max f_1(x) = C_{mH} \tag{5}$$

The tail wing handling efficiency:

$$C_{mH} = k_q C_{L\alpha H}((1-\varepsilon_\alpha)\alpha + \delta_H)\frac{S_H l_H}{Sc_A} \tag{6}$$

S_H is the area of the horizontal tail, l_H is the length of the tail force arm, $C_{L\alpha H}$ is the slope of the horizontal tail lift line, ε_α is the downwash at the horizontal tail, and ε_α is the velocity retardation coefficient at the horizontal tail.

$$\min f_2(x) = A_i^z \tag{7}$$

In the landing process of the Tail-sitter VTOL UAV, the Poincaré mapping point is obtained by the state of the previous moment of the touchdown collision. Based on this, the discrete time system $x_{k+1}^z = P^z(x_k^z)$ is defined as $\partial x_k^z = x_k^z - x^{z^*}$, Where x^{z^*} is a fixed point, so we can get the linearized model:

$$\partial x_{k+1}^z = A^z \partial x_k^z \tag{8}$$

where:

$$A^z = [A_1^z \; A_2^z] \tag{9}$$

And satisfy:

$$A_i^z = \frac{P^z(x^{z^*} + \Delta x_i^z) - P^z(x^{z^*} - \Delta x_i^z)}{2\Delta x_i^z} \tag{10}$$

When the absolute value of the A^z eigenvalue is greater than or equal to 1, the solution of the Tail-sitter VTOL UAV is unstable.

$$\max f_3(x) = S \tag{11}$$

Since the optimized landing gear parameters will be partially different from the bionic structure, in order to make the application of the obtained bionic data to the design of Tail-sitter VTOL UAV landing maneuvers, it is necessary to re-convert the similarity of the landing process.

Where $\Phi^H(t) = [\phi_1^H(t), \ldots, \phi_i^H(t), \ldots, \phi_N^H(t)]^T$ represents the joint angle of the bionic target cat at time t, and N is the number of joints. $p^R = \{\Phi^R(0), \ldots, \Phi^R(t), \ldots, \Phi^R(T)\}$ represent the joint angles of the Tail-sitter VTOL UAV landing gear at time t. $\Phi^R(T) = [\phi_1^R(t), \cdots, \phi_i^R(t), \cdots, \phi_N^R(t)]^T$ represent the joint angles of the Tail-sitter VTOL UAV landing gear, and the similarity is defined as follows:

$$S_i^{H-R}(t) = \varpi\left(1 + \frac{\phi_i^R(t) - \phi_i^H(t)}{\phi_{i-max}^R - \phi_{i-min}^R}\right)^{-1} + (1-\varpi)\left(1 + \frac{\dot{\phi}_i^R(t) - \dot{\phi}_i^H(t)}{\dot{\phi}_{i-max}^R - \dot{\phi}_{i-min}^R}\right)^{-1} \tag{12}$$

$$S^{H-R}(t) = \frac{\sum_{i=1}^{N} S_i^{H-R}(t)}{N} \tag{13}$$

$$S^{H-R} = \frac{\int_{t=0}^{T} S^{H-R}(t)dt}{T} \tag{14}$$

The ϕ_{i-max}^R and ϕ_{i-min}^R represent the maximum and minimum rotation range of the i-joint of the Tail-sitter VTOL UAV landing gear, respectively. The $S_i^{H-R}(t)$ represents the similarity between the *i*-joint of the Tail-sitter VTOL UAV landing gear at time *t* and the *i*-joint of the bionic object cat. ϖ is the transformation coefficient and ϖ. When ϖ tends to 1, it means that the action tends to be static. When ϖ tends to 0, it means that the action tends to be fast conversion. $S^{H-R}(t)$ are the average similarity of the whole joint at time t, and S^{H-R} is the average similarity within a period of trajectory time T. The above three indicators are represented by *S*, and the closer *S* is to 1, the higher the similarity is.

$$\begin{aligned} \text{st.} \quad & l_{imin} \le l_i \le l_{imax} \\ & \sum_{i=1}^{2} l_i = 1 \\ & 0 < A^z \le 1 \\ & 0 < S \le 1 \end{aligned} \tag{15}$$

$i = 1, 2$; the l_{imin} and l_{imax} are the minimum and maximum lengths of the i-th pole, respectively; the constraint $\sum_{i=1}^{4} l_i$ means that the sum of the four rod lengths is unchanged.

5 Tail-Sitter VTOL UAV Landing Test

To confirm the landing stability of Tail-sitter VTOL UAV and safeguard the structural integrity of the aircraft during the initial development phase, a series of drop tests were conducted on the UAV under varying working conditions. These tests were carried out using the drop test bench, which is illustrated in Fig. 7.

The primary componentry of the testing apparatus comprises of a profile outer frame, a vertical guide rail, a hanging basket, and a force measuring platform. The prototype is secured to the hanging basket by means of a connecting rod that is interposed with the vector surface. The hanging basket, which uses bearings to slide along the vertical guide rail, can adjust the drop height via the retraction of its rope, thereby simulating vertical landing motions. Moreover, the adjustable cross height structure of the hanging basket enables the attitude control of the prototype.

Through the drop shock test, as depicted in Fig. 8, the Tail-sitter VTOL UAV exhibited a capacity for adaptive landing across various attitudes, even in scenarios with ground inclinations of more than 20°.

Compared to the traditional landing method based on model building in Figs. 9 and 10, which performs inverse dynamics calculations based on the mission trajectory information to obtain the corresponding driving angle and monitors the motion trajectory using positive kinematics and sensors, and performs compensatory control

Fig. 7 Structure of Tail-sitter VTOL UAV drop test bench

Fig. 8 Landing test of Tail-sitter VTOL UAV landing gear

based on the deviation from the desired trajectory, as shown in Fig. 11, the pre-trained bionic technique has better stability performance.

The landing loads of the Tail-sitter VTOL UAV, as presented in Fig. 12, were acquired through the employment of sensors located at the landing gear joints of the Tail-sitter and the utilization of bottom force gauges. These measurements exhibit a significant resemblance to the landing dynamics of a cat, as simulated in the study.

As shown in Fig. 13, this section compares and analyzes the comparison of the handling ability of the Tail-sitter VTOL UAV in the air after the adoption of the free-tail technology through the flight test along the rectangular path at an altitude of 100–300 m. The comparison shows that the addition of the free-tail technology has a significant enhancement on the stability of the Tail-sitter VTOL UAV handling in the air and thus proves the great necessity of the research on the landing technology based on the free-tail technology (Fig. 14).

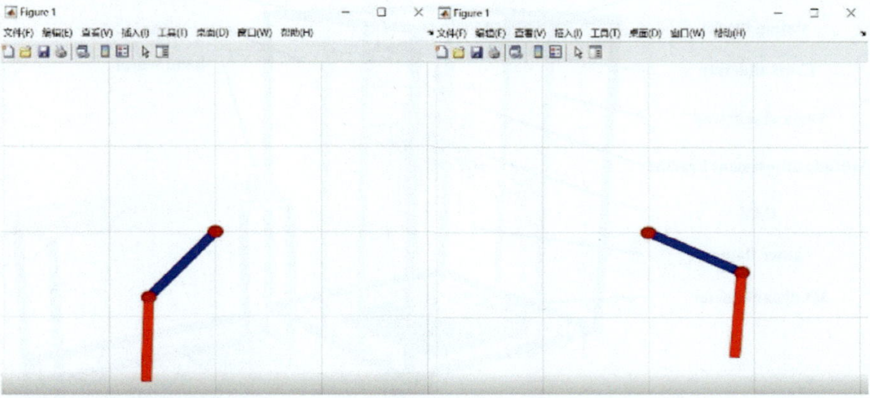

Fig. 9 Tail-sitter VTOL UAV landing gear landing simulation

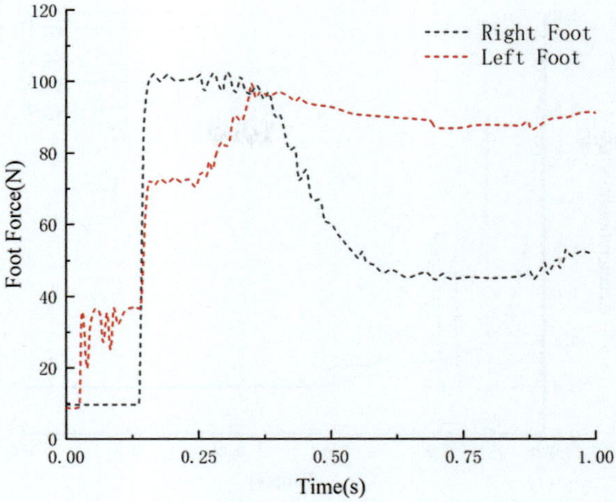

Fig. 10 Model-based test loads for landing gear landings

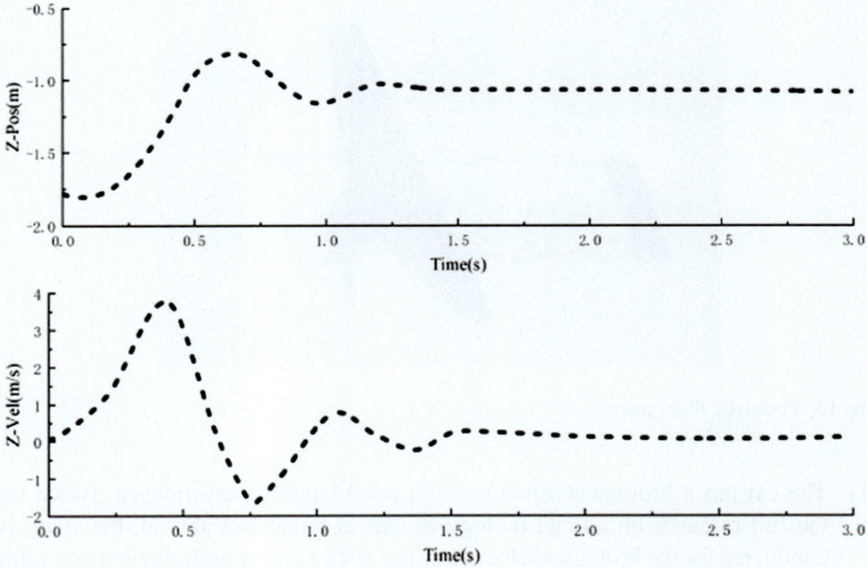

Fig. 11 Fuselage c.g. position and fuselage orientation for landing

6 Conclusions

This paper combines the landing device with the optimization of the overall layout of the UAV, and proposes a technical solution of Tail-sitter VTOL UAV based on Free-tail bionic technology.

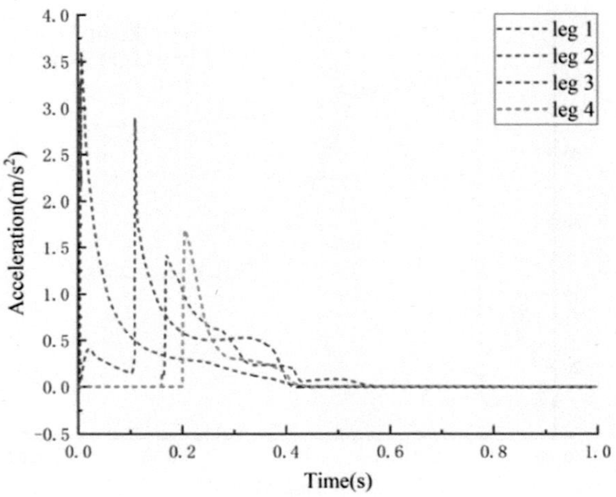

Fig. 12 Acceleration of each landing leg of landing gear

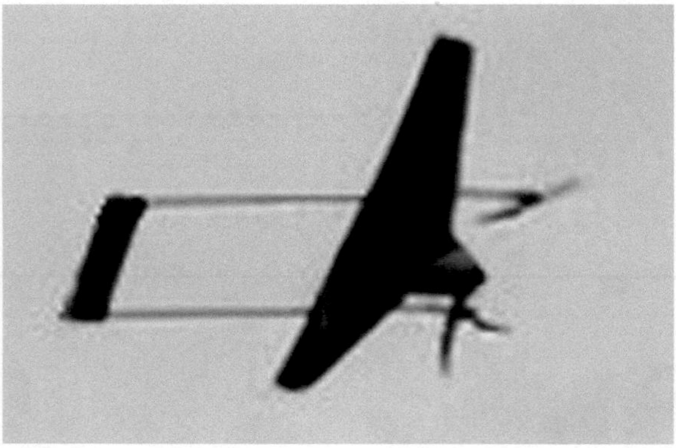

Fig. 13 Prototype flight photos

(1) The cat has a biological structure with good landing performance. Based on the full research on animal biology of cats at home and abroad, the study is conducted for the biological characteristics such as the morphological size ratio of the cat; through the analysis of the posture of the whole jumping process and the ground reaction force on the foot in different environments of the cat, the overall characteristics of the cat's jumping buffer and the overall buffering effect of the cat's front and hind legs on the cat during the jumping process are understood By analyzing the overall characteristics of the cat's jumping cushion and the different contributions of the cat's front legs and hind legs to the overall cushioning effect during the jumping process, and based on the motion position and coordinate information of its key skeletal nodes, we can use it as the learning target of Tail-sitter VTOL UAV landing gear for stable landing of the bionic cat.

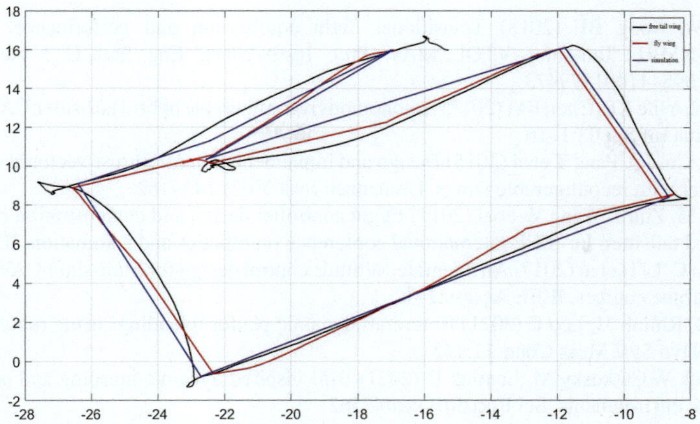

Fig. 14 Flight test trajectory

(2) Considering that the length of the landing gear in the technical scheme of the Free-tail bionic technology based Tail-sitter VTOL UAV involves the influence of both aerodynamic efficiency and landing stability, the multi-objective optimization of the landing gear length ratio requires a similarity conversion of its landing process again, and it is proved through the fallout test that the optimized Free-tail bionic technology based Tail-sitter VTOL UAV still has a better landing stability. The need for this technology for future research was demonstrated by flight test comparisons from both modes.

References

1. Alwateer M, Loke SW (2020) Emerging drone services: challenges and societal issues. IEEE Technol Soc Mag 39(3):47–51
2. Zong J, Zhu B, Hou Z, Yang X, Zhai J (2021) Evaluation and comparison of hybrid wing VTOL UAV with four different electric propulsion systems. Aerospace 8(9):256
3. Imumbhon JO, Alam MD, Cao Y (2021) Design and structural analyses of a reciprocating S1223 high-lift wing for an RA-driven VTOL UAV. Aerospace 8(8):214
4. Bacchini A, Cestino E (2019) Electric VTOL configurations comparison. Aerospace 6(3):26
5. Ma T, Wang X, Fu J, Hao S, Xue P (2022) Three-dimensional flight envelope for V/STOL aircraft with multiple flight modes. Aerospace 9(11):691
6. Zhong J, Wang C, Zhang H (2023) Transition control of a tail-sitter unmanned aerial vehicle with L1 neural network adaptive control. Chin J Aeronaut 36(7):460–475
7. Wang H, Su S, Qiu X et al (2022) The hovering stability of the Egretta tail-sitter VTOL UAV. Int J Aerosp Eng:1–12
8. Kubo D, Suzuki S (2012) Tail-sitter vertical takeoff and landing unmanned aerial vehicle: transitional flight analysis. J Aircr 45(1):292–297
9. Xia X, Yang M, Chen G et al (2021) Transition flight control and simulation of a novel tail-sitter UAV with varying fuselage shape. IEEE Access (99):1–1

10. Tang W, Song BF (2018) Transitional flight equilibrium and performance study for the X-NMRL Tail-sitter VTOL MAV. Proc Instit Mech Eng Part G J Aerosp Eng 233(8):095441001879473
11. Wang K, Yijie KE, Chen BM (2017) Autonomous reconfigurable hybrid tail-sitter UAV U-Lion. Sci China Inf Sci 03:1–16
12. Ang K, Cui JQ, Pang T et al (2015) Design and implementation of a thrust-vectored unmanned tail-sitter with reconfigurable wings. Unmanned Syst 3(02):143–162
13. Kuang M, Zhu J, Wang W et al (2017) Flight controller design and demonstration of a thrust-vectored tailsitter. In: IEEE international conference on robotics and automation. IEEE
14. Li H, Li C, Li H et al (2017) An integrated altitude control design for a tail-sitter UAV equipped with turbine engines. IEEE Access:1–1
15. Leon B, Rimoli JJ, Leo C (2021) Rotorcraft dynamic platform landings using robotic landing gear. J Dyn Syst Meas Contr 11:143
16. Roderick W,Cutkosky M, Lentink D (2021) Bird-inspired dynamic grasping and perching in arboreal environments. Sci Rob 6(61):eabj7562
17. Vukobratovic M (1973) How to control artificial anthropomorphic systems. IEEE Trans Syst Man Cybern
18. Nonaka S, Sugano K (2018) Model design of cat's skeleton robot based on muscle strength ratio estimation method. In: The proceedings of JSME annual conference on robotics and mechatronics (Robomec), pp 2P1–D10
19. Akpan ES, Dauda M, Kuburi LS et al (2021) Box-Behnken experimental design for the process optimization of catfish bones derived hydroxyapatite: a pedagogical approach. Mater Chem Phys 272:124916
20. Mccarty S (2014) Impact force patterns on a landing cat. Int Sch Bangkok J Sci 8(1)

Performance Reliability Assessment of Retraction Mechanism Based on Bayesian Theory

Yumeng Yan, Jiyuan Zhou, Yin Yin, Hong Nie, and Xiaohui Wei

Abstract The landing gear retraction mechanism incurs high test costs, leading to a shortage of test data and posing challenges in conducting performance reliability assessment. In response to this challenge, a mechanism performance reliability assessment method based on the Bayesian approach is proposed. This method aims to mitigate estimation errors resulting from limited test data and enhance the overall robustness of the assessment process. Firstly, a rigid-flexible coupling dynamic simulation model of the retraction mechanism is established, followed by the formulation of a kinematic accuracy reliability model. Subsequently, leveraging Bayesian theory and the motion characteristics of the retraction mechanism, the kinematic accuracy parameter estimation model that integrates prior and test samples is formulated. Finally, taking the retraction mechanism as research object, the estimated values of the kinematic accuracy parameter are obtained, and the performance reliability of the retraction mechanism is assessed. The results show that the proposed method yields consistently stable evaluation results and effectively diminishes the impact of uncertainties within the retraction mechanism on performance characterization.

Keywords Bayesian theory · Reliability · Retraction mechanism

1 Introduction

Restricted by factors such as test cost and test cycle, small sample tests are often the only feasible option for assessing landing gear retraction system [1]. The key influencing factors on the retraction performance are inevitably impacted by the uncertainties in the process of design, manufacture, and use [2, 3]. This makes it more challenging to extrapolate small-sample test outcomes to reliably characterize the underlying distribution of performance parameters. As a result, it can be tough to

Y. Yan · J. Zhou · Y. Yin (✉) · H. Nie · X. Wei
State Key Laboratory of Mechanics and Control for Aerospace Structures, Nanjing University of Aeronautics and Astronautics, Nanjing 210016, China
e-mail: yinyin@nuaa.edu.cn

Z. Liu et al. (eds.), *Advances and Challenges in Advanced Unmanned Aerial Systems*, Springer Aerospace Technology, https://doi.org/10.1007/978-981-99-8045-1_24

assess the accurate and stable reliability. Therefore, it is crucial to develop a stable and effective retraction mechanism performance reliability assessment method.

Targeting the issue of insufficient test samples in mechanism reliability analysis, prior studies have offered various approaches. Yin et al. [4] constructed the response surface based on the sample data collected from the mathematical model simulation and analyzed the influence of key parameters on the retraction mechanism reliability. Hu et al. [5] effectively reduced the computational cost of reliability analysis using the advanced adaptive Kriging model based on the multi-body dynamic simulation model of the adjusting mechanism of tail nozzle. However, these methods often relied on prior information that may not accurately capture the characteristics of new products, potentially affecting the accuracy of reliability estimation results.

Moreover, the classical interval estimation method relies on test data to estimate an interval of possible values of performance characteristic parameters. Li et al. [6] introduced an improved interval estimation method based on Bootstrap, providing parameter interval estimations for NC machine tools' reliability model that considered working conditions covariates. However, due to the inherent dynamics and uncertainty within the test process, small sample test data may not belong to the same population in essence, and the interval estimation results are at considerable risk [7].

In order to improve the analysis method, the Bayesian method is used to fuse the experimental information and prior information to achieve reasonable statistical inference of the kinematic accuracy parameters. This endeavor enhances both the stability and precision of mechanism kinematic reliability assessment outcomes under small-sample test conditions. Assuming the stress and strength random variables to follow an exponential distribution, K c et al. [8] calculated the estimators of stress and strength parameters by Bayesian method, thus estimating the reliability of the system. Similarly, Li et al. [9] proposed a non-probabilistic credible Bayesian reliability method and applied it to the structure reliability analysis of a supersonic vehicle.

This paper presents a novel method for assessing the reliability of mechanism performance, based on the Bayesian approach. In Sect. 2, a dynamic simulation model accounting for assembly deviation is established, followed by the development of retraction tests. The kinematic accuracy reliability model of the retraction mechanism is formulated. In Sect. 3, an innovative kinematic accuracy parameter estimation approach is introduced, involving the fusion of prior data and test data. In Sect. 4, the proposed method is applied to derive the kinematic accuracy parameters distribution for the retraction mechanism. The kinematic accuracy reliability is calculated.

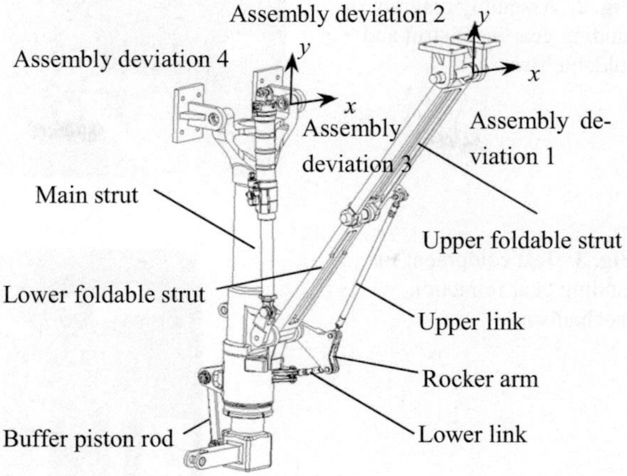

Fig. 1 Landing gear retraction mechanism

2 Retraction Mechanism Dynamics and Reliability Model

2.1 Function Definition and Dynamics Model of Retraction Mechanism

Based on the retraction principle, a dynamic simulation model for the retraction mechanism is established, as depicted in Fig. 1. The impact of assembly deviations on kinematic accuracy is taken into account by parameterizing assembly deviation. The retraction process is as follows: First, the actuator cylinder drives the main strut to rotate. This rotation subsequently propels the transmission component, comprising the lower foldable strut, upper foldable strut, upper link, rocker arm, and lower link. This motion prompts the rotation of the wheel on the buffer piston rod, achieving the desired rotational movement. The uncertain factors such as initial assembly deviation can lead to motion error in the retraction mechanism. When the landing gear reaches the designated angle during retraction, the buffer piston rod's motion error could exceed the permissible threshold. This could result in interference between the wheel and the cabin door, ultimately leading to failure in the kinematic accuracy of the retraction mechanism.

2.2 Dynamics Test of Retraction Mechanism

As shown in Fig. 2, by setting the adjustable fixture to simulate the assembly deviations in the main strut and foldable strut, the retraction mechanism test is carried out under various working conditions. The angle sensors are employed to actively monitor the motion trajectory of the landing gear in real-time. Subsequently, the gathered test data is transmitted to the dynamic signal testing system. During the

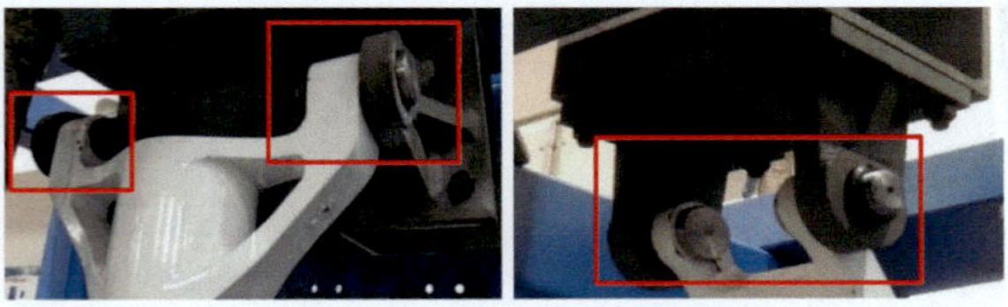

Fig. 2 Assembly diagram of landing gear main strut and foldable strut

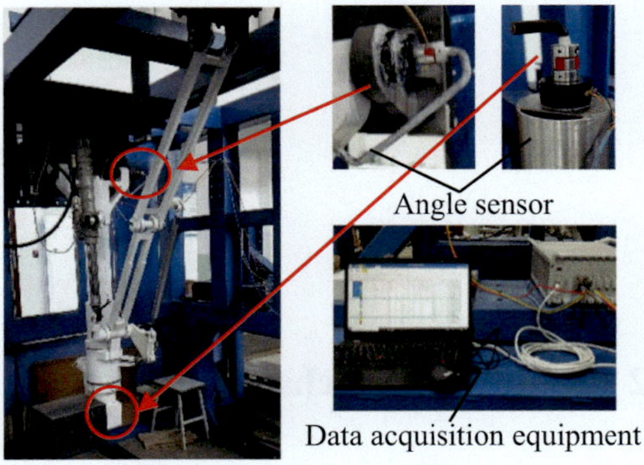

Fig. 3 Test equipment for landing gear retraction mechanism

test, the rotation angle of the buffer piston rod is meticulously documented when the landing gear attains the predetermined angle. Figure 3 showcases the retraction mechanism experimental prototype along with the data acquisition equipment.

2.3 Reliability Model of Retraction Mechanism

The performance function corresponding to the failure mode of the kinematic accuracy of the retraction mechanism is expressed as

$$Z(\boldsymbol{X}) = g(\boldsymbol{X}) = \Delta^* - \Delta \tag{1}$$

wherein, $\boldsymbol{X} = [x_1, x_2, ..., x_r]$ is made up of random variables that affect the kinematic precision function of retraction mechanism, the normal distribution is shown in Table 1. The position of the assembly deviation is shown in Fig. 1. Δ^* is expressed as the rotation limit error of the buffer piston rod. Δ is expressed as the rotation output error of the buffer piston rod, that is, the kinematic accuracy parameter.

It is assumed that the performance function Z obeys the normal distribution [10], Δ^* and Δ also obey the normal distribution. The kinematic accuracy reliability of retraction mechanism can be written as

Table 1 The distribution of random variable

Parameters/(mm)	Distributions	Parameters/(mm)	Distributions
Horizontal assembly deviation of upper foldable strut x_1	$N\left(\mu_{x_1} = 0, \sigma_{x_1}^2 = 0.02^2\right)$	Vertical assembly deviation of upper foldable strut x_2	$N\left(\mu_{x_2} = 0, \sigma_{x_2}^2 = 0.1^2\right)$
Horizontal assembly deviation of main strut x_3	$N\left(\mu_{x_3} = 0, \sigma_{x_3}^2 = 0.02^2\right)$	Vertical assembly deviation of main strut x_4	$N\left(\mu_{x_4} = 0, \sigma_{x_4}^2 = 0.1^2\right)$

$$P_r = P\left(\Delta^* - \Delta > 0\right) = \int_0^{\infty} \frac{1}{\sqrt{2\pi\left(\sigma_{\Delta^*}^2 + \sigma_{\Delta}^2\right)}} \exp\left[-\frac{(z - (\mu_{\Delta^*} - \mu_{\Delta}))^2}{2(\sigma_{\Delta^*}^2 + \sigma_{\Delta}^2)}\right] dz \tag{2}$$

3 Kinematic Accuracy Parameters Estimation Based on Bayesian Theory

The Bayesian method is used to fuse the prior data and test data to improve the stability of parameter estimation results of performance characterization Δ.

3.1 Bayesian Formula

According to the Bayesian formula, the parameter estimation expression of the mean μ_Δ and variance σ_Δ^2 of the kinematic accuracy parameter Δ is expressed as

$$\pi(\mu_\Delta, \sigma_\Delta^2 | \boldsymbol{T}) \propto p(\boldsymbol{T} | \mu_\Delta, \sigma_\Delta^2) \pi(\mu_\Delta, \sigma_\Delta^2) \tag{3}$$

wherein, $p(\boldsymbol{T} | \mu_\Delta, \sigma_\Delta^2)$ represents the likelihood function, $\pi(\mu_\Delta, \sigma_\Delta^2)$ represents the prior distribution, $\pi(\mu_\Delta, \sigma_\Delta^2 | \boldsymbol{T})$ represents the posterior distribution, $\boldsymbol{T} = [t_1, t_2, ..., t_n]$ consists of all test sample values.

3.2 Likelihood Function

Δ obeys normal distribution, the likelihood function that combines the test sample information and the overall information can be expressed as

$$
\begin{aligned}
p(\boldsymbol{T}|\mu_\Delta,\sigma_\Delta^2) &\propto \sigma_\Delta^2 \exp\left(-\alpha\sum_{j=1}^{n}(t_j-\mu_\Delta)^2\right)\\
&=\sigma_\Delta^{-n}\exp\left\{-\alpha\left[(n-1)s^2+n(\bar{t}-\mu_\Delta)^2\right]\right\}
\end{aligned}
\tag{4}
$$

in which, $\alpha = 1/2\sigma_\Delta^2$, $\bar{t}$ is the mean value of the test sample, and s^2 is the variance of the test sample.

3.3 Prior Distribution

When the test sample obeys the normal distribution and the mean and variance are unknown, the joint prior distribution of mean μ_Δ and variance σ_Δ^2 can be assumed to be normal-inverse gamma distribution, as shown in Eq. (5).

$$
\begin{aligned}
\pi(\mu_\Delta,\sigma_\Delta^2) &= \pi(\mu_\Delta\sigma_\Delta^2)\pi(\sigma_\Delta^2) = (\sigma_\Delta^2)^{-(\gamma_0+1)}e^{-(\beta_0/2\sigma_\Delta^2)}\\
&= (\sigma_\Delta^2)^{-((v_0+1)/2+1)}e^{-(v_0\sigma_0^2+k_0(\mu_\Delta-\mu_0)^2/2\sigma_\Delta^2)}
\end{aligned}
\tag{5}
$$

here, μ_0, k_0, v_0, and σ_0^2 are hyper parameters. They can be obtained by generating the prior mean sample $\mathbf{M}_{\Delta p} = \left[\mu_{\Delta p1}, ..., \mu_{\Delta pm}\right]$ and the prior variance sample $\boldsymbol{\Sigma}_{\Delta p}^2 = \left[\sigma_{\Delta p1}^2, ..., \sigma_{\Delta pm}^2\right]$.

The prior distribution of μ_Δ under condition σ_Δ^2 is normal distribution, the prior distribution of σ_Δ^2 is inverse gamma distribution. In order to ensure the rationality of the prior distribution hypothesis, the prior distribution should be tested for K-S goodness of fit.

3.4 Posterior Distribution and Point Estimation

The posterior distribution of mean μ_Δ and variance σ_Δ^2 can be rewritten as

$$
\pi(\mu_\Delta,\sigma_\Delta^2|\boldsymbol{T}) = (\sigma_\Delta^2)^{-(\gamma_n+1)}\mathrm{e}^{-(\beta_n/2\sigma_\Delta^2)} = (\sigma_\Delta^2)^{-((v_n+1)/2+1)}\mathrm{e}^{-((v_n\sigma_n^2+k_n(\mu_\Delta-\mu_n)^2)/2\sigma_\Delta^2)}
\tag{6}
$$

with posterior distribution parameters μ_n, k_n, v_n, and σ_n^2.

Using the maximum likelihood estimation method, the point estimation value can be derived as

$$
\widehat{\mu_\Delta} = \mu_n \text{ and } \widehat{\sigma_\Delta^2} = \frac{v_n\sigma_n^2}{v_n+3}
\tag{7}
$$

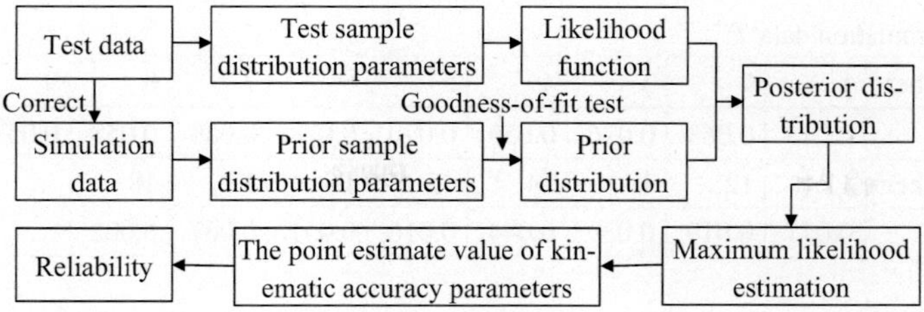

Fig. 4 Reliability estimation flow chart based on Bayesian

By substituting $\widehat{\mu_\Delta}$ and $\widehat{\sigma_\Delta^2}$ into Eq. (2), the kinematic accuracy reliability P_r is obtained.

3.5 Kinematic Accuracy Reliability Estimation Process Fusing Prior and Test Samples

The kinematic accuracy reliability estimation process is shown in Fig. 4.

4 Kinematic Accuracy Reliability Estimation of Retraction Mechanism

The feasibility and effectiveness of the proposed method are verified by estimating the kinematic accuracy reliability of the rotatable tire landing gear retraction mechanism.

4.1 Kinematic Accuracy Reliability Estimation of Retraction Mechanism

Based on the simulation test and the real test of the retraction mechanism, the test sample data and the simulation data are shown in Tables 2 and 3.

Train the Neural Network. The simulation data is used to train the neural network. The mean square error of Δ is 5.579×10^{-7}, less than 1×10^{-6}, so the neural network training is finished.

Table 2 Test sample data ***T***

Serial number	1	2	3	4	5	6	7	8	9	10
$T/(°)$	0.048	0.023	0.042	0.026	0.032	0.062	0.009	0.032	0.002	0.070

Table 3 Simulation data T^0

Serial number	1	2	3	4	5	6	7	8	9	10
$T^0/(°)$	0.043	0.063	0.026	0.059	0.070	0.078	0.080	0.058	0.051	0.045
Serial number	11	12	13	14	15	16	17	18	…	500
$T^0/(°)$	0.071	0.019	0.039	0.044	0.016	0.051	0.067	0.002	…	0.071

Generate Prior Sample Data. Using the Monte Carlo method and the random distribution information of $\boldsymbol{X}$, 80 groups of test points are generated and substituted into the neural network to generate prediction samples and calculate the mean $\mu_{\Delta pj}$ and variance $\sigma^2_{\Delta pj}$. The prior mean sample $\mathbf{M}_{\Delta p}$ and prior variance sample $\boldsymbol{\Sigma}^2_{\Delta p}$ are created by repeating above steps 3000 times, as shown in Table 4. The prior sample frequency distribution histograms are shown in Figs. 5 and 6.

It can be seen that the frequency distribution of prior mean samples is approximately normal distribution, while the frequency distribution of prior variance samples is approximately inverse gamma distribution.

Estimate Hyper Parameter. The prior sample data in Table 4 are used to obtain the estimated value of the hyper parameters. $\mu_0 = 0.039, k_0 = 82.775, v_0 = 99.606, \sigma_0^2 = 4.078 \times 10^{-4}$. The K-S goodness of fit test is carried out. The hypothesis distribution can be accepted as the distribution of the corresponding parameters.

Kinematic Reliability Estimation of Retraction Mechanism. The test sample data in Table 2 and the hyper parameters are brought into Eq. (7) to obtain the parameter

Table 4 Prior sample data

Serial number j	1	2	3	…	3000
Prior mean sample $\mu_{\Delta pj}$	0.036	0.034	0.039	…	0.038
Prior variance sample $\sigma^2_{\Delta pj}$	3.691×10^{-4}	4.861×10^{-4}	4.387×10^{-4}	…	3.329×10^{-4}

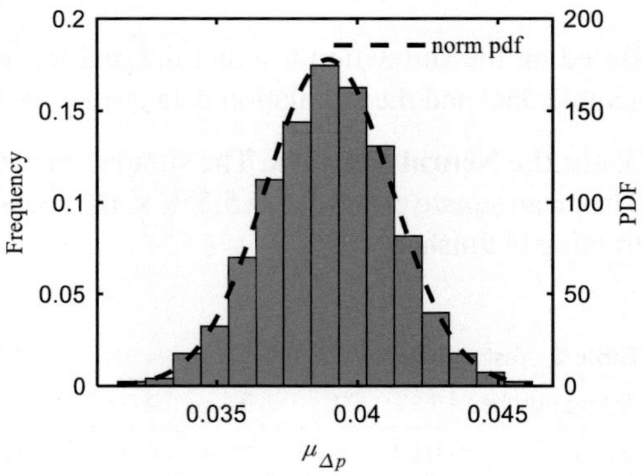

Fig. 5 Frequency distribution histogram of mean

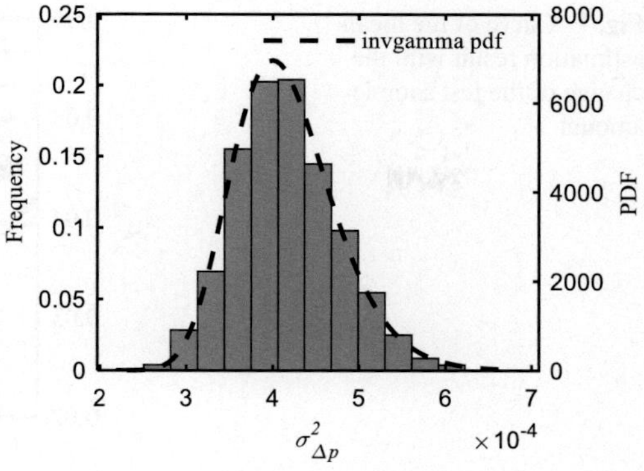

Fig. 6 Frequency distribution histogram of variance

values, in which $\widehat{\mu_\Delta} = 0.039$, $\widehat{\sigma_\Delta^2} = 7.978 \times 10^{-4}$. The reliability of the kinematic accuracy of the retraction mechanism is 0.999998.

4.2 Stationarity Analysis of Kinematic Accuracy Parameter Estimation

The classical interval estimation method aims to determine the confidence interval for parameter estimation through the analysis of experimental sample data. It is widely used to assess reliability indicators of products or systems, such as failure rates, mean time between failures, and more. This method provides a statistical foundation to quantify the uncertainty and robustness of estimated parameters in reliability calculations. Consequently, a comparison is drawn between the classical interval estimation method, with a confidence coefficient of 0.95, and the proposed approach. The study examines the variation trend in kinematic accuracy parameters μ_Δ and σ_Δ^2 of the retraction mechanism across test sample quantities ranging from 5 to 15. The calculation results are visually depicted in Figs. 7 and 8.

It is evident that the proposed method is significantly better than the classical interval estimation method in terms of the stability of estimation results for μ_Δ and σ_Δ^2 by fusing prior sample. The robustness of the proposed method under the condition of small sample is effectively verified. In addition, it's noteworthy that the proposed method exhibits a corrective influence on the estimation outcomes of distribution parameters when faced with the constraints of a small sample size.

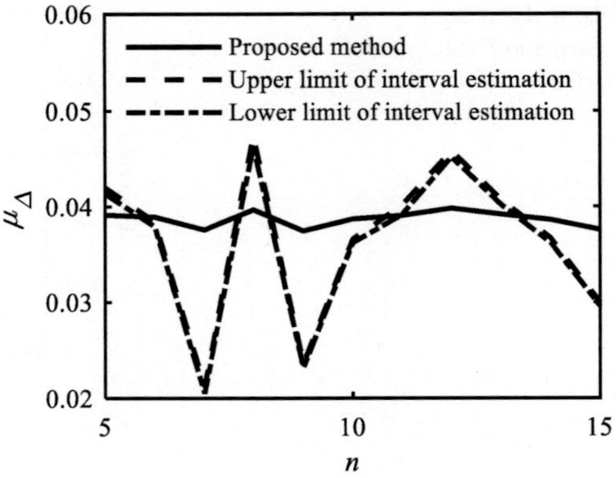

Fig. 7 Curve of the mean estimation result with the change of the test sample amount

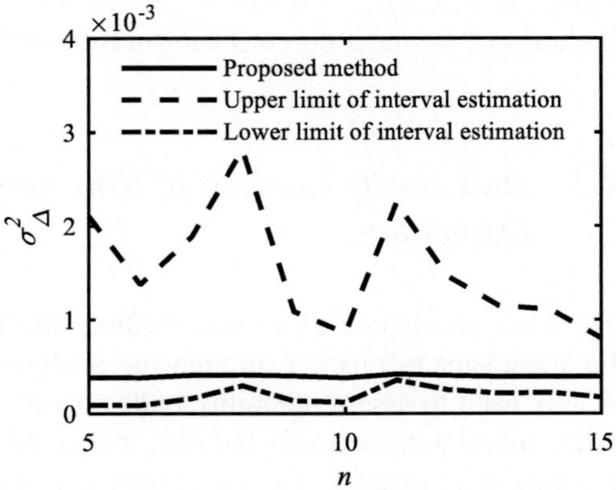

Fig. 8 Curve of variance estimation result with the change of test sample amount

5 Conclusion

In this paper, a novel method for assessing the mechanism performance reliability based on Bayesian theory is proposed. The method is applied to determine the kinematic accuracy reliability of the landing gear retraction mechanism. By comparing it with the classical interval estimation method, the proposed method exhibits a certain capacity to correct the estimation results for kinematic accuracy parameters through the incorporation of prior information. Furthermore, these results indicate a notably enhanced stability in the estimation outcomes. Therefore, the proposed method holds the potential to offer valuable insights into the assessment of performance reliability in similar mechanisms, particularly when constrained by limited availability of test samples.

Acknowledgements This study was supported by the National Natural Science Foundation of China (52172368, 52275114), the Natural Science Foundation of Jiangsu Province (BK20220135).

References

1. Liu J, Ren Y (2020) A general transfer framework based on industrial process fault diagnosis under small samples. IEEE Trans Ind Inf 17(9):6073–6083
2. Zhan ZH, Zhang XM, Zhang HD, Chen GC (2019) Unified motion reliability analysis and comparison study of planar parallel manipulators with interval joint clearance variables. Mech Mach Theory 138:58–75
3. Li XY, Chen WB, Kang R (2021) Performance margin-based reliability analysis for aircraft lock mechanism considering multi-source uncertainties and wear. Reliab Eng Syst Safe 205:107234
4. Yin Y, Nie H, Ni HJ, Zhang M (2016) Reliability analysis of landing gear retraction system influenced by multifactors. J Aircr 53(3):713–724
5. Hu HH, Wang P, Zhou HY (2022) Sequential reliability analysis for the adjusting mechanism of tail nozzle considering wear degradation. Machines 10(8)
6. Hongzhou L, Lixia S (2017) Reliability parameter interval estimation of NC machine tools considering working conditions. Math Probl Eng 2017:1–7
7. Ke Z, Zhou W (2019) Parameter estimation model of small test samples based on grey bootstrap method and unascertained rational number. Binggong Xuebao/Acta Armamentarii. 40(4):874–879
8. Kc S, Kumar M, Beer M (2021) Classical and Bayesian estimation of stress-strength reliability of a component having multiple states. Int J Qual Reliab Manag (UK):528–535
9. Li YL, Niu Z, Liu CH, Yan CL (2022) Credible Bayesian reliability model for structures with interval uncertain parameters. Structures 45:2151–2161
10. Zhang J, Xiao M, Gao L (2019) A new method for reliability analysis of structures with mixed random and convex variables. Appl Math Model 70:206–220

Control Allocation for Spin Interceptor with Aerodynamics and Lateral Thrust Blended

Xu Zheng, Haijin Di, and Jingfei Zhu

Abstract Focusing on the interception of high maneuverable targets such as new tactical ballistic missiles (TBM), a fast trajectory generation method is presented in this paper for interceptor with aerodynamics and lateral thrust blended. Firstly, the spin dynamics model is established in quasi-body coordinate system, and the guidance considering handover angle constraint between midcourse and terminal phrase is designed to capture accurately as well as achieve head-on collision killing. Secondly, the dynamic inverse controller is devised in the frame of feedback control idea, the control allocation is given for dual control interceptors. Finally, a combined algorithm considering by rough traversal search and iterative fine search is proposed to quickly obtain the launch azimuth, initial pitch angle and aiming angle of attack (AOA). The simulation results show that the launch window for reentry maneuverable TBM can be rapidly and effectively obtained less than 1 s through the combined search algorithm.

Keywords TBM · Dynamic inverse controller · Control allocation · Launch window

1 Introduction

Offensive and defensive confrontation has always been a hot area of research [1, 2]. The air rudder, as a single actuator of traditional anti-missile weapons, has been difficult to meet the requirements of guidance accuracy. Lateral thrust control can improve the rapid response capability of the missile, thereby effectively improving its maneuverability. Therefore, the use of aerodynamics and lateral thrust blended

X. Zheng (✉) · H. Di · J. Zhu
The 28th Research Institute of China Electronics Technology Group Corporation, Nanjing 210007, China
e-mail: zhengxu_hit@hotmail.com

J. Zhu
School of Computer Science and Engineering, Nanjing University of Science and Technology, Nanjing 210014, China

Z. Liu et al. (eds.), *Advances and Challenges in Advanced Unmanned Aerial Systems*, Springer Aerospace Technology, https://doi.org/10.1007/978-981-99-8045-1_25

has become the trend of anti-missile weapons [3, 4]. Interceptors with dual control actuator are divided into two types: orbit control type and attitude control type. The orbital control type is located near the center of mass, directly generating the overload for the interceptor (such as the 9M96E/2 interceptor in C-400). The attitude control type is located in front of the center of mass and is employed to generate overload and moment (such as the ERINT-1 interceptor in PAC-3). At present, more researches focus on interceptors in attitude control type [5, 6].

For the compound control interceptor with multiple actuators, a simple and reliable control allocation strategy needs to be designed to effectively improve the rapid response capability [7, 8]. Aiming at the problem of control allocation, literature [9] proposes a separate design scheme for inner and outer loops, using aerodynamic control as the inner loop and lateral thrust control as the outer loop to realize the control allocation of direct force and aerodynamic force. However, the influence of mutual interference is ignored. Literature [10] designs a neural network control distributor based on the use of convex optimization and non-linear programming to solve the control allocation problem of the asymmetry and saturation condition of the actuator. Literature [11] proposes to use a generalized inverse matrix method to abstract the control assignment problem into a parameterized input matrix zeroing problem under constrained conditions.

In view of the shortcomings of the existing composite control interceptor missiles that the launch window cannot be quickly determined, this paper proposes a combination method of ergodic coarse search and iterative accurate search. Then, a midcourse guidance law meeting the conditions of handover and capture is designed, and the direct force and aerodynamic actuator are allocated by the feedback control idea. Finally, aiming at the interception problem of the maneuverable TBM, the feasibility and effectiveness of the presented rapid generation model of launch window for composite control interceptor are demonstrated by simulation.

2 Dynamic Model of Spin Interceptor

The composite control interceptor flies to the predicted intercept point by inertial guidance system after launched. The interceptor rolls at a low speed and is controlled by an air rudder during the mid-flight. After the seeker intercepts the target, the spin speed is increased in order to facilitate attitude control, and the compound control combined of aerodynamics and lateral thrust is implemented at the end of interception to increase collision accuracy.

Due to the spin of the missile body, the dynamic model is generally established in quasi-projectile body coordinate system to facilitate the design of the control system. The spin interceptor dynamic model is as follows [8]

$$\begin{cases} \dot{\mathbf{x}}_1 = \mathbf{f}_1 + \mathbf{g}_1\mathbf{x}_2 + \mathbf{h}_1\mathbf{u}_1 + \mathbf{k}_1\mathbf{u}_2 \\ \dot{\mathbf{x}}_2 = \mathbf{f}_2 + \mathbf{h}_2\mathbf{u}_1 + \mathbf{k}_2\mathbf{u}_2 \end{cases} \tag{1}$$

where × 1 = $[\alpha^*\beta^*]$T, $\boldsymbol{x}_2 = [\,\omega_y^*\ \omega_z^*\,]^T$, $\boldsymbol{u}_1 = [\,\delta_y^*\ \delta_z^*\,]^T$, $\mathbf{u}_2 = [\,f_y\ f_z\,]^T$. α^* and β^* are respectively the quasi-attack angle and quasi-sideslip angle in quasi-projectile coordinate system; ω_y^* and ω_z^* are respectively the pitch and yaw rotational angular velocity in quasi-projectile system respect to launch coordinate system; δ_y^* and δ_z^* are respectively the virtual yaw and pitch rudder yaw angles in quasi-projectile coordinate system; δ_z^* and f_z are respectively the yaw and pitch directions lateral thrust in quasi-projectile coordinate system. The attitude angle and rudder deflection angle presented in missile body system and quasi-projectile system are as follows.

$$\begin{bmatrix} \alpha \\ \beta \end{bmatrix} = \begin{bmatrix} \cos\dot{\gamma}t & -\sin\dot{\gamma}t \\ \sin\dot{\gamma}t & \cos\dot{\gamma}t \end{bmatrix} \begin{bmatrix} \alpha^* \\ \beta^* \end{bmatrix}, \begin{bmatrix} \delta_z \\ \delta_y \end{bmatrix} = \begin{bmatrix} \cos\dot{\gamma}t & -\sin\dot{\gamma}t \\ \sin\dot{\gamma}t & \cos\dot{\gamma}t \end{bmatrix} \begin{bmatrix} \delta_z^* \\ \delta_y^* \end{bmatrix} \tag{2}$$

where $\dot{\gamma}$ refers to angular velocity of rotating projectile. In formula(1), the expression of each coefficient matrix is as follows.

$$\mathbf{f}_1 = \begin{bmatrix} -\frac{P\sin\alpha^*}{mV\cos\beta^*} - \frac{qSC_y^\alpha\alpha^*}{mV\cos\beta^*} + A \\ \frac{-P\cos\alpha^*\sin\beta^*}{mV} + \frac{qSC_z^\beta\beta^*}{mV} + B \end{bmatrix}, \mathbf{f}_2 = \begin{bmatrix} \begin{bmatrix} qSLm_y^\beta\beta^* + qSLm_y^{\overline{\omega}_y}\omega_y^*L/V \\ +(I_z - I_x)\omega_z^*\omega_x^* - I_z\omega_z^*\dot{\gamma} \end{bmatrix} / I_y \\ \begin{bmatrix} qSLm_z^\alpha\alpha^* + qSLm_z^{\overline{\omega}_z}\omega_z^*L/V \\ +(I_x - I_y)\omega_x^*\omega_y^* + I_y\omega_y^*\dot{\gamma} \end{bmatrix} / I_z \end{bmatrix} \tag{3}$$

$$A = \frac{g\sin\vartheta\sin\alpha^* + g\cos\vartheta\cos\alpha^*}{V\cos\beta^*},\ B = \frac{g\sin\vartheta\cos\alpha^*\sin\beta^* - g\cos\vartheta\sin\alpha^*\sin\beta^*}{V} \tag{4}$$

$$\mathbf{g}_1 = \begin{bmatrix} -\tan\vartheta\tan\beta^*\cos\alpha^* + \tan\beta^*\sin\alpha^* & 1 \\ \tan\beta^*\sin\alpha^* + \cos\alpha^* & 0 \end{bmatrix} \tag{5}$$

$$\mathbf{h}_1 = \begin{bmatrix} 0 & -\frac{qSC_y^{\delta_z}}{mV\cos\beta^*} \\ \frac{qSC_z^{\delta_y}}{mV} & 0 \end{bmatrix}, \mathbf{h}_2 = \begin{bmatrix} \frac{qSLm_y^{\delta_y}}{I_y} & \\ & \frac{qSLm_z^{\delta_z}}{I_z} \end{bmatrix} \tag{6}$$

$$\mathbf{k}_1 = \begin{bmatrix} -\frac{\cos\alpha^*}{mV\cos\beta^*} & 0 \\ \frac{\sin\alpha^*\sin\beta^*}{mV} & \frac{\cos\beta^*}{mV} \end{bmatrix}, \mathbf{k}_2 = \begin{bmatrix} 0 & -\frac{L_T}{I_y} \\ \frac{L_T}{I_z} & 0 \end{bmatrix} \tag{7}$$

In the formulas above, P is engine thrust, m is the mass of interceptor, V is the velocity value of interceptor; q is the dynamic pressure; S is characteristic area of interceptor body; C_y^α is the derivative of lift coefficient to AOA; C_z^β is the derivative of side force coefficient to side slip angle; g is the value of gravitational acceleration; ϑ is the pitch angle; $C_y^{\delta_z}$ is the derivative of lift coefficient to rudder deflection angle; $C_z^{\delta_y}$ is the derivative of side force coefficient to rudder deflection angle; L is the characteristic length of interceptor body; m_z^α and m_y^β are respectively the derivative of pitch moment coefficient to AOA and derivative of yaw moment coefficient to sideslip angle; $m_z^{\overline{\omega}_z}$ and $m_y^{\overline{\omega}_y}$ are respectively pitch damping torque coefficient and yaw damping torque coefficient; $m_z^{\delta_z}$ and $m_y^{\delta_y}$ are respectively the derivative of pitch moment coefficient to rudder yaw angle and the derivative of yaw moment coefficient to rudder yaw angle; I_x, I_y, I_z are respectively moments of inertia in rolling, pitching and yaw directions ($I_y = I_z$ for axisymmetric missiles); L_T is the distance from attitude control engine to projectile centroid.

3 Guidance and Control Law for Spin Interceptor

3.1 Design of Guidance Handover

The interceptor flies to the predicted encounter point in midcourse guidance phrase. To ensure that the seeker can capture the target during midcourse and terminal shift handover, the interceptor body direction shall be consistent with the line of sight (LOS) as much as possible, which can meet the terminal angle constraint. A midcourse guidance law considering the angle constraint of shift handover is obtained as follows.

$$a_{\varepsilon c} = K_1 V_m \dot{\varepsilon} + K_2 V_m \frac{\theta_f - \theta}{t_{go}} \tag{8}$$

$$a_{\eta c} = K_3 V_m \dot{\eta} + K_4 V_m \frac{\psi_{vf} - \psi_v}{t_{go}} \tag{9}$$

where K_1, K_2, K_3, K_4 are guidance coefficients; $\dot{\varepsilon}$ is the pitch rate in sight direction; $\dot{\eta}$ is yaw rate in sight direction; θ is the inclination angle; θ_f is the inclination angle at the predicted handover point; ψ_v is the trajectory yaw angle; ψ_{vf} is the trajectory yaw angle of at the predicted handover point; t_{go} is the predicted flight time of midcourse guidance; The first term in midcourse guidance law is the term obtained by proportional guidance, which aims to guide the interceptor to the target. The second is to rotate the trajectory inclination to reduce the terminal angle error.

3.2 Design of Control Allocation

Since the response speed of the projectile angular velocity is much faster than aerodynamic angle, the system dynamics can be divided into fast dynamics and slow dynamics by using the assumption of time scales. The fast dynamic refers to the angular velocity of projectile rotation, and the slow dynamic refers to the aerodynamic angle. Thus, the missile autopilot can be divided into two loop designs. The outer loop feedback is adopted to control the aerodynamic angle, generating angular velocity command and transmitting it to the inner loop for execution. The inner loop feedback is adopted to control the angular velocity, and transmitting the generated rudder deviation command to the missile body, as shown in Fig. 1.

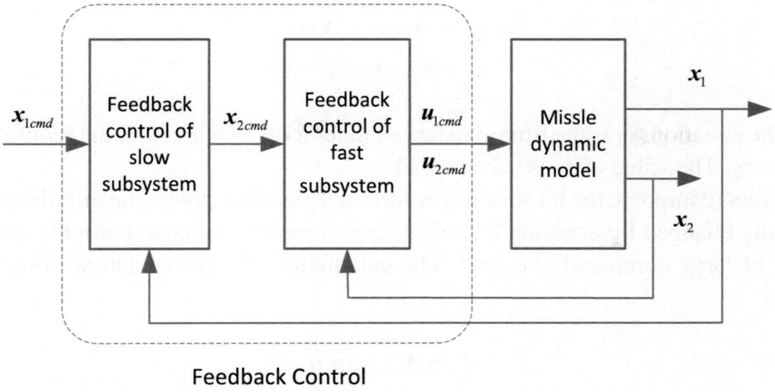

Fig. 1 Design of dynamic inverse controller

In Fig. 1, assume $\mathbf{x}_{1cmd}$ as $\mathbf{x}_{1cmd} = [\alpha_{cmd}\ \beta_{cmd}]^T$. The calculation formulas of converting command overload into command AOA and command sideslip angle given by guidance system are as follows.

$$\alpha_{cmd} = \frac{n_{yc} \cdot mg - qSC_y^{\delta_z}\delta_z}{qSC_y^{\alpha}} \tag{10}$$

$$\beta_{cmd} = \frac{n_{zc} \cdot mg - qSC_z^{\delta_y}\delta_y}{qSC_z^{\beta}} \tag{11}$$

When the command AOA and the command sideslip angle are obtained, the feedback control law of outer loop is firstly designed. The first formula in Eq. (1) obtains the control law in the outer loop slow subsystem.

$$\begin{aligned} \dot{\mathbf{x}}_{1cmd} &= (\mathbf{x}_{1cmd} - \mathbf{x}_1)/\tau_1 \\ \mathbf{x}_{2cmd} &= \mathbf{g}_1^{-1}(\mathbf{x}_{1cmd} - \mathbf{f}_1 - \mathbf{h}_1\mathbf{u}_{1pre} - \mathbf{k}_1\mathbf{u}_{2pre}) \end{aligned} \tag{12}$$

where $\mathbf{x}_{1cmd}$ is the command attitude angle, τ_1 is the time constant of outer loop, $\mathbf{x}_{2cmd}$ is the commanded attitude angular velocity, $\mathbf{u}_{1pre}$ and $\mathbf{u}_{2pre}$ are respectively the actuator control values in previous control cycle.

The control law of inner loop is correspondingly derived after giving the command angular velocity of outer loop. The actuator needs to be allocated with regard to includes air rudder and attitude control engine. Because the command overload at the hitting moment would be very large, control torque may not adequate enough if air rudder is single used for control. Therefore, lateral force is considered to introduce for auxiliary control under the condition of inadequate aerodynamic force.

When aerodynamic control is adopted, the control law of the inner loop is as follows.

$$\begin{aligned}\dot{\mathbf{x}}_{2cmd} &= (\mathbf{x}_{2cmd} - \mathbf{x}_2)/\tau_2 \\ \mathbf{u}_1 &= \mathbf{h}_2^{-1}(\dot{\mathbf{x}}_{2cmd} - \mathbf{f}_2)\end{aligned} \tag{13}$$

In the equation, τ_2 is the time constant of inner loop, and meeting the requirement of $\tau_1 > \tau_2$. The value of τ_2 is set as 0.024.

In order to improve the hit accuracy in terminal guidance phase, the missile attitude is rapidly changed by enabling lateral force to track the command attitude angle in facing of large command overload. The calculation formula of lateral force is as follows.

$$\begin{aligned}f_y &= NT_f \cos\eta \\ f_z &= NT_f \sin\eta\end{aligned} \tag{14}$$

where N ($N \leq 5$) is the number of engines consumed by current number, which can be determined according to the difference between command overload and current overload; T_f is the thrust of a single attitude control engine, whose value is set as 3111N in this paper; η is the angle of engine thrust direction in quasi missile-body system at the time of actual ignition.

4 Launch Window Design

The launch data $\boldsymbol{\eta} = [\,A_0\ \vartheta_0\ \alpha_1\,]^T$ of interceptor is selected as the design variable, among which A_0 is the launch azimuth angle, ϑ_0 is the initial launch pitch angle and α_1 is the aiming AOA. The determination of launch data at a certain time is carried out in two steps. Firstly, the launch data are traversed and roughly searched. After a group of launch data is selected, the optimal initial launch data can be found by taking the miss distance as terminal constraint index. Secondly, launch data with the minimum miss distance is taken as initial value, and the accurate launch data can be accurately derived by Newton iterative algorithm.

Under the condition of a certain interception time, the usual launch azimuth search method is traversal search by 360 degree. Obviously, this search algorithm will consume more calculation time and is not conducive to the interception operation. To overcome the shortcomings, an improved search algorithm is presented in this paper. The azimuth search towards the direction of target flight trajectory will significantly reduce the calculation time. Firstly, the initial launch azimuth is calculated according to the current target state in fixed-north coordinate system, and then the launch azimuth deviation value is obtained according to the rendezvous and interception time, which ceaselessly corrects the launch azimuth value. The correction formula of launch azimuth is as follows.

$$\begin{cases} A_0 = \arctan(z_{T0}/x_{T0}) \\ \Delta A = \arctan\left(\dfrac{z_{gT0} + v_{gzT0} \cdot t_f}{x_{gT0} + v_{gxT0} \cdot t_f}\right) \\ A = A_0 + \Delta A \end{cases} \tag{15}$$

where A_0 is the azimuth of the target in fixed-north coordinate system at the interception time; ΔA is the azimuth correction value predicted by rendezvous time; A is the corrected launch azimuth; x_{T0} and z_{T0} are respectively the x-direction and z-direction position of the target in fixed-north coordinate system; v_{gzT0} 、 v_{gzT0} are respectively the x-direction and z-direction velocity component of the target in launch coordinate system; t_f is the predicted interception and rendezvous time.

According to the azimuth search algorithm, the following gives the traversal coarse search process of launch data at a certain launch time.

Step 1 According to the first equation in (15), the initial launch azimuth is determined according to target position information at the beginning interception time.

Step 2 The predicted azimuth correction deviation ΔA is determined according to the second equation in (15).

Step 3 According to the third formula in (15), the launch azimuth angle is calculated and corrected from the initial launch azimuth angle and deviation value.

Step 4 The initial pitch angle ϑ_0 and aiming AOA α_1 are traversal searched. The section of ϑ_0 is set as $\vartheta_0 = [\,30°\ 50°\,]$ which is searched by every 1° The section of α_1 is set as $\alpha_1 = [\,-2°\ 2°\,]$ which is searched by every 0.5° The interceptor trajectory is calculated based on launch data.

Step 5 The miss distance is calculated at the rendezvous time according to the missile and target dynamics model.

Step 6 Compare the miss value calculated twice adjacently. If the miss distance is less than the last search value, go to **Step 7**. Otherwise, update the interception time and go to **Step 2**.

Step 7 : Save the launch time, launch azimuth, initial launch elevation angle and aiming AOA at the time of minimum miss distance, and go to **Step 8**.

Step 8 Determine whether the traversal is complete. If yes, end the search; otherwise, go to **Step 4**.

After deriving a set of optimal initial value by roughly search, the accurate launch data would be iteratively acquired with the constraint condition of minimum terminal miss distance.

5 Simulation Analysis

The PAC-3 is set as composite control interceptor in this paper, whose power unit adopts a high-performance single-stage solid rocket motor, employing inertial + terminal active radar guidance. After the interceptor is launched, it relies on the inertial guidance system to fly towards the predicted interception point. The mid-flight rolls at a speed of 30r/min and is controlled by an air rudder. After the seeker captures the target, the spin speed is increased to 180r/min. The compound control with lateral and aerodynamic force is adopted, and the KKV is released to perform collision killing to improve the interception accuracy. The parameters of the interceptor are given as follows. The engine thrust is 22960N, specific impulse is 2800 m/s, launch mass is 304 kg, fuel is 164 kg, projectile length is 4.63 m, projectile diameter is 255 mm, maximum range is 50 km, maximum firing height is 30 km, and maximum flight speed is 5 ~ 6 Ma.

It issues an incoming warning when ground-guided radar detects an incoming TBM. The target flight position and speed are respectively as $\mathbf{r}_T = [\,130000\ 90000\ -110000\,]^T$ m and $\mathbf{v}_T = [\,-900\ -100\ 700\,]^T$ m/s. The range of 68s to 92s since the beginning of detecting time is the launch window to intercept the incoming TBM, and the interception trajectory clusters of initial guidance, intermediate guidance and terminal guidance are derived less than 1 s by the launch data generated through rough and fine search, which are shown in Fig. 2. After calculation, the maximum angle between the direction of interceptor velocity and LOS at the shift handover is no more than 18o in all trajectory clusters, which can ensure that the seeker can effectively capture the target. The angle between the direction of interceptor speed and target speed shall not be less than 156 at the time of shift handover, and the "head-on interception" can be realized in the whole launch window.

Fig. 2 The interception trajectory cluster for TBM

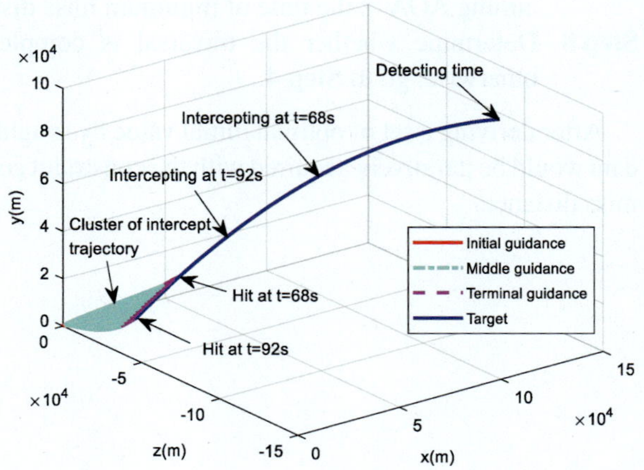

Fig. 3 The variation of quasi attack angle and quasi sideslip angle in terminal guidance phase

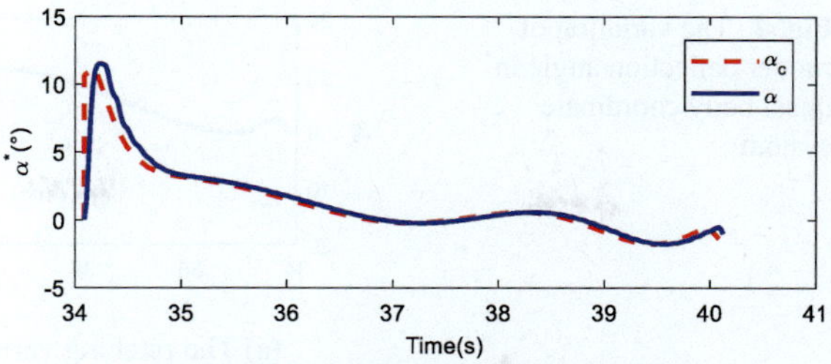

(a) The variation of quasi attack angle

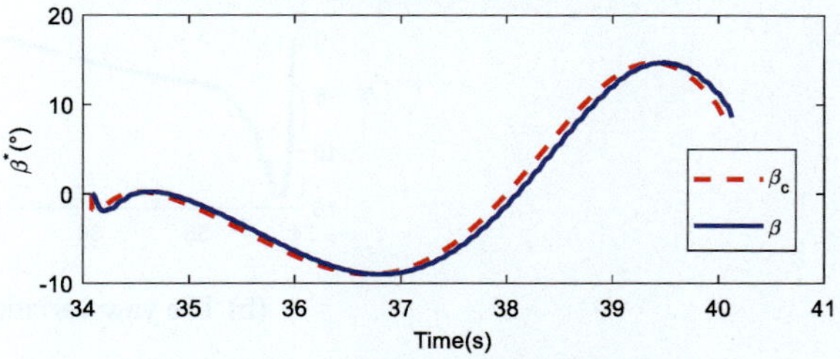

(b) The variation of quasi sideslip angle

The horizontal maneuverable acceleration is set to change according to the sinusoidal law, namely $\mathbf{a}_{zt} = 6 \times 9.8 \times \sin(t)$. The comparative changes of quasi attack angle and quasi sideslip angle, rudder deflection angle, attitude control engine consumption and missile target overload in the terminal guidance phase are represented in Figs. 3, 4, 5 and 6. In Figs. 3 and 4, the values of quasi angle of attack, quasi sideslip angle and rudder deflection angle become larger during the terminal phrase, and the control parameters(blue line) can track the guidance command(red line) well. In Fig. 5, 77 attitude control engines are consumed, with the total of 180 engines installed along the interceptor body by every 10 degrees, accounting for 42.8%. In Fig. 6, the interceptor overload increases obviously in order to ensure fast tracking, changing with the overload of maneuverable TBM. A large overload is formed in the last second to eliminate the LOS rotation rate. For this maneuverable TBM, the calculated miss distance is 0.362 m by employing aerodynamics and lateral thrust compound control technique. The maneuverable TBM cannot be intercepted if the aerodynamic rudder is individually used for attitude control. Therefore, the compound control can provide large available overload and possess high guidance accuracy.

Fig. 4 The variation of rudder deflection angle in quasi body coordinate system

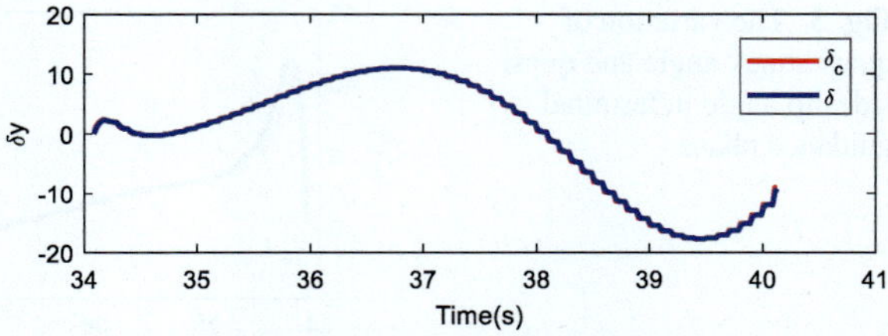

(a) The pitching variation of rudder deflection angle

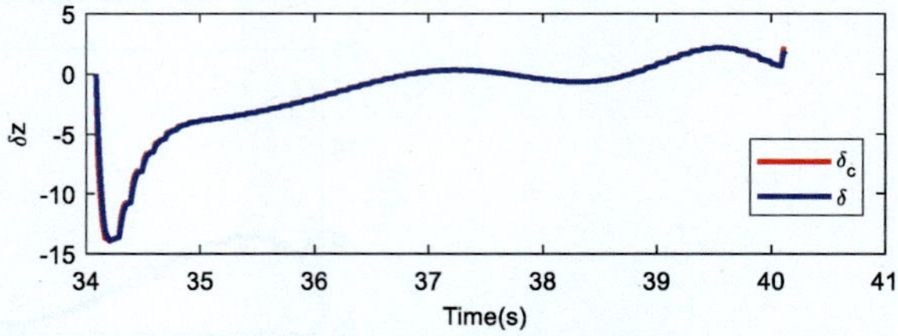

(b) The yaw variation of rudder deflection angle

Fig. 5 The consumption of attitude control engine

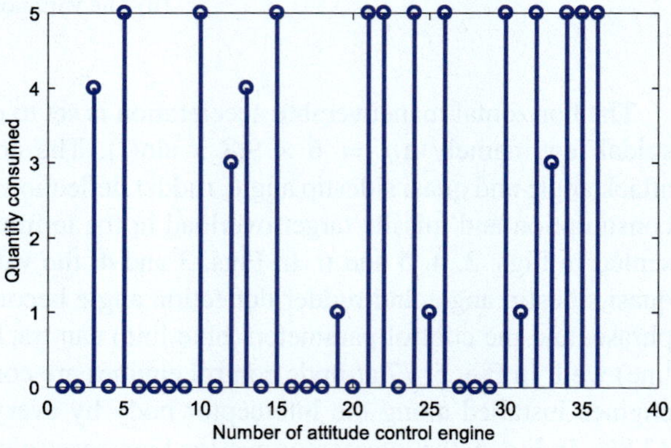

Fig. 6 The normal overload change of interceptor and TBM in terminal guidance stage

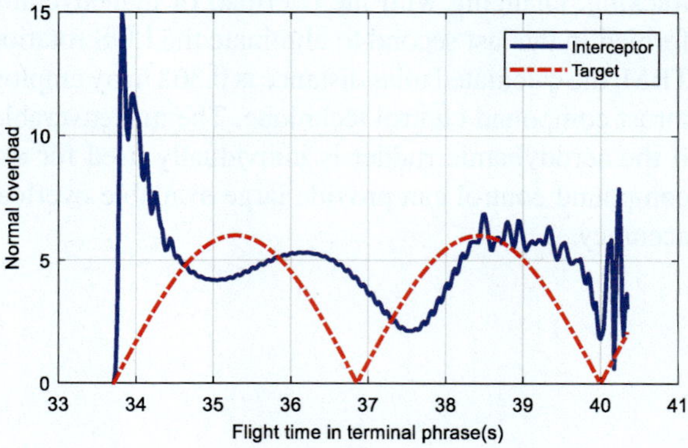

6 Conclusion

In this paper, a fast trajectory generation algorithm of interceptor is designed for dual controlled spin interceptor under the framework of quasi body coordinate system dynamic model, which is fairly instructive for engineering practice. The conclusions are as follows.

(1) The proposed method of traversal coarse search and iterative fine search can rapidly determine the interceptor command launch window, with the calculation of less than 1 s. The maneuverable TBM is simulated to verify the feasibility and effectiveness of the proposed method of launch window.
(2) The designed interceptor trajectory can meet the acquisition conditions of shift handover at the end of middle guidance, and realize the head-on interception. The collision killing is ensured by the dual control actuators with high guidance accuracy of 0.362 m.
(3) The dynamic inverse control based on feedback idea can superiorly realize the control allocation of lateral and aerodynamic force. The simulation results show that the interceptor can quickly track the command attitude angle and aerodynamic rudder deflection angle.

References

1. Xue YY, Hu GW (2018) Launch area solution for surface to air missile (SAM) based on deep leaning. Command Inf Syst Technol 9(4):48–52
2. Nan Y, Jiang L (2020) Midcourse penetration and control of ballistic missile based on deep reinforcement learning. Command Inf Syst Technol 11(4):1–9
3. Zhao QL, Dong XW, Song X (2019) Cooperative time-varying formation guidance for leader-following missiles to intercept a maneuvering target with switching topologies. Nonlinear Dyn 95(1):129–141
4. Bhavnesh P, Nilesh M, Talole SE (2017) Continuous-time predictive control-based integrated guidance and control. J Guid Control Dyn 40(7):1579–1595
5. Chen ZY, Yu JL, Dong XW (2021) Three-dimensional cooperative guidance strategy and guidance law for intercepting highly maneuvering target. Chin J Aeronaut 34(5):485–495
6. Gaudet B, Furfaro R, Linares R (2020) Reinforcement learning for angle-only intercept guidance of maneuvering targets. Aerosp Sci Technol 99(C):105746–105746
7. Zhang YA, Wu HL, Liu JM (2018) A blended control strategy for intercepting high-speed target in high attitude. Proc Instit Mech Eng Part G J Aerosp Eng 232(12):2263–2285
8. Xia C, Dong CY, Wang Q (2018) Adaptive integrated guidance and control back-stepping sliding mode design for blended control missile. Syst Eng Electr 40(10):2325–2333
9. Hong JH, Lee CH (2020) Nonlinear autopilot design for endo and exo-atmospheric interceptor with thrust vector control. IEEE Trans Aerosp Electron Syst 56(1):796–810
10. Chen M (2016) Constrained control allocation for over-actuated aircraft using a neuro-dynamic model. IEEE Trans Syst Man Cybern Syst 46(12):1630–1641
11. Kirchengast M, Steinberger M, Horn M (2018) Input matrix factorizations for constrained control allocation. IEEE Trans Autom Control 63(4):1163–1170

Trajectory Optimization for Spacecraft Close Proximity Based on Dual Quaternions

Qian Wang, Hongmin Zhai, and Shunli Li

Abstract The problem of trajectory optimization for spacecraft close proximity is addressed in this paper. The proposed algorithm aims to find a fuel-optimal trajectory while subject to line-of-sight constraint, glide slope constraint and control magnitude constraint, etc. Because line-of-sight constraint and glide slope constraint are coupled with the position and attitude motions of the spacecraft, and the dual quaternion can uniformly represent the relative position and attitude between coordinate systems. Therefore, dual quaternions were used to describe the motion of the spacecraft. Secondly, the sequential convex optimization is used to solve the fuel-optimal trajectory established by dual quaternions. Finally, the calculation time and energy consumption of the proposed method are given by numerical simulations.

Keywords Close proximity · Dual quaternions · Trajectory optimization

1 Introduction

This paper studied the trajectory optimization problem of spacecraft close proximity to a space target. There are many complex path constraints like line-of-sight constraint or collision avoidance constraint in the process of spacecraft close proximity to the target. Line-of-sight constraint is coupled with the position and attitude motions of the spacecraft. Dual quaternions can uniformly represent the relative position and attitude between coordinate systems, and the coupling constraint mentioned above can be uniformly represented by dual quaternions. In this way, dual quaternions

Q. Wang · S. Li (✉)
Harbin Institute of Technology, Harbin 150001, China
e-mail: lishunli@hit.edu.cn

Q. Wang
e-mail: 20B918102@stu.hit.edu.cn

H. Zhai
The Third Research Academy CASIC, Beijing 100074, China

Z. Liu et al. (eds.), *Advances and Challenges in Advanced Unmanned Aerial Systems*, Springer Aerospace Technology, https://doi.org/10.1007/978-981-99-8045-1_26

are used to establish the position and attitude integrated planning model, which can concisely express the dynamic model and coupling path constraints.

Quaternion as the foundation of dual quaternion have been put forward by Hamilton in 1843 [1]. But it has no practical applications for a long time. With the development of calculations, quaternions are increasingly used in flight control systems. Quaternion shows its superiority in the establishment of strapdown inertial system [2, 3], rigid body attitude control [4], attitude estimation [5, 6], etc.

The properties between quaternion and dual quaternion are inherited. But quaternions only represent rotation, while dual quaternions can represent not only rotation but also translation. Dual quaternions have been widely used for relative navigation [7], rendezvous and docking [8], etc. As mentioned above, the purpose of this paper is to find an optimal trajectory subject to the constraints such as the line-of-sight constraint, but line-of-sight constraint is coupled with attitude and position. Therefore, this paper referring to the method proposed by Lee [9] and Hazra [10] uses dual quaternions to establish the coupled rotational and translational models.

Our requirements for trajectory planning algorithm are feasibility, rapidity and optimality. The algorithm can converge to the trajectory with optimal objective as quickly as possible for any initial state. The obtained trajectory must satisfy the constraints at any time, and the algorithm can guarantee that the trajectory is global optimality. Convex programming is a high accurate nonlinear programming algorithm, which can efficiently solve the nonlinear optimal control problem by interior point methods [11]. The sequential convex optimization method is used in this paper to solve the optimal trajectory of the spacecraft close proximity to the space target, which requires convexification of the dynamics model represented by dual quaternion.

The paper is organized as follows. Section 2 first introduces the background of quaternion and dual quaternion, and then establishes the dynamic model and constraints of the spacecraft close proximity to the target, finally uses convex optimization to solve the optimal trajectory. Section 3 verified the efficiency of the proposed strategy through numerical simulations.

2 Dual Quaternion Method

2.1 Quaternion

The quaternion can be regarded as consisting of a vector part and a scalar part:

$$\boldsymbol{q} = \boldsymbol{q}_{1:3} + q_0 \tag{1}$$

$$\boldsymbol{q} = q_1\mathbf{i} + q_2\mathbf{j} + q_3\mathbf{k} \tag{2}$$

Sometimes a four-dimensional vector can be used to represent a quaternion

$$\boldsymbol{q} = [\boldsymbol{q}_{1:3}, q_0]^{\mathrm{T}} = [q_1, q_2, q_3, q_0]^{\mathrm{T}} \tag{3}$$

The conjugate quaternion is defined as $q^* = [-\boldsymbol{q}, q_0]^{\mathrm{T}} = [-q_1, -q_2, -q_3, q_0]^{\mathrm{T}}$. For unit quaternion, the conjugate quaternion is equal to its inverse quaternion $\boldsymbol{q}^*=\boldsymbol{q}^{-1}$. For two arbitrary quaternions, $\boldsymbol{q} = \left[\boldsymbol{q}_{1:3}^T, q_0\right]^{\mathrm{T}}, \boldsymbol{p} = \left[\boldsymbol{p}_{1:3}^T, p_0\right]^{\mathrm{T}}$, multiplication is defined as:

$$\boldsymbol{q} \otimes \boldsymbol{p}=\begin{bmatrix} q_0\boldsymbol{p}_{1:3} + p_0\boldsymbol{q}_{1:3} + \boldsymbol{q}_{1:3} \times \boldsymbol{p}_{1:3} \\ q_0 p_0 - \boldsymbol{q}_{1:3}^T\boldsymbol{p}_{1:3} \end{bmatrix} \tag{4}$$

The cross product is defined as:

$$\boldsymbol{q} \times \boldsymbol{p}=\begin{bmatrix} q_0\boldsymbol{p}_{1:3} + p_0\boldsymbol{q}_{1:3} + \boldsymbol{q}_{1:3} \times \boldsymbol{p}_{1:3} \\ 0 \end{bmatrix} \tag{5}$$

The quaternion product operation expressed in Eq. (3) can also be expressed as:

$$\boldsymbol{q} \otimes \boldsymbol{p}=[\boldsymbol{q}]_{\otimes}\boldsymbol{p}=[\boldsymbol{p}]_{\overset{\vee}{\otimes}}\boldsymbol{q} \tag{6}$$

where:

$$[\boldsymbol{q}]_{\otimes}=\begin{bmatrix} q_0 I_{3\times 3} + \boldsymbol{q}_{1:3}^{\times} & \boldsymbol{q}_{1:3} \\ -\boldsymbol{q}_{1:3}^T & q_0 \end{bmatrix} \tag{7}$$

Here $\boldsymbol{q}^{\times}$ denotes a cross product matrix.

2.2 Dual Quaternion

A dual number is defined as the sum of a primal part and a dual part [12], namely,

$$\hat{a} = a + \varepsilon a_0 \tag{8}$$

where ε is the dual unity, which verifies $\varepsilon \neq 0,\ \varepsilon^2 = 0$, while a and a_0 are real numbers. Dual quaternions can be thought of as replacing a and a_0 in dual numbers with two quaternions. We use dual quaternion to represent translation and rotation

$$\begin{aligned} \tilde{q} &= \boldsymbol{q}_r + \varepsilon \boldsymbol{q}_d \\ \boldsymbol{q}_d &= \frac{\boldsymbol{t}}{2} \otimes \boldsymbol{q}_r \end{aligned} \tag{9}$$

$\boldsymbol{q}_r$ is a unit quaternion representing the rotation, which represents the real part, $\boldsymbol{q}_d$ represents dual part. $\boldsymbol{t}$ is used to describe the translation. Due to $\varepsilon^2 = 0$, the multiplication of two dual quaternions can be given as

$$\tilde{q} \otimes \tilde{p} = q_r \otimes p_r + \varepsilon\left(q_r \otimes p_d + q_d \otimes p_r\right) \tag{10}$$

Like quaternion, dual quaternion can also be written as matrix form

$$\tilde{q} \otimes \tilde{p}=[\tilde{q}]_{\otimes}\tilde{p}=[\tilde{p}]_{\overset{\vee}{\otimes}}\tilde{q} \tag{11}$$

$$[\tilde{q}]_{\otimes}=\begin{bmatrix} [q_r]_{\otimes} & 0_{4\times 4} \\ [q_d]_{\otimes} & [q_r]_{\otimes} \end{bmatrix} \tag{12}$$

2.3 Dynamic Model

In order to describe the motions of the spacecraft relative to the target, reference frames are required. Target inertial frame I, centered at the target's center of mass, is fixed to the target, and axes fixed inertially. Body reference frame B is centered at the center of mass of the spacecraft, with axes along the principal moments of inertia of the spacecraft.

In Fig. 1, the general rigid body motion between the coordinate system I and the coordinate system B is represented by rotation $\boldsymbol{q}$ followed by translation (or $\boldsymbol{r}_I$ translated to $\boldsymbol{r}_B$ followed by rotation $\boldsymbol{q}$).The unit dual quaternion can be represented by $\boldsymbol{r}_I$ and $\boldsymbol{q}$ (or $\boldsymbol{q}$ and $\boldsymbol{r}_B$). $\boldsymbol{q}$ is the rotation quaternion from fixed coordinate I to coordinate system B.

$$\begin{aligned} \tilde{q} &= q + \varepsilon\frac{1}{2}r_I \otimes q \\ &= q + \varepsilon\frac{1}{2}q \otimes r_B \end{aligned} \tag{13}$$

The kinematic equation of dual quaternion is

$$\dot{\tilde{q}}=\frac{1}{2}\tilde{\omega}_I \otimes \tilde{q} = \frac{1}{2}\tilde{q} \otimes \tilde{\omega}_B \tag{14}$$

where $\tilde{\omega}_B = \left[\omega_B^T\ v_B^T\right]^T$, $\omega_B = \left[\omega_B^T\ 0\right]^T$, $v_B = \left[v_B^T\ 0\right]^T$, ω_B^T is the projection of the angular velocity of the coordinate system B relative to the coordinate system I in the coordinate system B. v_B^T denotes the translational velocity of the general rigid body. The dynamic equation considering position and attitude is as follows:

$$\begin{aligned} \boldsymbol{F} &= \mathrm{m}\dot{\boldsymbol{v}}_b + \boldsymbol{\omega}_b \times \mathrm{m}\boldsymbol{v}_b \\ \boldsymbol{T} &= \mathrm{J}\dot{\boldsymbol{\omega}}_b + \boldsymbol{\omega}_b \times \mathrm{J}\boldsymbol{\omega}_b \end{aligned} \tag{15}$$

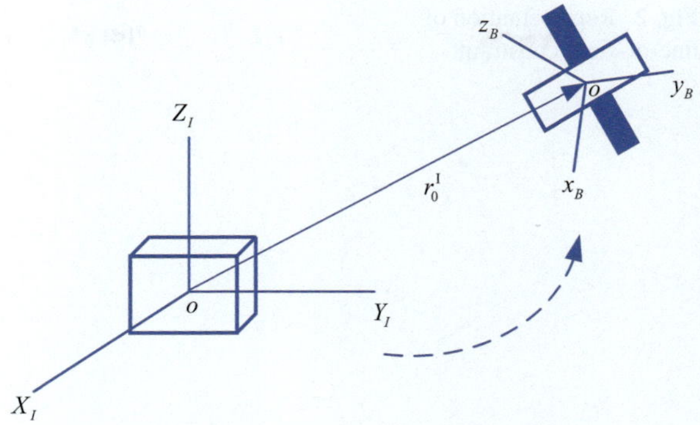

Fig. 1 Representation of reference frames O and B

The derivations of dynamic equations of the general rigid body in dual quaternion form can be given as:

$$\tilde{\boldsymbol{J}}\dot{\tilde{\boldsymbol{\omega}}}_B + \tilde{\boldsymbol{\omega}}_B \times \tilde{\boldsymbol{J}}\tilde{\boldsymbol{\omega}}_B = \tilde{\boldsymbol{F}} \tag{16}$$

where:

$$\tilde{\boldsymbol{J}}=\begin{bmatrix} \boldsymbol{0}_{3\times 3} & 0 & m\boldsymbol{I}_{3\times 3} & 0 \\ 0 & 0 & 0 & 1 \\ J & 0 & \boldsymbol{0}_{3\times 3} & 0 \\ 0 & 1 & 0 & 0 \end{bmatrix}, \quad \tilde{\boldsymbol{F}} = \begin{bmatrix} \boldsymbol{F} \\ 0 \\ \boldsymbol{T} \\ 0 \end{bmatrix} \tag{17}$$

2.4 Convex Optimization

Optimal Control Problem. In the process of the spacecraft approaching the target, some constraints need to be considered, such as the line-of-sight constraint, glide slope constraint, the control force constraint, etc. The line-of-sight constraint is to keep the target within the sight of the spacecraft at all times. The glide-slope is a cone with its vertex at the center of coordinate I, which guides the trajectory of the spacecraft to ensure no contact with the target before the end of the flight.

The representation of the LOS constraint is shown in Fig. 2. This constraint is represented in body frame *B*, and the mathematical interpretation of line-of-sight constraint is as shown below

$$-\boldsymbol{r}_B \cdot \boldsymbol{y}_B \geq \|\boldsymbol{r}_B\| \cos\theta \tag{18}$$

where y_B is a unit vector along the boresight of the sensor onboard the spacecraft in its body frame *B*. We express the line-of-sight constraint in the form of dual quaternion:

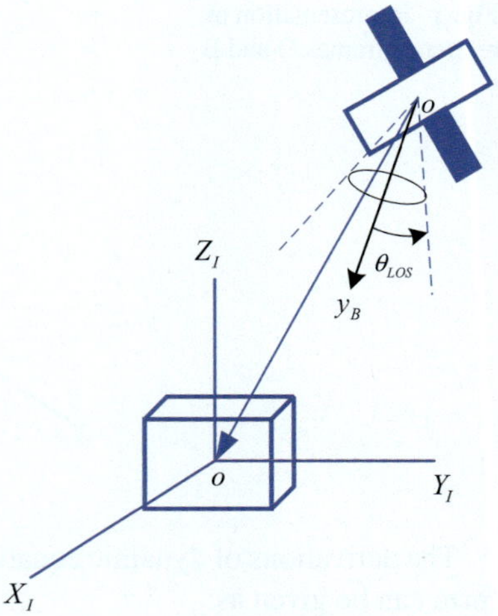

Fig. 2 Representation of line-of-sight constraint

$$f(\tilde{\boldsymbol{q}}) = \tilde{\boldsymbol{q}}^T M_H \,\tilde{\boldsymbol{q}} + \|2E_d\tilde{\boldsymbol{q}}\| \cos\theta \le 0 \tag{19}$$

where:

$$M_H = \begin{bmatrix} 0_{4\times4} & [\boldsymbol{y}_B]^T_{\overset{\vee}{\otimes}} \\ [\boldsymbol{y}_B]_{\overset{\vee}{\otimes}} & 0_{4\times4} \end{bmatrix},\ E_d = \begin{bmatrix} 0_{4\times4} & 0_{4\times4} \\ 0_{4\times4} & \boldsymbol{I}_{4\times4} \end{bmatrix}$$

The representation of the glide-slope constraint is shown in Fig. 3. This constraint is represented in fixed frame I, and the mathematical interpretation of glide slope constraint is as shown below

$$r_I \cdot z_I \ge \|r_I\| \cos\phi \tag{20}$$

z_I is the unit vector, and the direction is along the axis of the coordinate system I. $\boldsymbol{r}_I$ is the distance of the spacecraft relative to point, expressed in the coordinate system I. Treating glide slope constraint similar to line-of-sight constraint get its formulation in dual quaternion:

$$f(\tilde{\boldsymbol{q}}) = -\tilde{\boldsymbol{q}}^T M_G \,\tilde{\boldsymbol{q}} + \|2E_d\tilde{\boldsymbol{q}}\| \cos\phi \le 0 \tag{21}$$

The expression of E_d is the same, and:

$$M_G = \begin{bmatrix} 0_{4\times4} & [z_I]^T_{\otimes} \\ [z_I]_{\otimes} & 0_{4\times4} \end{bmatrix} \tag{22}$$

Fig. 3 Representation of glide slope constraint

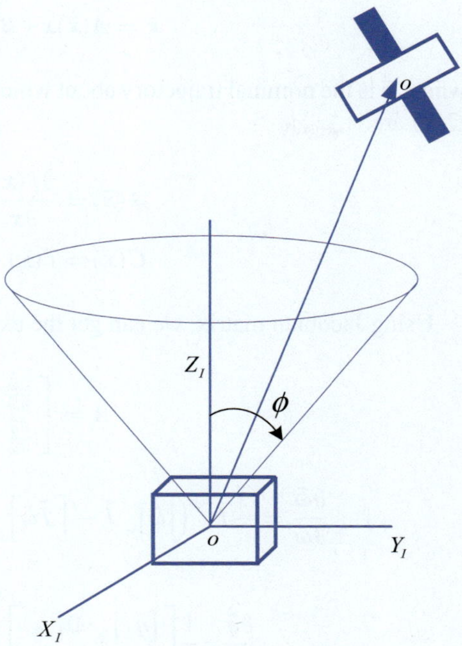

Some constraints are introduced above, then the problem of the spacecraft approaching the target can be regarded as a nonlinear optimal control problem with minimum fuel consumption.

Convex Optimization. The nonlinear optimal control problem needs to be convexized before being solved using convex optimization methods. First, the nonlinear equations must first be linearized. Because convex optimization method requires that equality constraints must be affine. Then, a zero-order hold approach was used to discretize nonlinear equation into a finite number of algebraic problems. Finally, sequential convex optimization is used to solve this problem.

We rewrite Eqs. (14) and (16), and omit subscript:

$$\begin{aligned} \boldsymbol{J}\dot{\tilde{\boldsymbol{\omega}}} + \tilde{\boldsymbol{\omega}} \times \boldsymbol{J}\tilde{\boldsymbol{\omega}} = \tilde{\boldsymbol{F}} \\ \dot{\tilde{\boldsymbol{q}}} = \frac{1}{2}\tilde{\boldsymbol{q}} \otimes \tilde{\boldsymbol{\omega}} \end{aligned} \tag{23}$$

Let $\boldsymbol{x} = \left[\tilde{\boldsymbol{\omega}}^T,\ \tilde{\boldsymbol{q}}^T\right]^T$, Eq. (23) can be written as:

$$\dot{\boldsymbol{x}} = f(\boldsymbol{x}) + B\tilde{\boldsymbol{F}} \tag{24}$$

where $B = \left[\left(\tilde{\boldsymbol{J}}^{-1}\right)^T\ 0_{8\times 8}\right]^T$. Equations (19), (21) and (24) must be linearized first so that can be converted in convex form. Linearizing Eq. (24) yields

$$\dot{\boldsymbol{x}} = A(\overline{\boldsymbol{x}})\boldsymbol{x} + B\,\tilde{\boldsymbol{F}} + C(\overline{\boldsymbol{x}}) \tag{25}$$

where $\overline{\boldsymbol{x}}$ is the nominal trajectory about which the equations are linearized. $A(\overline{\boldsymbol{x}})$ and $C(\overline{\boldsymbol{x}})$ are:

$$\begin{aligned} A(\overline{\boldsymbol{x}}) &= \left.\frac{\partial f(\boldsymbol{x})}{\partial \boldsymbol{x}}\right|_{\boldsymbol{x}=\overline{\boldsymbol{x}}} \\ C(\overline{\boldsymbol{x}}) &= f(\overline{\boldsymbol{x}}) - A(\overline{\boldsymbol{x}})\overline{\boldsymbol{x}} \end{aligned} \tag{26}$$

Using Jacobian matrix, we can get the expression of $A(\overline{\boldsymbol{x}})$:

$$A = \left.\begin{bmatrix} \frac{\partial \dot{\tilde{\omega}}}{\partial \tilde{\omega}} & \frac{\partial \dot{\tilde{\omega}}}{\partial \tilde{q}} \\ \frac{\partial \dot{\tilde{q}}}{\partial \tilde{\omega}} & \frac{\partial \dot{\tilde{q}}}{\partial \tilde{q}} \end{bmatrix}\right\| \tag{27}$$

$$\frac{\partial \dot{\tilde{\boldsymbol{\omega}}}}{\partial \tilde{\boldsymbol{\omega}}} = -\tilde{\boldsymbol{J}}^{-1}\left(\left[\tilde{\omega}\right]_{\times}\tilde{\boldsymbol{J}} - \left[\tilde{\boldsymbol{J}}\tilde{\boldsymbol{\omega}}\right]_{\times}\right), \quad \left[\tilde{\omega}\right]_{\times} = \begin{bmatrix} [\boldsymbol{\omega}]_{\times} & 0_{4\times4} \\ [v]_{\times} & [\boldsymbol{\omega}]_{\times} \end{bmatrix} \tag{28}$$

$$\frac{\partial \dot{\tilde{\boldsymbol{q}}}}{\partial \tilde{\boldsymbol{\omega}}} = \frac{1}{2}\begin{bmatrix} [q_1]_{\otimes} & 0_{4\times4} \\ [q_2]_{\otimes} & [q_1]_{\otimes} \end{bmatrix}\begin{bmatrix} \boldsymbol{I}_{3\times3} & 0 & \boldsymbol{0}_{3\times3} & 0 \\ 0 & 0 & 0 & 0 \\ \boldsymbol{0}_{3\times3} & 0 & \boldsymbol{I}_{3\times3} & 0 \\ 0 & 0 & 0 & 0 \end{bmatrix} \tag{29}$$

$$\frac{\partial \dot{\tilde{\boldsymbol{q}}}}{\partial \boldsymbol{q}} = \frac{1}{2}\begin{bmatrix} [\boldsymbol{\omega}]_{\overset{\vee}{\otimes}} & 0_{4\times4} \\ [v]_{\overset{\vee}{\otimes}} & [\boldsymbol{\omega}]_{\overset{\vee}{\otimes}} \end{bmatrix} \tag{30}$$

M_H, M_G in Eqs. (19) and (21) are positive semi-definite and cannot be solved directly by convex optimization method, they need to be linearized. Linearize Eqs. (19) and (21) using Taylor expansion at the nominal dual quaternion $\overline{\tilde{\boldsymbol{q}}}$:

$$\begin{aligned} f(\tilde{q}) &= f(\overline{\tilde{\boldsymbol{q}}}) + f'(\overline{\tilde{\boldsymbol{q}}})\left(\tilde{q} - \overline{\tilde{q}}\right) \le 0 \\ f'(\overline{\tilde{\boldsymbol{q}}}) &= 2M_H\overline{\tilde{\boldsymbol{q}}} + \frac{4E_d\overline{\tilde{\boldsymbol{q}}}}{\left\|2E_d\overline{\tilde{\boldsymbol{q}}}\right\|}\cos\theta \\ f'(\overline{\tilde{\boldsymbol{q}}}) &= -2M_G\overline{\tilde{\boldsymbol{q}}} + \frac{4E_d\overline{\tilde{\boldsymbol{q}}}}{\left\|2E_d\overline{\tilde{\boldsymbol{q}}}\right\|}\cos\phi \end{aligned} \tag{31}$$

According to the definition of $\boldsymbol{x}$, $\tilde{\boldsymbol{q}} = [\,\boldsymbol{0}_{8\times8}\ \mathbf{I}_{8\times8}\,]\boldsymbol{x}$. To convert the continuous-time optimal control problem into a finite-dimensional parameter optimization problem, a zero-order hold approach was used to discretize the trajectory at K points. Let t_f be the final time of the trajectory, we define the time at index k:

$$t_k = \frac{k}{K-1} t_f, \quad k \in [0, K) \tag{32}$$

$u(t)$ can be written in terms:

$$u(t_k) = u[k] \tag{33}$$

we express the discrete-time dynamics as follows:

$$\boldsymbol{x}[k+1] = A[k]\boldsymbol{x}[k] + B[k]\tilde{\boldsymbol{F}}[k] + C[k] \tag{34}$$

The state transition matrix of A, B and C from t_k to t_{k+1} are:

$$\begin{aligned} A[k] &= e^{A(\overline{x}(t_k))\Delta t} \\ B[k] &= \int_0^{\Delta t} e^{A(\overline{x}(t_k))\tau} \, B \, d\tau \\ C[k] &= \int_0^{\Delta t} e^{A(\overline{x}(t_k))\tau} \, C(\overline{x}(t_k)) \, d\tau \end{aligned} \tag{35}$$

where $\Delta t = t_{k+1} - t_k$.

Equation (31) is discretized as:

$$f(\tilde{\boldsymbol{q}}) = f(\bar{\tilde{\boldsymbol{q}}}[k]) + f'(\bar{\tilde{\boldsymbol{q}}}[k])\left(\tilde{\boldsymbol{q}}[k] - \bar{\tilde{\boldsymbol{q}}}[k]\right) \le 0 \tag{36}$$

Sometimes linearization may cause the problem to be infeasible, even if the original problem is feasible. This infeasibility will obstruct the convergence of the problem. To prevent this, an additive relaxation term $\boldsymbol{s}_k$ was introduced to the inequality constraints:

$$f(\bar{\tilde{\boldsymbol{q}}}[k]) + f'(\bar{\tilde{\boldsymbol{q}}}[k])\left(\tilde{\boldsymbol{q}}[k] - \bar{\tilde{\boldsymbol{q}}}[k]\right) \le \boldsymbol{s}_k \tag{37}$$

In addition, to prevent artificial infeasibility, the discrete-time dynamics is relaxed through introducing the virtual control variable $\boldsymbol{v}$:

$$\boldsymbol{x}[k+1] = A[k]\boldsymbol{x}[k] + B[k]\tilde{\boldsymbol{F}}[k] + C[k] + \boldsymbol{v}[k] \tag{38}$$

We also should concern that the linearization of nonlinear problems may render the problem unbounded. To prevent this risk, we define a region about the linearization point, which is the current best solution, such that the best solution from linear approximation lies within it. This region is called trust region [13].

We define $\delta \boldsymbol{x}_k = \boldsymbol{x}_k - \overline{\boldsymbol{x}}_k$, and introduce a new variable η_k, trust region is defined by:

$$\delta x_k^T \delta x_k \leq \eta_k \tag{39}$$

We need to add virtual control et al. to the objective function to bound their size:

$$\min \sum_{k=0}^{K-1} \left\| \tilde{\boldsymbol{F}}[k] \right\|_2 \Delta t + \omega_{dyn} \sum_{k=0}^{K-1} \|\boldsymbol{v}_k\| + \omega_s \sum_{k=0}^{K} \|\boldsymbol{s}_k\| + \omega_\eta \sum_{k=0}^{K} \|\eta_k\| \tag{40}$$

where $\Delta t = t_{k+1} - t_k = t_f \big/ K - 1$, ω_{dyn}, ω_s, ω_η are constant coefficients.

Until now, the dynamics model and constraints have been written in convex form. Nonlinear optimal control problem can be written in:

$$\begin{aligned} &\text{Eq.}(40) \\ s.t.\ &\text{Eq.}(38),\ \text{Eq.}(37),\ \text{Eq.}(39) \\ &\left\| \tilde{\boldsymbol{F}}[k] \right\|_\infty \leq \tilde{F}_{\max} \quad k \in [0, K) \\ &\boldsymbol{x}[0] = \boldsymbol{x}_0,\quad \boldsymbol{x}[K] = \boldsymbol{x}_f \end{aligned} \tag{41}$$

The process of linearization usually requires a nominal trajectory $\overline{\boldsymbol{x}}$, which is then discretized into $\overline{\boldsymbol{x}}[k]$. To ensure the nominal trajectory is good estimate true trajectory, sequential convex programming method is used. This method is to iteratively solve Eq. (41) until the iteration stop condition is met. In the first iteration, an initial trajectory is obtained by solving Eq. (41) without considering the constraints. Only controls are left in the objective function. But in the following iterations, this initial trajectory is used as the initial nominal trajectory of Eq. (41 to be solved iteratively. This process will continue until the iteration stop condition is met:

$$\|\boldsymbol{x}_m[k] - \boldsymbol{x}_{m-1}[k]\|_\infty < \varepsilon \tag{42}$$

$\boldsymbol{x}_{m-1}[k]$ is equal to the nominal trajectory of the previous iteration, $\overline{\boldsymbol{x}}[k] = \boldsymbol{x}_{m-1}[k]$.$\varepsilon$ is a small amount, and iteration stops until Eq. (42) is satisfied.

3 Numerical Simulations

In this section, simulation results of spacecraft close proximity to the target are presented using the algorithm proposed in Sect. 2. Some simulation parameters are given in Table 1.

Table 2 gives the boundary of constraints, initial and terminal dual quaternions. Two constraints are represented in different frames, where line-of-sight constraint is represented in body frame B and glide-slop is represented in inertial frame I. The unite vectors $\boldsymbol{y}_B$ and $\boldsymbol{z}_I$ are constant in two reference frames.

Table 1 Simulation parameters

Parameters	Values
Mass	10(kg)
Moment of inertia	[4.0514; 3.6628; 3.1879] $\text{kg}\cdot\text{m}^2$
Initial time and terminal time	$t_0 = 0, \quad t_f = 200\text{s}$
Force and torque	$F_{\max} = 10\text{N}, \quad T_{\max} = 1\text{N}\cdot\text{m}$

Table 2 Simulation parameters in dual quaternion

Parameters	Values
Initial state and terminal state	$\boldsymbol{r}_0 = [2.5\ -20\ 52]^{\text{T}}\text{m}, \boldsymbol{r}_f = [0\ 0\ 1]^{\text{T}}\text{m};$ $\boldsymbol{v}_0 = [0\ 0\ 0]^{\text{T}}\text{m/s}, \boldsymbol{v}_f = [0\ 0\ 0]^{\text{T}}\text{m/s};$ $\omega_0 = [0\ 0\ 0]^{\text{T}}\text{rad/s}, \omega_f = [0\ 0\ 0]^{\text{T}}\text{rad/s}$
Line-of-sight	$\boldsymbol{y}_B = [0\ 0\ -1]^{\text{T}}, \quad \theta = 30°$
Glide-slope	$z_I = [0\ 0\ 1]^{\text{T}}, \quad \phi = 30°$
Initial and terminal dual quaternions	$q_0 = [0\ 0\ 0.5\ 0.866]^{\text{T}}, q_f = [0\ 0\ 0\ 1]^{\text{T}}$

The sequential convex programming (SCP) algorithm has a convergence tolerance of $\varepsilon = 10^{-2}$. The trajectories obtained by dual quaternion are shown in Fig. 4. The dotted line represents the trajectory without considering line-of-sight and glide-slope constraints and is used as nominal trajectory for Eq. (41). Trajectory represented by solid line considering these constraints obtained by Eq. (41). Two trajectories have a slight difference. Figures 6 and 8 show the translational velocities and forces. The changes of velocity and force in each direction are different. Among them, the maximum speed variation range and fuel consumption are in z direction, followed by the y direction, and the minimum fuel consumption is in the x direction.

Figures 7 and 9 show the angular velocities and torques. The range of angular velocity variation in z direction is the largest, which means that the fuel consumption is the largest. Whether it is velocity or angular velocity, the maximum change in z direction is because it is related to the given initial and terminal states, and the deputy always rotates around the z axis relative to the chief.

Figures 10 and 11 depict the changes of the two constraint angles of the spacecraft from the initial position to the terminal position. The two thick solid lines in figures represent the bound angle of line-of-sight constraint and glide-slope constraint. The dotted line in Fig. 11 is the trajectory taking the line-of-sight constraint into account and satisfy the iteration stop condition. The trajectory shown in Fig. 10 always satisfies the glide-slope constraint, indicating that the deputy is always located in the cone as shown in Fig. 3 when approaching the chief.

Fig. 4 Trajectories with dual quaternions

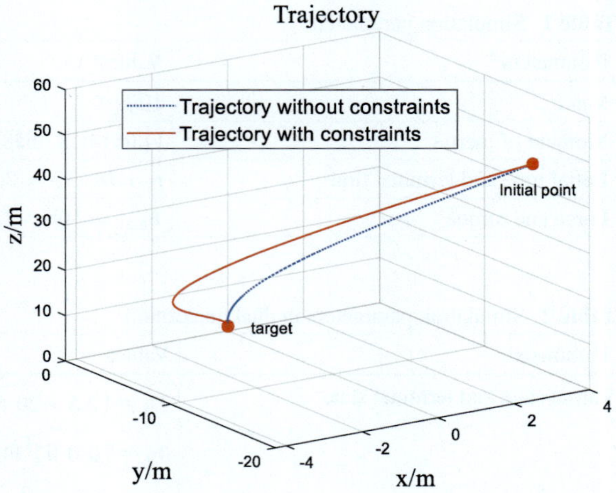

Figure 5 shows the change of the rotational quaternion of the body coordinate system B relative to the fixed coordinate system I. The quaternions shown in the figure satisfy $\|\boldsymbol{q}\|_2 = 1$ at each time.

Table 3 shows the objective-value and time cost calculated by dual quaternions. The objective value for rotation is 0.0471 $\text{kg} \cdot \text{m}^2/\text{s}$. The total objective value for position calculated by the dual quaternion is 1.0519 $\text{kg} \cdot \text{m/s}$. Compared with pseudospectral methods, sequential convex optimization has the lowest fuel consumption with the same convergence accuracy. In addition, because the algorithm proposed in this paper is sensitive to the initial guess of dual quaternions, different initial values may lead to different objective values.

Fig. 5 Dual quaternions

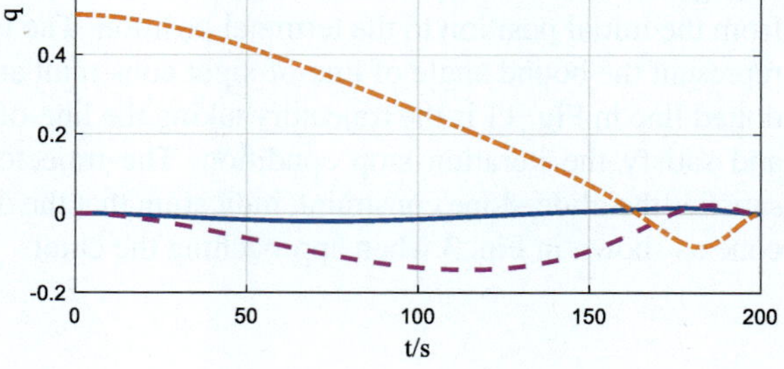

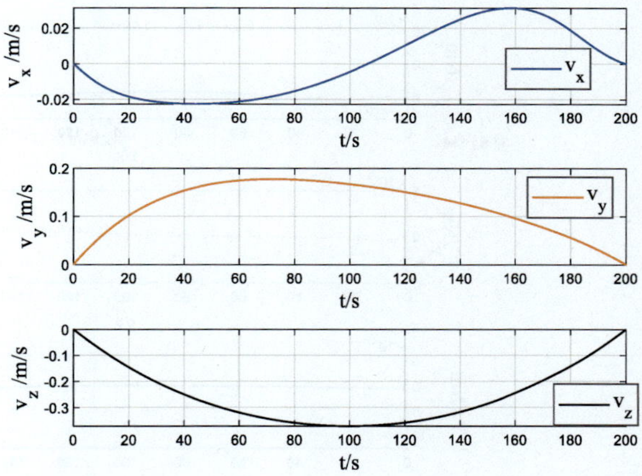

Fig. 6 The velocities of the spacecraft

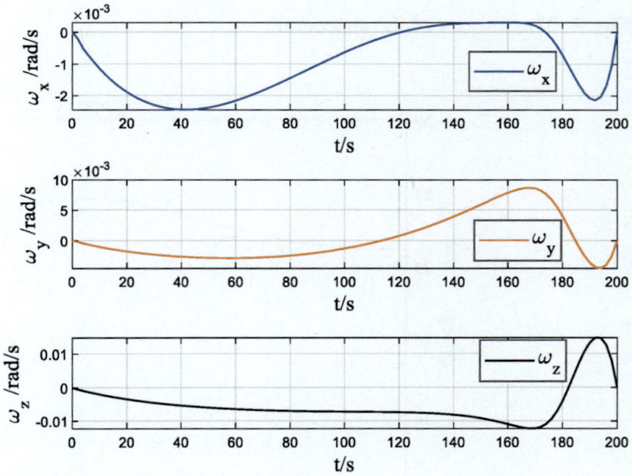

Fig. 7 The angular velocities

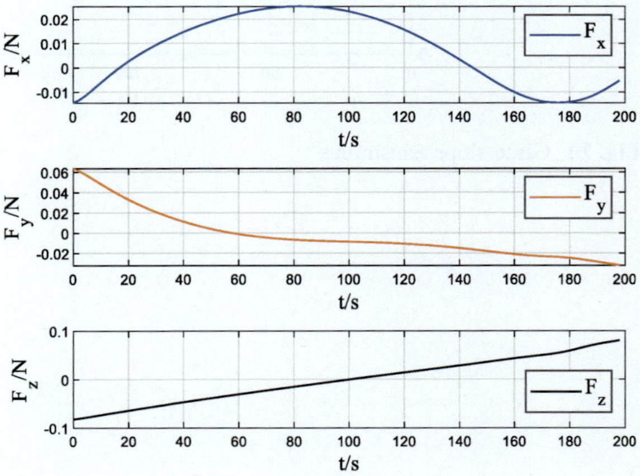

Fig. 8 The forces of the spacecraft

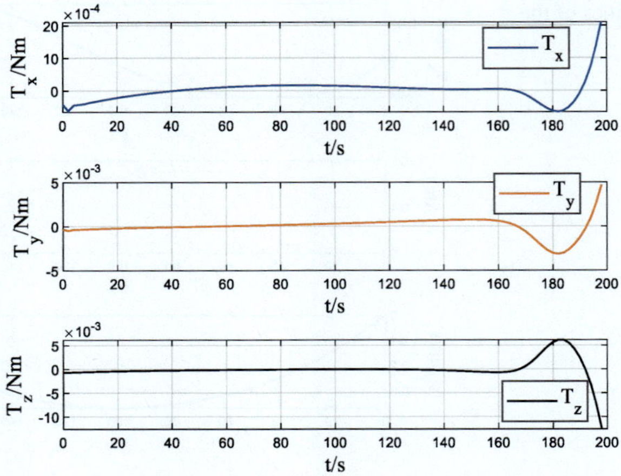

Fig. 9 The torques of the spacecraft

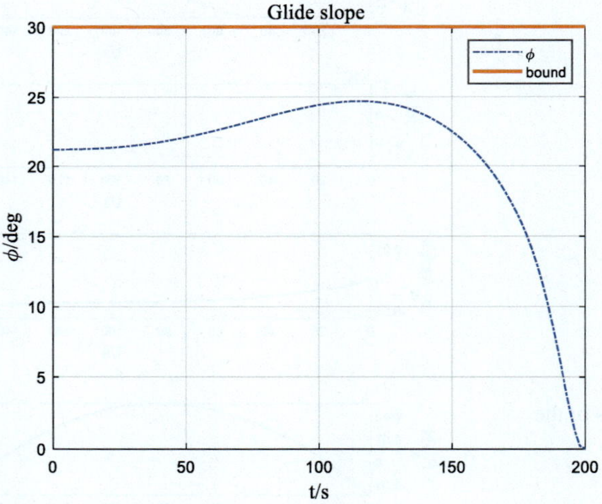

Fig. 10 Glide-slope constraints

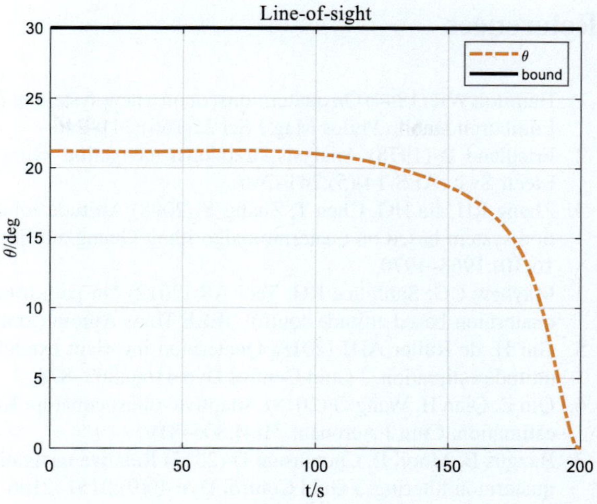

Fig. 11 Line-of-sight constraint

Table 3 Fuel consumption using dual quaternions

Method	Time–cost (s)	Objective-value for position $(\text{kg} \cdot \text{m/s})$	Objective-value for rotation $(\text{kg} \cdot \text{m}^2/\text{s})$
Dual quaternions	31.62	1.0519	0.0471

4 Conclusion

In this paper, the trajectory optimization problem of the spacecraft approaching a space target is studied. The sequential convex optimization method is used to solve the fuel-optimal trajectory subject to line-of-sight constraint, glide-slope constraint and control force amplitude constraint, etc. Since line-of-sight constraint is coupled with position and attitude of spacecraft, dual quaternions can combine the position and attitude into a unified state variable with 8 scalars. We used the dual quaternion to represent the dynamic model of the spacecraft. Finally, converting both the dynamics and constraints into a convex form which involves linearizing and discretizing them. Sequential convex optimization is applied to solve them iteratively. The simulation results show that convex optimization can efficiently solve the dual quaternions. Sequence convex optimization is sensitive to initial value when solving dual quaternion problem. In the future, we can find a more accurate algorithm that is insensitive to initial value, or find a better initial guess method.

References

1. Hamiton WR (1844) On quaternions; or on a new system of imaginaries in algebra. The London Edinburgh Dublin Philos Mag J Sci 25(166):241–246
2. Friedland B (1978) Analysis strapdown navigation using quaternions. IEEE Trans Aerosp Electr Syst (AES-14)(5):241–246
3. Zhang RH, Jia HG, Chen T, Zhang Y (2008) Attitude solution for strapdown inertial navigation system based on quaternion algorithm. Guangxue Jingmi Gongcheng/Optics Precis Eng 16(10):1963–1970
4. Mayhew CG, Sanfelice RG, Teel AR (2012) On path-lifting mechanisms and unwinding in quaternion-based attitude control. IEEE Trans Autom Control 58(5):1179–1191
5. Gui H, de Ruiter AHJ (2018) Quaternion invariant extended kalman filtering for spacecraft attitude estimation. J Guid Control Dyn 41(4):863–878
6. Qiu Z, Qian H, Wang G (2018) Adaptive robust cubature Kalman filtering for satellite attitude estimation. Chin J Aeronaut 31(4):806–819
7. Razgus B, Mooij E, Choukroun D (2017) Relative navigation in asteroid missions using dual quaternion filtering. J Guid Control Dyn 40(9):2151–2166
8. Büyükkoçak AT, Tekinalp O (2019) Safe spacecraft rendezvous using dual quaternions on time-dependent trajectories generated by model predictive control. In: 2019 AIAA scitech forum
9. Lee U, Mesbahi M (2017) Constrained autonomous precision landing via dual quaternions and model predictive control. J Guid Control Dyn 40(2):292–308
10. Hazra S (2019) Autonomous guidance for asteroid descent using successive convex optimisation. Delft University of Technology
11. Morgan D, Chung S, Hadaegh FY (2014) Model predictive control of swarms of spacecraft using sequential convex programming. J Guid Control Dyn 37(6):1725–1740
12. Wu Y, Hu X, Hu D, Lian J (2005) Strapdown inertial navigation system algorithms based on dual quaternions. IEEE Trans Aerosp Electron Syst 41(1):110–132
13. Szmuk M, Eren U, Açıkmeşe B (2018) Successive convexification for 6-DoF mars rocket powered landing with free-final-time. In: 2018 AIAA guidance, navigation, and control conference

Collaborative Multi-Drones Air Combat Threat Assessment and Credibility Study

Ruimin Pu and Meng Yu

Abstract With the rapid development of artificial intelligence technology, air combat is now developing in the direction of fully autonomous and collaborative operation. In a collaborative mode, drones need to assess the current situation of the environment when operating, and every drone will interact with each other for better information gathering and situational awareness. To this end, this study develops a novel index system for threat assessment based on seven factors that have impacts on the threat level of radiation sources. A hierarchical analysis means is introduced to quantify the drone threat assessment results. The consistency of threat assessment results among different drones is compared by combining the Jaccard coefficients and the sequence similarity values. Finally, exemplary simulations are carried out, results from which demonstrates the effectiveness of the proposed method when implementing in a drone collaborative exploration mission.

Keywords Credibility · Hierarchical analysis · Threat assessment

1 Introduction

In the modern air combat environment, weapons and air-to-air tactics are becoming more and more intelligent. In a complex and highly dynamic air combat scenario, multi-drones operating in a collaborative manner can create an operational advantage in terms of tactical flexibility and readiness of warfare sensing, thereby increasing the success rate of air combat. Although collaborative drone operation has been widely studied in recent years [1–3], there are still many problems in terms of collaborative-related technology, such as the reliability of information interaction between multiple drones [4], the task allocation problem of multi-drones operations [5, 6] and the accuracy of drone threat assessment in a highly dynamic and strong real-time operational environment [7–9].

R. Pu (✉) · M. Yu
Nanjing University of Aeronautics and Astronautics, Nanjing 210016, China
e-mail: p.ruimin_15@nuaa.edu.cn

Z. Liu et al. (eds.), *Advances and Challenges in Advanced Unmanned Aerial Systems*, Springer Aerospace Technology, https://doi.org/10.1007/978-981-99-8045-1_27

In an ever-changing battlefield, drones need to assess the current moment's operational posture in a timely manner, acquire as much information as possible, develop attack plans and make quick decisions based on the results of the threat assessment. Many methods for air combat threat assessment have been proposed, such as Bayesian networks [10, 11], fuzzy comprehensive evaluation [12], and information entropy [13]. In the case of collaborative operation, the judgment of the current situation on the battlefield varies due to the variability of the individual environment, thus the assessment results of the real-time situation by different aircraft in multiple drones often differ to some extent. The credibility of the threat assessment information provided by the remaining drones differs when each drone is making maneuvering decisions. How to quantify the credibility of each drone's threat assessment result remains an open problem. For this reason, this study proposes a method for comparing the consistency of drone threat assessment results using Jaccard coefficients and sequence similarity values as a way to represent the credibility of threat assessment results among drones. The research on the credibility of information among multiple drones can help drones develop more accurate tactics and make more rational decisions in complex operational environments.

In this paper, a new index system for threat assessment is established in Chapter "Multi-RISs Assisted UAV-Enabled Wireless Sensor Network", and the results of threat ranking are obtained using hierarchical analysis. Chapter "Research on Landing Attitude Control of Terrain Adaptive UAVs" proposes a method to measure the consistency of threat ranking between different drones using Jaccard coefficients and sequence similarity values, and analyzes the credibility of threat assessment between different drones based on the results of the measurements. Chapter "Optimal Event-Triggered Control of Quadrotor with Low Power Consumption and High Stability" presents a simulation example with threat assessment and credibility analysis, and presents the results in a visual format.

2 Principles of Drone Threat Assessment

2.1 Indicator System

(1) Type of target. In the modern air combat environment, the radiation sources detected by drones have different types. The types of radiation sources are divided into five types: NA(unknown), search radar, target-indicating radar, airborne fire-control radar and missile-borne radar, and if the information of the radiation source is missing or unknown, it is indicated by NA, the same as below. Each type of target has its own function, and the threat level varies for different types of targets, the threat level coefficient for each type of target is defined in the form of a discrete function, as expressed in (1).

$$M_{t\arg et\_type} = \begin{cases} 0 & NA \\ 0.40 & search\,radar \\ 0.50 & target - indicating\,radar \\ 0.75 & airborne\ fire - control\,radar \\ 1 & missile - borne\,radar \end{cases} \quad (1)$$

(2) Operation status of target. During operation each target has its own status, as the situation on the battlefield changes, the operation status will switch according to its pre-designed strategy if necessary, and the variation of operation status will have effects on its threat level. The main operation status of targets during air warfare are NA(unknown), search, tracking, midcourse guidance and terminal guidance. Generally, the tracking and guidance status poses a highest level of threat, while the search status has a lower threat level, and when the operation status of the target is NA, its threat level still cannot be neglected. The threat level coefficient of the target operation status is shown in (2).

$$M_{t\arg et\_state} = \begin{cases} 0.50 & NA \\ 0.30 & search \\ 1 & tracking \\ 1 & guidance \end{cases} \quad (2)$$

(3) Identification friend or foe. For targets locate within the detection range of the drone, their attribute identification (friend or foe) plays a decisive role in threat level assessment. The attribute of a target can be classified as NA, enemy, friend, neutrality and civilian. The threat level coefficients of the target's enemy attributes are shown in (3).

$$M_{relationship} = \begin{cases} 0.50 & NA \\ 1 & enemy \\ 0 & friend\,or\,civilian \\ 0.20 & neutrality \end{cases} \quad (3)$$

(4) Carrier frequency of target. The carrier frequency of the target is one of the crucial factors in analyzing the threat level of a target. The size of the carrier frequency determines the resolution and accuracy, and when the carrier frequency is in different ranges, the target has different operating characteristics and plays different roles on the battlefield, the threat level to our side is also different. The carrier frequency threat level coefficient of the target is shown in (4) (unit: GHz).

$$M_{carrier\_frequency}(x) = \begin{cases} 0.2 & x \le 4 \\ 0.5 & 4 < x \le 8 \\ 1 & x > 8 \end{cases} \quad (4)$$

(5) Pulse width of target. The pulse width of the target mainly reflects the minimum action distance resolution of the target, the width of the pulse width also has effects on the threat level of the target, for instance, the pulse width of the search status radar is relatively wide and the threat level is low, and the pulse width of the tracking status radar is relatively narrow and the threat level is high. The pulse width threat level coefficient of the target is shown in (5) (unit: μs).

$$M_{pulse\_duration}(x) = \frac{1}{1+x^2} \tag{5}$$

(6) Distance of target. Since the attack range between the enemy and us is limited, thus when the distance of the target is beyond the attack range, we consider that the target does not pose a threat, and when the distance of the target is within the attack range, the closer the distance, the greater the threat. The distance threat level coefficient of the target is shown in (6), and the threat level coefficient is 0.4 when the distance is NA.

$$M_{dis\tan ce}(x) = \begin{cases} 0.40 & NA \\ 1 - \frac{x}{d_{\max}} & x \le d_{\max} \\ 0 & x > d_{\max} \end{cases} \tag{6}$$

where x indicates the distance between the target and us, and $d_{\max}$ indicates the maximum enemy attack distance, which is set to 180 km.

(7) Range rate of target. The range rate of the target is a significant influencing factor of the threat level. The threat level coefficient for the range rate is formulated as an exponential function as shown in (7).

$$M_{velocity} = \begin{cases} 1 - e^{kv} & v \ge 0 \\ 1 - e^{-kv} & v < 0 \end{cases} \tag{7}$$

where v denotes the range rate between the target and us, and k is a constant factor taking the value of –0.005.

2.2 Assessment Method

When the drone conducts a threat assessment, each detected target is analyzed qualitatively and quantitatively using hierarchical analysis to determine the weights of the threat assessment indicators. The concrete procedure is presented as follows:

(1) Determining the judgment matrix

Experts compare the materiality of the seven indicators of type, operation status, identification friend or foe (IFF), carrier frequency, pulse width, distance, and range

Table 1 Judgment matrix

Indicator	Type	Operation status	IFF	Carrier frequency	Pulse width	Distance	Range rate
Type	1	7/5	7/5	7	7	7/4	7/4
Operation status	5/7	1	1	5	5	5/4	5/4
IFF	5/7	1	1	5	5	5/4	5/4
Carrier frequency	1/7	1/5	1/5	1	1	1/4	1/4
Pulse width	1/7	1/5	1/5	1	1	1/4	1/4
Distance	4/7	4/5	4/5	4	4	1	1
Range rate	4/7	4/5	4/5	4	4	1	1

rate of target according to the comparative quantification regulations of analytic hierarchy process (AHP), and obtain the judgment matrix T as shown in Table 1.

(2) Consistency test

In order to check the consistency of the constructed judgment matrix for the comparison of the importance of indicators, a consistency test is required. First, the single-order consistency index CI is calculated from the maximum eigenvalue of the matrix $\lambda_{\max}$ and the order of the matrix n, and then the total-order consistency index CR is calculated from the random consistency index RI. The calculation formulas are as follows.

$$CI = \frac{\lambda_{\max} - n}{n - 1} \tag{8}$$

$$CR = \frac{CI}{RI} \tag{9}$$

When the total-order consistency index CR < 0.1, the consistency test is passed, and if the consistency test is not passed, the judgment matrix needs to be reconstructed.

(3) Determining indicator weights

The weights of each index are calculated according to the judgment matrix using the arithmetic mean method or geometric mean method.

2.3 Assessment Result

The threat level coefficients of the seven indicators are multiplied with a certain confidence level according to the specific combat situation and then summed with their respective weights to obtain the threat value of the target. Then the result of ranking all the targets detected by a drone according to the threat value from the largest to the smallest is the result of the drone threat assessment.

3 Credibility Analysis of Threat Assessment Among Drones

When multiple drones collaborate to accomplish air combat missions, each drone has a different perspective and is in a different environment, and there is variability in the threat assessment of the current operational environment when multiple drones are completing the same attack mission. However, multiple drones need to interact with each other in real time while operating, making decisions and developing attack plans based on all the information collected. Therefore, it is necessary to investigate the similarity of different drones' threat assessment results during multi-drones collaboration air combat, which can reflect the credibility of the threat assessment information between different drones. The results of the threat assessment (i.e., threat ranking) of different drones are analyzed here mainly based on two values, the Jaccard coefficient and the sequence similarity value.

3.1 *Jaccard Coefficients*

The Jaccard coefficients can be used to measure the similarity between a finite set of samples. For a given two sets A and B, the ratio of the size of the intersection of set A and set B to the size of the union of set A and set B is the Jaccard coefficient, and the formula for quantifying the Jaccard coefficient J(A,B) is defined in (10).

$$J(A, B) = \frac{|A \cap B|}{|A \cup B|} = \frac{|A \cap B|}{|A| + |B| - |A \cup B|} \qquad J(A, B) \in [0, 1] \qquad (10)$$

The larger the Jaccard coefficient, the higher the similarity between samples and vice versa, and if the sets A and B are empty, then J(A,B) = 1.

3.2 *Sequence Similarity Value*

Since Jaccard coefficient can only be used to reflect the similarity of element composition in two sequences, in order to further measure the similarity of arrangement order between two sequences, the sequence similarity value is introduced here. The sequence similarity value S(a,b) is calculated mainly by using the idea of minor threat suppression, and the steps are as follows.

Suppose two sequences are given, and the two sequences are represented as vector a and b respectively. Firstly, the length of a and b is compared, the shorter length is denoted as len, and then the first three bits of a and b are respectively compared to see if they are the same. There are mainly the following four cases.

(1) If the first bit is different, take the sequence similarity value S(a,b) as 0.3.
(2) If the first bit is the same and the second bit is different, then the sequence similarity value S(a,b) is 0.4.
(3) If the first two bits are the same and the third bit is different, then the sequence similarity value S(a,b) is 0.5.
(4) If the first three bits are the same, the cosine of the angle between the two vectors is solved by (11), and then the sequence similarity value S(a,b) is calculated by (12).

$$\cos\theta = \frac{\boldsymbol{a} \cdot \boldsymbol{b}}{|\boldsymbol{a}||\boldsymbol{b}|} \tag{11}$$

$$S(\boldsymbol{a}, \boldsymbol{b}) = 0.7 + 0.3 * \cos\theta \qquad S(\boldsymbol{a}, \boldsymbol{b}) \in [0.7, 1] \tag{12}$$

4 Simulation Example

Suppose there are six drones, A, B, C, D, E, and F, flying in an exploration task, the detection range of drones is 120000 m and the mission goal is to strike one important location. There are 13 ground-radars in the simulation scenario, and it is assumed that A, B, C, and D are close together and arranged in a square formation, while E and F are relatively far away, and these six drones perform a real-time threat assessment of the operational environment during the mission and get different threat rankings.

The route maps of the six drones are shown in Fig. 1. Where, B and C fly one after the other, and the flight path in Fig. 1 shows overlap.

Threat assessment results of six drones for the current operational environment at 200 s are shown in Fig. 2, the vertical axis represents the current ranking of 13 radars in the scene, with 0 indicating they are not within the detection range.

The results of the threat assessment of the six drones were compared two-by-two every 100 s between 0 and 1000 s, and the corresponding Jaccard coefficients and sequence similarity values were calculated as shown in Figs. 3, 4, and 5.

From the above 15 sets of simulation results, we can see that, in general, the comparison results of Jaccard coefficients are more consistent, while the comparison results of sequence similarity values are more different. This indicates that when information is interacted between different drones, there is a higher credibility in the detection results of the threat targets (The composition of threat targets that drones can detect) and a lower credibility in the threat value assessment and ranking of the threat targets. In addition, the consistency of the threat assessment between A, B, C, and D, which are close to each other, is higher relative to E and F, which are farther

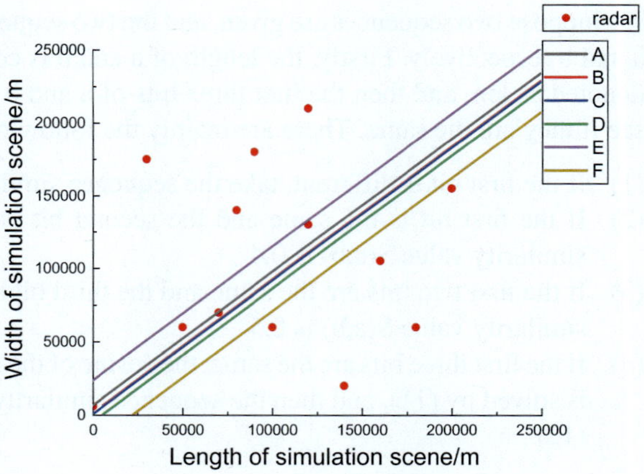

Fig. 1 Collaborative drones route map

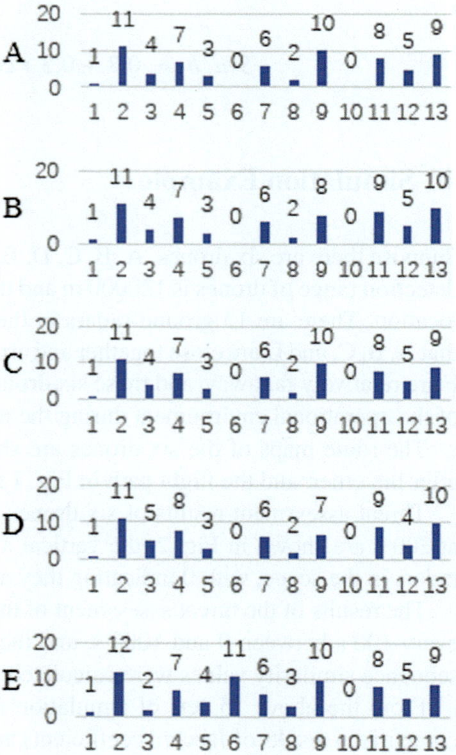

Fig. 2 Comparison of threat ranking results at 200 s

away from them, reflecting that the closer the relative distance between different drones when interacting with each other for information, the higher the credibility level.

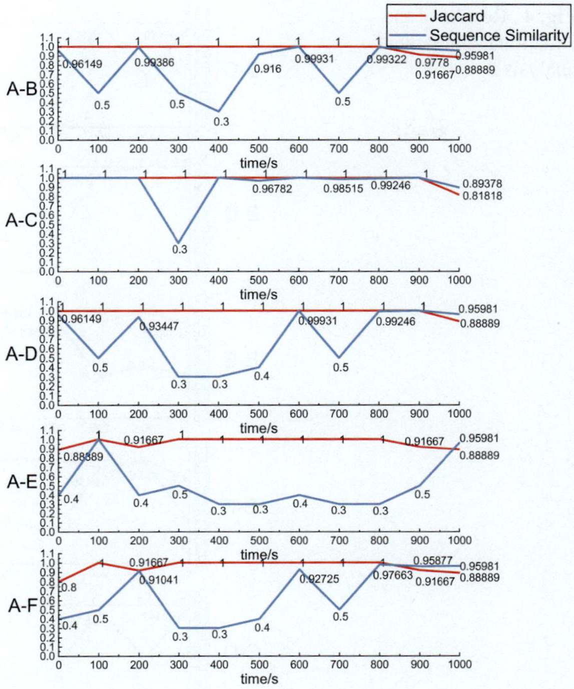

Fig. 3 Results of the threat assessment analysis 1

In summary, the indicators of type, operation status, attribution identification, carrier frequency, pulse width, distance and range rate of target are included all together to perform initial threat assessment, and a confidence level to each indicator is established, which completes the threat assessment. The entire threat assessment is computationally efficient, meaning that every drone in flying formation can carry out its own threat assessment separately and in the same time, share the assessment result with each other upon request. The calculation of Jaccard coefficients and sequence similarity values allows the variability of threat assessment results to be represented in a quantitative form, and the credibility analysis of multi-drones collaboration can benefit from the quantification of the variability of threat assessment results. The study of multi-drones collaboration credibility can provide trustworthy threat joint prediction trust intervals for subsequent task assignment and collaboration decision making.

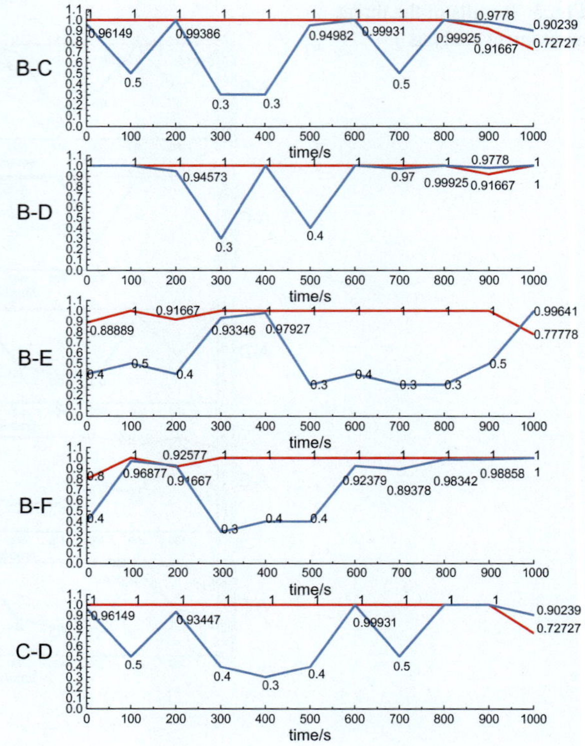

Fig. 4 Results of the situational assessment analysis 2

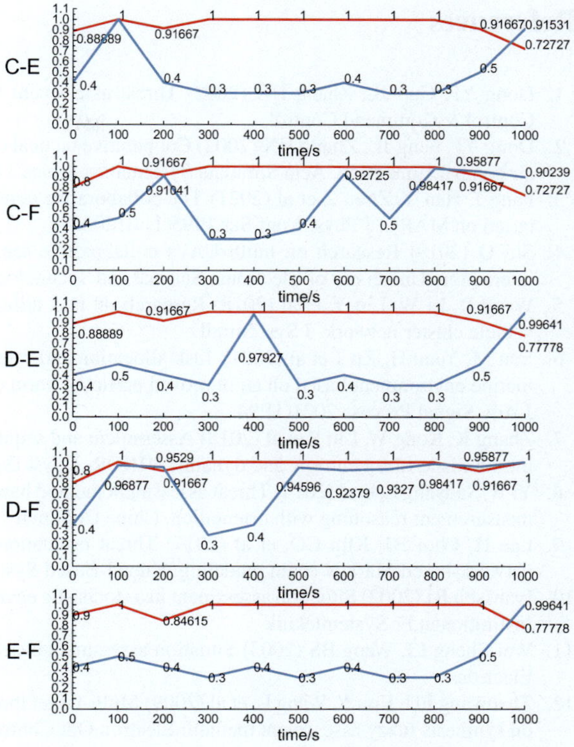

Fig. 5 Results of the situational assessment analysis 3

5 Conclusion

This paper establishes a drone threat assessment indicator system using seven factors: type, operation status, attribution identification, carrier frequency, pulse width, distance, and range rate of target and calculates the threat assessment results (i.e., threat ranking) through hierarchical analysis. Then a method is proposed to compare the consistency of threat assessment results among different drones using Jaccard coefficients and sequence similarity values. Finally, the scenario of multi-drones collaborative air combat is simulated and the consistency of the threat assessment results between different drones is compared. From the simulation results, we can see the credibility of the threat assessment information and the influencing factors between each other when multiple drones are flying together, which is a solid reference value for making reasonable mission allocation and collaboration decision when multiple drones are operating together in air combat.

References

1. Dong YF, Guo JL, Zhang HX (2002) Threat assessment for multi-aircraft-air combat. Fire Control & Command Control
2. Dong YF, Feng JL, Zhang HX (2002) Cooperative tactical decision methods for multi-aircraft air combat simulation. Acta Simulata Systematica Sinica 14(6):723–725
3. Fang J, Han Y, Zhou Z et al (2021) The collaborative combat of heterogeneous multi-UAVs based on MARL. J Phys: Conf Ser 1995(1):012023
4. Shi Q (2019) Research on multi-UAVs collaborative search system based on information interaction. University of Electronic Science and Technology
5. Wang R, Li W, Liu Y, et al (2018) Research on task allocation method of unmanned aerial vehicle cluster network. J Syst Simul
6. Yan M, Yuan H, Xu J et al (2021) Task allocation and route planning of multiple UAVs in a marine environment based on an improved particle swarm optimization algorithm. EURASIP J Adv Signal Process 2021(1):94
7. Zhang K, Kong W, Liu P et al (2018) Assessment and sequencing of air target threat based on intuitionistic fuzzy entropy and dynamic VIKOR. J Syst Eng Electron 29(2):6
8. Yi W, Sanyang L, et al (2014) Threat assessment method based on intuitionistic fuzzy similarity measurement reasoning with orientation. China Commun
9. Lee H, Choi BJ, Kim CO, et al (2017) Threat evaluation of enemy air fighters via neural network-based Markov chain modeling. Knowl-Based Syst 116(JAN.15): 49–57
10. Ivansson BJ (2002) Situation assessment in a stochastic environment using Bayesian networks. Institutionen Fr Systemteknik
11. Wei-Sheng LI, Wang BS (2003) Situation assessment based on Bayesian networks. Syst Eng Electron
12. Zhaowang FU, Kou Y, Wang L, et al (2009) Multi-target threat assessment of air combat based on synthesis fuzzy assessment method. Electron Opt Control
13. Tong Q, Tong ZX, Xiang-Dong GU, et al (2011) Target threat assessment for air combat based on information entropy. Math Pract Theory

A Novel Navigation Beneficial Path Planning Algorithm for Application in Lunar Surface Exploration

Huiting Wang, Meng Yu, and Ruimin Pu

Abstract The lunar south pole surface environment is characterized by sparse and homogeneous topographic features, poor optical texture and unstructured landforms, which brings great challenges to the lunar rover autonomous navigation. For high-precision autonomous navigation during lunar surface exploration, this paper proposes a novel path planning algorithm, the main purpose is to plan a travelable path that might be beneficial to improve the navigation accuracy. Firstly, the proposed method integrates the global auxiliary lines generated based on the A* algorithm into the Dynamic Window Approach (DWA). Then, a new cost function that takes into account the specialty of the lunar south pole environment and the accuracy impact factors of navigation is established. Finally, two kinds of simulations are conducted to verify the feasibility of the optimized path planning algorithm and whether it can improve the navigation accuracy. The simulation results indicated that the optimized path planning algorithm is more adaptable to complex environments and is effective in improving navigation accuracy.

Keywords Lunar rover · High-precision autonomous navigation · Path planning

1 Introduction

Up to now, more than 100 unmanned lunar explorations have been carried out worldwide, and over 20 lunar probes have been involved in lunar exploration missions. With the further exploration of the moon, scientists have discovered that there may be rich deposits of ice around the lunar polar region [1]. For this reason, many countries begin to pay attention to the exploration of the lunar polar region, especially the lunar south pole region. Compared to the front side of the moon, the environment of the lunar South Pole is more complicated and the rover might face with an extremely difficult operation situation. In order to ensure that the rover is able to carry out its

H. Wang (✉) · M. Yu · R. Pu
Nanjing University of Aeronautics and Astronautics, Nanjing, China
e-mail: wanghuiting@nuaa.edu.cn

Z. Liu et al. (eds.), *Advances and Challenges in Advanced Unmanned Aerial Systems*, Springer Aerospace Technology, https://doi.org/10.1007/978-981-99-8045-1_28

various exploration tasks, it is necessary to plan a safe and navigation-friendly path in an autonomous manner.

The path planning has been an important topic in the field of robotics ever since 1970s [2]. In general, path planning algorithms can be divided into global path planning algorithms and local path planning algorithms according to the extent of the awareness of environmental information [3]. Global path planning algorithms include the A* algorithm, the genetic algorithm and the fast random search tree algorithm. The fuzzy logic algorithm, the D* algorithm and the DWA algorithm all belong to local path planning algorithms. In recent years, path planning algorithms have been used in various exploration missions. Many researchers have also conducted research on path planning algorithms for complex lunar environments. Reference [4] proposed an algorithm to optimize the rolling window obstacle avoidance algorithm based on auxiliary line, which integrates global map information into the local path planning algorithm. The improved algorithm has a higher success rate and shorter planning path. Reference [5] demonstrated an algorithm that is more adapted to the complex lunar environments. The algorithm improves the cost function of the A* algorithm by weighted fusion of different environmental factors, such as lunar surface topography, illumination and communication. An optimized algorithm based on the D* algorithm is also proposed, which can ensure the safe travel of the rover in environments with unknown obstacles.

Compared to the environment on Earth, where there are plenty of tall buildings, roads and vegetations, the lunar south pole surface environment is characterized by sparse and homogeneous topographic features, poor optical textures, and unstructured landforms. These negative environmental factors make it difficult for sensors such as LiDAR or cameras to extract effective features, which further impact the accuracy of algorithms such as simultaneous localization and mapping [6] (SLAM) that rely on robust feature registration. To improve the navigation accuracy, a navigation-friendly path planning algorithm is introduced in this study. Section 2 provides an overview of DWA. A novel navigation beneficial path planning algorithm is introduced in the Sect. 3. In the Sect. 4, the impact of path planning on the navigation accuracy is evaluated via a numerical case study.

2 Local Path Planning Based on DWA

2.1 DWA Overview

DWA is a local path planning algorithm. It samples multiple sets of velocities in velocity space, and then simulates several sets of trajectories at these velocities for a certain period of time. The optimal travelable path will be selected by cost function, and the velocity and angular velocity of the rover can also be obtained.

All the simulations and descriptions in the paper are based on the three-degree-of-freedom kinematic model, assuming that the rover cannot move laterally and can

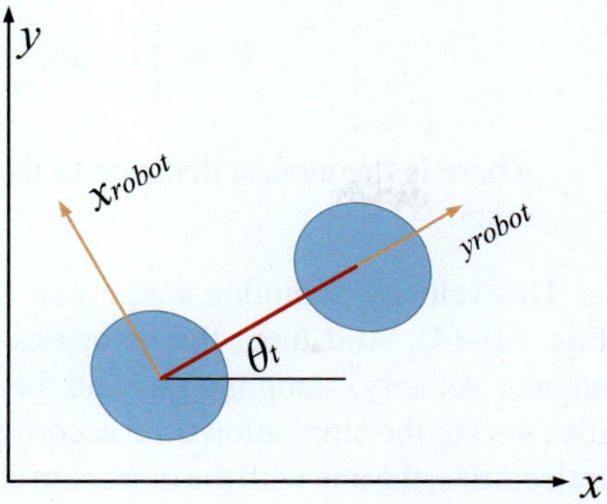

Fig. 1 Motion model of lunar rover

only move forward and rotate. Since the rover moves a short distance within adjacent moments, the paper regards the motion of the rover during the simulation time as uniform linear motion (as shown in Fig. 1), thus the position of the rover at the next moment can be expressed as:

$$\begin{cases} x = x + v\Delta t \cos(\theta_t) \\ y = y + v\Delta t \sin(\theta_t) \\ \theta_t = \theta_t + \omega \Delta t \end{cases} \tag{1}$$

where v and ω are the velocity and angular velocity of the rover at the current moment, θ_t is heading angle, and Δt is simulation time interval.

DWA can constrain the sampling velocities to a certain range depending on the restrictions of the rover's own performance and the limitations of environment.

(1) The rover has a maximum and minimum velocity limit:

$$V_m = \{v \in [v_{\min}, v_{\max}], \omega \in [\omega_{\min}, \omega_{\max}]\} \tag{2}$$

(2) The mobility of the rover is characterized by the performance of the engine, thus the velocity of the rover is limited by the maximum acceleration and deceleration:

$$V_d = \left\{ (v, \omega) | \begin{array}{c} v \in [v_c - \dot{v}_b \Delta t, v_c + \dot{v}_a \Delta t] \wedge \\ \omega \in [\omega_c - \dot{\omega}_b \Delta t, \omega_c + \dot{\omega}_a \Delta t] \end{array} \right\} \tag{3}$$

where $\dot{v}_a$ and $\dot{v}_b$ are the maximum acceleration and maximum deceleration of the rover, $\dot{\omega}_a$ and $\dot{\omega}_b$ are the maximum angular acceleration and the maximum angular deceleration. v_c is the velocity of the rover at the current moment.

(3) To guarantee the safety of the rover, it should stop before hitting an obstacle, therefore the velocity of the rover is limited:

$$V_a = \left\{ (v, \omega) | \begin{matrix} v \leq \sqrt{2 \cdot dist(v, \omega) \cdot \dot{v}_b} \wedge \\ \omega \leq \sqrt{2 \cdot dist(v, \omega) \cdot \dot{\omega}_b} \end{matrix} \right\} \quad (4)$$

where is the closest distance to the obstacle on the simulated trajectory of the rover.

The velocity sampling range can be obtained by finding the intersection of Eqs. (2)–(4). And then, the velocities are sampled according to the velocity (or angular velocity) sampling rate and the trajectories are simulated at different velocities during the simulation time according to Eq. (1). Finally, among several sets of trajectories, the one with the minimum cost function value is selected as the travelable path. The entire planning process of DWA is represented in Fig. 2.

The cost function of DWA consists of heading angle cost function, obstacle distance cost function and velocity cost function, which is computed as

$$G = \alpha \cdot H_t + \beta \cdot L_o + \gamma \cdot V \quad (5)$$

where α, β and γ are weight constants. H_t is the angle between the rover heading and the target. L_o is the distance between the position of the last moment on the simulated trajectory of the rover and the obstacle. V is the velocity of the rover at the current moment. To avoid overrepresentation of a factor in the cost function, H_t, L_o, and V have been normalized.

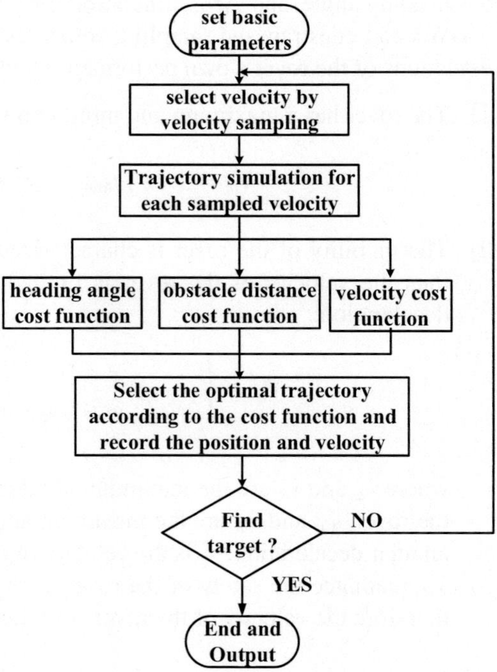

Fig. 2 Planning process of DWA

2.2 Application of DWA in Path Planning Case

The simulation parameters of path planning cases as shown in Table 1. In this exemplary case, obstacle maps are synthesized and demonstrated in Fig. 3. The black areas are obstacles in each figure, and the radius of each obstacle is set to 0.5 m.

The simulation results are presented in Fig. 3, where (a)–(c) are the path planning results in obstacle map 1 and (d) is the simulation result in obstacle map 2.

Figure 3a shows a successful application of DWA in case 1, where the green curves are the simulated trajectories to be selected and the blue curve is the planned path. In Fig. 3b, after modifying the weights of the heading cost function, the rover is too close to the obstacle to meet the safety requirements, which leads to the failure of

Table 1 Simulation parameters

Case	Start point	Target point	Weights
1	(0, 0)	(10, 9)	$\alpha = 0.05, \beta = 0.2, \gamma = 0.1$
2	(0, 0)	(10, 9)	$\alpha = 0.5, \beta = 0.2, \gamma = 0.1$
3	(0, 0)	(9, 9)	$\alpha = 0.05, \beta = 0.2, \gamma = 0.1$

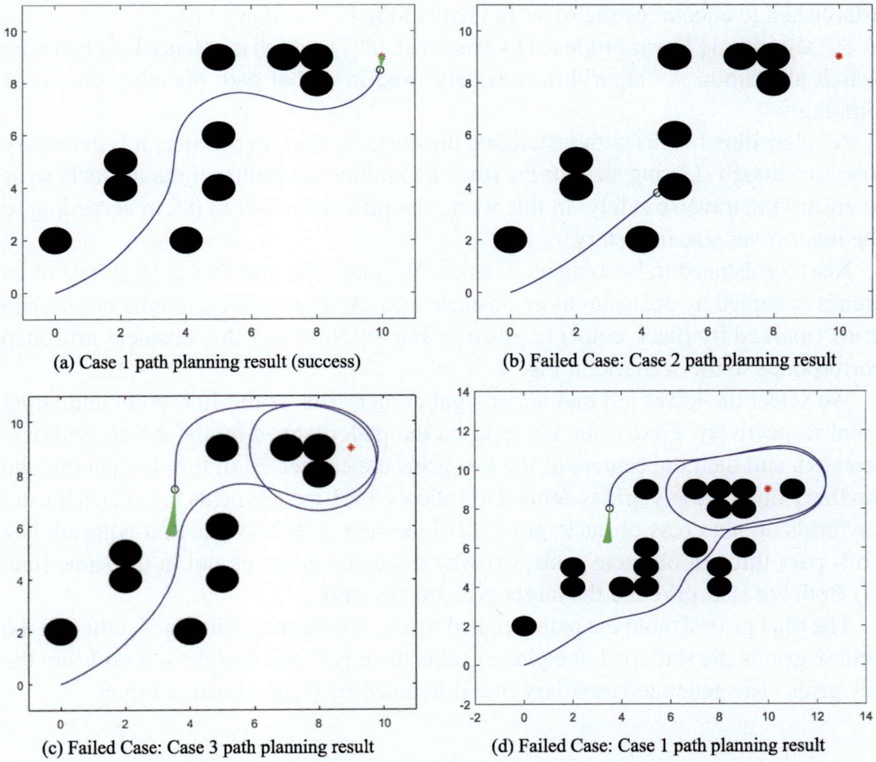

(a) Case 1 path planning result (success)

(b) Failed Case: Case 2 path planning result

(c) Failed Case: Case 3 path planning result

(d) Failed Case: Case 1 path planning result

Fig. 3 Path planning result in obstacle map 1 and obstacle map 2 respectively path planning. In Fig. 3c, changing the position of the target point, the rover gradually deviated from the target point after circling it once, and finally path planning failed. Applying Case 1 to a more complex obstacle map, as shown in Fig. 3d, it can be seen that the path planning failed to reach the final target spot.

In summary, DWA has the advantages of fast convergence to the target and smooth path generation, However, DWA is also susceptible to many factors, such as the selection of target point and the selection of cost function weights. In addition, DWA does not work well in more complex environments. For this reason, a method based on novel auxiliary line is integrated with traditional DWA, which is detailed in Sect. 3.

3 With Integrating Global Auxiliary Lines Path Planning

3.1 Algorithm Overview

In order to improve the reliability of DWA, this section employs the global auxiliary line constructed based on the A* algorithm for optimization [4]. First, an auxiliary line from the start point to the target point is constructed. Then, a reward function is established to encourage the rover to travel along the auxiliary line.

A* algorithm [7] was proposed by Hart et al. in [7], which is essentially a heuristic search algorithm. A* algorithm is widely used in global path planning due to its simplicity.

A* algorithm begins with rasterizing the obstacle map. In practice, it is necessary to leave enough steering size for the rover and inflate the radius of the obstacle so as to ensure the traverse safety. In this work, the grid size is set to 0.5 m according to the lunar rover size in history.

Next, grids need to be assigned a value. We mark the grid that at least half of its area is occupied by obstacles as an obstacle grid. An illustrative example of obstacle grids (marked by black color) is given in Fig. 4. Note that this obstacle grid map corresponds to the scenario in Fig. 4.

We select the lower left and upper right corner grids as the start point and target point respectively. Firstly, the key grid on the path planned by the A* algorithm is selected, and then the centers of the key grids are connected in turn to generate the auxiliary line. The key grid is defined as follows: (1) Line segments between adjacent key grids do not cross obstacle grids; (2) Line segments between non-adjacent key grids pass through obstacle grids; (3) Any three key grids are not in the same line; (4) Both the start grid and the target grid are key grids.

The blue grids denote the path planned by the A* algorithm in Fig. 5, wherein the yellow grid is the start grid, the green grid is the target grid and the red grids are the key grids. The generated auxiliary line is denoted by the red dash in Fig. 6.

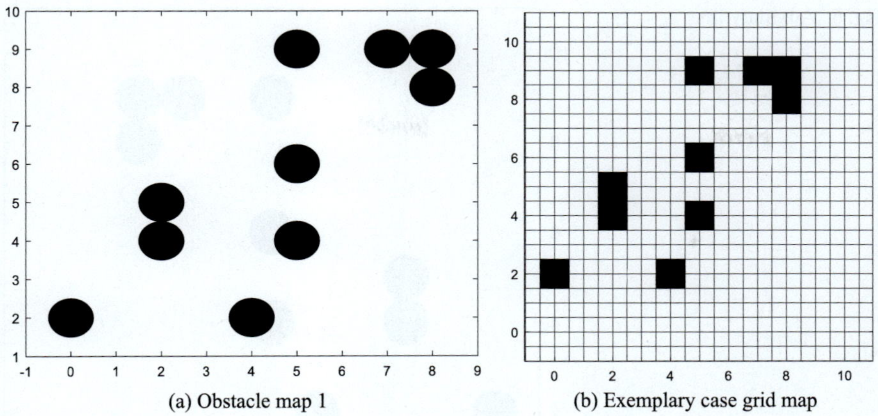

Fig. 4 Obstacle map and grid map

Fig. 5 Extraction of key grids

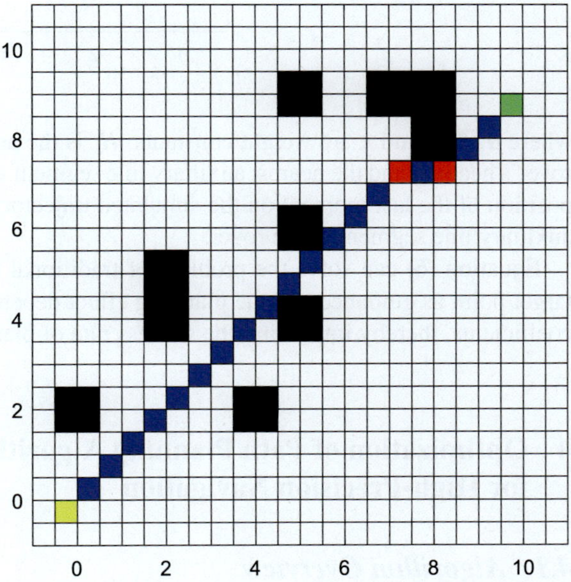

3.2 Simulation Result

To encourage the rover to walk along the auxiliary line, the angle and the distance between the rover and the auxiliary line are introduced as a cost value into Eq. (5). The optimized cost function is represented as:

$$G_e = \alpha \cdot H_s + \beta \cdot L_o + \gamma \cdot V + \lambda L_s \tag{6}$$

Fig. 6 Auxiliary line

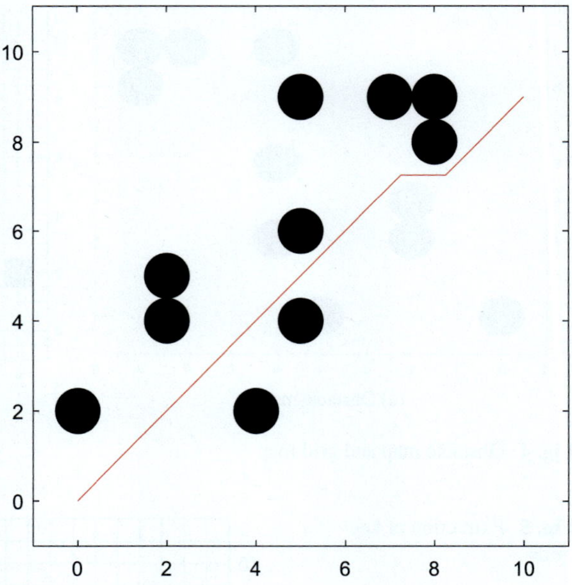

where α, β, γ and λ are weight constants. H_s is the angular difference between the rover's heading and the nearest auxiliary line segment. L_s is the distance between the position of the last moment on the simulated trajectory of the rover and the nearest auxiliary line segment to the rover.

Equation (6) can solve the problem of traditional DWA that only uses a single target point as guidance and the planning effect depends on the selection of weight coefficients, thereby improving the success rate of planning.

4 Optimization of Path Planning Algorithm for High-Precision Navigation

4.1 Algorithm Overview

SLAM is an effective method for providing location information for the lunar rover, which also affect the success rate of path planning. Therefore, it is important to improve the navigation accuracy in the path planning process of the lunar rover. Based on this, a new cost function that takes into account the specialty of the lunar south pole environment and the accuracy impact factors of navigation is established in this section.

The lack of adequate visual information on the south lunar pole surface limits the applications of optical sensors such as LiDAR or cameras, which further brings a greater challenge to obtain a sufficient number of visual features for navigation. In addition, due to the homogeneous topographic features of the lunar south pole, some visual features may have a high degree of similarity to each other, which brings much ambiguity to feature registration and localization estimation. However, the significant visual information and navigation landmarks at the lunar south pole can only be provided by the undulating obstacles such as rocks, craters, and slopes. Therefore, to plan a navigation-friendly path, the obstacle features should be utilized as much as possible while ensuring that the rover does not hit the obstacle. Specifically, the maximum and minimum thresholds of obstacle distance are added to obstacle distance cost function in Eq. (5), respectively. And then the weight of the obstacle distance cost function is increased when the distance between the rover and the obstacles exceeds the maximum threshold or is less than the minimum threshold. The new cost function is expressed as:

$$G_e = \alpha \cdot H_s + \beta \cdot L_{so} + \gamma \cdot V + \lambda L_s \tag{7}$$

where L_{so} is the modified obstacle distance cost function. When the lunar rover approaches obstacles while ensuring safety, the higher the score of L_{so}, which is because LiDAR can obtain more accurate ranging information and sufficient prominent navigation landmark features.

4.2 Simulation Result

There are two types of obstacle maps as shown in Figs. 7 and 8. The simulation parameters are demonstrated in Table 2.

For the path planning with the same start point and target point, Case 3 fails to plan reasonable path in the first place. In contrast, a smooth path can be successfully generated after introducing the auxiliary line method, as shown in Fig. 7a. As shown

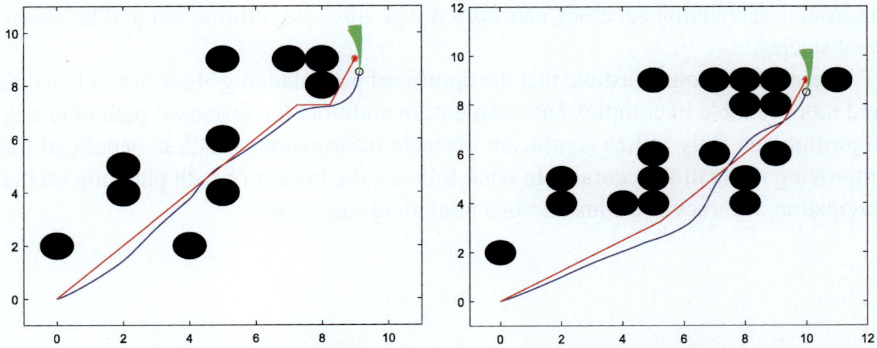

(a) Case 4 path planning result

(b) Case 5 path planning result

Fig. 7 Path planning result in obstacle map 1

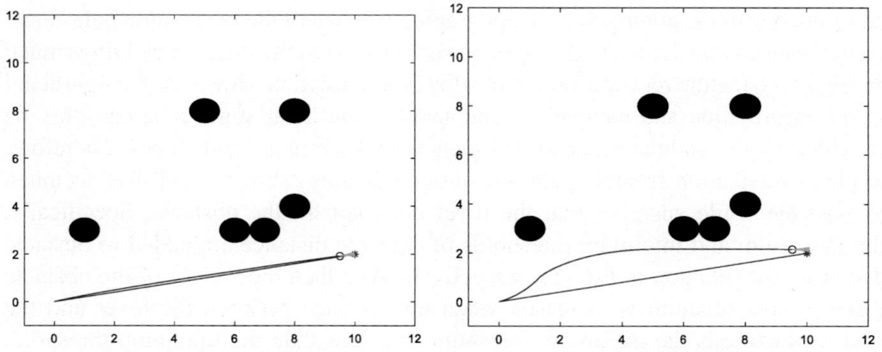

(a) Case 6 path planning result (b) Case 7 path planning result

Fig. 8 Path planning result in obstacle map 2

Table 2 Simulation parameters

Case	Start point	Target point	Weights
4	(0, 0)	(9, 9)	$\alpha = 0.2, \beta = 0.1, \gamma = 0.1, \lambda = 0.3$
5	(0, 0)	(10, 9)	$\alpha = 0.1, \beta = 0.1, \gamma = 0.1, \lambda = 0.3$
6	(0, 0)	(10, 2)	$\alpha = 0.7, \beta = 0.05, \gamma = 0.1, \lambda = 1$
7	(0, 0)	(10, 2)	$\alpha = 0.7, \beta = 0.05, \gamma = 0.1, \lambda = 1$

in Fig. 7b, the optimized path planning algorithm is also successfully applied in more complex environments.

To verify the application of the optimized path planning algorithm in scenarios with sparse environmental features, the scenario shown in Fig. 8 is set up. And then the fusion algorithm as show in Sect. 3 and the optimized path planning algorithm are applied in the scenario, the final simulation results are shown in Fig. 8a and b. From the figure, it can be seen that the path planned by the optimized algorithm based on improving navigation accuracy can fully utilize obstacle information and be closer to obstacles.

These simulations illustrate that the optimized path planning algorithm is feasible and more reliable in complex environment. In addition, the proposed path planning algorithm can fully utilize significant obstacle information, which is beneficial for improving navigation accuracy. In what follows, the impact of path planning on the navigation accuracy is evaluated via a numerical case study.

5 Numerical Simulation Analysis of Localization Accuracy Evaluation

5.1 Fundamental Setup

Reference [8] proposes an active re-observation algorithm, which could reduce the localization error of SLAM by revisiting the characteristic landforms with higher confidence. It relies on point cloud registration algorithms such as ICP (Iterative Closest Point [9]) to further accurately estimate the Localization of the lunar rover. The ability of LiDAR to detect significant navigation landmark feature information is also closely related to its suppression effect. Therefore, the algorithm proposed in the paper can be applied to the re observation process to further improve the navigation accuracy of the lunar rover.

Based on the above content, this section will apply the proposed algorithm to the fixed position detection task and re observation task of the lunar rover, and record the localization error of the lunar rover based on radar point cloud data to further evaluate the effectiveness of this algorithm in improving navigation accuracy.

5.2 Simulation Result

The point cloud map used to simulate the exploration environment of the rover at the lunar south pole is denoted in Fig. 9. The rover is equipped with a LIDAR in the simulation. The detection range of the LIDAR is set to 50 m. The travel path of the rover is also shown in Fig. 9, in which the red curve is the planned path and the green curve is the real travel path of the rover.

The simulation results are represented in Fig. 10. There are four sets of cases: (1) The lunar rover conducts fixed position detection task along the random path, the

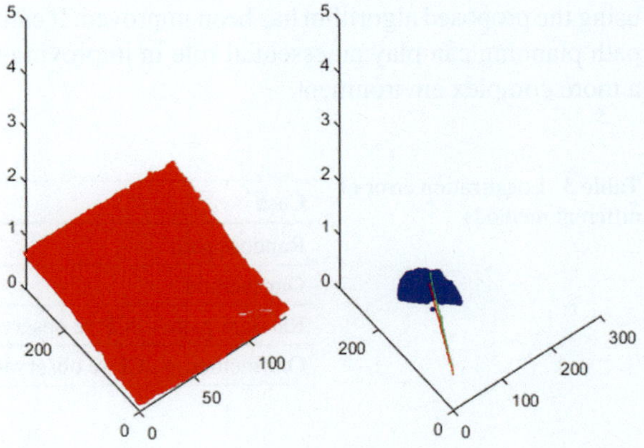

Fig. 9 The point cloud and the travel path

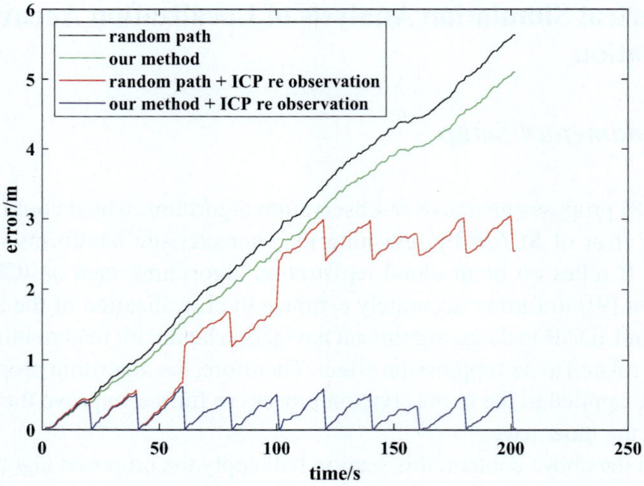

Fig. 10 Localization error of different methods

simulation result is shown in the black curve of Fig. 10; (2) The lunar rover conducts fixed position detection task along the path planned by the proposed algorithm, the simulation result is shown in the green curve of Fig. 10; (3) Localization estimation using ICP re observation every 20 s when the lunar rover conducts re observation task along the random path, the simulation result is shown in the red curve of Fig. 10.; (4) Localization estimation using ICP re observation every 20 s when the lunar rover conducts re observation task along the path planned by the proposed algorithm, the simulation result is shown in the blue curve of Fig. 10.

In Fig. 10, the vertical coordinate indicates the localization error, and the horizontal coordinate is the simulation time. The Mean Absolute Error (MAE) is demonstrated in Table 3.

As can be seen in Fig. 10 and Table 3, whether performing fixed position detection tasks or re observation tasks, the autonomous navigation accuracy of the lunar rover using the proposed algorithm has been improved. It can be expected that the proposed path planning can play an essential role in improving the accuracy of navigation in a more complex environment.

Table 3 Localization error of different methods

Case	MAE(m)
Random path	2.86
Our method	2.39
Random path + ICP re observation	1.75
Our method + ICP re observation	0.25

6 Conclusion

This paper proposes a novel path planning algorithm, where a method based on global auxiliary line is integrated with DWA to enhance the reliability of path planning. A new cost function that takes into account the specialty of the lunar south pole environment and the accuracy impact factors of navigation is established. Simulation result demonstrates that the proposed algorithm is feasible and reliable in terms of path planning and navigational performance improving. In future study, we will carry out a semi-physical experiment to verify the effectiveness of the path planning algorithm and provide a clearer picture for embedding navigation-friendly factors into on-board rover trajectory planning system.

References

1. Cheng A, Wang S, Li X, et al (2007) Progress and development trend of lunar north and south poles research. Adv Geophys 22(4): 1070—1074 (in Chinese)
2. Li X, Ma X, Wang X (2022) A review of path planning algorithms for mobile robots. Comput Measure Control 30(07): 9–19 (in Chinese)
3. Zhao Y (2017) Research on autonomous navigation technology of indoor robot based on laser sensor. Tianjin University of Technology (in Chinese)
4. Wang K, Song P, Liu S, Li H (2019) Research on lunar surface rover path planning based on A* algorithm optimization. Spacecraft Eng 28(01): 19–26 (in Chinese)
5. Wu W (2021) Research on the path planning method of lunar surface rover. Advisor: Xie Xiaomei. University of Electronic Science and Technology (in Chinese)
6. Ye T, Hongwei C, Fasheng W, Xingwen C, A review of SLAM algorithms for indoor mobile robots. Comput Sci 1–26 (in Chinese)
7. Hart PE, Nlsson NJ, Raphel B (1972) A formal basis for the heuristic determination of minimum cost paths. IEEE Trans Syst Sci Cybern 4(2): 28–29
8. Yu M, Research on autonomous visual navigation method for planetary landing and exploration mission. Harbin: Harbin Institute of Technology 20 (in Chinese)
9. Besl PJ, McKay HD (1992) A method for registration of 3D shape. IEEE Trans Pattern Anal Mach Intell 14(2):239–256

Design of Axial Force Sensor Based on Equal Strength Beam

Zhikang Li, Yuegang Tan, and Yi Liu

Abstract Axial force is an important indicator affecting the gas turbine's reliability. This work proposes a novelty annular force sensor based on an equal-strength beam for the axial force of bearing. The finite element method verifies that the sensor elastomer has a considerable strain-flat length and validates the sensor's measurement range. The static performance of the sensor is also studied, and the sensor has excellent linearity in the range of 0–3.5 t.

Keywords Axial force · Annular force sensor · Equal-strength beam · Strain-flat length

1 Introduction

The axial force of the main shaft of the gas turbine is an essential consideration in the design phase. The excessive axial force will lead to long-term overload of the machine and reduce its working life, and the gas turbine performance will not be fully effective conversely [1]. According to the measurement parameters, two main axial force measurement methods are used now, indirect and direct measurement [2, 3]. Direct measurement [4] directly measures the axial force of the main shaft by installing force sensors, which are straightforward and more accurate than indirect measurement. The disadvantage is that it needs to be taken up a particular space in the cavity.

At present, the main sensitive element used for elastomer measurement is resistance strain gauge, which is measured by using Huygens bridge with mature technology and high measurement accuracy, while the problems are complicated layout, many wire harnesses and susceptibility to electromagnetic interference [5]. Fiber Bragg Grating (FBG) [6] is an excellent sensing element with the advantages of anti-electromagnetic interference and wavelength division multiplex [7], and the available

Z. Li · Y. Tan · Y. Liu (✉)
School of Mechanical and Electronic Engineering, Wuhan University of Technology, Wuhan 430070, China
e-mail: Wgdliuyi2001@whut.edu.cn

Z. Liu et al. (eds.), *Advances and Challenges in Advanced Unmanned Aerial Systems*, Springer Aerospace Technology, https://doi.org/10.1007/978-981-99-8045-1_29

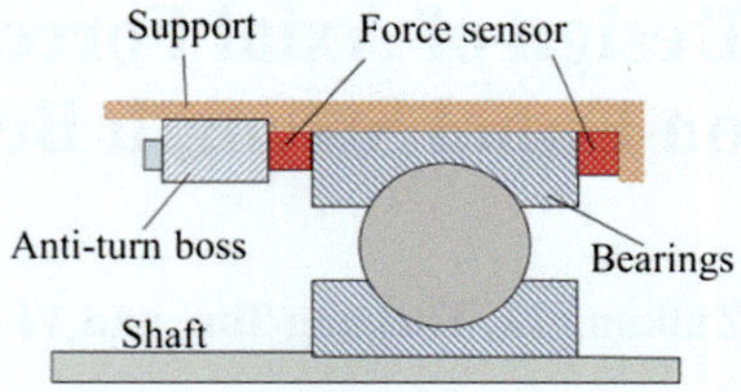

Fig. 1 Schematic of the installation position of the annual force sensor

length of FBG is 10 mm. Since the measurement result is the average strain of the grid length, the strain gradient of the existing conversion elastomer is large, resulting in the chirping of the FBG, which reduces the measurement accuracy. This paper proposes a conversion elastomer based on an equal-strength beam design for FBG measurements. The aim is to reduce the strain gradient in the measurement region.

In this paper, an Annular Force Sensor (AFS) was designed based on an equal-strength beam, which has excellent accuracy.

2 Design of Elastic Element Structure

2.1 *Elastic Structure Requirements*

The principle of axial force measurement with an annular force sensor is to measure the strain of the elastomer using an FBG. The AFS is fitted with an FBG sensor, installed between the end of the outer ring of the bearing and the housing support, and pre-calibrated before use. The signal is transmitted by wire to the acquisition system. The sensors are mounted before and after the bearings to measure the forward and backward axial forces, as shown in Fig. 1 [8].

2.2 *Elastic Elements*

As shown in Fig. 2, the traditional force sensor unit structure [9] is a two-point support structure, and strain-sensitive elements such as strain gauges or fiber Bragg gratings are installed on the back of the boss. A new structure is designed based on the equal-strength beam as the primary cycle unit of the force sensor. It reduces the measurement error due to strain gradient variation by forming a strain-flat region.

Usually, the space reserved for the sensor is limited in designing the gas turbine. High precision is also required under heavy loads. In this work, the height of the AFS is only 8 mm, the outer diameter is 280 mm, the measuring load range is 0–3.5 t, and the measuring accuracy must reach 2.5%.

The bending moment of the beam with fixed supports at both ends is in accordance with Eq. 1

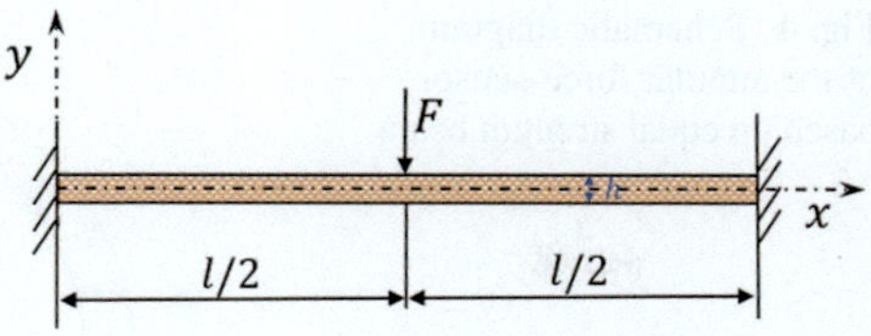

Fig. 2 Conventional force ring structural units

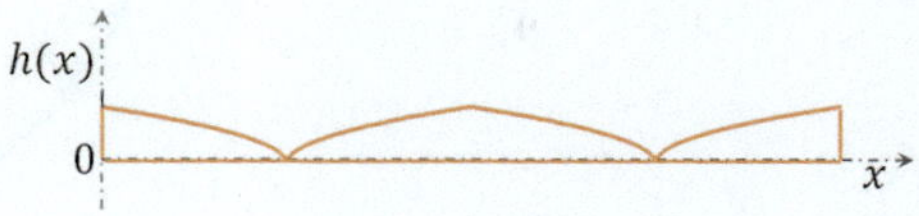

Fig. 3 The variable thickness of equal strength beam

$$M(x) = \begin{cases} \left|\frac{Fx}{2} - \frac{Fl}{8}\right|, & \left(0 \le x \le \frac{l}{2}\right) \\ \left|-\frac{F(l-x)}{2} + \frac{Fl}{8}\right|, & \left(\frac{l}{2} \le x \le l\right) \end{cases} \tag{1}$$

To comply with the equal strength, the variation law in the thickness of the beam with fixed supports at both ends obeys Eq. 2.

$$h(x) = \sqrt{\left|\frac{12Fx - 3Fl}{4[\sigma]b_0}\right|} \tag{2}$$

where $[\sigma]$ the force ring material's allowable stress, b is the force-measuring ring's width, the changing of $h(x)$ are shown in Fig. 3.

The minimum thickness is determined according to the material's shear strength after determining the contour of the variable curve. According to the engineering situation, a boss subject to the downward force is added $x = l/2$. Also, set two bosses on the opposite side at the place of $x = 0$ and $x = 1$.

The AFS unit is subjected to force on one side and fixed support on the other. It converts the vertical force into a transverse strain when the upper boss is subjected to pressure, and the back of the upper boss deforms. The FBG can be pasted on the bottom surface of the structure unit. Figure 4 is the schematic diagram annular force sensor based on equal strength beam.

2.3 Determination of the Structure of the Force-Measuring Ring

The AFS will be installed in the gas turbine, and the annular shape will be formed through the repetition of elastic elements, which guarantees the heavy load capacity. With the increase of boss supports, the bearing capacity will be enhanced, but the deformation and measurement accuracy of the sensor will be reduced. Limited by

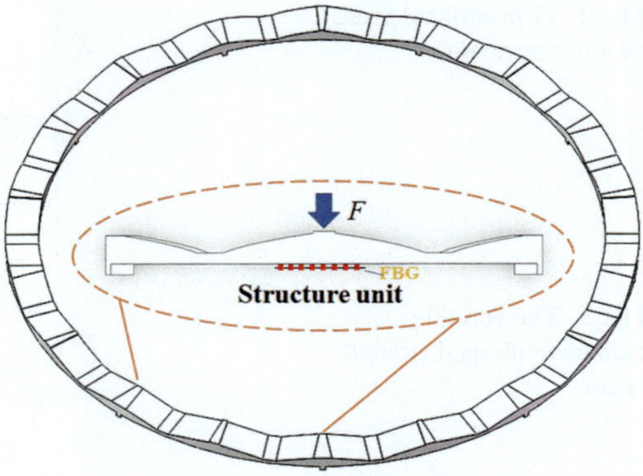

Fig. 4 Schematic diagram of the annular force-sensor based on equal strength beam

the strength and accuracy requirements, the number of structural units is determined to be nine.

3 Structure Simulation of Sensor

The finite element method (FEM) was used to perform a static simulation of the annular sensor through the ANSYS2022R1 program. The material of the sensor is high-strength chrome-nickel alloy steel. The density is 7.85×10^3 kg/m^3, the elastic modulus is 200 GPa, Poisson's ratio is 0.3, and the expansion coefficient is 15.4×10^{-6}.

According to the actual working situation, we apply a 3.5 t force to the boss of the thickness change surface. The upper boss is deformed downward under load, and the limit position is designed to protect the sensor from being destroyed under temporary overload.

The result of the structural simulation is shown in Fig. 5. The maximum stress of the force-measuring ring is 782 MPa, which meets the requirements of 835 MPa of alloy steel. The best sensitivity is obtained by pasting the fiber Bragg grating at the plane. If the stress results do not meet the strength requirements, the geometric parameters of the structure unit should be recalculated and modified.

A path is set at the flat surface's midline along the force ring's circumferential direction, as shown in Fig. 5b, to analyze the effect of the equal strength design. The strain change on the path is shown in Fig. 6, and it can be seen from the results that in the length range of 20 mm, the strain change is less than 2.3%, and there is good strain flatness.

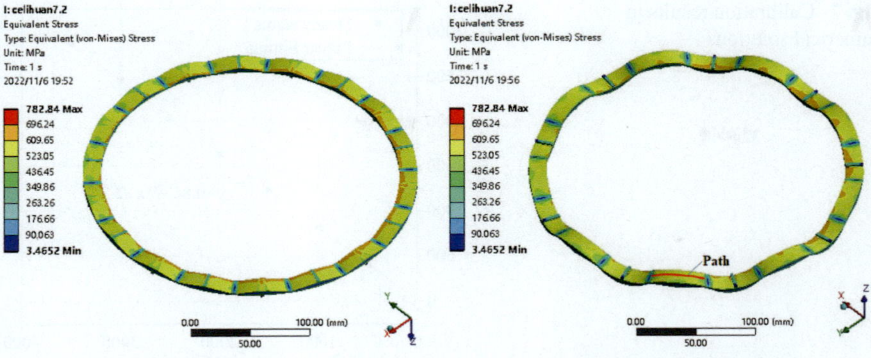

(a)Stress on the upper face (b)Stress on the lower face

Fig. 5 Equivalent stress

Fig. 6 The strain on the path

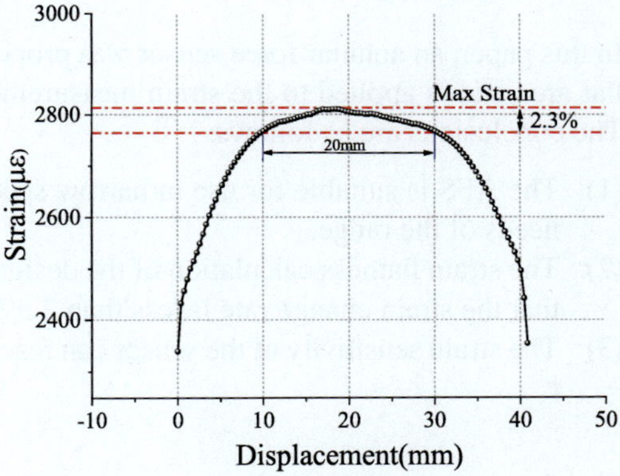

4 The Static Performance

The sensitive element used in force sensors is FBG, which is arranged at the place of the maximum strain value point. We define the measurement value as the mean over a long range of 10 mm maximum strain value. The measurement value of the sensor is 2860 under a load of 3.5 t. Figure 7 shows the calibration result in numerical solutions, the abscissa means the load, and the ordinate means the strain. The FEM results of the overall strain sensitivity.

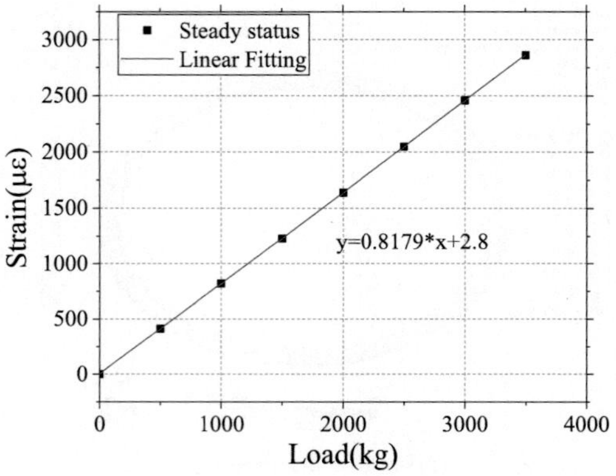

Fig. 7 Calibration results in numerical solutions

5 Conclusion

In this paper, an annular force sensor was proposed, which has a considerable strain flat area and is applied to the strain measurement used by the fiber Bragg grating. The conclusions are as follows:

(1) The AFS is suitable for use in narrow spaces and is easy to adjust to meet the needs of the range.
(2) The strain flatness calculation of the designed structure through the AFS shows that the strain change rate is less than 2.3% over a 20 mm long path.
(3) The strain sensitivity of the sensor can reach 0.8179 $\mu\varepsilon$/kg in the range of 0–3.5 t.

References

1. Sokolov NV, Maksimov TV, Khadiev MB, Sagbiev IR (2021) Influence of the axial force of the rotor of a centrifugal compressor on the operation of the thrust bearing. In: AIP conference proceedings, vol 2412, no 1. AIP Publishing LLC, p 030047
2. Bozzi L, Malavasi F, Garotta V (2011) Heavy-duty gas turbines axial thrust calculation in different operating conditions. In: Turbo expo: power for land, sea, and air, vol 54655, pp 945–954
3. Dorney D, Marcu B, Sargent S, Tran K (2003) Calculation of turbine axial thrust by coupled CFD simulations of the main flow path and secondary cavity flow in an SLI LOX turbine. In: 39th AIAA/ASME/SAE/ASEE joint propulsion conference and exhibit, p 4919
4. Bavassano F, Mantero M, Traverso R, Livermore-Hardy R, Blair B (2017) A system integration approach for heavy-duty gas turbine upgrades using improved rotor thrust predictions and application of advanced thrust bearing designs. In: Turbo expo: power for land, sea, and air, vol 50886. American Society of Mechanical Engineers, p V05BT15A012
5. Christenson J (2019) Chapter three—sensors and transducers. In: Segil J (ed) Handbook of biomechatronics. Academic Press, pp 61–93
6. Hill KO, Fujii Y, Johnson DC, Kawasaki BS (1978) Photosensitivity in optical fiber waveguides: application to reflection filter fabrication. Appl Phys Lett 32(10):647–649

7. Qi YB, Wu GF, Wang HX (2012) Development of fiber bragg grating sensing multiplexing mode. J Central South Univ (Science and Technology), 3058–3072
8. Yunfang O (2020) Measurement method for axial force of aero-engine special-shaped thrust bearing. J Aerospace Power (06): 1149–1156
9. Xue Y, Yansong Z, Daming Q (2021) Test technique for dynamic characteristic of rotor axial force on aero -engine. J Aerospace Power (09): 1804–1810

Stability of a Tractor with Implement by Employing Fluid-Based Inerters into Front Axle Suspension

Huachun Li and Xiaofu Liu

Abstract Vibration caused instability of the tractor with implements has always been severe during agricultural work. In this paper, the potential of applying fluid-based inerter into the front axle suspension of tractor to reduce the vibration was studied, where two inerter-based strut layouts were proposed and compared with the conventional layout towards the stability of ploughing depth was studied in pitch plane model, the performance benefits of each layout were optimised and analysed. Meanwhile, the optimum design parameters were identified, respectively. According to the optimization results, over a wide range of static suspension stiffness, the ploughing depth stability of both potential layouts has improved compared to the conventional layout, with a maximum performance improvement of 27.8% and 38.2% respectively.

Keywords Fluid-based inerter · Tractor · Front axle suspension · Stability · Plough

1 Introduction

The tractors have been widely used in agricultural works. Due the complex working condition, the tractors have to suffer long-term serious vibration during driving, which is harmful to the ride comfort of driver and stability of agricultural production. Improving the suspension performance is one of the most direct and effective approach to reduce the vibration. While the implements are usually mounted at the rear of tractor, which restrict the tuning of rear axle suspension to maintain the reliability of hitch system, the improvement of front axle suspension could be the most promising way to increase the stability of tractor.

Considering the cost and reliability, the tractor manufacturers still prefer to use passive suspensions instead of high performance active/semi-active suspensions.

H. Li · X. Liu (✉)
China Agricultural University, Beijing 100193, China
e-mail: lxf28@cau.edu.cn

Z. Liu et al. (eds.), *Advances and Challenges in Advanced Unmanned Aerial Systems*, Springer Aerospace Technology, https://doi.org/10.1007/978-981-99-8045-1_30

While most of the recent studies of passive suspension are mainly focused on parameter optimisation of the existing suspension structure [1–3], the improvement of suspension performance has been restricted by the physical limit of traditional vibration control units (e.g. springs or dampers) in certain structure. Until a novel element inerter has been introduced by Smith in 2002 [4], which fundamentally extends the range of vibration control and the theoretical performance benefits of incorporating inerter into suspension has been proved by numerous theoretical applications [5–7]. There are several ways to physically realise the inerter, where the fluid based form has been shown to be the most promising one for industrial production due to its durability, structural simplicity and the similarity with existing damper constructions [8]. Subsequently, Liu et al. accurately models the typical fluid-based inerter, reveal the coupling nonlinear properties, such as the inertance and inherent damping [9]. On the basis of such modelling methodology, further validate the generalisable model of Fluid-inerter Integrated Damping (FID) devices experimentally, where all the design parameters could be varied within pre-defined ranges, including the damping/inertance-dependent tube length and tube diameter [10].

In this paper, alternative layouts incorporate fluid-based inerter will be studied for the front axle suspension of tractors to explore the potential to improve the stability of agricultural implements, which dramatically influence the efficiency of agricultural production, however has been rarely considered in previous studies of front axle suspensions [11].

2 FID Device Model and Candidate Suspension Struts

A typical FID device usually consists of cylinder, piston rod, external tube and optional piston valve (Fig. 1). When the piston rod moves with relative velocity Δv, the pressure difference between chambers $\Delta p = p_1 - p_2$ are formed due to the flow paths with flow rate Q_1 through external tube and Q_2 through piston valve. The device terminal properties, mechanical inertance (b) is produced by the accelerated fluid in external tube, meanwhile, the inherent damping (c_h) is generated due to the friction between flowing fluid and tube inner wall; the optional piston valve could also provide an external damping (c_v).

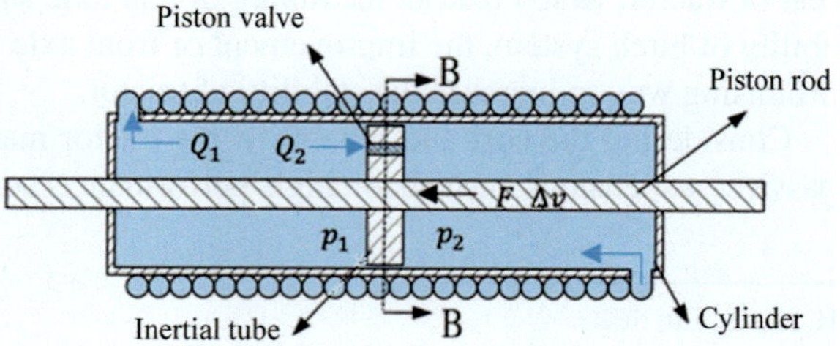

Fig. 1 The schematic plot of a typical FID device

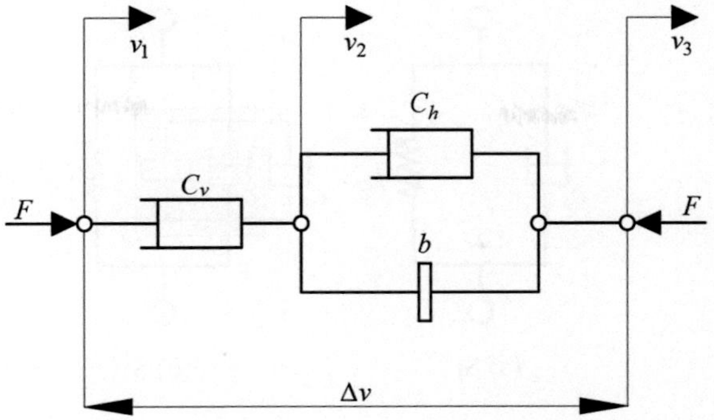

Fig. 2 Typical FID device equivalent mechanical network topology

Based on the hydraulic structure of FID device and the analogy between mechanical and hydraulic networks [6], the equivalent mechanical network of the typical FID device is shown in Fig. 2.

The models of inertance and inherent tube damping of FID device are built in previous studies [9, 10]:

$$b = \frac{\rho l (A_1)^2}{A_2} \tag{1}$$

$$c_h = f_c \frac{2\rho l A_1}{D_h} \left(\frac{A_1}{A_2} \right)^2 \tag{2}$$

where f_c is the Fanning friction factor, $f_c = 0.08\,\mathrm{Re}^{-\frac{1}{4}} + 0.012\sqrt{\frac{D_h}{2R}}$, Reynolds number $\mathrm{Re} = \frac{\rho D_h Q_1}{\mu A_2}$, ρ is fluid density, μ is fluid viscosity, l is the length of external tube, A_1 is the effective area of piston, A_2, $D_h$and$R$ are the cross-sectional area, inner diameter and bend radius of external tube, respectively. Besides, the external valve damping c_v is valve opening size related, which needs to be experimentally tested for certain valve setting. As Eqs. (1) and (2) shown, the independent design parameters tube length l and diameter D_h will be selected for optimisation of front axle suspension.

In this paper, three passive suspension layouts are studied.

Figure 3a is a conventional parallel spring-damper strut (S1), Fig. 3b is a fluid-based inerter (grey block) connected in parallel with strut S1, Fig. 3c is spring connected in parallel with a fluid-based inerter.

The admittance $Y(s)$ of each suspension strut layout is deduced as:

$$Y_{s1}(s) = k_d s^{-1} + c_s \tag{3}$$

$$Y_{s2}(s) = k_d s^{-1} + \frac{1}{(c_h + bs + c_s)^{-1}} \tag{4}$$

$$Y_{s3}(s) = k_d s^{-1} + \frac{1}{(c_h + bs)^{-1}} \tag{5}$$

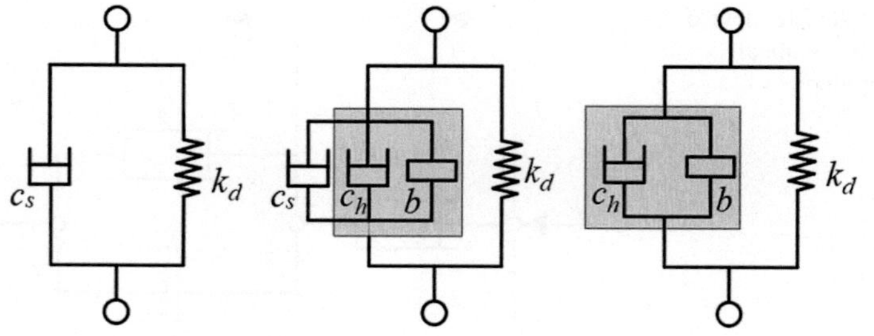

(a) S1 (b) S2 (c) S3

Fig. 3 Candidate suspension strut layouts

where s represents the complex variable in Laplace domain.

3 Pitch Tractor Model Optimisation

In order to analyse the tractor pitch plane vibration, assume the tractor structure is symmetrical about the centre surface and the left and right wheels drive on the same road conditions. Therefore, the 4 degree of freedom pitch-plane half tractor model is built in Fig. 4. The lower arm of implement hitch is equivalent to a parallel damping c_a and stiffness k_a connected. In this paper, the plough is chosen as the typical agricultural implement for stability study. The ploughing resistances of the soil during working are expressed as the horizontal direction resistance R_x and the vertical direction resistance R_y [12].

Where O_b is the tractor body centre of mass, O_a is the plough centre of mass, z_b is the vertical displacement at O_b, φ_b is the angular displacement at O_b, φ_a is the angular displacement at O_a, m_b is the tractor body mass, m_a is the plough mass, J_b is the tractor body moment of inertia, J_a is the plough moment of inertia, D is the hitch point of upper arm, L_b is the vertical distance from O_b to D, L_g is the vertical distance from O_b to the lower hitch arm, L_d is the horizontal distance from O_b to D, L_a is the horizontal distance from O_a to D, L_i is the vertical distance from O_a to D, L_f is the horizontal distance from O_b to front axle, L_r is the horizontal distance from O_b to rear axle, k_{ft} is front wheel stiffness, k_{rt} is rear wheel stiffness, c_{ft} is front wheel damping, c_{rt} is rear wheel damping, z_{uf} is the vertical displacement of the front axle unsprung mass, z_{rf} and z_{rb} are the front and rear wheel road inputs.

Establish the tractor half tractor pitching model equation:

$$\begin{gathered}\left(\hat{z}_{rb} - \hat{z}_b - \hat{\varphi}_b L_r\right) s c_{rt} + \left(\hat{z}_{rb} - \hat{z}_b - \hat{\varphi}_b L_r\right) k_{rt} - \left(\hat{z}_{uf} - \hat{z}_b - \hat{\varphi}_b L_f\right) s Y(s) \\ - m_u \hat{z}_{uf} s^2 = 0\end{gathered} \tag{6}$$

$$\left(\hat{z}_{uf} - \hat{z}_b + \hat{\varphi}_b L_f\right) s Y(s) - R_y + \left(\hat{z}_{rb} - \hat{z}_b - \hat{\varphi}_b L_r\right) s c_{rt} + \left(\hat{z}_{rb} - \hat{z}_b - \hat{\varphi}_b L_r\right) k_{rt}$$

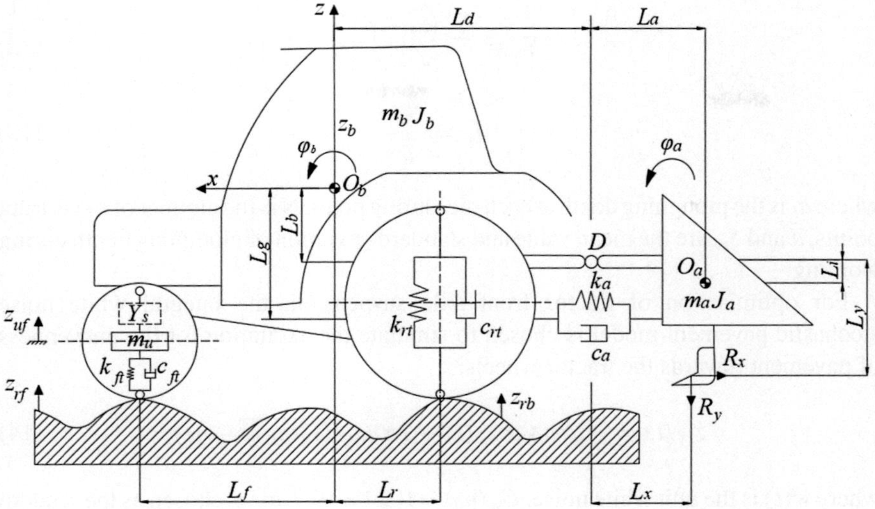

Fig. 4 Pitch plane half tractor model

$$-(m_a + m_b)\hat{z}_b s^2 = 0 \tag{7}$$

$$\widehat{F}_q = (\widehat{\varphi}_b L_g - \widehat{\varphi}_a(L_g - L_b))k_a + (\widehat{\varphi}_b L_g - \widehat{\varphi}_a(L_g - L_b))sc_a\# \tag{8}$$

$$\widehat{F}_q\left(L_g - L_b\right) + R_x L_x - R_y L_y - J_a\widehat{\varphi}_a s^2 = 0 \tag{9}$$

$$\left(\hat{z}_{rb} - \hat{z}_b - \widehat{\varphi}_b L_r\right)sL_r c_{rt} + \left(\hat{z}_{rb} - \hat{z}_b - \widehat{\varphi}_b L_r\right)L_r k_{rt} - \left(\hat{z}_{uf} - \hat{z}_b - \widehat{\varphi}_b L_f\right)sY(s)L_f$$
$$+\left(\widehat{F}_q + R_x\right)L_b - \left(m\hat{z}_b s^2 + R_y\right)L_d - \widehat{F}_q L_g - J_b\widehat{\varphi}_b s^2 = 0 \tag{10}$$

where F_q is the hitch lower arm force, and $\wedge$ denotes Laplace transform variable.

Due to inconsistent soil resistance and time delay of front and rear road inputs, the pitching motion of plough could alter the ploughing depth during working. The tillage quality based on uneven ploughing depth hardly meet the agronomic requirements and therefore influence the crop yield.

According to the standard of ploughing depth, the stability of the plough operation can be measured by the coefficient of depth variation (J_p), the smaller variation brings better stability of tractor with plough in pitch plane. J_p is defined as [13]:

$$\overline{\overline{a}} = \frac{\sum a_i}{N} \tag{11}$$

$$S_m = \frac{\sum \left(a_i - \overline{\overline{a}}\right)^2}{N - 1} \quad (12)$$

$$J_p = \frac{S_m}{\overline{\overline{a}}} \quad (13)$$

where a_i is the ploughing depth at each measuring point, N is the number of measuring points, $\overline{\overline{a}}$ and S_m are the mean value and standard deviation of ploughing depth during working.

For optimisation of tractor front axle suspension, the integral white noise stochastic pavement model is chosen to simulate the excitation for the unevenness of pavement towards the tractor wheels:

$$\dot{z}_{rf}(t) = -2\pi n_{00} u z_{rf}(t) + 2\pi n_0 \sqrt{G_q(n_0)u} \cdot w(t) \quad (14)$$

where $w(t)$ is the unit white noise, $G_q(n_0) = 1024 \times 10^{-6}\,\text{m}^2$ is chosen as the random pavement unevenness coefficient, $n_0 = 0.1\,\text{m}^{-1}$ is selected to be the variance of the random pavement unevenness excitation signal, $n_{00} = 0.011\text{m}^{-1}$ is the lower cut-off space frequency, and u = 10 km/h is assumed for the tractor driving speed during working. The road excitation to front and rear wheels are assumed to be the identical but with delay factor of τ.

The suspension assembly room is restricted by the tractor structure, thus the corresponding limit of design parameters are set to be $0.002\,\text{m} \le D_h \le 0.023\,\text{m}$, $0.245\,\text{m} \le l \le 35\,\text{m}$ according to the CF700 tractor front axle suspension arrangement space [2]. The range of suspension static stiffness is set as $40000\,\text{N} \cdot \text{m}^{-1} \le k_d \le 330000\,\text{N} \cdot \text{m}^{-1}$, which covers majority of tractors on the market. The other model parameters are listed in Table 1.

In order to explore the potential of stability improvement of the tractor front axle suspension, the damping c_s of struts S1, S2 and the FID suspension design parameters tube length l and tube diameter D_h of struts S2, S3 are optimised by using the global patternsearch with local fminsearch algorithm in MATLab to minimise the coefficient

Table 1 The main simulation parameters for pitch plane half tractor model

Description	Value	Description	Value
Helix radius R (m^2)	0.04	Piston area A_1 (m^2)	1.0×10^{-3}
Fluid viscosity μ (Pa · s)	0.0365	Fluid density ρ (kg/m^3)	874
Tractor body mass m_b (kg)	2700	Plough mass m_a (kg)	600
Unsprung mass m_u (kg)	146	Resistance R_x&R_y (N)	895 and 929
Tractor body moment of inertia J_b (kg · m^2)	3610	Plough moment of inertia J_a (kg · m^2)	184
Front wheel damping c_{ft} (N · s/m)	3361	Front wheel stiffness k_{ft} (N/m)	703,834

of depth variation (J_p) in half tractor pitch model. The results are plotted in Fig. 5, in which Fig. 5a shows the optimum value of J_p for each strut layout, Fig. 5b further illustrates up to 27.8 and 38.2% improvement of J_p can be achieved by struts S2, S3, compared to conventional layout S1, with the maximum improvement appears around static stiffness of 0.1 MN/m. The corresponding optimum damping value and inertance are plotted in Fig. 5c and d, the damping value of S3 is always larger than S2, due to an external damping is connected parallel with fluid-based inerter in S2, which could provide extra damping c_s. It is also noted that when employing a fluid-inerter in strut as S3, even more performance benefits could be obtained without an external damping valve required.

To verify the performance benefits, the optimum parameters of external tube diameter $D_h = 0.01$ m and tube length $l = 32$ m are adopted for tractor front axle suspension with layout S3 at static stiffness of $0.07\,\mathrm{MN \cdot m^{-1}}$. The corresponding plough jump values of S1 and S3 are simulated and plotted in Fig. 6. The results show that the fluid-inerter involved suspension S3 could effectively reduce the plough

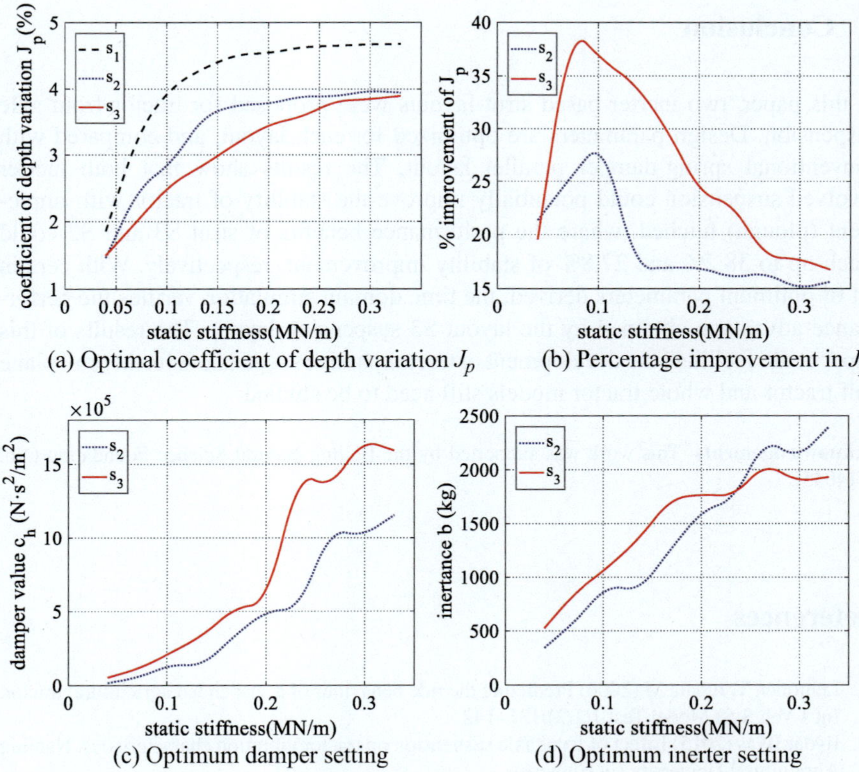

Fig. 5 The optimum coefficient of depth variation J_p and corresponding percentile improvements, optimum damping value, optimum inertance for alternative strut layouts in pitch plane tractor model

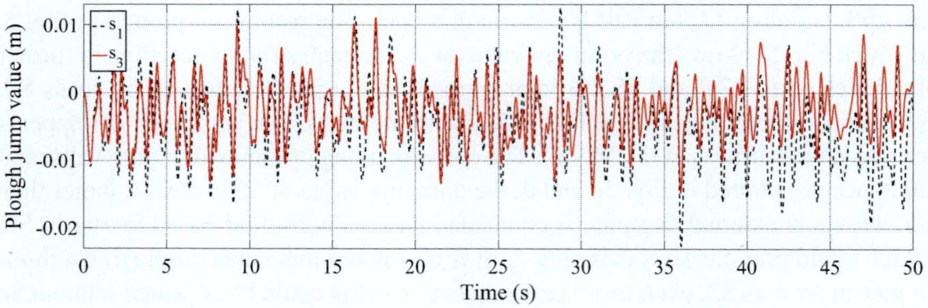

Fig. 6 Plough jump value corresponding to the suspension S1 and S3 in the time domain

jump and improve the stability of tractor with implement compared with conventional suspension S1.

4 Conclusion

In this paper, two inerter-based strut layouts were proposed for tractor front axle suspension. Design parameters are optimized for each layout, and compared with conventional spring-damper parallel layout. The results show that both inerter involved suspension could potentially improve the stability of tractor with implement (plough) hitched, where the performance benefits of strut S3 and S2 could reach up to 38.2% and 27.8% of stability improvement, respectively. With certain set of optimum parameters derived, the time domain simulation verifies the performance advantage obtained by the layout S3 suspension layout. The results of this paper could facilitate the development of tractor passive suspension, further roll plane half tractor and whole tractor models still need to be studied.

Acknowledgements This work was supported by the Beijing Natural Science Foundation (No. 3214045).

References

1. Lehtonen T, Juhala M (2006) Predicting the ride behaviour of a suspended agricultural tractor. Int J Veh Syst Model Test 1(1/2):131–142
2. Ilydar Ilyas (2015) Effect of front axle suspension on tractor vibration characteristics. Nanjing Agricultural University (in Chinese)
3. Yuan J, Fan J, Zhou Y (2018) Optimal design of tractor front axle suspension parameters. Noise Vibr Control 38(1): 8 (in Chinese)
4. Smith MC (2002) Synthesis of mechanical networks: the inerter. IEEE Trans Automatic Control 47(10):1648–1662

5. Scheibe F, Smith MC (2009) Analytical solutions for optimal ride comfort and tyre grip for passive vehicle suspensions. Veh Syst Dyn 47(10):1229–1252
6. Wang FC, Hsieh MR, Chen HJ (2011) Stability and performance analysis of a full-train system with inerters. Veh Syst Dyn 50(4):545–571
7. Jiang JZ, Matamoros-Sanchez AZ, Zolotas A, Goodall R, Smith M (2015) Passive suspensions for ride quality improvement of two-axle railway vehicles. Proc Inst Mech Eng Part F 229(3):315–329
8. Swift S, Smith MC, Glover A, Papageorgiou C, Gartner B, Houghton NE (2013) Design and modelling of a fluid inerter. Int J Control 86(11):2035–2051
9. Liu X, Jiang JZ, Titurus B, Harrison A (2018) Model identification methodology for fluid-based inerters. Mech Syst Signal Process 106:479–494
10. Liu X, Titurus B, Jiang JZ (2019) Generalisable model development for fluid-inerter integrated damping devices. Mech Mach Theory 137:1–22
11. Liu X, Qiao N, Zheng X, et al (2014) Development and test of 1LF-550 amplitude adjustable hydraulic tilting plow. Xinjiang Agric Mechanization (06): 17+47 (in Chinese)
12. Zheng E, Cui S, Yang Y et al (2019) Simulation of the vibration characteristics for agricultural wheeled tractor with implement and front axle hydropneumatic suspension. Shock Vib 2019:1–19
13. B/T 14225-2008, Spade plow (in Chinese)

High Precision Measurement of High Frequency Dynamic Strain of Aerospace Structures with FBG Strain Sensors

Hongyang Li, Feng Chen, Jiangbo Yuan, Meili Wang, Yu Pan, Wei Xu, Pengfei Xue, and Li Wang

Abstract Fiber Bragg grating (FBG) strain sensors are widely used in strain measurement of aerospace structures. FBG strain sensors have received widespread attention as an important strain testing method. With the continuous development of the aerospace manufacturing industry, the dynamic properties of new spacecraft materials have been rapidly improved, and dynamic strain testing has extremely high requirements for FBG strain sensors. The measurement characteristics of FBG sensors often change with the frequency of the measured strain signal. Therefore, current dynamic strain measurements cannot accurately obtain the strain magnitude of aerospace structures. In this paper, a high-precision measurement and error analysis method with FBG sensors is proposed to obtain high-frequency dynamic strains of aerospace structures. The measurement errors under different amplitude and frequency of dynamic strain are obtained. Experiment results show that the maximum error is only 36 $\mu\varepsilon$. Accurate and reliable measurement of high-frequency dynamic strain of aerospace structures using FBG strain sensors is realized.

Keywords Aerospace structures · FBG strain sensors · Dynamic strain testing · Measurement errors

1 Introduction

With the development of optical fiber sensing technology, Fiber Bragg grating (FBG) strain sensors are widely concerned as new strain measurement means [1]. FBG strain sensors have the advantages of light weight, small size, strong anti-interference ability relative to the traditional metal-strain sensors, which are widely used in the field of strain measurement of aerospace structures, FBG sensors applied in the aircraft strain measurement is shown in Fig. 1 [2, 3].

Surface-paste FBG strain sensor can measure the strain on the surface of the aerospace structures without destroying the structures. It is the most common used

H. Li (✉) · F. Chen · J. Yuan · M. Wang · Y. Pan · W. Xu · P. Xue · L. Wang
Science and Technology on Space Physics Laboratory, Beijing, China
e-mail: Hongyangli@buaa.edu.cn

Z. Liu et al. (eds.), *Advances and Challenges in Advanced Unmanned Aerial Systems*, Springer Aerospace Technology, https://doi.org/10.1007/978-981-99-8045-1_31

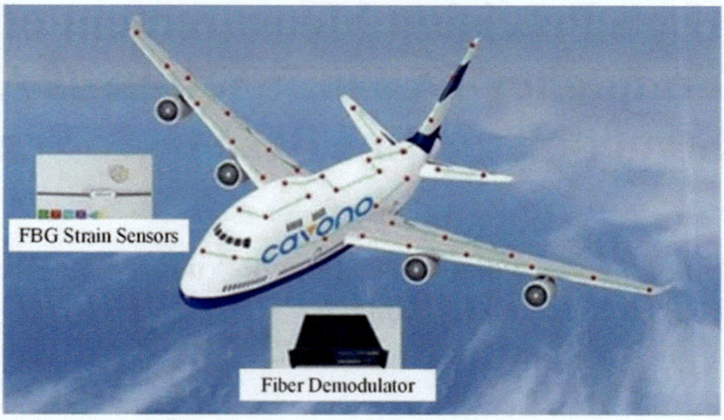

Fig. 1 FBG sensors applied in the aircraft strain measurement

FBG sensor with simple installation process, easy replacement, high measurement accuracy and low testing cost.

The main object of traditional strain testing is static strain [4], but with the continuous development of aerospace manufacturing industry, industrial design departments have rapidly increased their requirements for dynamic performance of materials. The traditional static strain test means has been unable to meet the demand of high speed mechanical structure test. Engineering researchers urgently need to understand the real-time health of materials at high speed running [5]. By grasping the operation data and spectrum information of mechanical system, they analyze the fatigue damage of engineering structure in dynamic strain environment, and provide experimental data support to improve the dynamic performance of the structure [6]. Therefore, dynamic strain testing based on FBG strain sensors has gradually become a hotspot in scientific research field and industrial sectors [7, 8]. Different from the static strain testing, the dynamic strain testing has a very high requirement for the dynamic performance of FBG sensors. The external high frequency excitation is a very severe testing for the FBG sensor in the field of fatigue resistance, installation reliability and high frequency response capability. In addition, due to the dynamic strain process and transfer mechanism is complex, the dynamic strain testing and analysis is usually inaccurate. The measurement characteristics of the FBG sensor usually change with the measured signal frequency, so the current dynamic strain measurement cannot accurately know the strain value of the aerospace structures [9], which greatly limits the application of FBG sensor in aerospace structure dynamic strain testing.

In this paper, the critical technologies of high precision measurement for dynamic strain of aerospace structures with surface-bonded FBG strain sensors are researched by theoretical analysis of optical fiber measurement and dynamic calibration experiments.

2 The Theoretical Analysis of FBG Sensor Measurement

FBG is the most widely used one in the engineering field. During the engraving process, the optical fiber is placed under the laser source, and the intensity of the laser source changes spatially periodically due to the interference, resulting in a permanent periodic modulation of the optical refractive index of the fiber core.

Wideband light sources penetrate the fiber core at a specific angle, transmission along optical fiber. When wideband light reaches the grating, most of the light passes through the grating continue to spread forward, and the light whose wavelength corresponds to the modulation period is reflected back by the grating. The wavelength of reflected light is called the Bragg wavelength λ_B. The relationship between Bragg wavelength λ_B and the period of the modulation cycle is:

$$\lambda_B = 2n_{eff}\Lambda \tag{1}$$

where n_{eff} is the refractive index of the fiber core and Λ is the modulation period of the fiber. Differentiation operations on both sides of Eq. (1):

$$d\lambda_B = 2\Lambda dn_{eff} + 2n_{eff}d\Lambda \tag{2}$$

Therefore, when the fiber grating is subjected to axial forces, both the modulation period and the refractive index will change, which leads to wavelength change of the reflected light. According to the fiber theory, the relationship between Bragg wavelength change and strain of fiber grating is:

$$d\lambda_B = \left\{\frac{n_{eff}^2}{2}[(p_{11} + p_{12})\nu - p_{12}] + 1\right\}\varepsilon \tag{3}$$

p_{11} and p_{12} are the axial and radial elasto-optic coefficient respectively, which describe the change of the fiber core refractive index caused by the deformation in the axial and radius directions of the fiber structure. ν is the Poisson ratio of the fiber core. The effective elastic optical coefficient p_e is set as:

$$p_e = \frac{n_{eff}^2}{2}[(p_{11} + p_{12})\nu - p_{12}] \tag{4}$$

According to the formula (4) and (3), the Bragg wavelength of FBG sensor can be described as:

$$\Delta\lambda_B = (1 - p_e)\lambda_B\varepsilon \tag{5}$$

The effective elastic optical coefficient p_e depends on the material characteristics of the FBG sensor. When the grating of FBG sensors is formed, its effective elastic

Table 1 The sensitivity coefficient of FBG strain sensors with different Bragg wavelengths

Bragg wavelength (nm)	1350	1400	1450	1500	1550	1600
Sensitivity coefficient (pm/με)	1.053	1.092	1.131	1.170	1.209	1.248

optical coefficient and central wavelength are both constant. Thus, a linear relationship between the Bragg wavelength$\Delta\lambda_B$ and the axial strain ε of the optical fiber grating is obtained according to Eq. (5). In this paper, the refractive index of the fiber core n_{eff} is 1.456, and Poisson ratio $v = 0.17$. p_{11} is 0.12, and p_{12} is 0.27. Therefore, p_e is 0.22 according to the formula (4). Thus, the sensitivity coefficient of the FBG strain sensor is related to the Bragg wavelength. The sensitivity coefficient of FBG strain sensors with different Bragg wavelengths are shown in Table 1.

As shown in Table 1, the sensitivity coefficient of FBG strain sensors with different Bragg wavelengths from 1350 to 1600 nm varies in the range of 1.053–1.248 pm/με. As it is varied for the Bragg wavelength, the fiber core material, the engraving process as well as the annealing process, there is a large difference for sensitivity coefficient of different FBG strain sensors. It is necessary to calibrate the sensitivity coefficient of FBG strain sensors before using them in the dynamic strain measurement of aerospace structures.

3 Dynamic Strain Calibration Experiments

A series of rectangular strain beams with different natural frequencies were designed, as shown in Fig. 2. The strain beam adopts a symmetric structure to ensure the symmetric vibration and avoid the influence of transverse force on vibration. The upper surface of the strain beam provides sufficient installation space for the FBG strain sensor. The strain beam is connected to the platform vibrator by the screws, and the pretightening force limits the displacement of the rectangular strain beams relative to platform vibrator. Thus, the vibration of the platform vibrator is applied to the fixed end of the strain beam to realize the excitation of the strain beam.

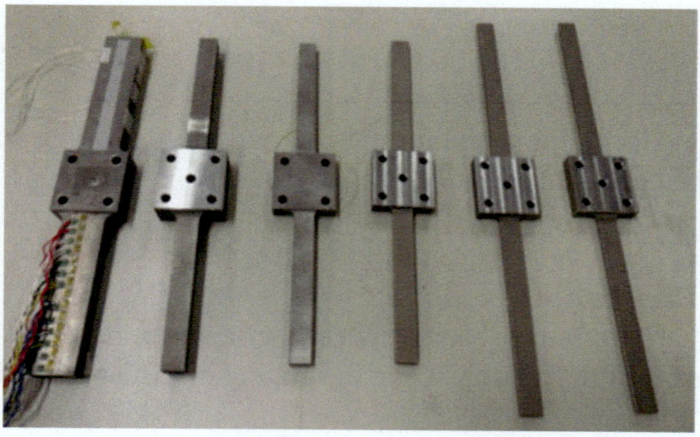

Fig. 2 Rectangular strain beams with different natural frequencies

Dynamic strain calibration experiment was carried out, using strain beams with different natural frequencies. Loctite 1C-LV glue is used to install FBG sensor on different strain beams with the same adhesive layer parameters. The parameters of FBG sensor and adhesive layer are shown in Table 2.

Seven sensors are installed on the lower surface of the strain beam from the root to the end on both sides of the central axis, and the installation position of each sensor is recorded. The central wavelength of the sensor is 1345 nm, and the installed strain beam is placed at room temperature for 20 h to ensure that the adhesive layer is completely solid. The strain beam installed with FBG strain sensors is as shown in Fig. 3.

Reflective glass beads are pasted on the upper surface of the strain beam. As the strain beam is symmetrical, the strain amplitudes on the upper and lower surfaces of the rectangular strain beam are equal. As shown in Fig. 4, install the strain beam on the dynamic strain calibration device with the side of the reflective glass bead facing upward. Use a torque wrench to tighten all positioning screws with the same preload to ensure a reliable connection between the strain beam and the platform vibrator. In the experiment, a SmartScan CAN fiber demodulator was used, and the sampling frequency was set to 25 kHz. The optical fiber connectors of each FBG sensor is connected to demodulation channel of the demodulator, and the demodulator is connected to the calibration data processing system. The demodulator is turn on, which is allowed to warm up for a period of time.

Start the platform vibrator and adjust the excitation frequency to the required calibration frequency to make the strain beam vibrate. Start the laser vibrometer, and

Table 2 The parameters of FBG sensor and adhesive layer

Parameter	Elastic modulus	Poisson's ratio	Geometric size
The Loctite 1C-LV	0.98 Gpa	0.25 ~ 0.35	10 × 2 × 1 mm
Fiber grating	71.10 Gpa	0.25	Φ250 μm

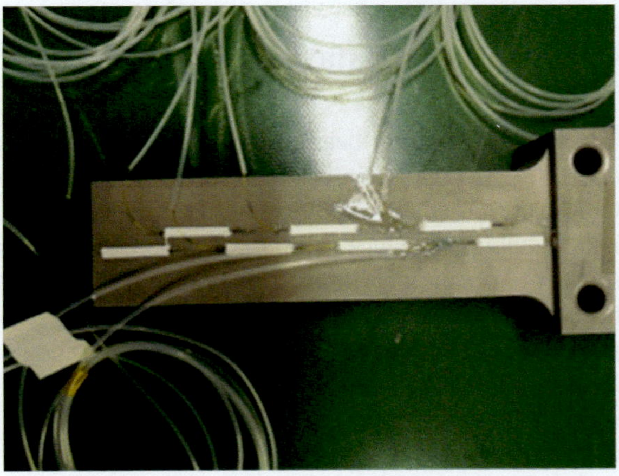

Fig. 3 The strain beam installed with FBG strain sensors

Fig. 4 Dynamic calibration experiment of FBG strain sensor

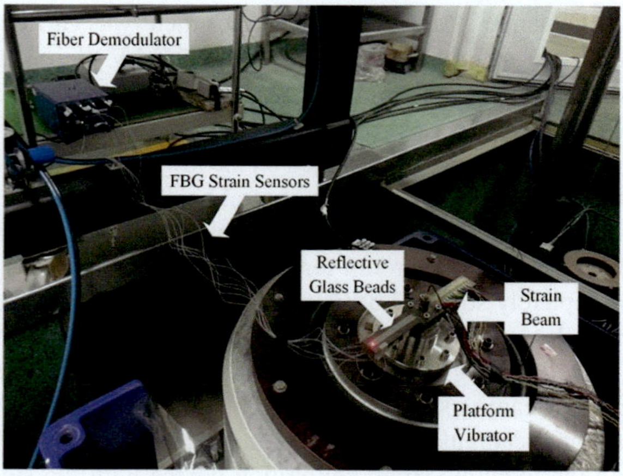

adjust the laser irradiation position on the reflective glass beads at the end of the strain beam. The vibration amplitude of the strain beam was measured, and the strain on the surface of the strain beam can be obtained by the single point traceability method. The strain amplitude on the strain beam can be set by adjusting the output power of the platform vibrator. When the dynamic strain amplitude is constant a period of time latter, it indicates that the strain beam has reached a stable vibration state. Subsequently, the calibration data acquisition system was started. The strain traceability data and the wavelength change of the FBG sensors were recorded. During the experiment, multiple FBG strain sensors were installed at different positions of the strain beam from the root to the end, and the wavelength change data of the FBG sensors at different positions were obtained. The strain traceability data at the installation location of the FBG strain sensor was calculated according to the single point traceability method. Further, the curves obtained by strain traceability and measured by the FBG sensor were comparative analyzed.

4 Experiment Data Analysing

Three types of strain beams were selected with natural frequencies of 500, 800, and 1000 Hz. The strain beams were installed on the vibration table. The vibration table was started, and the excitation acceleration was set. When the strain beam is in a stable resonance state, the dynamic strain was measured using FBG strain sensors to obtain the vibration amplitude of each point on the surface of the strain beam. At the same time, the strain traceability curve obtained according to the single point traceability method is obtained. Figure 5a shows the surface strain distribution curves of the 500 Hz strain beam under acceleration excitation of 10 m/s^2, 35 m/s^2, and 100 m/s^2, respectively; Fig. 5b shows the surface strain distribution curves of an 800 Hz strain beam under acceleration excitation of 10 m/s^2, 35 m/s^2, and

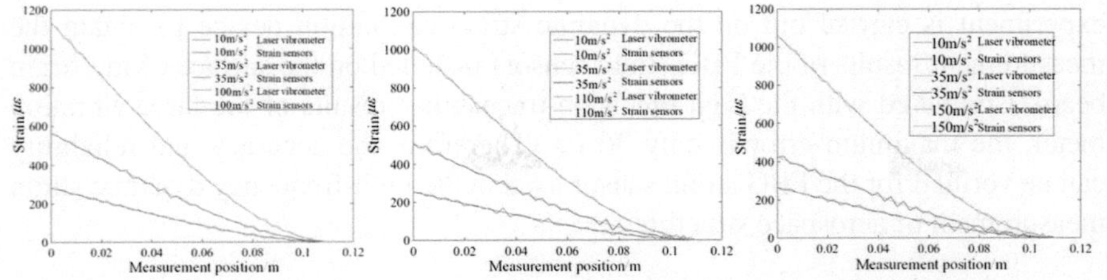

(a) Strain on 500 Hz beam (b) Strain on 800 Hz beam (c) Strain on 1000 Hz beam

Fig. 5 Comparison between measurement results of FBG strain sensors and strain traceability results by laser vibrometer

110 m/s^2, respectively; Fig. 5c shows the surface strain distribution curves of a 1000 Hz strain beam under acceleration excitation of 10 m/s^2, 35 m/s^2, and 150 m/s^2, respectively. The experiment results shows that the strain at the root of the strain beam is the largest, and gradually decreases from the root to the free end of the strain beam. The strain at the free end is approximately 0. For a strain beam, the strain at each point increases as the excitation acceleration increases. For strain beams with natural frequencies of 500, 800, and 1000 Hz, the strain at the root reaches 1000 when the excitation acceleration on the vibration table is 100 m/s^2, 110 m/s^2, and 150 m/s^2, respectively. In addition, the measurement results of the FBG strain sensors are basically consistent with the strain traceability results obtained by the single point traceability method. The stability of strain data obtained by the single point traceability method is significantly higher than that measured by the FBG strain sensor. This is mainly due to the fact that the error of each measurement point of the FBG strain sensor will be introduced into the interpolation results of the strain measurement curve. However, the single point traceability method is based on the data from one single measurement point, which only introduces the measurement error of that point. In experiments, the point near the root of the strain beam is usually taken as the measurement point of FBG sensors, as the strain value of that point is large and the relative error is small.

Compare the difference between the dynamic strain measurement results with the FBG strain sensors and the results obtained by the single point traceability method with the laser vibrometer, and the measurement error of the FBG strain sensor is obtained. The measurement error is randomly distributed, and the maximum error is only 36 $\mu\varepsilon$, which satisfied the high frequency dynamic strain measurement requirement of aerospace structures.

5 Conclusion

In this paper, the measurement theory and traceability method of high frequency dynamic strain of aerospace structures using FBG strain sensors is analyzed, and the dynamic measurement mechanism is obtained for FBG strain sensors. Traceability experiment is carried out on the dynamic strain calibration device to obtain the measurement results of the FBG strain sensors installed on the surface of the strain beam. Compared with the high-precision traceability results of the laser vibration meter, the maximum error is only 36 $\mu\varepsilon$. Therefore, the accuracy and reliability can be verified for the FBG strain sensor used in the high-frequency dynamic strain measurement of aerospace structures.

References

1. Yi J, et al (2012) Spatial shape reconstruction using orthogonal fiber Bragg grating sensor array. Mechatronics 22(6): 679–687
2. Roveri N et al (2016) Carcaterra A. OPTYRE a new technology for tire monitoring: evidence of contact patch phenomena. Mech Syst Signal Process 66–67:793–810
3. Sun G et al (2018) 3D shape sensing of flexible morphing wing using fiber Bragg grating sensing method. Optik – Int J Light Electron Optics 156:83–92
4. Umesh S, et al (2016) Fiber Bragg grating based bite force measurement. J Biomech 49(13): 2877–2881
5. Kang LH, et al (2007) Estimation of dynamic structural displacements using fiber Bragg grating strain sensors. J Sound Vibr 305(3): 534–542
6. Songa X et al (2018) Dynamic input estimation and shape sensing for a nonlinear beam based on distributed fiber bragg grating sensor network. Optik 156:896–905
7. Billon A, et al (2015). Qualification of a distributed optical fiber sensor bonded to the surface of a concrete structure: a methodology to obtain quantitative strain measurements. Smart Mater Struct 24(11): 115001
8. Im J, et al (2014) Aluminum-thin-film packaged fiber Bragg grating probes for monitoring the maximum tensile strain of composite materials. Appl Optics 53(17): 3615–3620
9. Panciroli R et al (2016) Live monitoring of the distributed strain field in impulsive events through fiber Bragg gratings. J Fluids Struct 61:60–75

ADRC Flight Control System Design for Flying Wing UAV with High Aspect Ratio

Shi Jiayu, Li Yuchao, and Ma Fei

Abstract The flying wing unmanned aerial vehicle (UAV) with high aspect ratio is taken as the research object in this study. Based on the theory of auto-disturbance-rejection control (ADRC) and combined with the characteristics of multi-rudder configuration of flying wing layout, the attitude controller of flying wing UAV is designed. The results show that the flight controller has good control performance, which can meet the performance requirements, and can effectively deal with the coupling effects between lateral, directional and longitudinal of the flying wing UAV. Compared with the conventional flight control methods, the controller designed in this paper has fewer parameters to be set, and a set of control parameters can be used in the whole flight envelope. The process is relatively simple. It provides a reference for further engineering applications.

Keywords Flying wing · ADRC · Flight envelope · Coupling effects

1 Introduction

The high aspect ratio flying wing UAV removes the tail, it has a very wide and short fuselage, and the fuselage is fused with the wings [1]. Therefore, it has the advantages of high lift-drag ratio, outstanding stealth performance and large available space of the fuselage. Now it has become the mainstream layout of high-altitude long-endurance UAV, but there are some shortcomings in the flying wing UAV itself. As a controlled object, it has characteristics of strong nonlinear, multi-variable, time-varying and coupling, so designing its flight control system is particularly difficult [2].

At present, the design of UAV flight control law in engineering practice is mainly achieved through gain scheduling for multiple flight state points. This traditional method has designed many effective and reliable control laws in the past, but there are many shortcomings: the controller is designed based on the small perturbation linearization mathematical model of equilibrium state points, which in essence

S. Jiayu (✉) · L. Yuchao · M. Fei
Chinese Flight Test Establishment, Xi'an, China
e-mail: 616575093@qq.com

Z. Liu et al. (eds.), *Advances and Challenges in Advanced Unmanned Aerial Systems*, Springer Aerospace Technology, https://doi.org/10.1007/978-981-99-8045-1_32

only has the local stability in the equilibrium point neighborhood, but cannot guarantee the global stability in the whole envelope. Due to the limited robustness of each state point, the traditional method also needs to select as many state points as possible within the flight envelope for repeated design, which increases the complexity and workload of controller design. The flight envelope of UAV is limited to the linear range of aerodynamic parameters and the potential performance of UAV cannot be fully utilized. This controller can only control the single input single output system (SISO), but cannot do anything for the multiple input multiple output system (MIMO). Since there is no ability to estimate the disturbance, when the system is subjected to constant disturbance, the speed and effect of the controller's dynamic response are greatly affected. Therefore, this method becomes inadequate when facing the flying wing UAV, and more advanced control methods need to be introduced.

2 Basic Technology of ADRC Flight Control

2.1 Six Degrees of Freedom Mathematical Model of Flying Wing UAV

The flying wing UAV with high aspect ratio is a very complex nonlinear dynamic system, so it is necessary to have an precise model before control system designing, which will directly affect the reliability of the flying wing UAV flight simulation results. The rudder configuration of the flying wing UAV is shown in Fig. 1.

According to the conversion relationship between commonly used coordinate systems of aircraft [3], based on kinematics and dynamics model theories, the rigid-body six-degree of freedom model [4] of flying wing UAV is obtained through derivation, as shown in formulas 1–12.

$$\frac{d\alpha}{dt} = -p\cos\alpha\tan\beta + q - r\sin\alpha\tan\beta + \frac{g\cos\mu\cos\gamma}{V\cos\beta} \tag{1}$$

$$\frac{d\beta}{dt} = p\sin\alpha - r\cos\alpha + \frac{g\sin\mu\cos\gamma}{V} - \frac{T\cos\alpha\sin\beta}{MV} + \frac{Y\cos\beta}{MV} \tag{2}$$

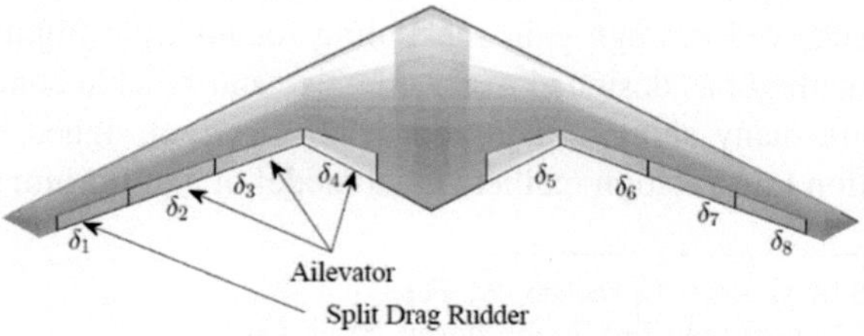

Fig. 1 Flying wing UAV rudder configuration

$$\frac{d\mu}{dt} = p\frac{\cos\alpha}{\cos\beta} + r\frac{\sin\alpha}{\cos\beta} - \frac{g\cos\gamma\cos\mu\tan\beta}{V} + \frac{L}{MV}(\sin\mu\tan\gamma + \tan\beta) + \frac{Y}{MV}\tan\gamma\cos\beta\cos\mu + \frac{T\sin\alpha}{MV}(\sin\mu\tan\gamma + \tan\beta) - \frac{T\cos\alpha}{MV}(\cos\mu\tan\gamma\sin\beta) \quad (3)$$

$$\frac{dp}{dt} = \frac{I_{xz}(I_{xx} - I_{yy} + I_{zz})pq + [I_{zz}(I_{yy} - I_{zz}) - I_{xz}^2]qr}{I_{xx}I_{zz} - I_{xz}^2} + \frac{I_{zz}L + I_{xz}N}{I_{xx}I_{zz} - I_{xz}^2} \quad (4)$$

$$\frac{dq}{dt} = \frac{1}{I_{yy}}[M + (I_{zz} - I_{xx})pr + I_{xz}(r^2 - p^2)] \quad (5)$$

$$\frac{dr}{dt} = \frac{[I_{xx}(I_{xx} - I_{yy}) + I_{xz}^2]pq - I_{xz}(I_{xx} - I_{yy} + I_{zz})qr}{I_{xx}I_{zz} - I_{xz}^2} + \frac{I_{xz}L + I_{xx}N}{I_{xx}I_{zz} - I_{xz}^2} \quad (6)$$

$$\frac{dV}{dt} = \frac{1}{m}(T\cos\alpha\cos\beta - D - mg\sin\gamma) \quad (7)$$

$$\frac{d\chi}{dt} = \frac{T}{mV\cos\gamma}[\sin\alpha\sin\mu - \cos\alpha\sin\beta\cos\mu] + \frac{C}{mV\cos\gamma}\cos\mu + \frac{L}{mV\cos\gamma}\sin\mu \quad (8)$$

$$\frac{d\gamma}{dt} = -\frac{T}{mV}[-\sin\alpha\cos\mu - \cos\alpha\sin\beta\sin\mu] - \frac{C}{mV}\sin\mu + \frac{L}{mV}\cos\mu - \frac{g}{V}\cos\gamma \quad (9)$$

$$\frac{dx_g}{dt} = V\cos\gamma\cos\chi \quad (10)$$

$$\frac{dy_g}{dt} = V\cos\gamma\sin\chi \quad (11)$$

$$\frac{dz_g}{dt} = -V\sin\gamma \quad (12)$$

2.2 *Structure of ADRC Controller*

In view of the shortcomings of traditional controllers, ADRC is a new practical technology which has successfully absorbed the achievements of modern control, enriched and developed the mature control strategy of PID controller based on error

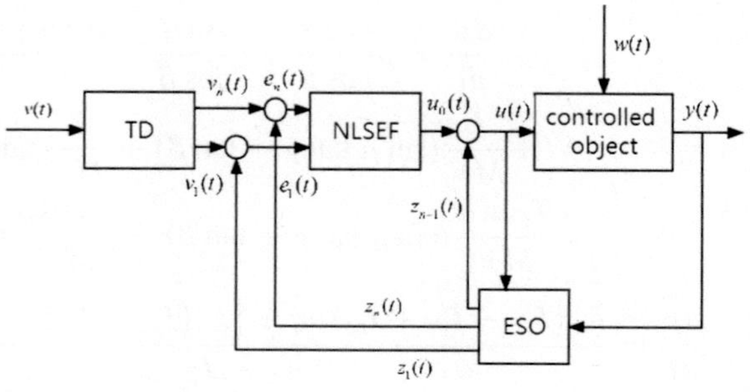

Fig. 2 ADRC controller structure

to eliminate error, and successfully applied the nonlinear effect to get good control effect. The problem of low control precision of PID controller is solved. The ADRC does not rely on the mathematical model, but treats the nonlinear uncertainties of the system as disturbances for observation, estimation and compensation. Then, the nonlinear error feedback is used to transform the design of the nonlinear control system into the linear control system design, which blurs the boundary between linear system and nonlinear system. It directly targets at the full nonlinear model of UAV, and designs a globally stable controller with strong robustness, avoiding the design process of traditional gain scheduling.

The classical ADRC system consists of three parts: nonlinear tracker (TD), extended state observer (ESO) and nonlinear error feedback control law (NLSEF), as shown in Fig. 2. The tracking differentiator limits the change rate of the input signal according to a certain law, so that it changes smoothly and continuously without a wide range of pulsation. The nonlinear state feedback is to convert the feedback information of the state observer and the output information of the tracking differentiator into a simple series system, and finally obtain a modular nonlinear state error feedback model. ESO is the key of ADRC controller. Its function is to treat the imprecise modeling part of the system model and the external interference as the system interference to observe, estimate and compensate, which can make the controlled object linearized.

3 Implementation of ADRC Flight Control

3.1 Time Scale Separation of State Variables

From the previous passage, the flying wing UAV model used in this study contains 5 input variables and 12 output variables. Obviously, the number of outputs is greater than the number of inputs, that is, the system itself cannot be decoupled. To solve this problem, ADRC control theory is combined with singular perturbation theory (i.e., time scale separation of state variables [5]). These state variables were divided into different parts [6, 7] according to different response speeds, and the approximate

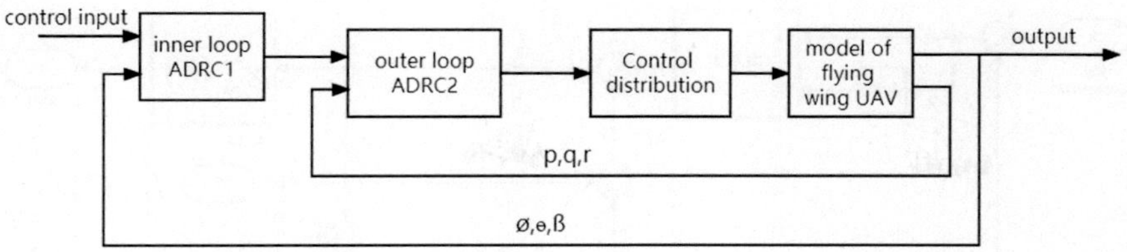

Fig. 3 Flying wing UAV ADRC attitude controller

solution of the system was obtained. When the control law of the faster loop is considered, the state variable of the slower loop is treated as constant [8].

3.2 Algorithm of Attitude Control

The ADRC attitude controller consists of two loops: the inner loop is the fast loop, which is used to control the angular rates (p, q, r) of the flying wing UAV (input: angular rate; output: torque); the outer loop is the slow loop, which is used to control the attitude angles (ϕ, θ, β) of the flying wing UAV [9] (input: attitude angle; output: angular velocity). The structure diagram of flying wing UAV ADRC attitude controller in this study is shown in Fig. 3.

1. Design of inner loop control law (ADRC2)

For the inner loop design [10, 11], the differential equation of state variables x_1 can be written as formula (13):

$$\begin{bmatrix} \dot{p} \\ \dot{q} \\ \dot{r} \end{bmatrix} = \begin{bmatrix} f_p(\overline{x}) \\ f_q(\overline{x}) \\ f_r(\overline{x}) \end{bmatrix} + g_{pqr}(\overline{x}) \begin{bmatrix} l \\ m \\ n \end{bmatrix} \tag{13}$$

$$\begin{bmatrix} f_p(\overline{x}) \\ f_q(\overline{x}) \\ f_r(\overline{x}) \end{bmatrix} = f_{pqr,1}(\overline{x}) + f_{pqr,2}(\overline{x}) \tag{14}$$

$f_{pqr,1}(\overline{x})$ is the known item model, as shown in formula (15), and $f_{pqr,2}(\overline{x})$ is the sum of the unmodeled part of the model and the external interference, see formula(16), which is the namely sum interference. They are derived from the flying wing UAV system model.

$$f_{pqr,1}(\overline{x}) = \begin{bmatrix} \frac{I_{xz}(I_{xx}-I_{yy}+I_{zz})pq+[I_{zz}(I_{yy}-I_{zz})-I_{xz}^2]qr}{I_{xx}I_{zz}-I_{xz}^2} \\ \frac{1}{I_{yy}}\left[(I_{zz}-I_{xx})pr+I_{xz}\left(r^2-p^2\right)\right] \\ \frac{[I_{xx}(I_{xx}-I_{yy})+I_{xz}^2]pq-I_{xz}(I_{xx}-I_{yy}+I_{zz})qr}{I_{xx}I_{zz}-I_{xz}^2} \end{bmatrix} \tag{15}$$

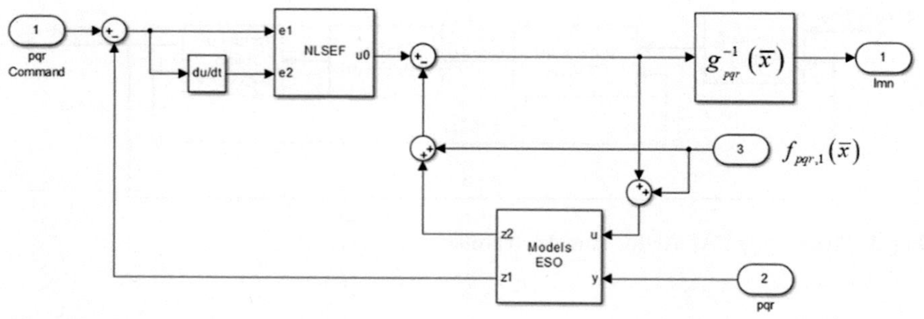

Fig. 4 Structure of inner loop ADRC2

$$g_{pqr}(\overline{x}) = \begin{bmatrix} \frac{I_{zz}}{I_{xx}I_{zz}-I_{xz}^2} & 0 & \frac{I_{xz}}{I_{xx}I_{zz}-I_{xz}^2} \\ 0 & \frac{1}{I_{yy}} & 0 \\ \frac{I_{xz}}{I_{xx}I_{zz}-I_{xz}^2} & 0 & \frac{I_{xx}}{I_{xx}I_{zz}-I_{xz}^2} \end{bmatrix} \tag{16}$$

The design of fast variable controller (ADRC2) is shown in Fig. 4.

The NLSEF in this controller adopts a structure containing first-order differential terms, which has good control effect on the model containing rudder-module. The ESO uses the most basic second-order structure, in which the control commands are p_c, q_c, r_c, and the output variables are l, m, n. Some core design parameters are shown in Table 1.

2. Design of outer loop control law (ADRC1)

Similar to the design of the inner loop control law, the differential equation of state variables x_2 can be written as formula (17):

$$\begin{bmatrix} \dot{\phi} \\ \dot{\theta} \\ \dot{\beta} \end{bmatrix} = \begin{bmatrix} f_\phi(\overline{x}) \\ f_\theta(\overline{x}) \\ f_\beta(\overline{x}) \end{bmatrix} + g_{\phi\theta\beta}(\overline{x}) \begin{bmatrix} p \\ q \\ r \end{bmatrix} \tag{17}$$

Table 1 NLSEF parameters in the fast loop controller

NLSEF	Parameter values of each loop		
	p	q	r
β_1	3	4	2
β_2	2	2	1
δ	1	1	1
α_1	0.5	0.5	0.5
α_2	1.25	1.25	1.25

$$\begin{bmatrix} f_\phi(\overline{x}) \\ f_\theta(\overline{x}) \\ f_\beta(\overline{x}) \end{bmatrix} = f_{\phi\theta\beta,1}(\overline{x}) + f_{\phi\theta\beta,2}(\overline{x}) \tag{18}$$

$$f_{\phi\theta\beta,1}(\overline{x}) = \begin{bmatrix} 0 \\ 0 \\ \frac{g\sin\mu\cos\gamma}{V} + \frac{Y\cos\beta}{V} - \frac{T\cos\alpha\sin\beta}{MV} \end{bmatrix} \tag{19}$$

$$g_{\phi\theta\beta}(\overline{x}) = \begin{bmatrix} 1 & \sin\phi\tan\theta & \cos\phi\tan\theta \\ 0 & \cos\phi & -\sin\phi \\ \sin\alpha & 0 & -\cos\alpha \end{bmatrix} \tag{20}$$

The outer loop controller (ADRC1) calculates the attitude angular rates (p, q, r) with the attitude angles (ϕ, θ, β). The output variables of the outer loop controller are the control commands of the inner loop controller (i.e. ADRC2). The outer loop controller adopts the principle of time scale separation, which can ignore the dynamic characteristics of the fast variables. It uses NLSEF without differential module and second order ESO, and the structure of the outer loop controller is similar to that of the inner loop controller.

4 Test and Verification

In order to demonstrate the control capability of the established ADRC controller of flying wing UAV, a digital simulation platform is used to verify whether the ADRC controller can make the controlled object meet the performance requirements. The simulation step is set as h = 0.001 s, and the tracking and holding verification of pitch angle, roll angle, yaw angle and height were carried out at different state points. The simulation results are shown in Fig. 5.

It can be seen from the above digital test results that under different requirements, the ADRC attitude controller can always make the flying wing UAV accurately track the commands within a certain period of time, the response curves are very smooth. The whole response process of pitch Angle, sideslip Angle and height had no overshoot and no oscillation. The response of roll angle was slightly overshoot. By calculation, the steady-state error of this point did not exceed 0.5°, and the adjustment time did not exceed 8 s, which met the evaluation conditions. During the whole test, the controller operated smoothly. In this process, the controller runs smoothly and has good dynamic response characteristics. Therefore, the nonlinear flight controller of the flying wing UAV designed based on the ADRC control theory has good control performance, and can better deal with the nonlinear characteristics of the longitudinal and lateral-directional coupling of the flying wing UAV, and meets the requirements of flight performance. Theoretically, the controller can make the flying wing UAV satisfy the performance requirements in longitudinal, directional, lateral and altitude control.

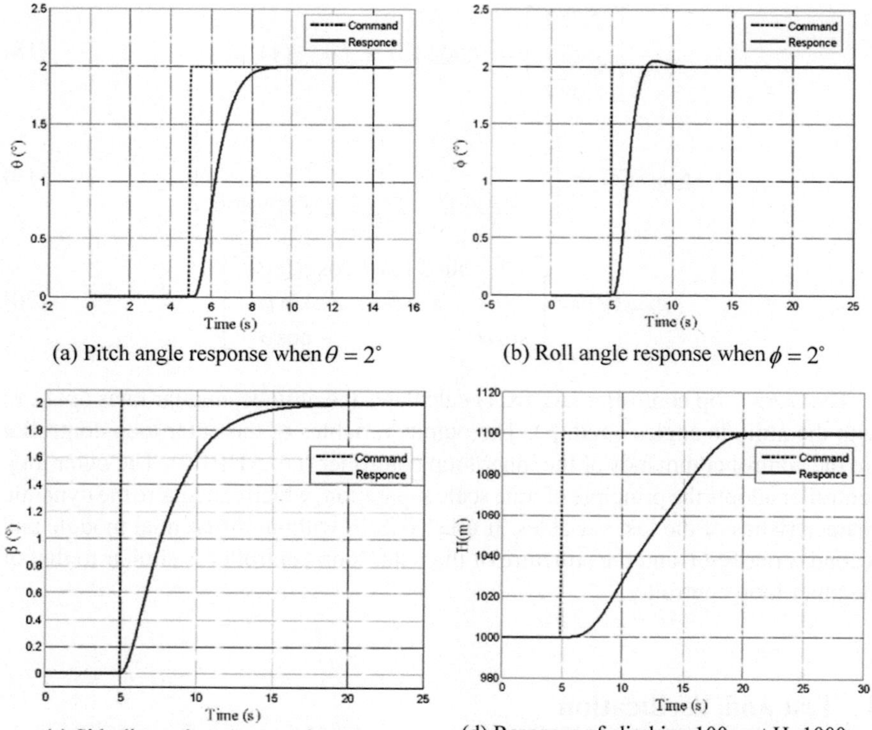

(a) Pitch angle response when $\theta = 2^{\circ}$

(b) Roll angle response when $\phi = 2^{\circ}$

(c) Sideslip angle response when $\beta = 2^{\circ}$

(d) Response of climbing 100m at H=1000m

Fig. 5 Digital simulation results at different state points

5 Conclusion

In this study, a rigid-body six-degree-of-freedom model of flying wing UAV with high aspect ratio is established. Based on the ADRC control theory, the nonlinear flight control system of flying wing UAV is designed. The tracking and holding tests of pitch angle, roll angle, yaw angle and altitude were carried out at different state points, the following conclusions are drawn: The designed controller can make the flying wing UAV meet the requirements of flight control performance in longitudinal, directional, lateral and altitude control, and has good dynamic response characteristics and good tracking accuracy. The controller has the advantages of simple structure, fewer parameters to be set and can be used in the whole envelope, which is suitable for application in engineering practice.

References

1. Donghong Z (2018) Research on automatic landing technology of UAV with high aspect ratio. Nanjing University of Aeronautics and Astronautics, Nanjing, pp 68–93
2. Yi Y, Li Z (2019) Unmanned aerial vehicle and future combat. National Defense 36–39
3. Zhang T, Li Q, Zhang C (2018) Development trend of intelligent unmanned autonomous systems. Unmanned Syst Technol (01): 11–22
4. Zhenping F (2005) Flight dynamics of aircraft. Beihang University Press, Beijing, pp 20–22
5. Wei S, Yanzhao Y (2020) Aerodynamic modeling and flight simulation of Tailseat UAV with flying wing layout. Comput Simulation 7:55–60
6. Rogalski T, Nowak D (2019) Control system for aircraft take-off and landing based on modified PID controllers. MATEC Web Conf (12)
7. Guerrero-Sanchez ME, Lozano R (2021) Nonlinear control strategies for a UAV carrying a load with swing attenuation. Appl Math Model (Mara): 709–722
8. Deslich J, Flick P, Meckstroth CM (2021) Evaluating the effectiveness of compliant leading edge control surfaces on an oblique flying wing for directional control. AIAA SciTech Forum and Exposition, pp 1364–1379.
9. Jingqing H (2009) Active disturbance rejection control technology. National Defense Industry Press, Beijing, pp 280–281
10. Rong Z, Wenjie H, Wen T (2018) Applicability and tuning of linear active disturbance rejection control. Control Theory Appl 11:1654–1660
11. Zuo L, Zhang X, Li Z et al (2023) Design of UAV control law based on active disturbance rejection method. J Beijing Univ Aeronautics Astronsutics 34–36

A Prestrained Bistable Composite Gridded Structure

Chenmin Zhao, Bing Wang, Xiayu Chen, Xinyu Lin, Chenglong Guan, and Shuncong Zhong

Abstract Bistable composite structures have attracted growing interest in morphing applications to aerospace industry. Here, we device a novel prestrained bistable composite gridded structure, consisting pairs of prestrained composite strips on both sides and oriented in 90°. This is achieved by employing the elastic fibre prestressing (EFP) technique, where a plain-weave carbon prepreg was stretched in two directions at a constant strain level, and the tensile strain was maintained throughout the curing process to produce a prestrained composite trip. Upon load removal, recovery from the prestrained carbon fibres generates compressive stresses and interacts with thermal residual stresses, which in turn changing the in-plane stress level within a composite structure and induce out-of-plane deflections. Therefore, the bistability is generated from the pairs of the oriented prestrained composite strips, their deflections give opposite cylindrical configurations to the mid-plane. Here, we presented further details on the biaxial fibre straining rig; samples were produced with different prestrain levels, in order to reveal the underlying mechanisms from the fibre prestraining. These results provide valuable insights for the design of aerospace deployable structures.

Keywords Prestrain · Bistable · Composite · Gridded structure

1 Introduction

Morphing structures play an important role in development of aerospace engineering, especially for shape-changing or deployable structures. Bistable composite structures have attracted growing interest in morphing mechanical design owning to their superior performance in reducing structural weight, and complexities [1]. Since they

C. Zhao · B. Wang (✉) · X. Chen · X. Lin · C. Guan · S. Zhong
Fujian Provincial Key Laboratory of Terahertz Functional Devices and Intelligent Sensing, School of Mechanical Engineering and Automation, Fuzhou University, Fuzhou 350108, P.R. China
e-mail: b.wang@fzu.edu.cn

Institute of Precision Instrument and Intelligent Measurement and Control, Fuzhou University, Fuzhou 350108, P.R. China

Z. Liu et al. (eds.), *Advances and Challenges in Advanced Unmanned Aerial Systems*, Springer Aerospace Technology, https://doi.org/10.1007/978-981-99-8045-1_33

are bistable, a continuous energy input is not essential during the shape-changing [2]. Bistable composite technology is a proven technique, structures are generally characterised into two categories based on the curvature directions of their stable shapes, namely (i) opposite-sense bistable composite structures[3], and (ii) equal-sense bistable composite structures [4]. For (i), the curvatures of both the stable shapes are in opposite direction, i.e., the Gaussian curvature is negative; the design principle is mainly on introducing unsymmetric internal stress levels in order to induce out-of-plane deflections. Whilst for (ii), the Gaussian curvature is positive; the design principle is mainly on employing the geometric curvature effects, also known as bistable composite tape-spring structure [5].

The opposite-sense bistable composite structures are usually produced by exploiting the thermal residual stresses, where unsymmetric internal stress levels are achieved through either unsymmetric layup designs, or by employing external stresses to alter the in-plane stress levels. Plain-weave carbon composites show well-balanced properties and widely applied in aerospace industries. For their plain-weave nature, carbon fibre yarns are curled and waved in both orthogonal directions [6]. The curling and interlocking of the yarns restrict the shrinkage of the substrate [7], making their plain-weave laminates free from unsymmetric in-plane stresses, and thereby restricting the layup design freedom on bistability. Plain-weave Fibre prestressing technique is effective in adjusting the internal stress levels within a composite. Commonly applied fibre prestressing methods include elastic fibre prestressing (EFP) [3] and viscoelastic fibre prestressing (VFP) [8].

Since elastic fibre-based woven composites have become the mainstream and widely used in industries, here we investigated the feasibilities on applying EFP to carbon woven composite, in order to produce bistable composite structures. A bespoke biaxial fibre straining rig was developed to apply a constant strain level on both directions of the plain-weave carbon prepreg fibre; gridded composite structures were then manufactured with different prestrain levels, and subsequently tested to reveal further insight into the underlying mechanisms. The developed bistable gridded structure serves as a basic shape-changing unit, which can be beneficial for designs and applications to morphing wings, as well as adaptive structures.

2 Experimental

Table 1 shows the plain-weave carbon composite laminate, supplied by Weihai Guangwei Composites Co., Ltd., China. The curing temperature used is 120 °C and the curing time is 100 min.

Table 1 Mechanical properties of plain-weave carbon composite laminate

E_{11} (GPa)	E_{22} (GPa)	G_{12} (GPa)	ν_{12}	t (mm)
52	52	3.5	0.075	0.25

A bespoke biaxial fibre straining rig was developed to apply different strain levels on both directions of the plain-weave carbon prepreg, see Fig. 1a. It consisted of two fixtures, an electronic-controlled heating plate (300 × 300 × 40 mm in dimension), and two sets of straining modules. Each of the straining module composed of an electric linear actuator, a load cell, a set of prepreg fixture, and an electric linear actuator to ensure stable straining speed. The fibre straining process was monitored by using a force transducer with a range of 0–500 kg and an accuracy of ±0.03%. The heating plate was made of aluminum alloy with good heat efficiency, ensuring uniform heat transfer during the curing process. A digital temperature control box was applied to provide real-time feedback on temperature.

To produce a bistable composite gridded structure, four carbon prepreg strips were cut into designated sizes and clamped on the biaxial fibre straining rig as in Fig. 1a. A designated strain was applied on both fibre directions of the uncured prepreg through aforementioned straining modules. The whole setup was then subjected to heat treatment, where temperature was heated at a constant speed, and then kept at 120 °C for 100 min. The fully cured structure was naturally cooled to room temperature, the tensile strain was then removed to produce the bistable gridded structural sample. Figure 1b shows the two stable configurations of a prestrained bistable composite gridded structure, both of the stable curvatures were monitored to reveal prestraining effects.

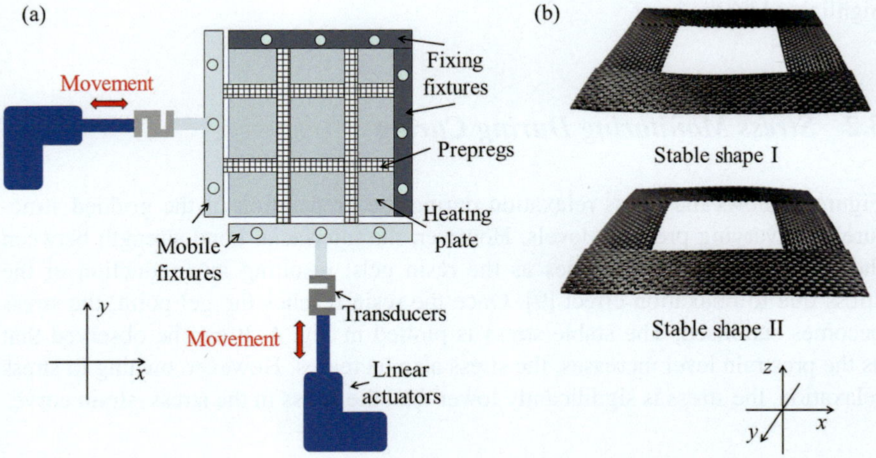

Fig. 1 **a** Schematic of the bespoke biaxial fibre straining rig; **b** stable shapes of the manufactured prestrained bistable gridded structure

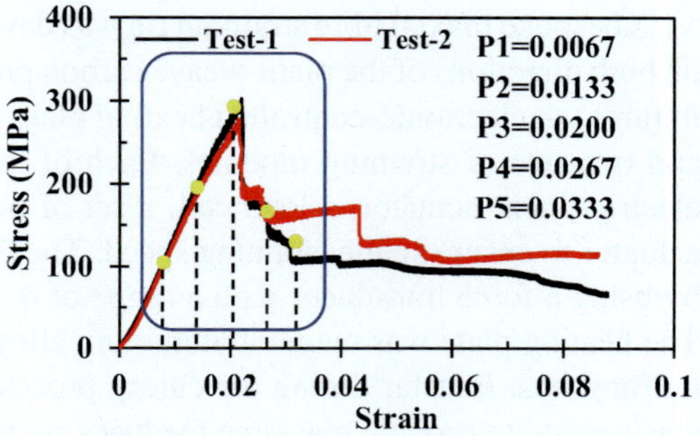

Fig. 2 Stress–strain curve of a plain-weave carbon prepreg in tension

3 Results and Discussion

3.1 *Elastic Fibre Prestraining Level*

To produce a prestrained bistable composite gridded structure, range of the applied prestrain level needs to be controlled and determined first. Tensile tests were conducted on plain-weave carbon fibre prepreg. The results are showed in Fig. 2, giving a mean ultimate stress value of 290.4 ± 17.5 MPa, and ultimate strain value of 0.0218 ± 0.0003. Based on these results, the appropriate level of prestrain is highlighted by the inset.

3.2 *Stress Monitoring During Curing*

Figure 3 shows the stress relaxation during the preparation of the gridded structure with varying prestrain levels. However, the interfacial bond strength between the fibres and matrix increases as the resin gels, resulting into reduction of the stress due to relaxation effect [9]. Once the resin reaches the gel point, the stress becomes stabilised. The stable stress is plotted in Fig. 4. It can be observed that as the prestrain level increases, the stress also increases. However, owning to stress relaxation, the stress is significantly lower than the stress in the stress–strain curve.

3.3 *Effect of Prestrain on the Curvature of a Gridded Structure*

The curvature of the prestrained gridded structure in the x and y directions is shown in Fig. 5. The figure demonstrates that as the prestrain increases, the curvature in both the x and y directions increases, and subsequently decreases. The decrease in

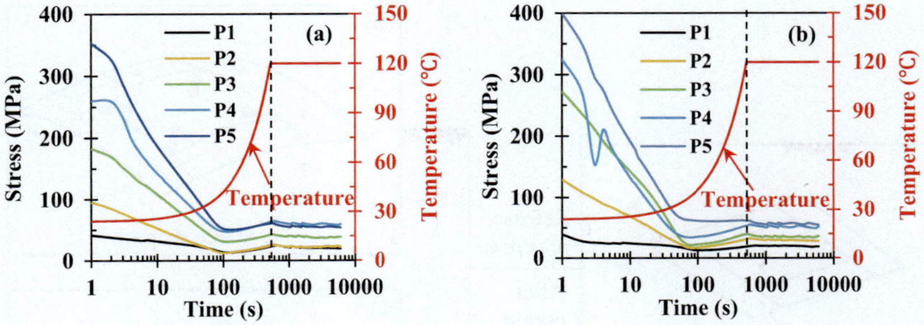

Fig. 3 Monitoring of stresses during curing of prestrained gridded structures, **a** *x*-directional stress; **b** *y*-directional stress

Fig. 4 Stable temperature during curing of prestrain gridded structure

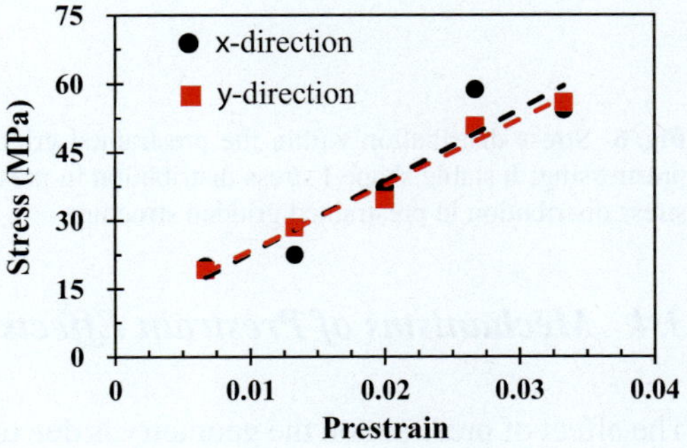

curvature may be attributed to an uneven distribution of in-plane stresses, possibly due to some fibre damage caused by prestrain exceeding the ultimate strain.

Fig. 5 Effect of prestrain on the curvature of the gridded structure

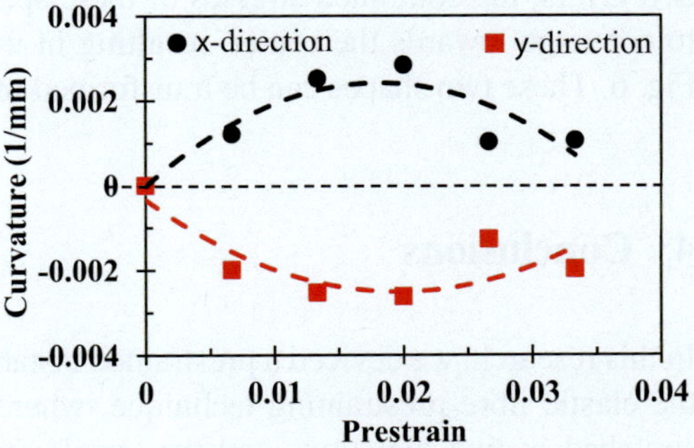

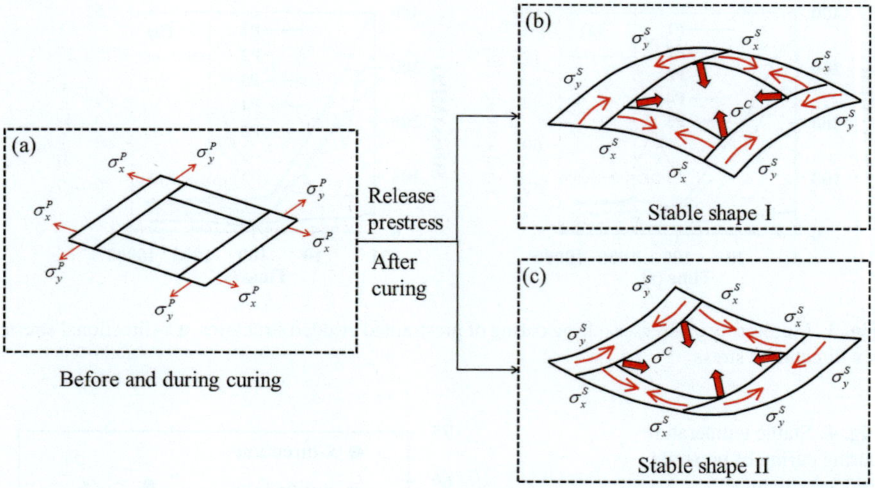

Fig. 6 Stress distribution within the prestrained gridded structures. **a** stress distribution during prestressing; **b** stable shape I stress distribution in prestrained gridded structure; **c** stable shape II stress distribution in prestrained gridded structure

3.4 Mechanisms of Prestrain Effects on Gridded Structure

The effect of prestrain on the geometry is due to the release of prestrain after curing, which causes spring-back strains in the opposite direction of the prestrain. These stresses interact with the thermal residual stresses in the gridded structure and regulate the in-plane stresses of the grided structure. When the spring-back stresses more than the thermal residual tensile stresses, compressive stresses are induced in the interior of the structure. Moreover, since the corners of the gridded structure are cured in two layers, the combined stresses of these spring-back stresses cause the structure to converge towards the center, resulting in the geometric configuration shown in Fig. 6. These two shapes can be transformed into each other.

4 Conclusions

In this research, we deviced a prestrained bistable composite gridded structure using the elastic fibre prestraining technique, where a plain-weave carbon prepreg was stretched in two directions and the tensile strain was maintained throughout the curing process. Upon prestrain removal, recovery from the prestrained carbon fibres generates compressive stresses and interacts with thermal residual stresses, which in turn changing the in-plane stress level within a composite structure. The main findings include:

(1) During the preparation of the prestrained gridded structure, significant stress relaxation was observed in the plain-weave carbon fibre composite. However, this effect was reduced after the onset of curing, which led to increased strength of the interfacial bond between the fibres and the matrix, and ultimately stabilising the internal stress.

(2) As the prestrain level increases, the curvature of the gridded structure also increases, it is observed there is an optimal level to maximise the bistability.

References

1. Emam SA, Inman DJ (2015) A review on bistable composite laminates for morphing and energy harvesting. Appl Mech Rev 67:1–15
2. Li H, Dai F, Du S (2012) Numerical and experimental study on morphing bi-stable composite laminates actuated by a heating method. Compos Sci Technol 72:1767–1773
3. Daynes S, Diaconu CG, Potter KD, Weaver PM (2010) Bistable prestressed symmetric laminates. J Compos Mater 44:1119–1137
4. Zhang Z, Li Y, Yu X, Li X, Wu H, Wu H, Jiang S, Chai G (2019) Bistable morphing composite structures: a review. Thin-Walled Struct. 142:74–97
5. Daton-Lovett AJ (1996) An extendible member
6. Naik NK, Tiwari SI, Kumar RS (2003) An analytical model for compressive strength of plain weave fabric composites. Compos Sci Technol 63:609–625
7. Xiong JJ, Shenoi RA, Gao J (2009) An analytical model to predict residual thermal stress in 2D orthogonal plain weave fabric composites. Int J Solids Struct 46:1872–1883
8. Fancey KS (2000) Investigation into the feasibility of viscoelastically generated pre-stress in polymeric matrix composites. Mater Sci Eng A 279:36–41
9. Hu H, Li S, Wang J, Zu L, Cao D, Zhong Y (2017) Monitoring the gelation and effective chemical shrinkage of composite curing process with a novel FBG approach. Compos Struct 176:187–194

Multidisciplinary Multi-Point Optimization Design of Transport Aircraft Wing

L. I. Yuchao and Q. I. Chanying

Abstract In traditional aircraft design, the interaction between aerodynamic and structural discipline is not fully considered in the preliminary design stage, so it is difficult to obtain the optimal design. In this context, this paper takes the transport aircraft wing as the research object to develop an aerodynamic/structural multi-disciplinary multi-point optimization design method for transport aircraft wing in the preliminary design stage. In this study, the optimization method of Kriging surrogate model is used to carry out multi-point optimization design of transport aircraft wing under various typical speed. During optimization design, using the full-velocity potential equation for aerodynamic calculation, and Ansys for structural finite element analysis to optimize the design with the lift-to-drag ratio and structural weight as the goals. The effectiveness of the method was preliminarily verified.

Keywords Wing optimization design · Aerodynamic/structural · Kriging surrogate model · Wing weight · Lift-to-drag ratio

1 Introduction

Aerodynamic/structural/fuel/propulsion/weight/control disciplines influence each other in wing design. The aerodynamic/structural influence is particularly important. For transport aircraft, due to its large aspect ratio, the wing will produce large bending and torsion deformation under the action of aerodynamics, and the aerodynamic properties of such deformed wing will change greatly, and will also produce flutter–that is, wing static/dynamic aeroelastic problem. In traditional aircraft design, aeroelasticity is not fully considered in the preliminary design stage, and it is difficult

L. I. Yuchao (✉)
Chinese Flight Test Establishment, Xi'an, China
e-mail: lyc-si@163.com

Q. I. Chanying
Xi'an 5th Military Representative's Institute of Chinese Air Force Equipment, Xi'an, China

Z. Liu et al. (eds.), *Advances and Challenges in Advanced Unmanned Aerial Systems*, Springer Aerospace Technology, https://doi.org/10.1007/978-981-99-8045-1_34

to obtain an optimal design, so multidisciplinary optimization design with surrogate model is required [1–3].

In this paper, we propose to take the transport aircraft wing as the research goal, and adopt the optimization method based on the Kriging surrogate model to develop the multidisciplinary multi-point optimization design method of transport aircraft wing in the preliminary design stage.

2 Optimization Method

The aerodynamic/structural optimization design of the transport aircraft wing adopts an optimization framework based on the Kriging surrogate model [4], in addition to the estimated surface, Kriging also provides a measure of error or uncertainty of the estimated surface, however, the accuracy will be limited if the number of observations sampled is small, the data has limited spatial scope, or the data is in fact not sufficiently spatially correlated. In this research, the observations sample is evenly distributed in the parameter design space, and considerably large to ensure the optimization is in enough spatial scope, so the Kriging surrogate model is adopted [5, 6].

The multi-objective optimization design uses the Latin hypercube sampling to generate the initial sample points, input the shape of the wing structure corresponding to the sample points for aeroelastic analysis, and then, establish a Kriging surrogate model according to the sample points and response values, use genetic algorithm to find the optimal result via repeated iteration until convergence or reaching maximum test points, and finally solve the Pareto solution set [7].

3 Optimization Design Model

3.1 Aerodynamics and Structure Model

In this study, the aerodynamic discipline adopts the method of full-velocity potential equation plus viscosity correction to analyze the aerodynamic characteristics of wing-body combination. The design parameters selected for the preliminary design of the wing of the high subsonic transport aircraft are as follows: cruising altitude 10,000 m, design gross take-off weight 54t, cruising speed 0.76 Ma, wing reference area 105 m^2, using an internal and external two-segment wing structure. In the optimization design, the influence of wing surface aerodynamic forces on structural deformation and the influence of wing structure deformation on its surface aerodynamic distribution are considered. In order to bring the wing design closer to the actual situation, the wing-body combination was selected as the aerodynamic discipline analysis object, so that the aerodynamic interference of the fuselage on the wing could be taken into account. As shown in Fig. 1, X, Y, and Z are the fuselage coordinate system, X

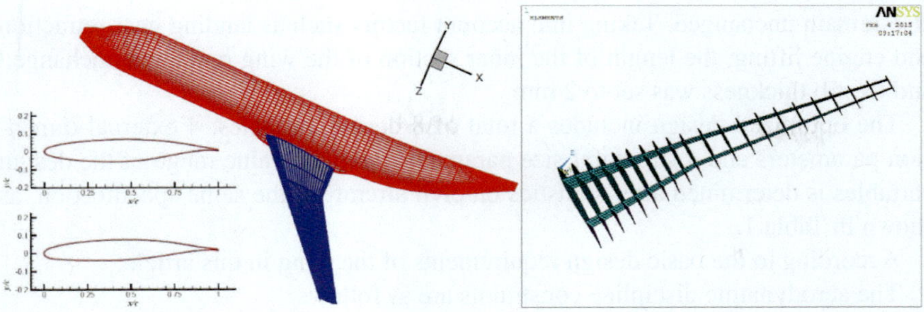

Fig. 1 Aerodynamics and structure model

direction is the axial direction of the fuselage, Z direction is the wingspan direction, and Y is determined according to the right-hand rule. This optimized design adopts a supercritical airfoil, with eight different airfoils configured from the wing root to the wingtip, and Fig. 1 shows the airfoil at the junction of the inner and outer wing (the top one) and the wingtip (the bottom one).

The transport aircraft wing generally adopts a monoblock wing structure, which helps to improve the efficiency of structural bending and torsion, and is beneficial to improve the local stiffness of the skin, ensuring the aerodynamic shape and improving the aeroelastic characteristics.

According to the above analysis, this paper selects the double-bar monoblock wing structure layout, the front beam is arranged at 20% of the chord direction of the wing, the rear beam is arranged at 60% of the chord direction, the front and rear beams are I-beams, the upper and lower edge strips and long trusses of the beam are pressed into the wing skin to increase the internal space of the wing to place the landing gear, and the rear beam has a turn in the inner section of the wing. A total of 21 ribs along the wingspan; The upper and lower skins are made of shell units. The wing structure is shown in Fig. 1. In aerodynamic/structural calculation, aerodynamic load are applied to the structural grid by load conversion method.

3.2 Wing Optimization Design Model

In the actual aircraft design, the designer hopes to carry out a multi-objective optimization to give an optimization solution set, that is, the Pareto solution set, so that the designer can have a comprehensive understanding of the possible design solutions, so as to better make trade-off and compromise, improving design efficiency.

In this paper, the wing of the wing-body combination is designed, and the wing-body combination is selected for aerodynamic analysis, while the structural discipline only needs to analyze the wing structure which is fixed at the root. The length of the fuselage, the position of the rectifier package, and the fuselage data are unchanged, so the relative position of the wing to the fuselage and the wing root chord length also remain unchanged. Taking into account factors such as landing gear retraction and engine lifting, the length of the inner section of the wing remained unchanged and the rib thickness was set to 2 mm.

The optimized design includes a total of 8 design variables: 4 external dimension parameters and 4 structural size parameters, and the value range of the design variables is determined after statistics on civil aircraft of the same specification, as shown in Table 1.

According to the basic design requirements of the wing in this article:

The aerodynamic discipline constraints are as follows:

(1) **Lift**: If the total weight of the aircraft is 54t, the wing must provide lift that can balance the weight in the cruising state, that is, L≥54t.
(2) **Wing area**: The wing reference area is 105 m^2, in order to prevent large changes in wing loading, the wing area generated by the design cannot deviate too much from the reference value, therefore, the area deviation set in this paper is $\pm 5\ m^2$, that is $100\ m^2 \leq S_{wing} \leq 110\ m^2$.

The constraints of structural disciplines are as follows:

(1) **Strength**: The maximum stress σmax of the wing structure σmax ≤ 412,000,000/1.5 calculated by Ansys finite element simulation. Consider a safety factor of 1.5.
(2) **Stiffness**: Finite element analysis by Ansys. Maximum deformation of the wing: $\delta_{max} \leq 1$ m.

In a multi-objective optimization design, the weight and lift-to-drag ratio are each objective function, as shown in the following equation, and an optimal solution solution set is obtained by optimizing the design

$$\mathrm{Obj}(1) = \mathrm{Weight}/100 \quad (1)$$

$$\mathrm{Obj}(2) = -\mathrm{L}/\mathrm{D} \quad (2)$$

Table 1 Number range of design parameters

Design parameter	Span/m	Taper ratio	Torsion angle(°)	Sweep back/ mm	Front bar web/ mm	Back bar web/ mm	Lower skin thickness/ mm	Upper skin thickness/ mm
Symbol	B	TR	θ	Λ	T_{FS}	T_{BS}	T_{LS}	T_{US}
Min	26	0.2	−3	25	4	4	4	4
Max	34	0.4	−1	35	8	8	12	12

4 Optimization Design Result

Multi-point design is to optimize the design in multiple flight states, which is more comprehensive and closer to the real flight environment of the aircraft, and can obtain more engineering optimization results. In this chapter, the multi-point optimization design of the wing will be carried out at 0.74, 0.76, and 0.78 Ma.

By investigating the speed of mainstream transport aircraft, it is found that its flying speed is about 0.74~0.80 Ma, so this multi-point optimization design selects 0.74, 0.76 and 0.78 Ma three flying speeds for multi-point design, representing cruising speed and non-design state speed respectively.

Under different Ma number, the maximum value of stress and deformation is selected by the structural discipline and the minimum value of lift is selected by the aerodynamic discipline as the final state variable, so as to ensure that the design results in any flight state will not fail.

During the whole flight, the aircraft is in the cruise state most of the time, so we select the cruise state, that is, 0.76 Ma as the main design state, and the objective function of this state is used as the final objective function of the multi-point design.

In this optimization, the Kriging surrogate model is used, 100 sets of initial sample points are generated using Latin hypercubes sampling, and genetic algorithm is used to perform the optimization: the population size is 50, the maximum algebra is 300, the maximum number of sample points is 300, and the pareto solution set is finally obtained after 4 cycles, as shown in Fig. 2.

Figure 2 shows the two wing shapes corresponding to the optimization results, here named pareto-1 and pareto-2, Final_response represents the response results of this group of design variables synthesized in three design states. The corresponding

Fig. 2 Pareto solution set for multipoint design

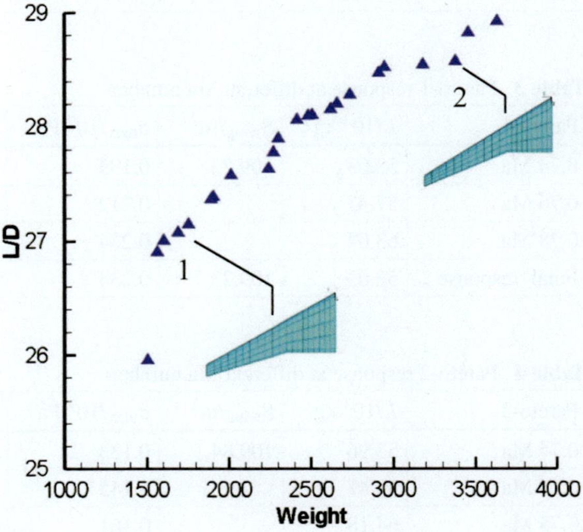

sample point data, the state function and objective function response at the three speeds are shown in the Tables 2, 3 and 4.

Through data comparison, it is not difficult to find that the aspect ratio of pareto-2 is about 16% larger than that of pareto-1, and its wing shape is relatively narrow and long, so it has relatively excellent aerodynamic characteristics, however, in order to meet the structural stress and maximum deformation constraints, it is necessary to increase the structural strength, so its structural weight is relatively large. pareto-1 has smaller span and lower lift-to-drag ratio, because of its small wingspan, the requirements for wing structural strength and stiffness are lower than that of pareto-2, so the structural weight is lighter.

The distribution of surface pressure of the wings of pareto-1 and pareto-2 is shown in Fig. 3: the wing surface pressure of pareto-1 and pareto-2 has similar distribution rules under the same Ma number; And the maximum lift-drag ratio is reached both under the main design condition, whose Ma number is 0.76, as shown in Fig. 3.

By comparing the aerodynamic data, it is found that with the increase of Ma number, the lift and drag increase and the lift-to-drag ratio decreases. The lift and drag characteristics of pareto-2 were better than pareto-1 under the same Ma number, which ultimately led to better lift-to-drag characteristics than pareto-1. From 0.74 to 0.76 Ma, the lift-to-drag ratio does not change much, and the 0.76 to 0.78 Ma lift-to-drag ratio decreases significantly, mainly because when the speed reaches 0.78 Ma, the drag increases significantly.

Table 2 Design variable values corresponding to pareto-1 and pareto-2

	B/m	TR	$\theta/°$	$\Lambda/°$	T_{FS}/mm	T_{BS}/mm	T_{LS}/mm	T_{US}/mm
Pareto-1	31.1	0.2	−1	27.4	4.00	7.06	4.11	4.70
Pareto-2	33.4	0.2	−1	29.7	4.04	8.00	10.84	8.80

Table 3 Pareto-1 response at different Ma number

Pareto-1	$L/10^3$ kg	S_{wing}/m^2	$\sigma_{max}/10^9$ Pa	δ_{max}/m	Weight/kg	L/D
0.74 Ma	53.03	109.23	0.193	0.83	**1689**	**27.19**
0.76 Ma	57.47		0.212	0.90		**27.07**
0.78 Ma	63.04		0.234	0.99		**25.41**
Final_response	53.03	109.23	0.234	0.99	**1689**	**27.07**

Table 4 Pareto-2 response at different Ma number

Pareto-2	$L/10^3$ kg	S_{wing}/m^2	$\sigma_{max}/10^9$ Pa	δ_{max}/m	Weight/kg	L/D
0.74 Ma	53.96	109.84	0.133	0.72	**3379**	**28.53**
0.76 Ma	58.47		0.145	0.78		**28.57**
0.78 Ma	64.18		0.161	0.85		**27.42**
Final_response	53.96	109.84	0.161	0.85	**3379**	**28.57**

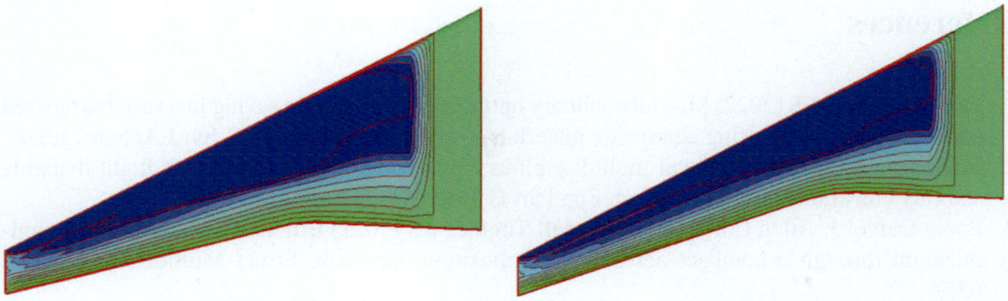

pareto-1 pareto-2

Fig. 3 Wing upper surface pressure distribution of pareto-1 and pareto-2 under 0.76 Ma

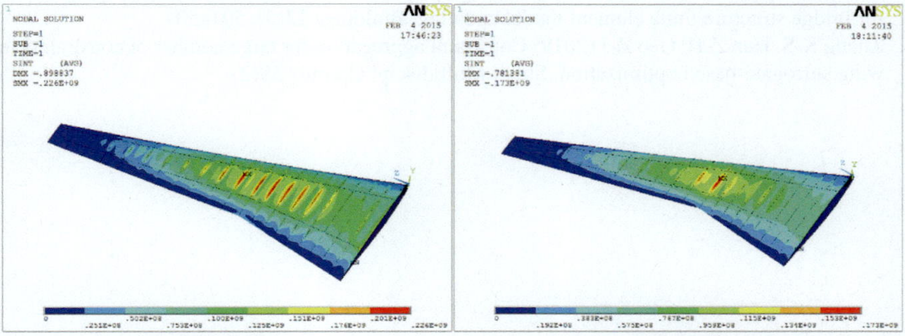

pareto-1 pareto-2

Fig. 4 Wing surface stress cloud of pareto-1 and pareto-2 under different 0.76 Ma

The surface stress cloud of the pareto-1 and pareto-2 wings is shown as Fig. 4: For the same wing, with the increase of Ma number, the stress of the wing structure gradually increases, and the maximum stress is concentrated at the turning point of the trailing edge of the wing, and the stress gradually decreases from the turn to the wingtip and to the wing root. Compared with pareto-2, pareto-1 is lighter in weight, smaller in size, and larger in structural stress and deformation.

5 Conclusion

Taking the transport aircraft wing as the research object, this paper constructs an aerodynamic, structural, and optimization design model, and sets weight and lift-to-drag ratio as objective function to carry out a multi-point optimization design based on the Kriging surrogate model, and obtains an optimization solution set. The effectiveness of the proposed method is preliminarily verified, which can provide designers with more choices when weighing the pros and cons and comprehensively integrating the performance of all aspects.

References

1. Elelwi M, Pinto FS (2022) Multidisciplinary optimization for weight saving in a variable tapered span-morphing wing using composite materials—application to the UAS-S4. J Actuat 11(5)
2. Mieloszyk J (2020) Numerical multidisciplinary optimization of aircraft with flight dynamic stability constraints. Proc Inst Mech Eng Part G J Aerospace Eng
3. Poole Daniel J, Allen Christian B, Rendall Thomas CS (2022) Efficient aeroelastic wing optimization through a compact aerofoil decomposition approach. Struct Multidiscipl Optimiz 65(3)
4. Shi Y, Mader CA, Martins JRRA (2021) Natural laminar flow wing optimization using a discrete adjoint approach. Struct Multidiscipl Optimiz 1–22
5. Han Z-H, Zhang Y, Song C-X, Zhang K-S (2017) Weighted gradient-enhanced kriging for high-dimensional surrogate modeling and design optimization
6. Wu J, Cheng F, Zou C (2022) Swarm intelligent optimization conjunction with Kriging model for bridge structure finite element model updating. Buildings **12**(5), 504–504
7. Zhang K-S, Han Z-H, Gao Z-J (2019) Constraint aggregation for large number of constraints in wing surrogate-based optimization. Struct Multidiscipl Optimiz 59(2)

Temperature Analysis of Machine Tool Based on Numerical and Testing Methods

Wang Changfa, Peng Liang, Hu Yi, Chen Zhenlei, and Ye Feng

Abstract Thermal issues constitute one of the primary factors affecting the machining accuracy of CNC machine tools, with temperature fluctuations being a critical element in controlling machine tool errors. In this study, a specific type of CNC machine tool was chosen as the research subject. A comprehensive analysis of the heat generated by each heat source during the machine tool's operation was conducted. A temperature field test was performed, accompanied by a corresponding transient finite element simulation, covering an 8-h machine tool operation period. When comparing the simulation results with the test findings, it was observed that the temperature field error in the simulation was within 5%. This leads to the conclusion that the transient temperature field numerical simulation method presented in this paper can accurately predict the actual temperature field of the machine tool. This, in turn, establishes a robust foundation for subsequent analyses of thermal error compensation in the machine tool.

Keywords Machine tool · Heat source · Numerical simulation · Temperature field · Thermal error

1 Introduction

The rapid development of electronic information technology has led the machine tool industry to the era of mechatronics with digital manufacturing as the technical core, of which the typical representative is the CNC machine tool [1]. The factors mainly affecting the machining accuracy of CNC machine tool workpieces include the thermal errors of the machine guide, tool, workpiece, spindle, and the structural errors of the CNC machine tool [2]. In precision machining with machine tools,

W. Changfa · P. Liang · H. Yi · C. Zhenlei (✉)
Faculty of Maritime and Transportation, Ningbo University, Ningbo, China
e-mail: chenzhenlei@nbu.edu.cn

Y. Feng
Fotile Group Ningbo, Fotile Kitchenware Co., Ltd, Ningbo, China

Z. Liu et al. (eds.), *Advances and Challenges in Advanced Unmanned Aerial Systems*, Springer Aerospace Technology, https://doi.org/10.1007/978-981-99-8045-1_35

thermal deformation-induced machining errors can make up a significant portion, ranging from 40 to 70%, of the total machining errors [3–6]. Liang et al. [7] developed a heat-flow-solid finite element model for the spindle system to analyze its relationship with the machine's temperature field, confirming the model's effectiveness in predicting spindle system thermal characteristics. In 1975, Okushima et al. [8] employed finite element analysis to calculate the entire machine tool's temperature field distribution and established a mathematical model linking machine temperature variations to spindle thermal displacement. American scholar Bernd Bossmans [9] computed spindle temperature fields using the finite difference method and assessed the impact of bearing heat sources on spindle heat distribution. Wang et al. [10] analyzed the overall thermal characteristics of the XK7 CNC milling machine using finite element analysis software, outlining the influence of various cooling conditions on spindle bearing temperature rises. Zhao et al. [11] emphasized that understanding the machine tool's thermal characteristics is a prerequisite for calculating its thermal error, with these characteristics providing input values for subsequent thermal error modeling.

Previous thermal error models primarily addressed thermal errors by directly affecting individual sub-models of the machine tool through heat source heat flow density [12]. However, comprehending the overall machine tool's temperature field distribution patterns is fundamental to studying machine tool thermal errors. Therefore, it is essential to reconstruct the temperature variation and temperature field distribution of the entire machine tool, serving as the foundation for subsequent thermal error deformation investigations.

This paper takes a certain type of CNC machine tool as the research object, and proposes a simulation analysis method for the temperature field of the whole machine based on CFD analysis. In this paper, the temperature field of the machine tool running continuously for 8 h is tested and simulated. The simulation results show that the maximum error with the measurement result is within 5%, which lays a solid foundation for the later research on the thermal displacement compensation of the machine tool.

2 Mathematical Model of Machine Tool Temperature Field

The heat transfer within each component of the machine tool adheres to both the first law of thermodynamics and the heat conduction equation. Presented below is the three-dimensional unsteady heat conduction model incorporating an internal heat source, which serves as the basis for this paper's analysis.

$$\rho c \frac{\partial t}{\partial \tau} = \lambda \nabla^2 T + \dot{\Phi} \tag{1}$$

In the formula, ρ is the density of the heat conducting object, c is the specific heat capacity of the heat conducting object (*J/(kg K)*), $t = t(x, y, z, \tau)$ is the object temperature (K), τ is time (s), $\dot{\Phi}$ is the term of internal heat source (*W/m*^{3}). $\dot{\Phi}$ satisfies $\dot{\Phi} = \dot{\Phi}_1 + \dot{\Phi}_2 + \dot{\Phi}_3 + \cdots$.

The air flow in the constant temperature chamber satisfies the basic equations of fluid mechanics, and it is an incompressible gas with a stable air flow rate. The basic control equations for its flow are as follows:

Continuity equation:

$$\nabla \cdot U = 0 \tag{2}$$

Above, U is the flow velocity vector (*m/s*), $U = ui + vj + wk$;

Conservation of momentum equation:

$$\rho(U \cdot \nabla)U = -\nabla p + \mu \nabla^2 U \tag{3}$$

In the above integral formulation of this equation, ρ is the fluid density (*kg/ m*^{3}); p is pressure forces (*Pa*); μ is the fluid dynamic viscosity (*Pa s*).

Conservation of energy equation:

$$\rho c_p(U \cdot \nabla)T = \lambda \nabla^2 T \tag{4}$$

In the above integral formulation of this equation, c_p is specific heat capacity at constant pressure (*J/kg/K*); λ is thermal conductivity (*W/m/K*);T is the temperature (*K*).

In this context, due to the small temperature gradient between each sub-model of the machine tool, the consideration of radiation heat transfer for each sub-model is omitted.

3 Heat Source Calculation of Machine Tool Temperature Field

3.1 Calculation of External Heat Source of Machine Tool

The primary external heat source for the machine tool primarily involves convective heat exchange between the airflow within the constant temperature chamber and the machine tool's surface. In essence, this analysis focuses on the heat exchange between the air and the machine tool surface under constant temperature conditions (Fig. 1).

Based on the above analysis, the machine tool's external heat source is described as the third type of boundary condition for heat transfer, outlined as follows:

$$\rho c \frac{\partial t}{\partial \tau} = \lambda \nabla^2 T + \dot{\Phi} \tag{5}$$

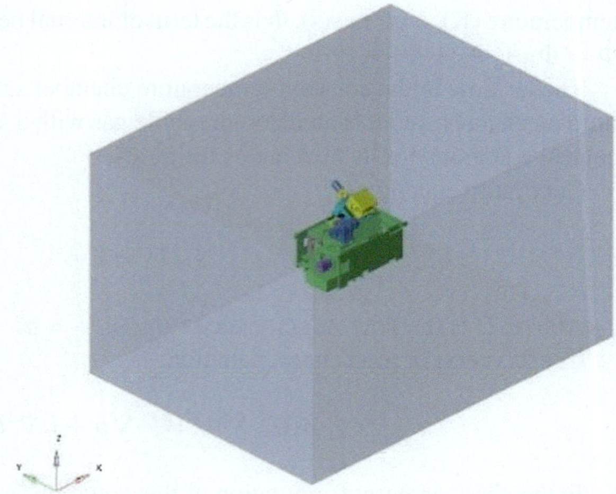

Fig. 1 Machine tool model in a constant temperature room environment

In the above formula, $\lambda(\frac{\partial t}{\partial n})_{wall}$ is the heat transfer through the wall (W/m^2); λ is the thermal conductivity of the machine tool ($W/m/K$),h is the convective heat transfer coefficient ($W/m^2/K$); A is the surface area of the machine tool in contact with the air; t_w is the wall machine tool's temperature.

3.2 Evaluating the Machine Tool's Internal Heat Generation

Studies have shown that [13, 14], the internal heat sources of machine tools are mainly divided into bearings, screw nut heat sources, motor heat sources, and hydraulic station heat sources (Fig. 2).

Among them, the heat source of the machine tool bearing is mainly the heat generated by the friction of the spindle bearing and the heat generated by the external load moment, and the heat source of the screw nut is mainly the heat.

The heat source heat of machine tool spindle bearing can be calculated by the following formula:

$$Q = \frac{2\pi n M}{60} \tag{6}$$

In the above formula, M is friction torque ($N{\cdot}m$, same below); n is the rotational speed (rpm).

Machine tool spindle bearings use 7000 °C angular contact ball bearings [15]. Bearing heat source $M = M_0 + M_1$ in formula (7). M_0 is the friction torque caused by the viscosity of the lubricant, M_1 is the applied load moment. Among them $M_0 = 10^{-7}(v{\cdot}n)^{2/3}D_m^{\ 3}, f_0$ depends on bearing type and lubrication method. This article takes $f_0 = 5$, bearing pitch $D_m = 34\ mm$, spindle rotation speed $n = 2500\ rpm$,

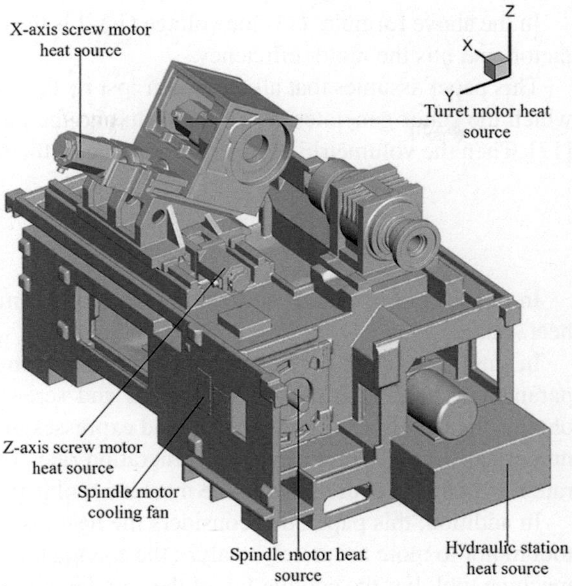

Fig. 2 Heat source distribution of machine tool

kinematic viscosity of lubricant under working condition $\nu = 150m^2/s$. According to this formula, $M_0 = 0.10N \cdot M$. $M_1 = f_1 P_1 D_m$, where f_1 is a factor related to the type of bearing and the load it is subjected to, $f_1 = 3.9 \times 10^{-4}$, bearing friction moment load [16] $P_1 = 2232\ N$. Calculated to get $M_1 = 0.030\ N \cdot M$.

For the screw nut heat source, it is mainly the heat generated by the friction between the X-axis screw nut and the Z-axis screw nut. At this time, $M = M_D + M_P$ in formula (7). Where $M_D = FDP/(2\pi\eta)$ is the driving torque and MP = FP $(1-\eta^2)/(2\pi\eta)$ is the resistance torque. Axial forced FD = μ mg · cos θ- mg · sin θ on the screw nut. F_P is the axial preloading force of the screw nut. For the heat source of the screw nut of the X-axis and the Z-axis, the screw lead $P = 10$mm in the formula, the transmission efficiency $\eta = 0.98$, and the axial force θ on the screw nut in the formula is 30°.

For the X-axis screw, the nut bears a mass of 158kg, the friction coefficient μ is 0.6, and the axial preloading force $F_P = 1440N$ of the screw nut. According to this calculated, the X-axis screw $F_D = 32N$, $M_D = 0.05N \cdot M$, $M_P = 0.093N \cdot M$ are obtained.

For the Z-axis screw, the nut bears a mass of 450kg, rail friction coefficient μ is 0.02, and the axial preloading force $F_P = 1290N$ of the screw nut. According to this, the Z-axis screw $F_D = 90N$, $M_D = 0.15N \cdot M$, MP = $0.083N \cdot M$ are obtained.

The power loss with respect to the heat source of the motor is calculated by the following formula:

$$P = \sqrt{3}UIcos\varphi(1 - \eta) \tag{7}$$

In the above formula, U is the voltage (V), I is the current (A), $cos\varphi$ is the power factor, and η is the motor efficiency.

This paper assumes that all the power lost by the motor is converted into heat, of which the stator generates 2/3 of the heat, and the mover generates 1/3 of the heat [17]. Then the volumetric heat generation rate of the motor is:

$$q = \frac{P}{V} \tag{8}$$

In the above formula, P is the power loss of the motor (W), V is the volume of heat source.

In summary, based on the analysis and calculation of machine tool motor power parameters, bearing friction heat sources and screw nut heat sources, this paper obtains the heat of each heat source, and expresses the effect of heat sources in the numerical form of volumetric heat generation rate. The volumetric heat generation rate for each heat source within the machine tool is provided in Tables 1 and 2.

In addition, this paper also considers the heat dissipation system of the machine tool motor to more accurately analyze the change law of the temperature field of the machine tool. For the cooling fan of the spindle motor of the machine tool, through the fan characteristic curve, the pressure conditions of the three key points of 0 m^3/min, 3 m^3/min, and 6.3 m^3/min at 60 Hz in Fig. 3 are used to simulate the working condition of the fan, so as to achieve the cooling effect of the spindle motor and restore the operating condition of the cooling system of the machine tool.

Table 1 Volumetric heat generation rate of motor and hydraulic station heat source

Heat source	Spindle motor	X-axis screw motor	X-axis screw motor	Turret motor	Hydraulic station
Heat generator rete (W/m3)	5900	10,500	6000	8500	4800

Table 2 Volume heat generation rate of frictional heat source

Heat source	Heat generator rete (W/m^3)	Heat source	Heat generator rete (W/m^3)
Bearing 1	40,625	Z-axis screw nut heat source	26,438
Bearing 2	108,512	X-axis screw bearing heat source 1	121,388
Bearing 3	18,900	X-axis screw bearing heat source 2	25,490
X-axis screw nut heat source	29,948	Z-axis screw bearing heat source 1	10,546

Fig. 3 Fan characteristic curve

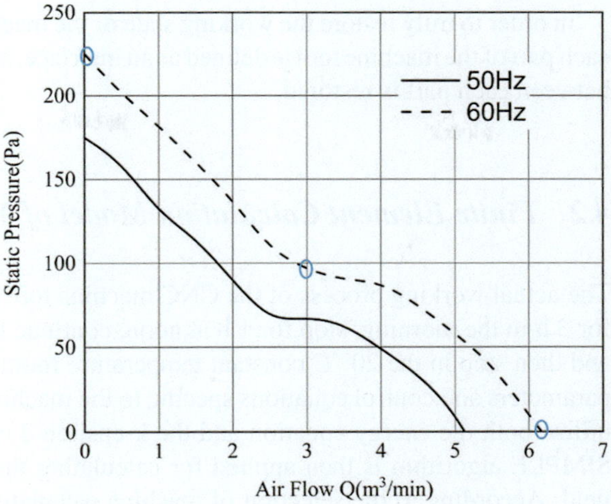

4 Simulation Analysis of Machine Tool Temperature Field

4.1 Machine Tool Mesh Model

CNC machine tool is a high-tech, high-precision machining center specially used for processing precision machinery. It is widely used in automobile, aviation and other machinery industries, and is an indispensable key equipment for the machinery manufacturing industry. Taking a certain type of CNC machine tool as the research object, Hyper Mesh is used to generate the tetrahedral mesh of the machine tool. Figure 4a is the tetrahedron grid of the machine tool, the number of which is 6046628. For calculation efficiency and convergence purpose, the tetrahedral grid is converted into a polyhedral grid through Fluent, and the number of converted polyhedral grids is 2356165, which is 61% less than the number of tetrahedral grids, as shown in Fig. 4b.

Fig. 4 Machine tool mesh model

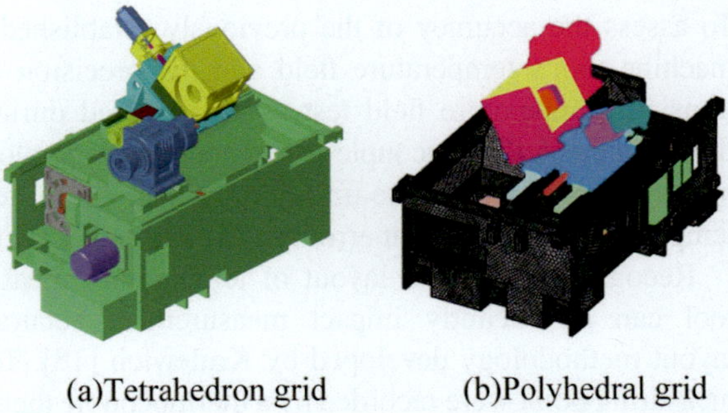

In order to truly restore the working state of the machine tool, the contact between each part of the machine tool is defined as an interface, and the interaction relationship between each part is restored.

4.2 Finite Element Calculation Model of Machine Tool

The actual working process of the CNC machine tool studied in this paper is to run for 3 h in the morning, stop for 1 h at noon, continue to run for 4 h in the afternoon and then stop in the 20 °*C* constant temperature room. Considering the operational parameters and control equations specific to the machine tool, Fluent is employed to utilize both the energy equation and the k-epsilon 2 equation model. The transient SIMPLE algorithm is then applied for calculating the machine tool's temperature field. According to the selection of machine calculation equations and models, set the boundary conditions of the machine tool. The boundary conditions of the volume heat generation rate of each structure shown in Tables 1 and 2 are used. In addition, according to the calculation and analysis method of formula (6) [18], the convective heat transfer coefficients of the spindle, the lead screw and the tool bed are set as 28.2, 16.5 and 5.6 $W/(m^2/K)$, respectively.

When the machine tool motor stops running at noon, the heat generation rate of each motor heat source is set to 0, but at this time the machine tool spindle motor fan continues to run. According to the fan characteristic curve in Fig. 3, the flow velocity is obtained through the fan inlet and outlet air volume, and the pressure difference between the fan inlet and outlet is calculated, so that the intake-fan model is defined by the pressure difference in Fluent to simulate the operation of the machine tool spindle motor fan.

5 Machine Tool Temperature Field Test and Simulation Result Verification

To assess the accuracy of the previously established mathematical model for the machine tool's temperature field and the precision of simulation results, an 8-h transient temperature field test was conducted during the machine tool's operational phase. A thermocouple thermometer was employed for temperature measurements, offering a response time shorter than that of temperature step changes. The temperature measurement error ranged within 0.1 °C for 50% of the recorded time.

Recognizing that the layout of temperature measuring points on the machine tool can significantly impact measurement accuracy, this study adopted the layout methodology developed by Krulewich [18]. Temperature readings for each measuring point were recorded by a thermocouple thermometer for every hour of the machine tool's operation. The temperature measuring point arrangement is illustrated in Fig. 5.

In addition, in order to avoid the uncertainty of external environmental factors in the operating state of the machine tool. The temperature of the machine tool is measured in a constant temperature room, and the ambient temperature is 20 °*C*. The thermostatic chamber and temperature measuring points are shown in Fig. 6.

Figure 7 displays the temperature field calculation outcomes for the machine tool's simulation model. In regions of the machine tool substructure with concentrated heat sources, temperatures are notably elevated, featuring steep temperature gradients.

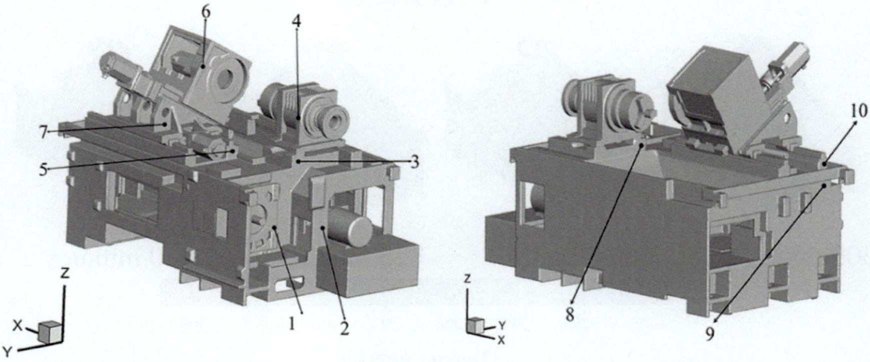

Fig. 5 Schematic diagram of distribution of temperature measuring points

Fig. 6 Schematic diagram of temperature measuring points

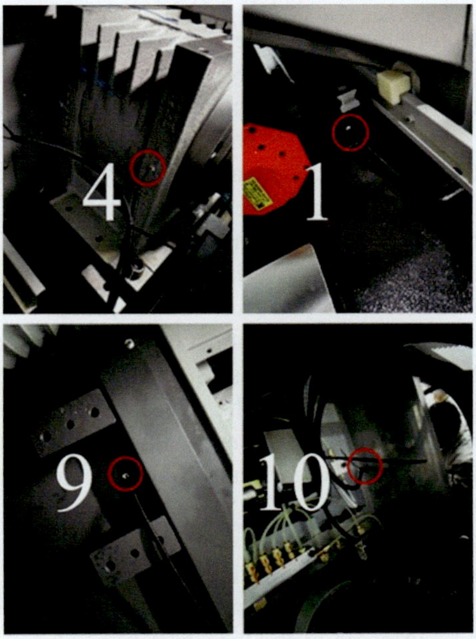

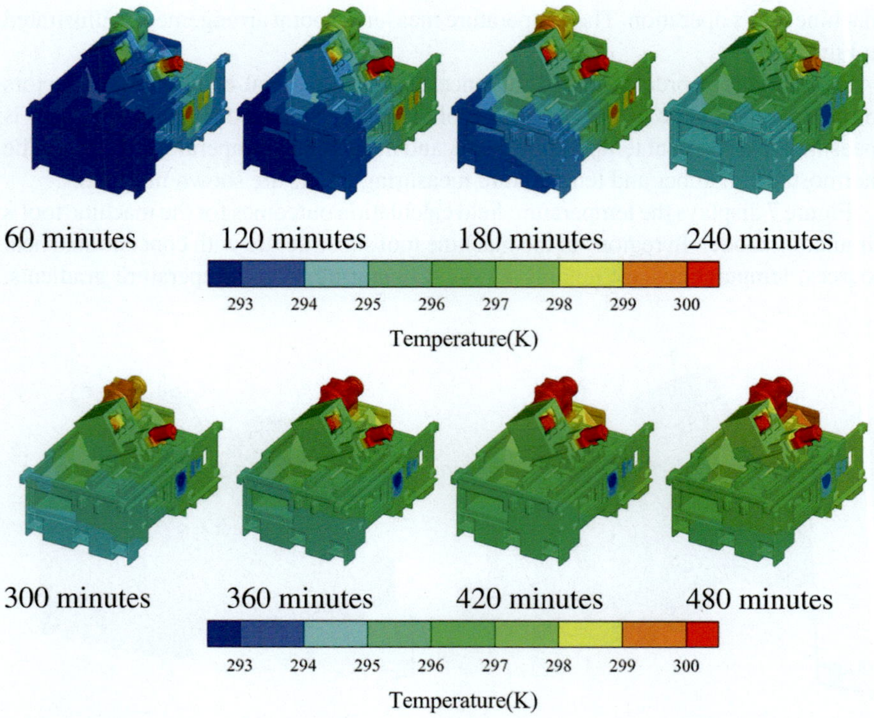

Fig. 7 Temperature cloud map of the machine tool at different times

Contrasting the temperature field cloud diagrams at 180 and 240 min, as presented in Fig. 7, reveals a decrease in motor heat source temperature during the machine tool's midday pause, contributing to an overall cooling of the machine tool.

In addition, it is noted that the temperature of the Z-axis screw of the machine tool does not change much, and the temperature gradient along the length direction decreases. It can be seen that the heat exchange of each heat source of the machine tool has a delay effect on the heat transfer on the Z-axis screw, that is, a pseudo-hysteresis effect. This effect verifies the need to consider the effect of heat lag transfer for the lead screw thermal error compensation.

Referring to the temperature-cloud diagram of the machine tool, the guide rail's temperature undergoes a substantial elevation during operation. Initially, in the early stages of operation, the temperature gradient along the machine tool guideway is minimal. However, after running for a duration of 360–480 min, a significant increase in guide rail temperature occurs, accompanied by a more pronounced temperature gradient.

Figure 8 presents a comparative analysis of the test and simulation results at each representative measuring point on the machine tool.

According to the comparative analysis of the measured temperature values of each measuring point and the simulation value, it can be seen that the temperature change

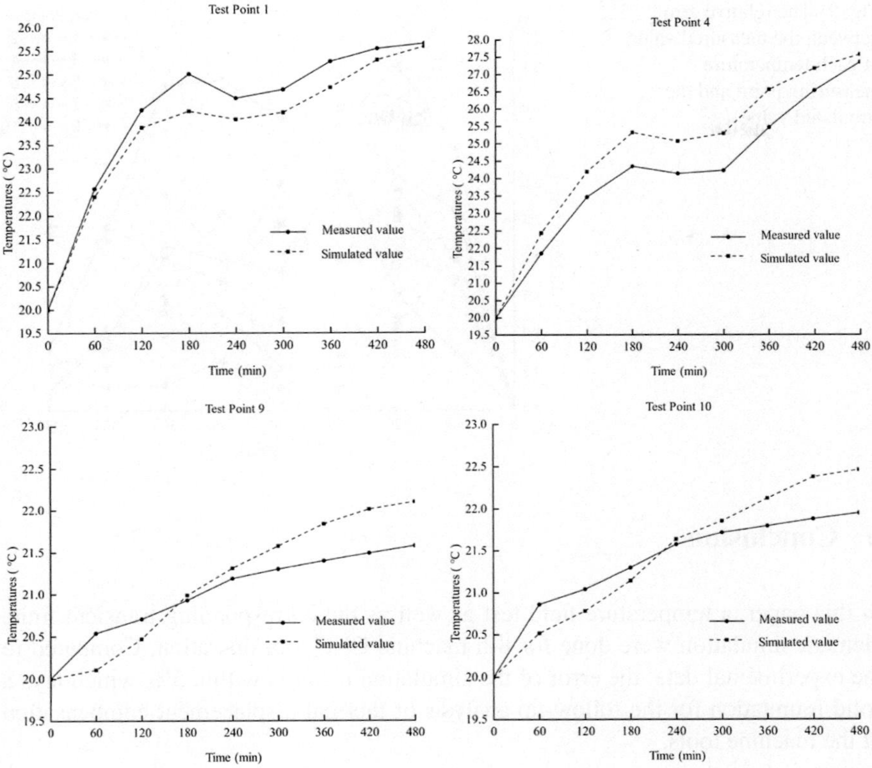

Fig. 8 Comparison of the temperature measurement value of each measuring point with the simulated value

trend of each measuring point on the time axis is similar. Stop at noon, the volume heat generation rate of the heat source is 0, so that the temperature near the heat source drops faster, while the temperature at the farther away from the heat source still rises slowly. Combining the temperature nephogram and temperature curve, it can be concluded that the pseudo-hysteresis effect of the machine tool screw is obvious, and this effect should be fully considered for the subsequent thermal error research. Moreover, through the temperature trend analysis of the machine tool at 360–480 min, the temperature values of each measuring point tend to be balanced, indicating that the time for each substructure of the machine tool to reach thermal equilibrium is relatively long.

Figure 9 illustrates the relative errors at each measuring point, computed from both the simulation and test temperature data. The relative temperature error for all measuring points remains below 5%, with a decreasing trend observed between 360 to 480 min. This trend indicates that the simulation model effectively captures the long-term thermal balance between machine tool substructures.

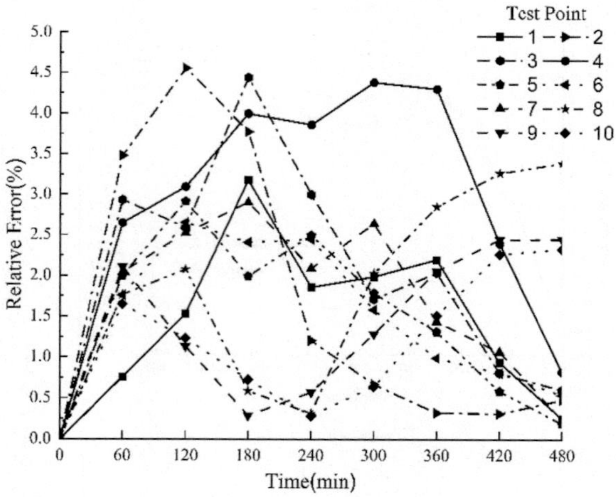

Fig. 9 The relative error between the measured value of each temperature measuring point and the simulated value

6 Conclusion

In this paper, a temperature field test as well as the corresponding transient finite element simulation were done for 8-h machine CNC tool operation. Compared to the experimental data, the error of the simulation result is within 5%, which lays a solid foundation for the follow-up analysis of thermal displacement compensation of the machine tools.

Acknowledgements We would like to thank the Zhejiang Province Public Welfare Technology Application Research Project (LGG22E080020), Healthy & Intelligent Kitchen Engineering Research Center of Zhejiang Province and 2025 Major Programs on Science Technology Innovation of Ningbo (2020Z06) for their support for this research.

References

1. Wu YX (2020) What is the future direction of machine tools. Woodwork Mach (03):46
2. Liu G (2017) Research on screw deformation prediction model based on real-time data of machine tools. Huazhong University of Science and Technology
3. Ramesh R, Mannan MA, Poo AN (2000) Error compensation in machine tools-a review: part II: thermal errors. Int J Mach Tools Manuf 40(9):1257–1284
4. White AJ, Postlethwaite SR, Ford DG (1970) Measuring and modelling thermal distortion on CNC machine tools. WIT Trans Eng Sci
5. Bryan JB (1990) International status of thermal error research. CIRP Ann 39(2):645–656
6. Mayr J, Jedrzejewski J, Uhlmann E (2012) Thermal issues in machine tools. CIRP Ann 61(2):771–791
7. Liang Z, Mohan L, Xingxing Z (2020) A new thermal error control method for spindle system of high precision computer numerical control machine tools. J Shanghai Jiao Tong University 54(11):1165–1171

8. Okushima K, Kakino Y (1976) An analysis of methods used in minimising thermal deformations of machine tools. In: Proceedings of the sixteenth international machine tool design and research conference. pp 195–201
9. Bossmanns B, Tu JF (1999) A thermal model for high speed motorized spindles. Int J Mach Tools Manuf 39(9):1345–1366
10. Wang J, Weng Z, Yao C (2004) Application of ANSYS in the analysis of thermal characteristics on CNC milling machines. J Zhejiang University Technol (03):13–17
11. Zhao W, Liu H, Hu S (2021) Research on the thermal characteristics of machine tools. Sci Technol Eng 21(07):2563–2574
12. Hongyang DU, Tao TAO, Ruisheng HOU (2021) A first-order autoregressive modeling method for axial thermal error of machine tool spindle. J Harbin Inst Technol 53(07):60–67
13. Haitao Z, Jianguo Y, Jinhua S (2007) Simulation of thermal behavior of a CNC machine tool spindle. Int J Mach Tools Manuf 47(6):1003–1010
14. Lee J-H, Yang S-H (2002) Statistical optimization and assessment of a thermal error model for CNC machine tools. Int J Mach Tools Manuf 42(1):147–155
15. Wang Z (1996) Practical bearing manual, vol 2. Shanghai Science and Technology Literature Press
16. Aihua ZHU, Chengjiu ZHU, Weihua Z (2008) Analysis on calculation of friction torque of rolling bearings. Bearing (07)
17. Yang S, W Tao (1998) Heat transfer, 3rd edn. Higher Education Publisher, Xi'an
18. Krulewich DA (1998) Temperature integration model and measurement point selection for thermally induced machine tool errors. Mechatronics 8(4), 395–412

Research on Sensorless Control of Permanent Magnet Synchronous Motor Based on Adaptive Gate Filter

Shuping Meng, Zhishu Xu, Lijuan Zhang, Dikai Jiang, Sijia Cao, and Zhipeng Zheng

Abstract Permanent magnet synchronous motor (PMSM) sensorless control system is widely used in some special environments, such as water immersion, oil immersion and high temperature and so on. A novel improved adaptive gate filter without attenuation and phase lag is proposed to solve the problems of the low-pass filter when the classical sliding mode observer (SMO) control algorithm is applied to PMSM sensorless control. Meanwhile, the saturation function is adopted to replace the traditional switching function, which effectively improves the problems of discontinuous jitter at the switch interface. The experimental results show that the proposed method can effectively eliminate the influence of filter and the chatting and effectively improve position tracking ability. Compared with the classical algorithm, the amplitude attenuation of the observed back EMF signal is reduced from about 34% to about 4%, and the high frequency buffeting component caused by the sliding mode structure is effectively filtered. As a result, the estimated rotor position error decreases from $-52°\sim-35°$ to $-2°\sim2°$. When applied to the sensorless speed servo control of PMSM, the speed fluctuation is -30r/min ~ $+30$r/min and the motor runs smoothly.

Keywords Permanent magnet synchronous motor · Adaptive gate filter · Saturation function · Sliding mode observer · Sensorless control

1 Introduction

With the development of new power electronics technology, computer control technology and new material technology, control system with PMSM as control object is widely used in aerospace, industrial production and other fields [1–3]. However, some water immersion, oil immersion and high temperature applications limit the

S. Meng (✉) · Z. Xu · L. Zhang · D. Jiang · Z. Zheng
Beijing Institute of Precision Mechatronics and Controls, Beijing, China
e-mail: Mengshuping1008@sina.com

S. Cao
Laboratory of Aerospace Servo Actuation and Transmission, Beijing, China

Z. Liu et al. (eds.), *Advances and Challenges in Advanced Unmanned Aerial Systems*, Springer Aerospace Technology, https://doi.org/10.1007/978-981-99-8045-1_36

use of position sensors. PMSM sensorless control has become the research direction for more and more scholars [4, 5].

The PMSM sensorless control method generally can be divided into two categories. The first type is based on the mathematical model of the motor [6–11]. The main closed-loop algorithm for medium to high speed operation include flux estimation method, model reference adaptive method, Kalman filtering method, sliding mode observer method, etc. Such methods is relatively easy to implement and of high reliability at high speed areas. However, at low speeds, the back EMF is difficult to detect due to its small amplitude. The signal is prone to interference and the accuracy decreases, so this type of method is even unable to use at zero speed. The second type is based on the saliency, such as high frequency signal injection method, etc., [12–14]. This type of method does not require motor accurate parameters and can be applied to extremely low speed operation. This type of method has good control effect at low speed areas. Therefore, combining these two type methods is an effective way to achieve speed servo control of PMSM in entire motor operating range without speed sensors.

The SMO method has the advantages of simple calculation, strong robustness when parameter changes or external disturbances occur. It has been widely applied due to its advantages. The classical SMO sensorless control algorithm [15, 16] uses a low-pass filter to filter the buffeting signal, which results estimated error. The phase delay requires a phase compensation, while the compensation angle changes with the motor speed. By using the estimated back EMF as feedback variable into the sliding mode observer and a speed-adaptive cut-off frequency filter, the chattering problems of traditional SMO is improved and the angle compensation caused by the filter becomes a fixed value [17]. However, the signal amplitude of back EMF will become very small at low speed due to the filter. In addition, the back EMF signal may contain other harmonic content.

On the other hand, due to the need for real-time and accurate frequency information in both speed controllers and flux observation, frequency tracking is an important aspect. At present, the widely used methods are Phase Locked Loop (PLL) or Frequency Locked Loop. Reference [18] points out that the application of PLL will reduce the dynamic performance, and proposes a PLL based on tangent function to solve the instability problem. However, the tangent function has singular points at $\pm \pi$, making it difficult to ensure the global convergence of PLL. Reference [19] constructed a frequency adaptive algorithm by adding an additional flux observer and utilizing observation errors to avoid singularity problems. FLL does not require the phase tracking link in PLL, and directly utilizes the error feedback of sine input for frequency tracking. However, the above schemes do not consider the rapid change of the rotor speed, so it is difficult to accurately track speed frequency in real-time when the motor speed changes frequently (such as continuous acceleration and deceleration).

Aiming at weakening the chattering phenomenon, amplitude attenuation and phase shift of the classical SMO, an adaptive gate filter is proposed to obtain the effective back EMF signal. With the adaptive gate filter applied in SMO algorithm, the back EMF signal can be obtained with almost no distortion and phase shift is eliminated and the accuracy of position estimation is improved. At the same time, the saturation function worked as the control function and a phase-locked loop is adopted to weaken the chattering phenomenon of the classical SMO. Finally, experimental research was conducted on an SPMSM to verify the improved strategy in this paper.

2 Classical SMO Sensorless Algorithm

2.1 Design of Classical SMO

The mathematical model of SPMSM in two-phase stationary coordinate system (α-β) is as follows:

$$\begin{bmatrix} \dot{i}_\alpha \\ \dot{i}_\beta \end{bmatrix} = \frac{1}{L_s}\begin{bmatrix} -R_S & 0 \\ 0 & -R_S \end{bmatrix}\begin{bmatrix} i_\alpha \\ i_\beta \end{bmatrix} + \frac{1}{L_s}\begin{bmatrix} 1 & 0 \\ 0 & 1 \end{bmatrix}\begin{bmatrix} u_\alpha - e_\alpha \\ u_\beta - e_\beta \end{bmatrix} \tag{1}$$

where, u_α, u_β are respectively the voltage, i_α, i_β are respectively the current, *Rs* and *Ls* are the resistance and inductance, e_α, e_β are respectively the back EMF component.

The back EMF e_α, e_β satisfies:

$$\begin{pmatrix} e_\alpha \\ e_\beta \end{pmatrix} = \psi_f \omega_e \begin{pmatrix} -\sin\theta_e \\ \cos\theta_e \end{pmatrix} = \lambda \begin{pmatrix} -\sin\theta_e \\ \cos\theta_e \end{pmatrix} \tag{2}$$

where, ω_e is the electrical angular velocity of the rotor and ψ_f is the rotor flux. From Eq. (2), it can be seen that the rotor position can be estimated through the back EMF.

The SMO utilizes the current signals measured through current sensor to form a closed-loop disturbance observer. Select the stator current as the sliding surface, $s(x) = \begin{bmatrix} \tilde{i}_\alpha \\ \tilde{i}_\beta \end{bmatrix} = \begin{bmatrix} \hat{i}_\alpha - i_\alpha \\ \hat{i}_\beta - i_\beta \end{bmatrix}$, where, the symbol with superscript "^" and " ~ " respectively represent the estimated values and the error values.

Establish a state observer using stator current as state variables, and obtain the mathematical model of the observer based on state Eq. (1):

$$\begin{bmatrix} \dot{\hat{i}}_\alpha \\ \dot{\hat{i}}_\beta \end{bmatrix} = \frac{1}{L_s}\begin{bmatrix} -R_s & 0 \\ 0 & -R_s \end{bmatrix}\begin{bmatrix} \hat{i}_\alpha \\ \hat{i}_\beta \end{bmatrix} + \frac{1}{L_s}\begin{bmatrix} 1 & 0 \\ 0 & 1 \end{bmatrix}\begin{bmatrix} u_\alpha - z_\alpha \\ u_\beta - z_\beta \end{bmatrix} \tag{3}$$

where, $\hat{e}_\alpha$, $\hat{e}_\beta$ are respectively the estimated value of back EMF, z_α, z_β are respectively the components of sliding mode control function. The expression for the equivalent control component z_α, z_β is as follows: $\begin{bmatrix} z_\alpha \\ z_\beta \end{bmatrix} = k{*}u(s) = k\begin{bmatrix} \mathrm{sgn}\left(\hat{i}_\alpha - i_\alpha\right) \\ \mathrm{sgn}\left(\hat{i}_\beta - i_\beta\right) \end{bmatrix}$.

Where, k represents the gain coefficient parameter and sgn(x) represents the sign function.

Subtracting Eq. (1) from Eq. (3), we can obtain the equation of current estimating error as

$$\begin{bmatrix} \dot{\tilde{i}}_\alpha \\ \dot{\tilde{i}}_\beta \end{bmatrix} = \frac{1}{L_s}\begin{bmatrix} -R_s & 0 \\ 0 & -R_s \end{bmatrix}\begin{bmatrix} \tilde{i}_\alpha \\ \tilde{i}_\beta \end{bmatrix} + \frac{1}{L_s}\begin{bmatrix} 1 & 0 \\ 0 & 1 \end{bmatrix}\begin{bmatrix} \hat{e}_\alpha - z_\alpha \\ \hat{e}_\beta - z_\beta \end{bmatrix} \tag{4}$$

When entering the sliding mode $\hat{e}_\alpha = z_\alpha, \hat{e}_\beta = z_\beta$. Where, $\hat{e}_\alpha, \hat{e}_\beta$ are the estimated value of the $\alpha\beta$ axis component of back EMF.

When reaching the sliding surface, the observer state will maintain equilibrium and stable operation near the sliding surface. The SMO reaches a stable state, $\dot{\tilde{i}}_\alpha = 0, \dot{\tilde{i}}_\beta = 0, \tilde{i}_\alpha = 0, \tilde{i}_\beta = 0$, and $\hat{e}_\alpha = z_\alpha, \hat{e}_\beta = z_\beta$ equation is valid. The expression for estimating the back EMF in a stable sliding mode is:

$$\begin{bmatrix} \hat{e}_\alpha \\ \hat{e}_\beta \end{bmatrix} = \frac{\omega_c}{s + \omega_c}\begin{bmatrix} z_\alpha \\ z_\beta \end{bmatrix} \tag{5}$$

where, ω_c represents the cut-off frequency.

The actual control of sliding mode is a discrete high-frequency switching signal. Traditional observers use a switching function, and the high-frequency signal generated during switching is filtered out by a low-pass filter. This method reduces system chattering to some extent, but it causes phase lag of the extended back EMF, requiring phase compensation to obtain the actual motor rotor position for closed-loop control, greatly increasing the complexity of the system, and cumbersome calculations can generate significant computational noise. If the cut-off frequency is fixed, the phase lag will change with the change of motor speed, and the estimated motor position error will also change with the change of motor speed. It is necessary to establish the corresponding table of phase compensation angle with the motor speed, and use the table lookup method to compensate. To solve this problem, the cut-off frequency of the improved filter is selected as the estimated electrical angular velocity, that is, under different motor speeds, the amplitude of the effective component is uniformly attenuated to 0.707 times the original amplitude, and a 45° phase lag is uniformly generated [17].

The position error signal of the filtered back EMF signal after heterodyne (equivalent to PD link in PLL) processing is described as

$$\begin{aligned} \varepsilon &= -\hat{e}_\alpha \cos\hat{\theta}_e - \hat{e}_\beta \sin\hat{\theta}_e \\ &= \lambda \sin\theta_e \cos\hat{\theta}_e - \lambda \cos\theta_e \sin\hat{\theta}_e \\ &= \lambda \sin\left(\theta_e - \hat{\theta}_e\right) \end{aligned} \tag{6}$$

When the error between the estimated and the actual rotor position is very small, $\sin\left(\theta_e - \hat{\theta}_e\right) \approx \theta_e - \hat{\theta}_e = \Delta\theta_e$.. The error is input into the phase-locked loop (PLL), and then the rotor position is obtained using the closed-loop phase detection function of the PLL.

2.2 Observer Stability Analysis

The stability of the above SMO is analyzed according to the reachability condition of sliding mode structure. Take positive definite function as

$$V = \frac{1}{2}s^T s = \frac{1}{2}\left(\tilde{i}_\alpha^2 + \tilde{i}_\beta^2\right) \tag{7}$$

If the system is stable, it is required $\dot{V}$ to be negative.

$$\begin{aligned}
\dot{V} &= s^T s \\
&= \tilde{i}_\alpha \cdot \dot{\tilde{i}}_\alpha + \tilde{i}_\beta \cdot \dot{\tilde{i}}_\beta \\
&= \tilde{i}_\alpha\left\{-\frac{R_s}{L_s}\tilde{i}_\alpha - \frac{1}{L_s}\hat{e}_\alpha - \frac{1}{L_s}k\mathrm{sgn}\left(\tilde{i}_\alpha\right)\right\} + \tilde{i}_\beta\left\{-\frac{R_s}{L_s}\tilde{i}_\beta - \frac{1}{L_s}\hat{e}_\beta - \frac{1}{L_s}k\mathrm{sgn}\left(\tilde{i}_\beta\right)\right\} \\
&= -\frac{R_s}{L_s}\left(\tilde{i}_\alpha^2 + \tilde{i}_\beta^2\right) - \frac{1}{L_s}\left\{\tilde{i}_\alpha\left(\tilde{e}_\alpha + k\mathrm{sgn}\left(\tilde{i}_\alpha\right)\right)\right\} - \frac{1}{L_s}\left\{\tilde{i}_\beta\left(\tilde{e}_\beta + k\mathrm{sgn}\left(\tilde{i}_\beta\right)\right)\right\} \\
&= V_1 - V_2
\end{aligned} \tag{8}$$

where,

$$\left\{\begin{aligned}
V_1 &= -\frac{R_s}{L_d}\left(\tilde{i}_\alpha^2 + \tilde{i}_\beta^2\right) \\
V_2 &= \frac{1}{L_d}\left\{\tilde{i}_\alpha\left(\hat{e}_\alpha - k\mathrm{sgn}\left(\tilde{i}_\alpha\right)\right)\right\} + \frac{1}{L_d}\left\{\tilde{i}_\beta\left(\hat{e}_\beta - k\mathrm{sgn}\left(\tilde{i}_\beta\right)\right)\right\}
\end{aligned}\right. \tag{9}$$

Equations (8, 9) shows that V_1 is negative constantly. When $k > \max\left(|e_\alpha|, |e_\beta|\right)$, the system converges and the observed current values converge to the actual current values. Then, the results observed by the observer characterize the rotor position.

3 Adaptive SMO Sensorless Algorithm

In the SPMSM speed control system, some non-ideal factors such as phase lag, loop delay, and high-frequency chattering signals can affect the observation accuracy. In order to improve the shortcomings of traditional SMO, an improved adaptive sliding mode observer (ASMO) is proposed. The structural diagram of the improved ASMO is shown in Fig. 1.

In Fig. 1, an adaptive gate filter is adopted to obtain the effective signal from the equivalent control component z_α, z_β based on the traditional SMO. This adaptive gate filter can filter out the observed harmonic components of the back EMF, but it does not produce filtering phase shift or signal amplitude attenuation, which improves system accuracy while hardly affecting system performance. Attention here, a saturation function is adopted in the equivalent control function $u(s)$, which enables the system to operate stably without chattering over a wide speed range [17].

The implementation principle block diagram of the adaptive gate filter is shown in Fig. 2. The equivalent control signals of the alpha axis and beta axis adopt a completely consistent signal processing mechanism, as shown as ANF block in the figure. Here, we will only use alpha axis component as an example for detailed derivation. The adaptive gate filter takes the equivalent control signal and the estimated motor electrical angle position as inputs, and after processing with sine cosine and difference calculation, it can effectively extract the components of the same frequency as the motor electrical angle from the equivalent control signal, namely the effective back EMF component. The result obtained by removing the effective component from the equivalent control signal is high-frequency noise in the equivalent control signal. In Fig. 2, the values suffixing (s) represent frequency domain variables, while the corresponding values suffixing (t) represent time domain components. $\mathcal{L}()$ represents the Laplace operator.

In Fig. 2, $\hat{\theta}_e = \hat{\omega}_e t$, represents the estimated value of the motor position. The effective fundamental back EMF can be undistorted calculated by using the feedback of the estimated position. This eliminates the need to adjust the cut-off frequency of the filter according to the speed, reducing the complexity of the algorithm.

Since the input signals multiplied with $\sin\hat{\theta}_e$ or $\cos\hat{\theta}_e$ achieve a DC signal superimposed by a high frequency signal, a low-pass filter is needed to separate out the DC

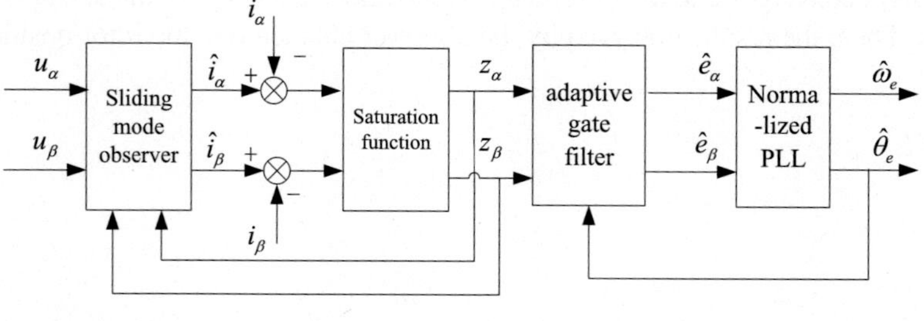

Fig. 1 Improved adaptive sliding mode observer

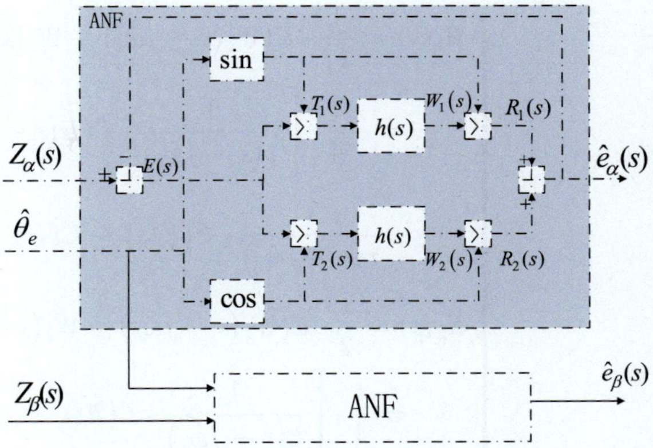

Fig. 2 Block diagram of adaptive gate filter algorithm

signal which contains the amplitude and phase information of the fundamental back EMF. The $h(s)$ meets the following formula (10). We can figure out $h(s) = {}^{1}/_{\tau s}$.

$$g(s) = \frac{h(s)}{1 + h(s)} = \frac{1}{\tau s + 1} \tag{10}$$

The derivation process of filter expression $G(s)$ in Fig. 2 is shown as follows.

$$\begin{cases} T_1(s) = \mathcal{L}\big(e(t)\cdot\sin(\hat{\omega}_e t)\big) = \mathcal{L}\left(e(t)\cdot\dfrac{e^{j\hat{\omega}_e t} - e^{-j\hat{\omega}_e t}}{2j}\right) \\ \quad = \dfrac{1}{2j}\mathcal{L}\big(E\big(s - j\hat{\omega}_e\big) - E\big(s + j\hat{\omega}_e\big)\big), \\ T_2(s) = \mathcal{L}\big(e(t)\cdot\cos(\hat{\omega}_e t)\big) = \mathcal{L}\left(e(t)\cdot\dfrac{e^{j\hat{\omega}_e t} + e^{-j\hat{\omega}_e t}}{2}\right) \\ \quad = \dfrac{1}{2}\mathcal{L}\big(E\big(s - j\hat{\omega}_e\big) + E\big(s + j\hat{\omega}_e\big)\big). \end{cases} \tag{11}$$

$$\begin{cases} W_1(s) = G(s)\cdot T_1(s) = \dfrac{1}{2j}\cdot\dfrac{1}{\tau s}\cdot\mathcal{L}\big(E\big(s - j\hat{\omega}_e\big) - E\big(s + j\hat{\omega}_e\big)\big), \\ W_2(s) = G(s)\cdot T_2(s) = \dfrac{1}{2}\cdot\dfrac{1}{\tau s}\cdot\mathcal{L}\big(E\big(s - j\hat{\omega}_e\big) + E\big(s + j\hat{\omega}_e\big)\big). \end{cases} \tag{12}$$

$$\begin{cases} R_1(s) = \frac{1}{2j} \cdot \mathcal{L}\big(W_1\big(s - j\hat{\omega}_e\big) - W_1\big(s + j\hat{\omega}_e\big)\big) \\ \quad = -\frac{1}{4} \cdot \left[\frac{1}{\tau\big(s - j\hat{\omega}_e\big)} \cdot \mathcal{L}\big(E\big(s - j2\hat{\omega}_e\big) - E(s)\big) \right. \\ \quad \left. -\frac{1}{\tau\big(s + j\hat{\omega}_e\big)} \cdot \mathcal{L}\big(E(s) - E\big(s + j2\hat{\omega}_e\big)\big) \right], \\ R_2(s) = \frac{1}{2} \cdot \mathcal{L}\big(W_2\big(s - j\hat{\omega}_e\big) + W_2\big(s + j\hat{\omega}_e\big)\big) \\ \quad = \frac{1}{4} \cdot \left[\frac{1}{\tau\big(s - j\hat{\omega}_e\big)} \cdot \mathcal{L}\big(E\big(s - j2\hat{\omega}_e\big) + E(s)\big) \right. \\ \quad \left. +\frac{1}{\tau\big(s + j\hat{\omega}_e\big)} \cdot \mathcal{L}\big(E(s) + E\big(s + j2\hat{\omega}_e\big)\big) \right]. \end{cases} \tag{13}$$

$$\begin{aligned} Y(s) = R_1(s) + R_2(s) &= \frac{1}{2} \cdot \frac{1}{\tau\big(s - j\hat{\omega}_e\big)} \cdot E(s) + \frac{1}{2} \cdot \frac{1}{\tau\big(s + j\hat{\omega}_e\big)} \cdot E(s) \\ &= E(s) \cdot \frac{1}{\tau} \cdot \frac{s}{s^2 + \hat{\omega}_e^2} \end{aligned} \tag{14}$$

$$G(s) = \frac{Y(s)}{X(s)} = \frac{Y(s)}{E(s) + Y(s)} = \frac{Y(s)/E(s)}{1 + Y(s)/E(s)} = \frac{s}{\tau(s^2 + \omega^2) + s} \tag{15}$$

From Formula (15), frequency response curve of $G(s)$ can be obtained as shown in Fig. 3, assuming that, $\hat{\omega}_e = 125.6$ *rad/s*.

From Formula (15) and Fig. 3, the proposed adaptive gate filter algorithm of which the scheme shown in Fig. 3 acts as an adaptive gate filter. The characteristic of this filter is that it can separate the corresponding frequency signal without attenuation and phase shift at the desired frequency point. Meanwhile, the filter frequency varies automatically and need no adjustment to extract the effective back EMF at different

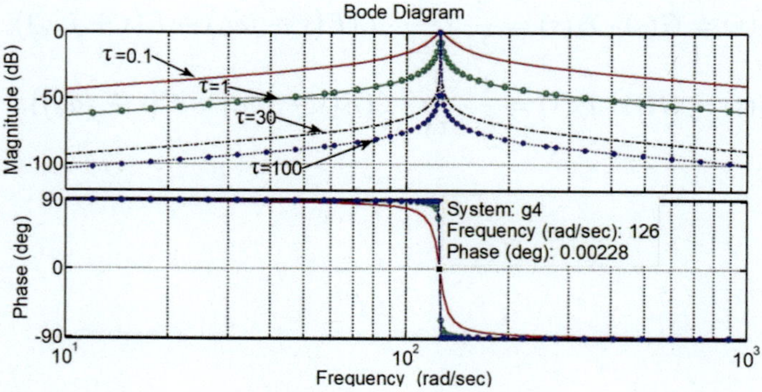

Fig. 3 Frequency response curve of $G(s)$

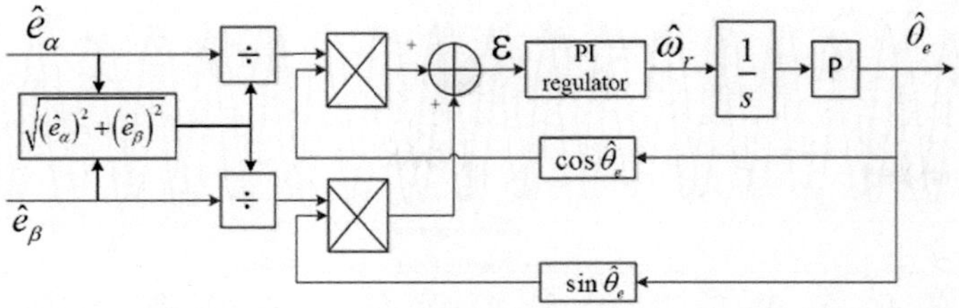

Fig. 4 Principle diagram of PLL

motor speed. The parameter τ is critical to the performance of the filter. The larger τ is, the deeper the filtering level is, which results the higher estimation accuracy and the slower convergence speed.

The obtained back EMF signal is processed by heterodyne method to obtain the error signal, and then the rotor position signal is obtained from the error signal through normalized PLL processing, as shown in Fig. 4.

4 Experimental Verification

To verify the effectiveness of the proposed estimation position correction strategy, an SPMSM sensorless experimental system of which the rated power is 1 kW was built. The switching frequency and current sampling frequency of the inverter are both 10 kHz. The speed loop period and current loop period are respectively 500us and 100us. The test motor parameters are shown in Table 1. The measured position data through a position sensor is only used for comparison with the estimated rotor position. The classical SMO algorithm and the improved adaptive SMO algorithm based on this paper are experimentally studied.

Table 1 Test SPMSM Parameters

Parameter	Symbol	Value
Rated line voltage	*Udc*	270
Rated power	*P*	1 kW
Rated speed	*n*	3000 r/min
Pole-pairs	n_p	5
Stator inductance	L_d	3.65 mH
Stator resistance	*R*	0.6Ω
Rotor flux	φ	0.071793 Wb
Moment of inertia	J	1.36*10–4 Kgm^2
Damping coefficient	B	1.15*10^{-4} NmS/rad
Line back EMF coefficient	Ce	0.46 v.s/rad

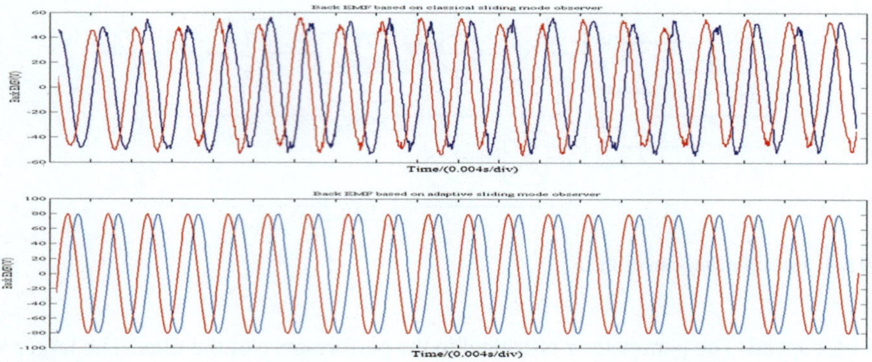

Fig. 5 Comparison curve of observed back EMF

Given step speed command $n_{ref} = 3000$ r/min and the switching gain parameter $k = 90$. The comparison curves are shown in Fig. 5. The back EMF signal using the classical algorithm has an obvious buffeting signal with an amplitude of about *55* V, which is about *34%* attenuated compared with the theoretical opposite EMF amplitude (*3000/9.55*0.46/1.732* = *83.43* V). While the improved ASMO algorithm can effectively filter the high-frequency buffeting signal, the amplitude of the back EMF is about 80 V, which is about 4% less than the theoretical opposite EMF amplitude.

So the advantage of the proposed scheme over the adaptive slide mode observer proposed in literature [17] can be summarized as follows. The amplitude of the back EMF has almost no attenuation based on the adaptive gate filter, while the output signal amplitude of the filter will be reduced to 0.707 times the amplitude of the pending signal based on the variable cut-off frequency first-order filter. In addition, the back EMF based on the adaptive gate filter has less harmonic content than that based on the variable cut-off frequency first-order filter.

The comparison curves of the estimated speeds observed by the classical SMO algorithm and the improved ASMO algorithm are shown in Fig. 6. It can be seen that the speed fluctuation range estimated by the classical SMO algorithm is about $\pm$ 30 r/min, while the speed fluctuation range estimated by the ASMO algorithm is about $\pm$ 20 r/min.

The comparison curves of the estimated rotor position and rotor position errors observed by the classical SMO algorithm and the improved ASMO algorithm are shown in Figs. 7 and 8 respectively. The error of the rotor position estimated by the classical SMO algorithm is about $-52°$ ~ $-35°$ compared with that of the real rotor position, and the lag angle is about 45°. Moreover, under different rotational speeds, the error fluctuation range changes. However, the error of the rotor position estimated by the improved ASMO algorithm is about $-$ 2°–2°, The lag angle between the estimated rotor position and the actual rotor position is about 2°.

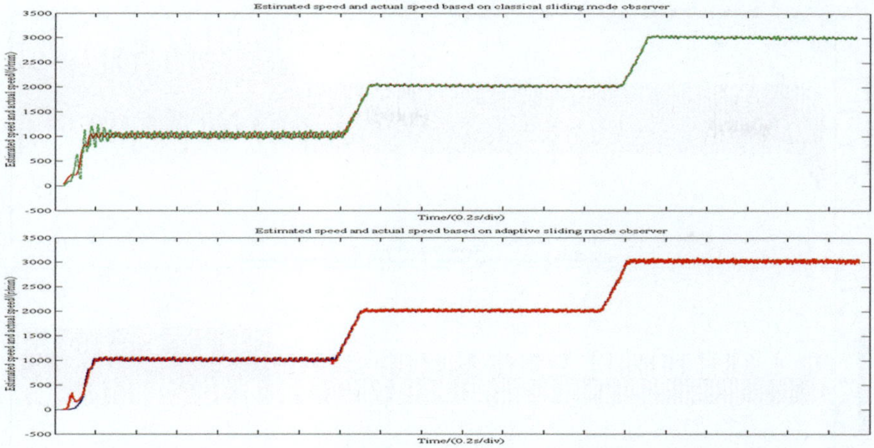

Fig. 6 Comparison curve of estimated speed

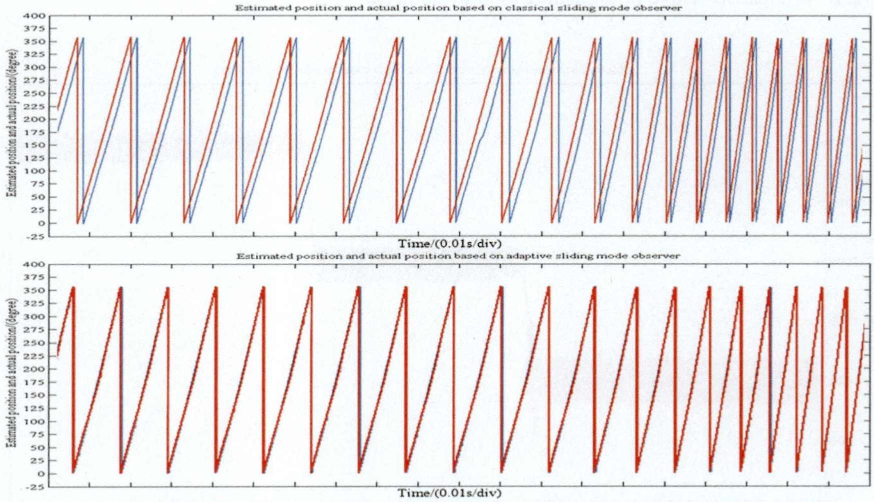

Fig. 7 Comparison curve of estimating rotor position

Apply the rotor position and speed obtained from the two algorithms separately to achieve closed-loop speed control of the motor. The rotor speed curves are shown in Figs. 9 and 10 respectively。

From Figs. 9 and 10, we can find that the motor can both operate stably when two algorithm are applied to sensorless speed control. Under sensorless operating conditions, the motor can achieve normal acceleration or deceleration operation and forward or reverse switching operation without speed commands. When the classical SMO algorithm is used, the motor speed fluctuation range is large about

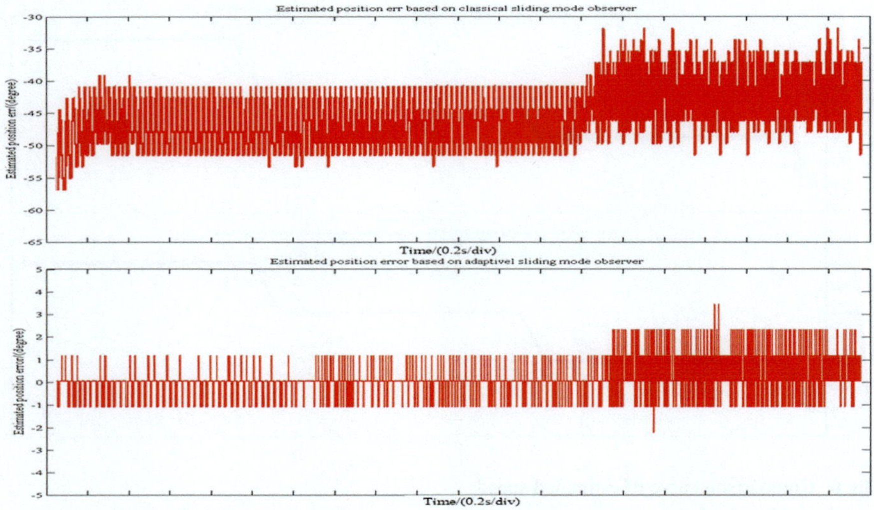

Fig. 8 Comparison curve of estimating rotor position err

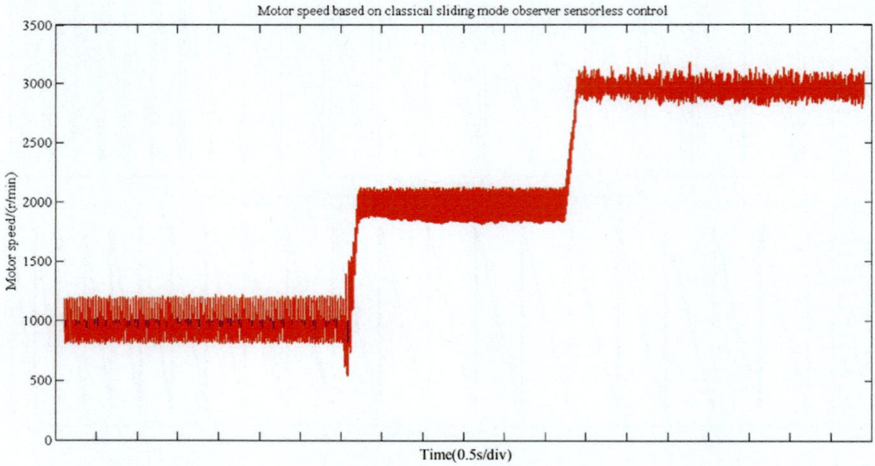

Fig. 9 The classical SMO algorithm used for closed-loop control (with rotor position compensation)

−120 r/min ~ +120 r/min, which is mainly due to the low position estimation accuracy. When the improved ASMO algorithm is used, the motor speed fluctuation range is small about *30 r/min ~ +30 r/min.* Comparing the running results of these two algorithms, it can be found that the improved adaptive ASMO algorithm can achieve higher precision speed servo control.

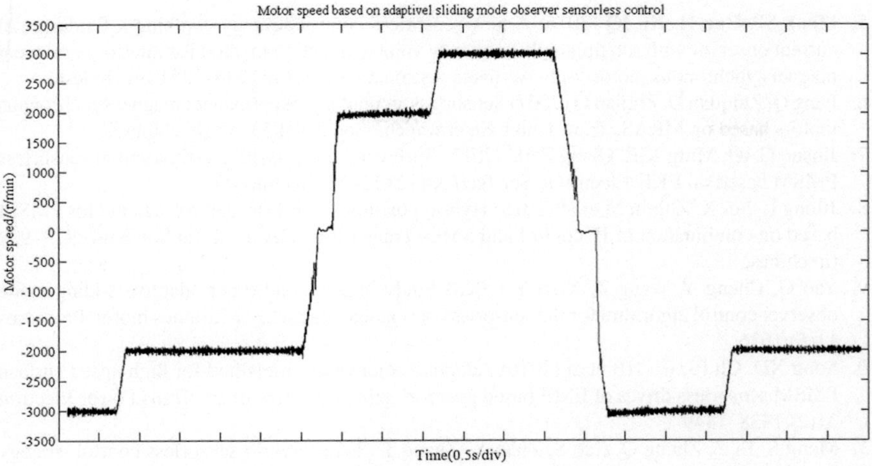

Fig. 10 SMO algorithm based on adaptive gate filter used for closed-loop control

5 Conclusion

In this paper, an improved ASMO algorithm using adaptive gate filter combined with saturation function as control function is proposed for the high-frequency chattering and phase shift of classical SMO. Based on the analysis of classical SMO algorithm, the adaptive gate filter algorithm is theoretically analysed and experimentally studied. From the experimental results, we can conclude that compared with the classical SMO algorithm, the improved ASMO algorithm can effectively filter the high-frequency buffeting signal and extract the back EMF with small amplitude attenuation and small phase lag, which requires no compensation in practical application. With the improved ASMO, the estimated rotor speed and rotor position accuracy is significantly improved. It has a small speed fluctuation range when applied to the PMSM closed-loop sensorless speed control.

References

1. Dutta R, Rahman MF (2008) Design and analysis of an interior permanent magnet(IPM) machine with very wide constant power operation range. IEEE Trans Energy Convers 23(1):25–33
2. Chen L, Sun XD, Jiang HB et al (2014) High performance control of a permanent magnet synchronous motor for electric vehicle applications. J Comput Theoret Nano sci 11(3):706–710
3. Dong C, Xia Y, Sun D, et al (2021) Analysis of loss on permanent magnet servo motor with high speed. Micromotors 54(3):5 (in chinese)
4. Gaolin W, Yang R, Dianguo X, et al (2013) DSP-based control of sensorless IPMSM drives for wide-speed-range operation. IEEE Trans Industr Electron 60(2):720–727

5. Meng SP, Guo H, Xu JQ (2016) A new sensorless control strategy combing a fundamental current observer with rotating high-frequency voltage injection method for interior permanent magnet synchronous motor. Acta Aeronaut Astronaut Sin 37(4):1336–1351 (in chinese)
6. Fang Q, Zhiquan D, Zhijian Q (2007) Sensorless technology of permanent magnet synchronous motors based on MRAS. Trans China Electro-techn Soc 22(4):53–58 (in chinese)
7. Jinsuo GAO, Ming XIE, Qiang ZHU (2017) Speed and rotor position estimation of sensorless PMSM based on EKF. Electronic Sci Tech 30(12):55–58 (in chinese)
8. Jilong L, Fei X, Zhiqin M et al (2018) Hybrid position sensorless, control scheme for PMSM based on combination of IF control and SMO. Trans China Electro techn Soc 33(4):919–929 (in chinese)
9. Yao G, Cheng Y, Wang Z, Xiao Y (2023) Study on a second-order adaptive sliding-mode observer control algorithm for the sensorless permanent magnet synchronous motor. Processes 11(6):1636
10. Song XD, Ch FJ, Ch HB et al (2016) Adaptive compensation method for high-speed surface PMSM sensorless drives of EMF-based position estimation error. IEEE Trans Power Electron 31(2):1438–1449
11. Meng S, Li Z, Zheng Q, Cao S, Zhao X, Zhang T (2022) Hybrid sensorless control strategy for PMSM based on combination of IF control and adaptive sliding mode observer. Springer Science and Business Media LLC
12. Luo X, Tang QP, Shen AW et al (2016) PMSM sensorless control by injecting HF pulsating carrier signal into estimate fixed frequency rotating reference frame. IEEE Trans Industr Electron 63(4):2294–2303
13. Wang GL, Yang L, Zhang GQ et al (2017) Comparative investigation of pseudorandom high frequency signal injection schemes for sensorless IPMSM drives. IEEE Trans Power Electron 32(3):2123–2132
14. Liu JM, Zhu ZQ (2014) Sensorless control strategy by square-waveform high-frequency pulsating signal injection into stationary reference frame. IEEE J Emerg Select Top Power Electron 2(2):171–180
15. Wen D, Deliang L, Zhanqiang L (2012) Position sensorless control of PMSM using SMO with two-stage filter. Electric Mach Control 16(11):1–10 (in chinese)
16. Siyi C, Youguo P (2016) Soft switching SMO for PMSM sensorless control. Trans China Electrotechn Soc 31(12):108–117 (in chinese)
17. Ding J, Wang A, Zhang T, Zhang H (2021) Research on sensorless control of permanent magnet synchronous motor based on adaptive sliding mode observer. In: 2021 4th International conference on circuits, systems and simulation (ICCSS). pp 195–201
18. Shuyi L, Weidong Z (2017) An adaptive sliding-mode observer with a tangent function-based PLL structure for position sensorless PMSM drives. Int J Electr Power Energy Syst 88:63–74
19. Shuaichen Y, Xiaoxian Y (2020) A modified flux slidingmode observer for the sensorless control of PMSMs with online stator resistance and inductance estimation. IEEE Trans Power Electron 35(8):8652–8662

High-Sensitivity and Low-Noise Interrogation for Spindle Fault Monitoring

Jie Wang, Yue Gang Tan, and Yi Liu

Abstract To reduce the error, the stiffness of the high-precision machine tool spindle is often remarkably high, and the strain, when subjected to force, is extremely small. It is difficult for traditional strain monitoring methods to monitor the failure of the spindle. In this paper, enhanced sensitivity of the fiber Bragg grating (FBG) interrogation system is proposed, which can increase the sensitivity of traditional FBG strain sensors at high bandwidth (0–4 kHz). The system measurement sensitivity reached 96.9 mV/με, and the minimum strain reached 0.1 με. High-stiffness spindle misalignment faults and bearing outer ring faults can be identified.

Keywords Spindle · Strain monitoring · Sensitivity · FBG Strain Sensors

1 Introduction

Reducing error has always been the goal pursued by high-precision machine tools. The primary source of machine tool error occurs when the spindle is deformed by force. The machine tool spindle design stiffness is extremely high to reduce deformation. Therefore, the strain induced by force when the faults occur is minimal. The traditional strain sensor is difficult to get the fault signal. The machine tool spindle fault monitoring usually is used, oil sensors [1], thermography [2], and acceleration vibration sensors [3–6] as a signal acquisition. For the acceleration detection method, vibration noise is introduced from other system components, which can reduce the signal-to-noise ratio and affect signal identification [7].

References [7–9] used the narrowband slope interrogation method of FBG to measure the strain size of gears and was able to identify the fault characteristics of the teeth at the wheel. Reference [10] used the narrowband slope interrogation method of FBG to measure the fault characteristics of the spindle and the fault size. Despite these improved sensitivities, it is still difficult to meet the measuring requirements for

J. Wang · Y. G. Tan · Y. Liu (✉)
School of Mechanical and Electronic Engineering, Wuhan University of Technology, Wuhan 430070, China
e-mail: Wgdliuyi2001@whut.edu.cn

Z. Liu et al. (eds.), *Advances and Challenges in Advanced Unmanned Aerial Systems*, Springer Aerospace Technology, https://doi.org/10.1007/978-981-99-8045-1_37

high-stiffness spindles. In addition, there was randomness in setting the position of the working wavelength. This leads to inconsistency between the sensitivity obtained at the calibration platform and the sensitivity used by the equipment at the actual work site.

This paper proposed a method to enhance the sensitivity of the measurement system. The theoretical analysis found system sensitivity at the point of maximum sensitivity. The sensitivity of the interrogation system was calibrated by cantilever beam experiments to determine the relationship between the location of the sensitivity and the operating wavelength. Experimental tests are conducted on a faulty spindle with high stiffness using the interrogation system.

2 Principle

The schematic diagram of the narrowband slope interrogation system is shown in Fig. 1a. The setup includes a laser, optical circulator, FBG, photodetector (PD), and data acquisition (DAQ). The FBG reflects a specific wavelength of light. The reflected light passes through port 2 of the optical circulator, and the PD receives the light intensity from port 3, converting the light intensity into a voltage value. Therefore, the relationship between the DAQ output voltage and the FBG reflection peak can express by Eq. (1).

$$U(\lambda) = \alpha R(\lambda) \tag{1}$$

where $\alpha = G_{\mathrm{loss}}.G_{PD}.P_{source}$, G_{loss} is the optical path loss. G_{PD} is the enlargement factor of the DAQ, which is also a constant. Due to the very tight bandwidth of the laser output wavelength, the optical power at the corresponding wavelength is a continuous value P_{source}. $R(\lambda)$ is the spectral reflection peak function.

Set the output wavelength of the system light source to the maximum sensitivity point. When the FBG is affected by strain, the reflection spectrum move left and right,

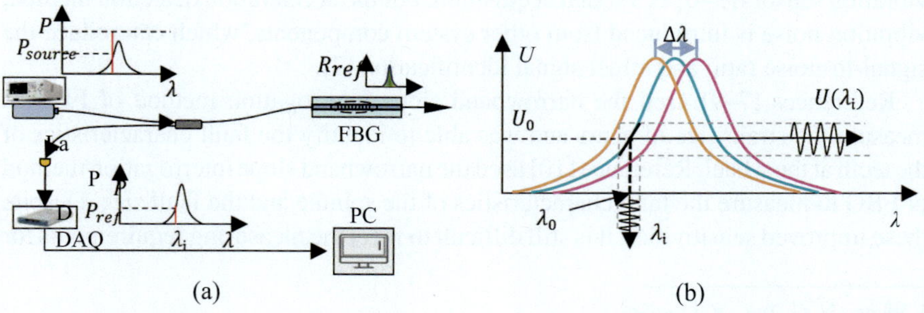

Fig. 1 **a** Schematic diagram of the FBG narrowband slope interrogation system. **b** Relationship between light intensity voltage and wavelength

resulting in the signal output change, as shown in Fig. 1b. The measured voltages are expanded by first-order Taylor at the maximum sensitivity operating point.

$$U(\lambda) = U(\lambda_0) + U'(\lambda_0)(\lambda - \lambda_0) \tag{2}$$

3 Experiments and Discussion

The narrowband slope interrogation device is shown in Fig. 1. The various setup used are as follows: the laser is the TSL-550 model of Santec Corporation, the laser line width is 40 MHz after the laser Coherence mode is turned on Coherence mode is turned off, and the laser linewidth is 10 kHz. PD is PDA10D2 of THORLABS Corporation, the bandwidth is 25 MHz, the DAQ (NI instrument) model is NI-6351, and the maximum sampling rate is 1.25 M/s. The ADC is 16 bits. The traditional electrical acceleration vibration sensor is an LMS system with a sampling frequency of up to 102.4 k.

3.1 Static Experiments

The purpose of the static experiment is to obtain the sensitivity of the interrogation system. The sensitivity was calibrated using the experimental setup within Fig. 2. The experiment obtained the wavelength position of the optimal operating point λ_0=1542.99 nm. Keeping the cantilever beam stable, the weights were regulated from 0 to 10 g, maintaining a step length of 2 g. The weight was applied to the loaded area of the cantilever beam. The response curve of the integration system was characterized.

Figure 3 shows the variation of the system when receiving different loads, using electrical sensors as a reference. It can be seen that the strain variation measured with strain gauges is about 1.6$\mu\varepsilon$ when the end load is 2 g each in Fig. 3a. Meanwhile, the voltage variation of the interrogation system is about 170 mV. From the partially enlarged detail in Fig. 3a, it can be known that the fluctuation is about 10 mV when the end load is 2 g and 8 g. The linear fitting curve is shown in Fig. 3b. It shows that the actual data points match the fitted curve very well. In considering the uncertainty of the cutoff of the fitting curve, the expanded uncertainty was calculated under a confidence interval of 0.9990 as 96.9 mV/$\mu\varepsilon$.

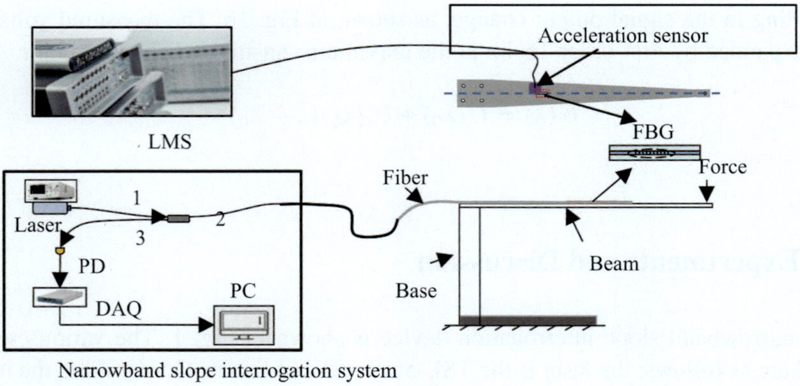

Fig. 2 Cantilever beam experimental setup

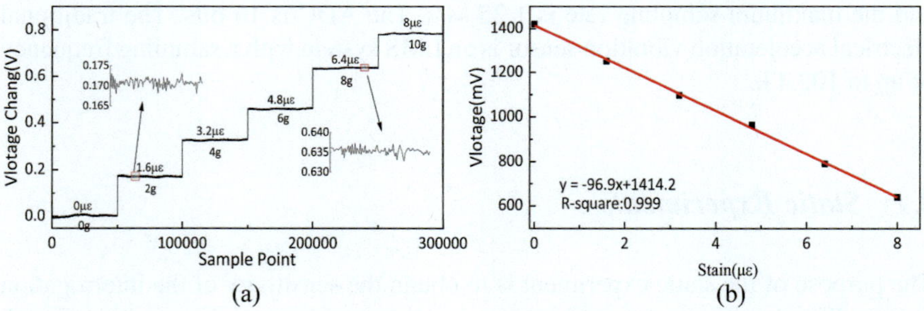

Fig. 3 **a** Cantilever beam static loading test results. **b** The sensitivity curve characterized

3.2 *Spindle Fault Diagnosis Experiments*

Spindles, such as bearing damage, and lubrication failure, will also appear after running. There are many common forms of failure, such as spindle axis misalignment and bearing failure. References [2, 3], the characteristic frequency expression of the outer ring failure when the rolling bearing fails is as follows:

$$f_{bw} = 0.5Z(1 - \frac{d}{D}\cos\alpha)f_r \tag{3}$$

In Fig. 4a, where D is the diameter of the bearing, d is the diameter of the rolling element, α is the contact angle, Z is the roller number, f_r is the relative rotational frequency of the inner and outer rings, In Eq. 3, f_{Bw} is the fault frequency of the outer ring.

Failures of spindles often have multiple failures occurring at the same time. In this paper, only bearing damage and spindle misalignment are considered. According

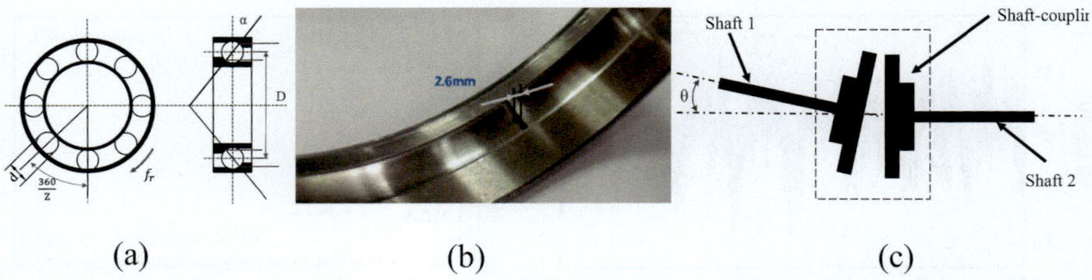

Fig. 4 **a** angular contact bearing diagram. **b** 2.6 mm size inner ring failure. **c** angular misalignment failure

to literature [11], the manifestation of axial misalignment faults is often shown as frequency multiplication in the frequency domain diagram.

To verify that the interrogation system can measure the effective signal in the case of high spindle stiffness. In Fig. 4b, a destructive notch is created on the bearing to simulate the occurrence of an outer ring failure. When assembling the bearing, let the shaft 1 and the shaft 2 have a certain angle, which is used to simulate the misalignment of the spindle in Fig. 4c. The bearing model used in this experiment is the HRB7012 AC angular contact matching bearing, which has a 2.6 mm notch on the inner wall of the outer ring of the bearing in Fig. 5. The parameters of the bearing are shown in Table1.

The vibration signals obtained by the FBG sensor and the piezoelectric sensor using the narrowband slope interrogation system show differences, as depicted in Fig. 6a and c. Figure 6a clearly exhibits periodicity, while Fig. 6c does not show any clear periodic results. By transforming the signals using a fast Fourier transform (FFT) and analyzing the frequency-domain signals in Fig. 6b and d, the frequencies of the motor can be obtained under the power supply frequency of 40 Hz. The experimental results in Fig. 6b reveal distinct characteristic frequencies, such as 40, 80.4, 120.4, 310.6, 621.2, 931.2, and 1241.6 Hz. Using Eq. (3) and Table 1, the theoretical failure frequency of the outer ring is determined to be 313.78 Hz under the rotational frequency of 40 Hz. Therefore, the 40 Hz signal represents the motor rotation

Fig. 5 Testing device for simulating high-stiffness spindle failure

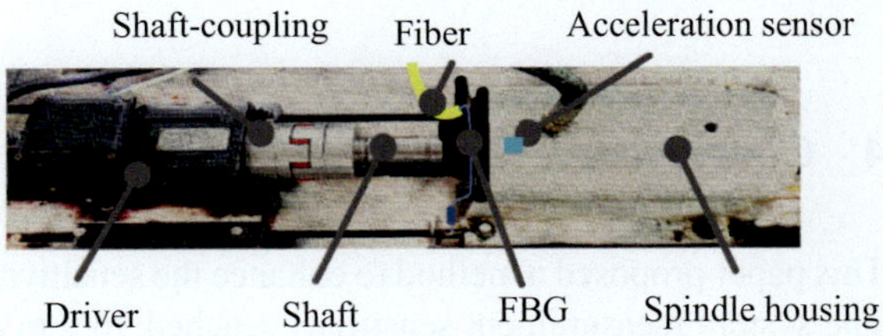

Table 1 Bearing parameters

Parameters	*D*	*d*	α	Z
Value	77.5 mm	10.98 mm	25°	18

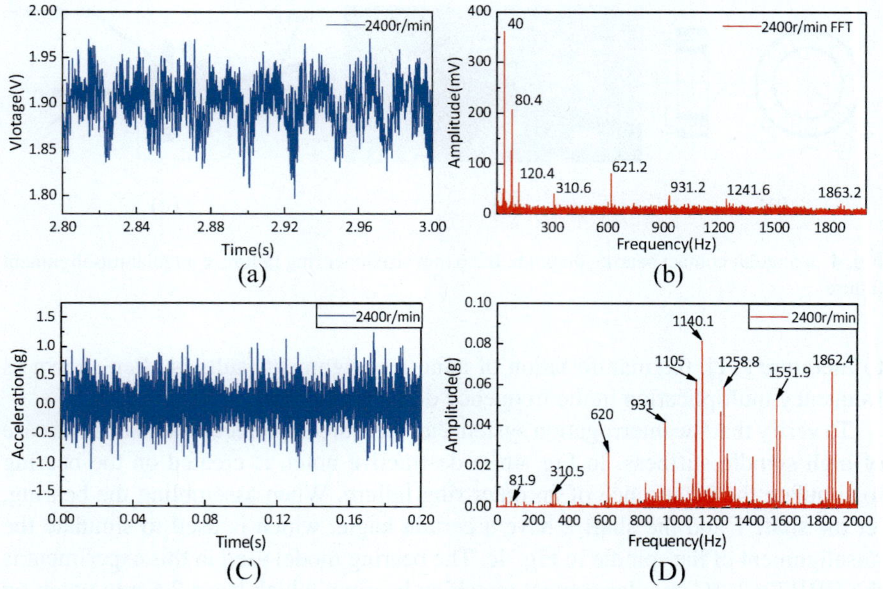

Fig. 6 Bearing failure signal measurement. **a** and **b** the narrowband slope interrogation system measure. **a** Time domain diagram **b** Frequency domain diagram **c** and **d** the piezoelectric accelerometers measure. **c** Time domain diagram **d** Frequency domain diagram

frequency, and the 310.6 Hz signal is the practical failure frequency. A comparison of the theoretical and practical measurements indicates a slight deviation, which may be attributed to manufacturing accuracy errors and non-uniform motor speed. The frequencies of 80.4 and 120.4 Hz are multiples of the motor rotation frequency, while 310.6, 580.6, 621.2, 931.2, 1241.6, and 1863.2 Hz are the failure frequencies caused by spindle misalignment, which results in frequency multiplication. The frequency-domain signal in Fig. 6d has too many frequencies, low signal-to-noise ratio, and lacks sufficient contrast between signals, making it difficult to accurately identify fault frequencies.

4 Conclusion

This paper proposed a method to enhance the sensitivity of the measurement system. The system measurement sensitivity reached 96.9 mV/$\mu\varepsilon$, and the minimum strain reached 0.1$\mu\varepsilon$. The experimental results show that the interrogation system has a better discriminatory ability for outer ring faults and spindle misalignment. In this paper, we verify the feasibility of enhancing the sensitivity of this sensor from theoretical analysis and experiments. Due to the high sensitivity of this method, it may have broad application prospects in some weak strain fields.

References

1. Mihara Y, Someya T (2002) Measurement of oil-film pressure in engine bearings using a thin-film sensor. Tribol Trans 45(1):11–20. https://doi.org/10.1080/10402000208982516
2. Touret T, Changenet C, Ville F, Lalmi M, Becquerelle S (2018) On the use of temperature for online condition monitoring of geared systems—a review. Mech Syst Signal Process 101:197–210. https://doi.org/10.1016/j.ymssp.2017.07.044
3. Chaudhari YK, Gaikwad JA, Kulkarni JV (2014) Vibration analysis for bearing fault detection in electrical motors. In: 2014 First international conference on networks & soft computing (ICNSC2014). Guntur, Andhra Pradesh, India, pp 146–150. https://doi.org/10.1109/CNSC.2014.6906681
4. Wang P, Zhao Y, Tian B, Li Y, Liu Y (2015) A MEMS accelerometer for spindle monitoring system. In: 10th IEEE international conference on nano/micro engineered and molecular systems. Xi'an, China, pp 545–549. https://doi.org/10.1109/NEMS.2015.7147488
5. Postel M, Aslan D, Wegener K, Altintas Y (2019) Monitoring of vibrations and cutting forces with spindle mounted vibration sensors. CIRP Ann 68(1):413–416. https://doi.org/10.1016/j.cirp.2019.03.019
6. Zhang X, Niu H, Hou C, Di F (2020) An edge-filter FBG interrogation approach based on tunable Fabry-Perot filter for strain measurement of planetary gearbox. Optic Fiber Technol 60:102379. https://doi.org/10.1016/j.yofte.2020.102379
7. Song H, Song E, Liu Y, Wang B, Wang Q, Wang Z (2023) Miniature FBG vibration sensor with high performance and low angle dependence for two-dimensional vibration measurement. IEEE Trans Instrument Meas 72:1–11. https://doi.org/10.1109/TIM.2022.3223061
8. Niu H, Zhang X, Hou C (2017) An approach for the dynamic measurement of ring gear strains of planetary gearboxes using fiber bragg gratings. Sensors (Basel) 17(12). https://doi.org/10.3390/s17122872
9. Huo Z, Zhang Y, Francq P, Shu L, Huang J (2017) Incipient fault diagnosis of roller bearing using optimized wavelet transform based multi-speed vibration signatures. IEEE Access 5:19442–19456. https://doi.org/10.1109/access.2017.2661967
10. Mao K, Wu Y (2011) Fault diagnosis of rolling element bearing based on vibration frequency analysis. Presented at the 2011 Third international conference on measuring technology and mechatronics automation
11. Lees AW (2007) Misalignment in rigidly coupled rotors. J Sound Vib 305(1–2):261–271. https://doi.org/10.1016/j.jsv.2007.04.008

Design of Control System for Precision Parts Cleaning Equipment Based on PLC

Yehui Ding, Yanzhi Guan, and Ziliang Pang

Abstract The traditional cleaning process of precision parts is completed by manual operation and ultrasonic machine. In order to solve problems such as high labor intensity, low efficiency, inaccurate solvent injection and organic solvents are harmful to human health, etc. A precision parts cleaning equipment with ultrasonic cleaning technology as the core was designed. The automatic control system for equipment takes Siemens PLC as the core, uses mechanical arm to cooperate to grasp parts, uses weighing method to inject cleaning solvent, and uses Siemens HMI to realize human–machine interaction function. A simple and effective filtering algorithm is designed to ensure more accurate cleaning solvent injection. The field test shows that the equipment can adapt to the field cleaning environment, the control system has reasonable functions and can clean according to the corresponding process, which greatly reduces manual participation and improves the cleaning efficiency.

Keywords ET200SP PLC · Analog quantity processing · Ultrasonic cleaning

1 Introduction

Precision parts include bearings and shafting parts. Precision parts are not only important components of some precision instruments, but also widely used in all walks of life. Precision parts play an important role in research, military defense, aerospace, medical equipment and precision instrument. Precision parts need to be cleaned before use, and their surface cleanliness is a necessary condition for a normal operation of parts. The surface cleanliness of the cleaned precision parts will be greatly improved, and the life and performance of the parts can be guaranteed. There are many cleaning methods for precision parts. Xue [1] compares the four cleaning methods for bearings. It is concluded that the best cleaning method for bearings is using ultrasonic wave, and the cleanliness of bearing surface is the highest after

Y. Ding (✉) · Y. Guan · Z. Pang
North China University of Technology, Beijing, China
e-mail: Skies_hui@126.com

Z. Liu et al. (eds.), *Advances and Challenges in Advanced Unmanned Aerial Systems*, Springer Aerospace Technology, https://doi.org/10.1007/978-981-99-8045-1_38

ultrasonic cleaning. Therefore, ultrasound cleaning technology is more and more used in precision parts cleaning.

At present, cleaning of precision parts mainly depends on the manual operation and ultrasonic cleaning machine. Problems such as low efficiency of manual cleaning, inaccurate solvent injection and new contamination may be introduced during solvent injection. In addition, the volatilization of some organic cleaning solvents also brings harm to the health of workers. Using the green, the safe and flexible design concept and the mature automatic control technology, the precise parts automatic cleaning equipment based on PLC is designed. The control system of the equipment takes Siemens PLC as the core and programming according to the process flow to realize the detection and transmission of mechanical action and the electric signal. HMI realizes human–machine interaction and can set parameters and can display real-time status of the whole equipment. The application of this equipment can effectively solve the problems existing in the current traditional manual cleaning process.

2 Control System Analysis

Precision parts cleaning equipment mainly consists of mechanical system and control system. The overall structure is shown in Fig. 1. The control system of cleaning equipment consists of PLC control unit, servo movement unit, mechanical arm unit, solvent injection unit, waste liquid collection unit, ultrasonic cleaning unit and HMI human–machine interaction. All components are centrally controlled by the PLC control unit. The control system structure is shown in Fig. 2.

Fig. 1 Integral structure

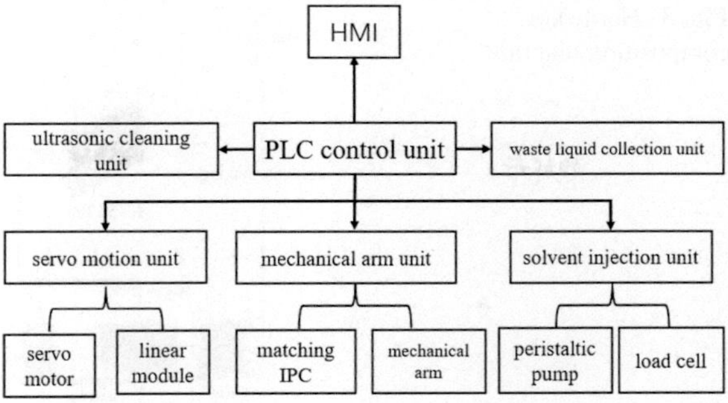

Fig. 2 Control system structure

3 Hardware Composition of the Control System

3.1 Hardware Configuration of Control System

The control system uses an industrial three-level control structure composed of management, decision-making and executive levels. (1) At management level, mainly consists of PC, which communicates with PLC via Ethernet. Hardware configuration, network communication configuration, programming and monitoring debugging are carried out on PC by TIA Portal V15.1 software. (2) At the decision-making level, Siemens ET200SP PLC is the core decision-making unit, which makes logical judgment on signals such as external buttons, sensors and network communication equipment, and provides corresponding signals to the executive level. (3) At the execution level, AUBO collaborative robot arm, peristaltic pump, electromagnetic pinch valve, AC contactor, ultrasonic cleaner, servo linear module and sensor detection are the main components. AUBO collaborative arm communicates with PLC via Modbus TCP [2]. Sensor detection mainly includes photoelectric sensor, magnetic proximity switch and weighing sensor. The load cell converts the parameter signal into 0–10 V voltage signal which is sent to CPU by PLC analog module for data processing and logic operation. The servo motor is controlled by Siemens V90 servo driver and powers the movement of the arm's root with linear module [3]. The HMI consists of Siemens KTP1200, which adjusts the process parameters and displays the status of the automatic control system [4]. ET200SP PLC communicates with Siemens touch screen by PN. The hardware composition of the control system is shown in Fig. 3. V90 servo driver is AC servo driver. The power density and acceleration characteristics of the ac servo drive are bing recognized as being superior to those of the dc counterpart. The ac servo family consists of permanent-magnet synchronous, dc brushless, induction, and switched reluctance motor drivers [5].

According to the control requirements of cleaning equipment, a total of 120 DI points, 40 DQ points and 7 AI analog quantities are required. SIMATIC ET200SP series PLC is selected, CPU model is 1510SP-1 PN, PM207 5A power module. BA 2xRJ45 module provides an expansion interface for PN communication of V90 servo driver and HMI. In DI module, DI 16 × 24VDC ST is used for signal input of

Fig. 3 Hardware composition diagram

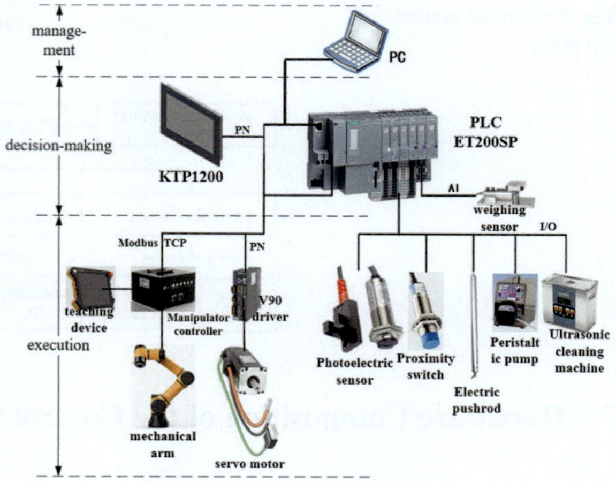

equipment operation button, digital sensor and AC contactor status. In DQ module, DQ 16 × 24VDC/0.5A ST is used for the output of digital control signals such as equipment AC contactor, electromagnetic relay and indicator light. In AI module, AI 4×I 2-,4-wire ST is used for analog input of load cell signal in the equipment.

The cleaning equipment control system designed in this paper has the following advantages: (1) The control system development software is TIA Portal V15.1. Compared with the previous S7-200 PLC development software, it is more flexible and efficient to complete the project, and the development platform function is more complete, especially the on-line diagnosis function, which brings convenience to the commissioning work. (2) KTP1200 touchscreen is used on-site to improve the flexibility of system parameter setting and enhanced personalization of equipment status display. (3) Industrial level three-level control structure with clear division of labor at each level and coordination between units enables the control system to operate efficiently and stably.

3.2 System Core Unit Analysis

Cleaning Solvent injection Device: The detailed structure is shown in Fig. 4. It mainly includes peristaltic pump, 5000 mL solvent bottle, weight sensor and tray.

Two stainless steel pagoda straight connections are installed at the bottle opening of the solvent bottle. One of the pagoda straight connections is connected to the hose of the peristaltic pump at the top and to the glass tube at the bottom. The bottom opening of the glass tube leads to the bottom of the solvent bottle for easy liquid extraction. Cleanliness of solvents is guaranteed by inserting glass tubes into solvent bottles. The connection hose is made of fluororubber with excellent heat and corrosion resistance. Fluororubber reduces the risk of connection hoses being corroded by organic solvents.

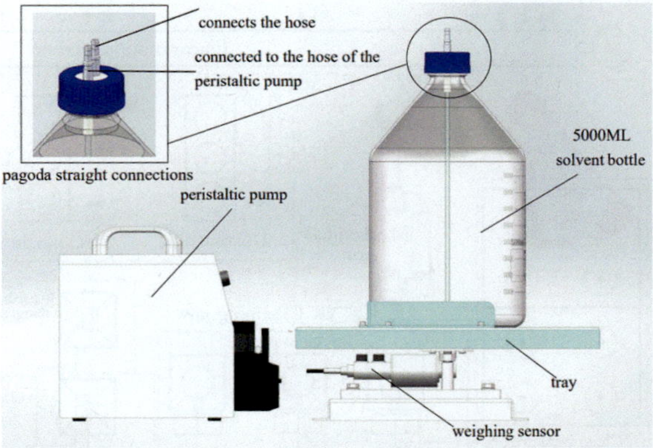

Fig. 4 Solvent injection device

Peristaltic pump as power source for injection solvent. Cleaning solvents in the transfer process, peristaltic pump will not contact with the solvent, to maximize the cleanliness of the solvent. New hoses can be replaced quickly when they are damaged or contaminated.

The top opening of the other pagoda straight-through joint connects the fluororubber hose, which connects the normally closed electromagnetic pinch valve not far from the bottle opening. When the peristaltic pump is injected with solvent, the electromagnetic pinch valve loosens the fluororubber hose to make the inside of the bottle communicate with the atmosphere. When the peristaltic pump does not work, the electromagnetic pinch valve clamps the fluororubber hose to isolate the solvent inside the bottle from the atmosphere. Thus, the solvent can be avoided from volatilizing out of the bottle, which can save the solvent and reduce the safety risk caused by solvent volatilization.

The measuring range of the weighing sensor is 10 kg, and the accuracy is 0.1%. It can accurately weigh the solvent, thus realizing the closed-loop control of solvent injection. The load cell is a resistance strain type cantilever beam load cell, which has the advantages of high accuracy, low drift, wide range, simple structure, wide application range, strong anti-bias load capability, and can be used in relatively poor environments.

4 Program Design of Control Systems

4.1 Cleaning Process Flow of Control System

The working areas for cleaning equipment include beaker area, solvent injection area, ultrasonic cleaning area, parts to be cleaned area, parts drying area, waste liquid collection area, beaker cleaning area. Its layout is shown in Fig. 5.

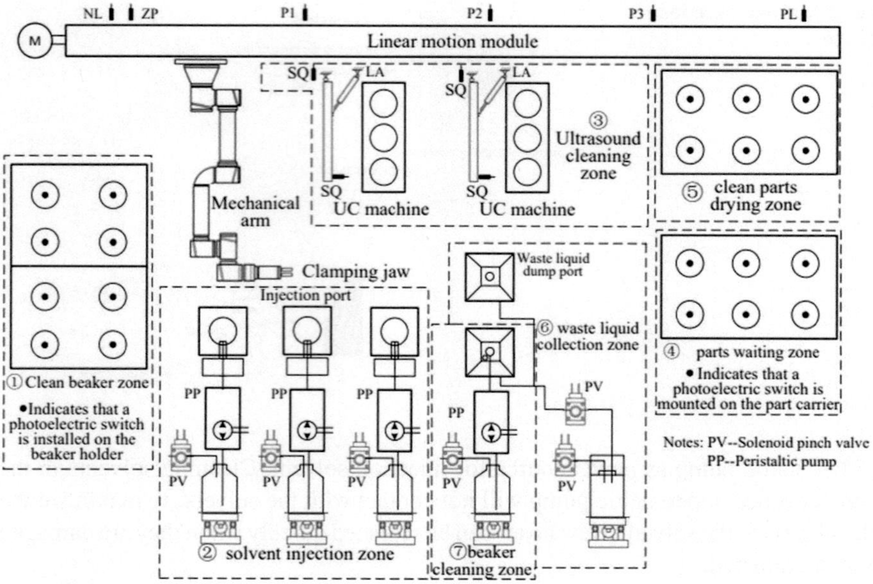

Fig. 5 Working area distribution

According to the cleaning process requirements of precision parts, the process flow of the automatic cleaning system is as follows: First, the mechanical arm grabs the beaker to the cleaning solvent fill port. After the solvent is filled, place the beaker in the ultrasonic cleaning machine (transfer ultrasonic energy with water between beaker and ultrasonic machine). Then the mechanical arm clamps the precision parts into a beaker injected with solvent. Then close the cover of the ultrasonic machine. During the cleaning process, the cover of the ultrasonic cleaning machine is closed and record time. After the cleaning time is completed, the cover of the ultrasonic cleaning machine is opened. The mechanical arm takes the parts and puts them in the specified drying area of clean parts. Then take the beaker out of the ultrasonic machine and pour the waste liquid in the beaker into the waste liquid recovery device in the waste liquid collection area. Then transfer the beaker to the beaker cleaning area to clean the beaker, and put the beaker back into the beaker area after cleaning. In this way, an ultrasonic cleaning of a precision part with a cleaning solvent is completed. The working area passed in the cleaning process of parts is ①–⑦ in Fig. 5. Similarly, automatic cleaning of multiple parts can be realized. Different kinds of cleaning solvents can be added by using different peristaltic pumps to avoid mutual pollution of different solvents. This equipment provides three sets of solvent injection devices (there are three common cleaning solvents). It can realize independent cleaning of three different solvents and cleaning in sequence of solvents.

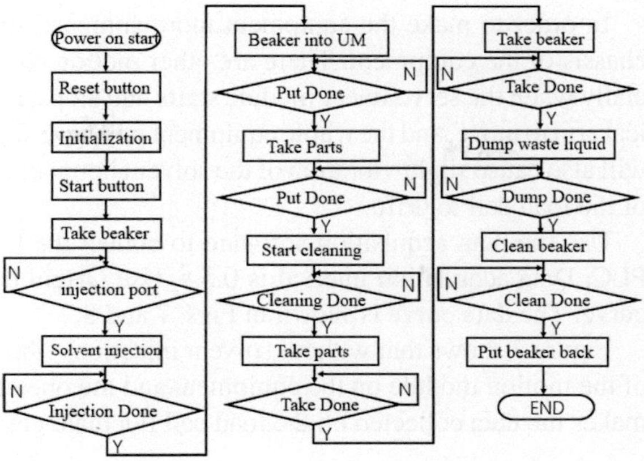

Fig. 6 Process flow chart

4.2 *Programming*

Before PLC programming, the control flow chart shall be designed according to the needs of the control system. The system control process flow diagram is shown in Fig. 6. According to the process requirements of cleaning and the action flow of the mechanical arm, the modular programming idea of PLC is applied in this system. The program is divided into the following modules: Main procedure, beaker clamping procedure, solvent injection procedure and part clamping procedure. Modular programming, simple and clear program structure, independent and interrelated program modules, and convenient program upgrade and maintenance. Through modular programming, the stability and reliability of PLC program operation have been greatly improved, and the equipment commissioning progress has also been greatly improved [6].

4.3 *Analog Quantity Processing Method of Weighing Module in the System*

The load cell uses a 0–10 V voltage type sensor. The AI module transmits the 0-10 V analog signal to the PLC CPU for processing, and the processed signal will be converted into a number between 0 and 27,648. Therefore, it is necessary to standardize the collected signal data of the weighing sensor to calibrate the measuring range of the weighing sensor. Convert the data collected by the load cell into 0–5000 g. Use $V = M/\rho$ Calculate volume. Store the volume value in the program WSS_value_R(self defined variable name) variable, which is the actual volume of solvent in the solvent bottle.

In order to make the equipment more compact, the load cell is installed on the chassis of the equipment. There are other motion modules on the equipment, especially when the servo linear module starts and stops, it drives the root of the mechanical arm to move, and the whole equipment will have slight vibration. These vibrations will also cause slight vibration of the solvent bottle on the load cell, causing the value of the load cell to drift.

Use the data acquisition software to collect the load cell data displayed on the PLC. Data acquisition interval is 0.5 s. Use Origin drawing software to draw data curve. The data curve is shown in Figs. 7 and 8.

Figure 7 shows that without solvent injection, vibration caused by the movement of the motion module on the equipment and the operation of the peristaltic pump, it makes the data collected on the load cell fluctuate greatly, the maximum fluctuation

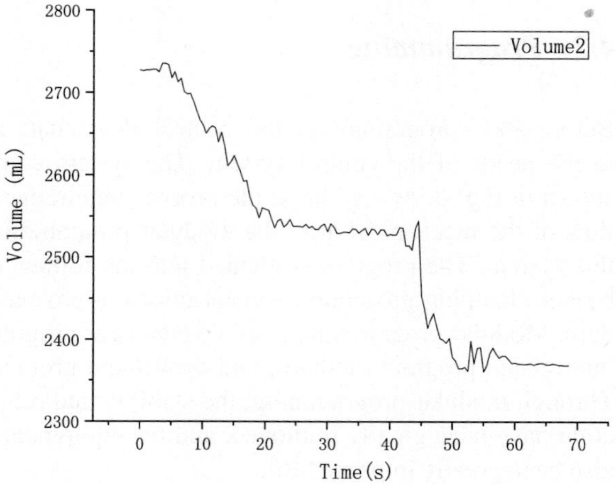

Fig. 7 Without solvent injection

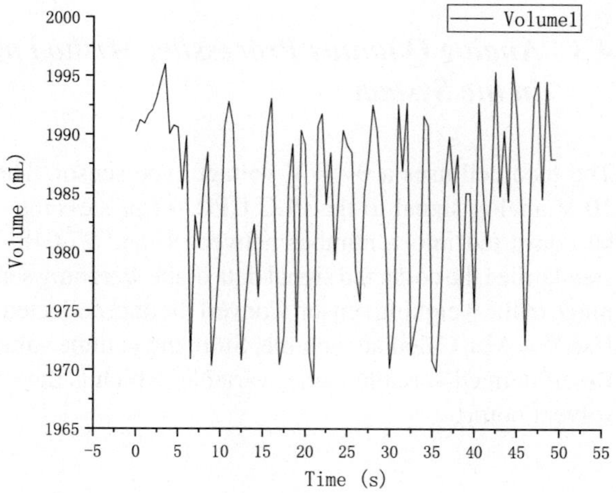

Fig. 8 Solvent injection curve

is about 25 mL. This seriously affects the accuracy of solvent injection. Figure 8 shows that without injection function, the data collected by normal solvent injection is as shown in the figure, the zigzag shape of the curve is very serious. This seriously affects the accuracy of solvent injection.

In order to reduce the solvent injection error caused by the above fluctuations, the analog quantity is filtered. There are many methods of analog filtering. Humin Yan [7] elaborated three common filtering algorithms: arithmetic average method, sliding average method and strong anti-interference average method. The system adopts arithmetic mean filtering algorithm. The algorithm formula:

$$\overline{x} = \frac{1}{n}\sum_{i=1}^{n} x_i \tag{1}$$

In the formula (1), the greater the value of n, the better the smoothness of the numerical curve, but the sensitivity will be reduced. The value adopted by the system through many experiments is n = 10.

After processing with arithmetic mean filtering algorithm, the collected data are plotted into curves, as shown in Figs. 9 and 10.

Figure 9 shows that without liquid injection, the weight figure drifts around 3 mL due to the vibration caused by the movement of other modules and the operation of the peristaltic pump. Compared with Fig. 7, the smoothness of the curve is improved. Figure 10 shows that after the introduction of filtering function, the smoothness of the curve is also improved compared with Fig. 8 under the condition of solvent injection.

After adding filtering algorithm to the analog quantity in the program, although the error fluctuation is relatively small, it will also cause other effects on the control. Volume quantitative judgment is used in the system as the starting condition of the peristaltic pump. Because of small fluctuations, the peristaltic pump will start

Fig. 9 Without solvent injection (filtering)

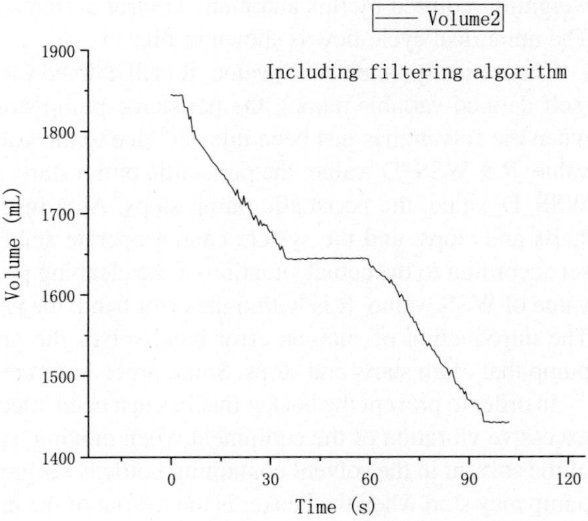

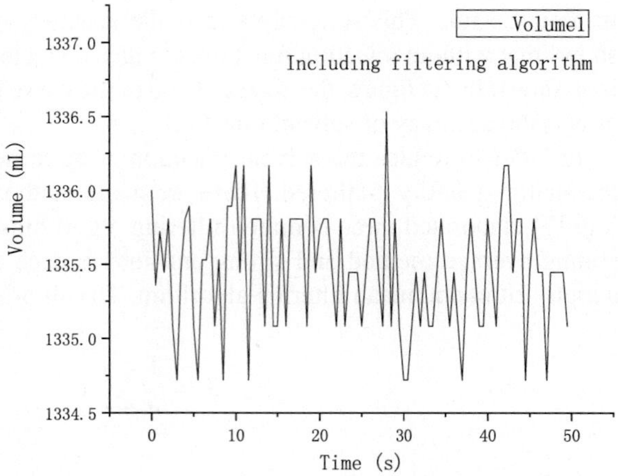

Fig. 10 Solvent injection curve (filtering)

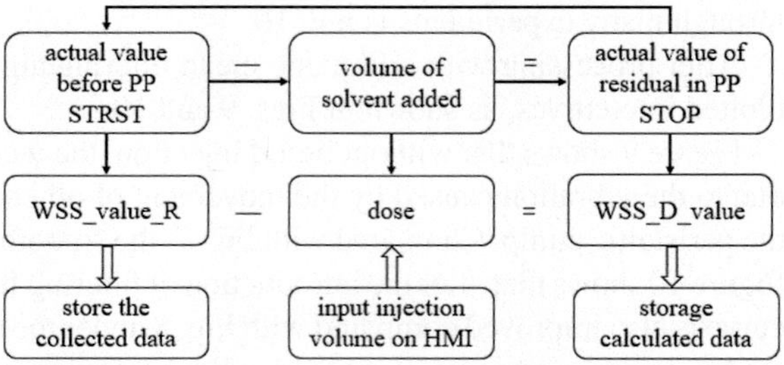

Fig. 11 Numerical cycle flow

repeatedly at the end of injection. In order to avoid repeated starting of peristaltic pump near the critical value, threshold filtering function (error band) is added to the program, so that the system operation can avoid the impact of small fluctuations. The weighing required by this automatic control system is a dynamic D-value process. The numerical cycle flow is shown in Fig. 11.

Theoretically, when WSS_value_R (self defined variable name) = WSS_D_Value (self defined variable name), the peristaltic pump stops. Especially at the moment when the solvent has just been injected, due to the volume fluctuation, when WSS_value_R > WSS_D_value, the peristaltic pump starts again; when WSS_value_R < WSS_D_value, the peristaltic pump stops. As a result, the peristaltic pump often starts and stops, and the system cannot operate stably. The ± 3 mL error band is set according to the actual situation of the cleaning process requirements. When the value of WSS_value_R is within the error band, the peristaltic pump stops working. The introduction of such an error band solves the unstable state of the peristaltic pump that often starts and stops. Some procedures are shown in Fig. 12.

In order to prevent the beaker that has just been injected with solvent from causing excessive vibration of the equipment when moving, so that the amount of vibration of the solvent in the solvent containing bottle is far greater than 3 mL, the peristaltic pump may start when the beaker is moved out of the injection port. For this purpose,

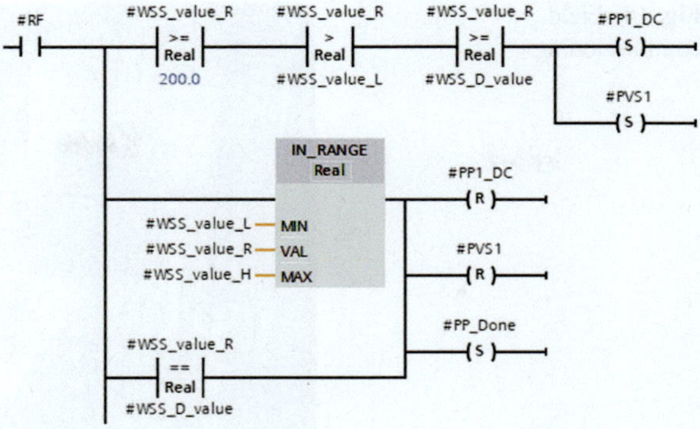

Fig. 12 Some procedures

a safety procedure section is added using the photoelectric sensor at the solvent injection port. When the beaker is removed, no matter what the solvent value is, the peristaltic pump cannot be started.

4.4 *HMI Screen Design*

The control system of precision parts cleaning equipment integrates many functions on HMI interface, realizing human–computer interaction. Some interfaces are shown in Fig. 13. According to the system control requirements, the automatic and manual control interface is developed. The screen can realize the system start, stop and other functions.

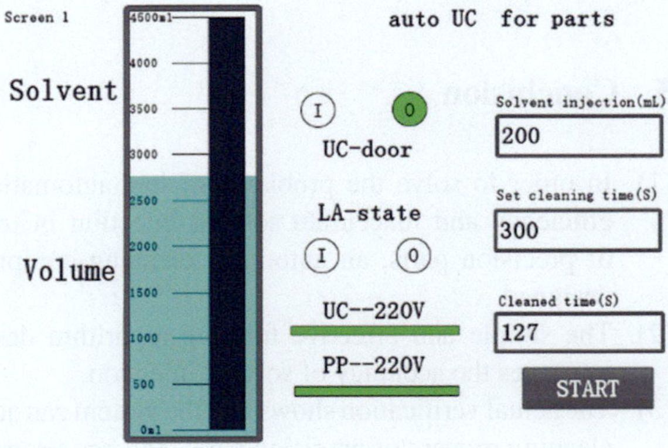

Fig. 13 Some interfaces

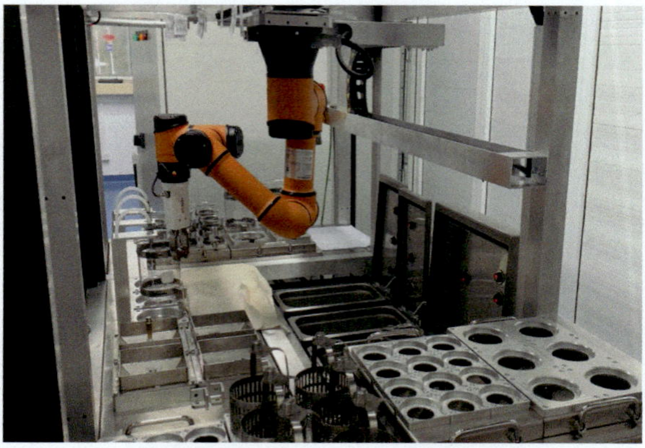

Fig. 14 Field commissioning

4.5 *Field Test*

The on-site commissioning was carried out according to the process flow of solvent injection of 200 mL and ultrasonic cleaning of 5 min. Field commissioning diagram is shown in Fig. 14. The accuracy of linear module in repeated positioning can meet the requirements of the equipment. During the field debugging, the accuracy of the end of the mechanical arm can still accurately grasp the parts in the multi-point repeated positioning, meet the operating requirements of the equipment. The experiment shows that the automatic cleaning can be realized according to the corresponding process, instead of manual cleaning. This equipment can clean up to 18 parts at a time, and the efficiency of cleaning 18 parts is increased by about 20% compared with manual cleaning.

5 Conclusion

(1) In order to solve the problems of low automation, high labor intensity, low efficiency and inaccurate solvent injection in traditional ultrasonic cleaning of precision parts, an automatic cleaning equipment for precision parts was designed.
(2) The simple and effective filtering algorithm designed in the system greatly improves the accuracy of solvent injection.
(3) The actual verification shows that the system can achieve precise control over the cleaning process of precision parts, and has strong practicability. It can replace manual work for all-weather work, and improve the cleaning capacity.
(4) The design of this equipment provides a new technical scheme with strong practicability for the traditional ultrasonic cleaning department of precision parts.

References

1. Xue F (2015) Discussion on application of bearing ultrasound cleaning technology. J Harbin Bear 36(02):28–29+34
2. Tamboli S, Rawale M, Thoraiet R, et al (2015) Implementation of modbus RTU and modbus TCP communication using siemens S7–1200 PLC for batch process. In: 2015 International conference on smart technologies and management for computing, communication, controls, energy and materials (ICSTM). IEEE, pp 258–263
3. Yang F, Li H (2021) Design and realization of automatic warehouse based on S7–1500PLC. Electr Sci Eng 3(1):6
4. Feng T, Guangpeng L (2021) Design of intelligent feeding control system based on S7–1200 PLC. E3S Web Conf EDP Sci 245:01051
5. Krishnan R (1987) Selection criteria for servo motor drives. IEEE Trans Indus Appl (2):270–275
6. Fan W, Yang Y, Wang D (2021) Design of forklift frame welding production line system based on PLC. Manuf Automat 43(07):123–128
7. Yan H, Zhang Y (2007) Research on digital filtering algorithm of analog sampling in PLC control system. Dev Innov Mach Electr Prod (04):136–137

High-Altitude Airship Propeller Fluid–Structure Interaction Solver

Dongchen Wang, Bifeng Song, and Dong Xue

Abstract This paper presents a study on fluid–structure interaction (FSI) solver for propellers of composite materials, which are commonly used on high-altitude airships and require FSI consideration for better propulsion optimal design. The FSI solver consists of an aerodynamic solver and a structural solver. The aerodynamic solver is a RANS (Reynolds Average Navier–Stokes) solver based on an in-house code named ROTNS. The structural solver is based on the secondary development of a commercial finite element software using Patran Command Language (PCL). The general framework of the FSI solver is built using C++ language and a weak coupling approach. To validate the accuracy of the solver for propulsion optimal design, wind tunnel test, ground static thrust test and structural deformation test were conducted. The numerical results show good agreement with the experimental results.

Keywords High-altitude airship propeller · Fluid–structure interaction · Reynolds average Navier–Stokes

1 Introduction

Solar-powered stratospheric airships have attracted considerable attention for years. This paper investigates the effect of different working conditions on propeller performance. The working conditions include climbing, cruising and descending modes, which correspond to different altitudes and speeds [1]. At altitudes ranging from 0 to 20 km, the lift-to-drag ratio of propeller varies significantly with working conditions, with a difference of more than an order of magnitude between maximum and minimum values [2, 3]. In order to fulfill the demands for lightweight and efficient propulsion systems in high-altitude airships, composite materials have been employed for propellers, resulting in an elongated and slender configuration. These factors collectively give rise to a notable fluid-structural coupling phenomenon [4, 5].

D. Wang · B. Song · D. Xue (✉)
Northwestern Polytechnical University, Xi'an, China
e-mail: xuedong@nwpu.edu.cn

Z. Liu et al. (eds.), *Advances and Challenges in Advanced Unmanned Aerial Systems*, Springer Aerospace Technology, https://doi.org/10.1007/978-981-99-8045-1_39

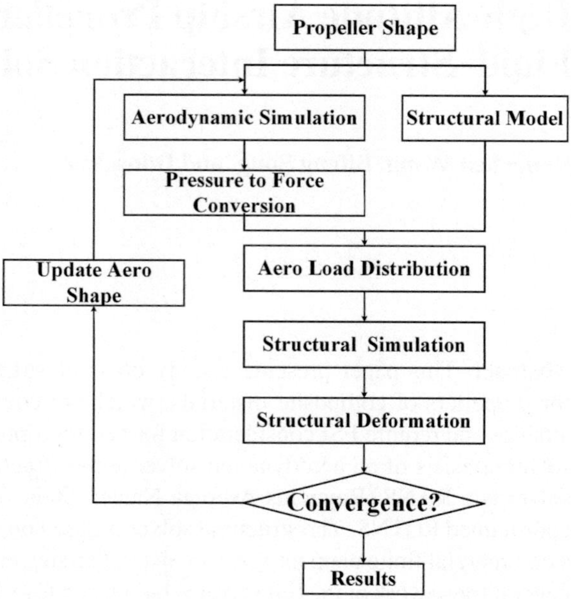

Fig. 1 Fluid-structural coupling workflow

Numerous studies have been conducted on propeller fluid-structural coupling solution in the last decade. The majority of them employed commercial software to compute the deformation of propeller and obtained satisfactory accuracy. A representative study is Chen's paper published in 2013. It implemented the propeller fluid-structural coupling solution procedure using ANSYS software and achieved an accuracy of 5% [6]. Nonetheless, commercial software solution procedure also entails some limitations. For example, researchers are unable to monitor the parameter variations during the iterative solution process in real time. Moreover, the input and output parameters and formats of commercial software are also constrained.

To avoid these limitations, this paper constructs a partially-independent-developed propeller fluid–structure coupling solution process. For the aerodynamic aspect, an in-house Reynold averaged Navier–Stokes flow solver denoted as ROTNS [7–9] is employed, while the structural aspect utilizes a custom development derived from a structural software. The schematic representation of the fluid-structural solver workflow can be observed in Fig. 1.

2 Shape Generation

In this manuscript, we elucidate the configuration of a propeller blade through the incorporation of 12 distinct airfoil sections [10]. Each of these sections is defined by parameters including its radial distance from the rotational center, airfoil geometry,

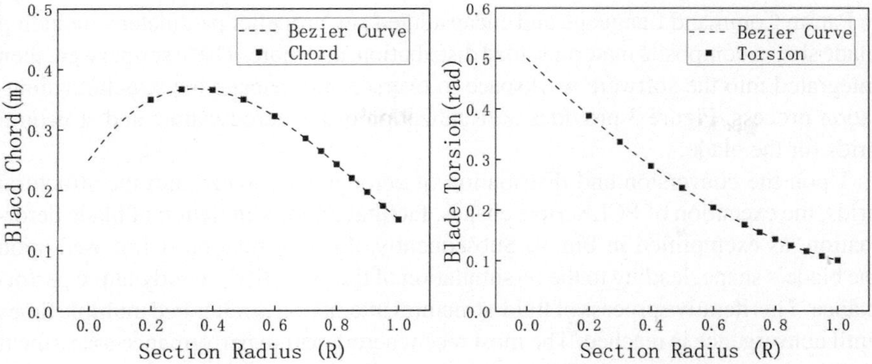

(a) Chord figure (b) Torsion figure

Fig. 2 Chord and torsion figure based on Bézier curves

chord length, and torsion angle. These 12 airfoil sections are positioned at various radii, denoted as 'r' in the context of this study.

The chord length and torsion angle of the airfoil section located at radial position 'r' are determined through the utilization of a Bézier curve. Bézier curves are parametric curves that were invented in 1962 by the French engineer Pierre Bézier for designing automobile bodies [11]. They are widely used in computer graphics and animation today [12, 13]. Within this study, a Bézier Curve of third-degree is employed for the generation of chord length and torsion angle for the section situated at radial position *r*:

$$\mathrm{Chord(r)} = \mathrm{C_1(1-r)^3 + 3C_2r\,(1-r)^2 + 3C_3r^2(1-r) + C_4r^3} \tag{1}$$

$$\mathrm{Torsion(r)} = \mathrm{T_1(1-r)^3 + 3T_2r\,(1-r)^2 + 3T_3r^2(1-r) + T_4r^3} \tag{2}$$

As depicted in Fig. 2 the utilization of the Bézier Curves method has achieved a seamless transition in the blade's chord length and torsion angle.

3 Simulations

The aerodynamic simulation was executed utilizing ROTNS, a finite volume method solver designed for solving the governing equations. The turbulence modeling adopted the Baldwin-Lomax model. Spatial discretization was achieved using the Jameson central scheme, while time marching was carried out employing the Lower–Upper Symmetric Gauss–Seidel implicit scheme.

On the other hand, the structural simulation was conducted through a script-based automation framework dedicated to structural analysis. These scripts were composed in Patran Command Language and encapsulated key propeller parameters, including blade shape, composite materials, load distribution, and more. These scripts were then integrated into the software workspace to oversee and manage the structural simulation process. Figure 3 provides a visualization of the aerodynamic and structural grids for the blade.

Upon the conversion and distribution of aerodynamic loads onto the structural grids, the execution of PCL scripts ensues, facilitating the simulation of blade deformation, as exemplified in Fig. 4. Subsequently, this deformation is imposed upon the blade's shape, leading to the re-simulation of the propeller's aerodynamic performance. This iterative process of fluid-structural interaction is reiterated multiple times until convergence is reached. The most recent aerodynamic performance assessment is ultimately accepted as the final result.

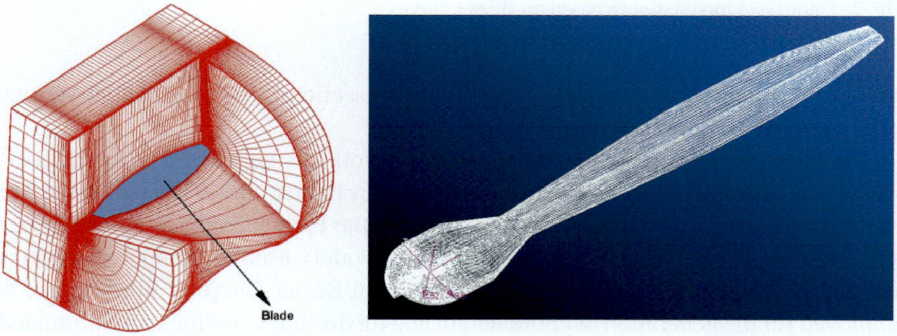

(a) Aero blade grid figure (b) Structural blade grid figure

Fig. 3 Blade grid figure

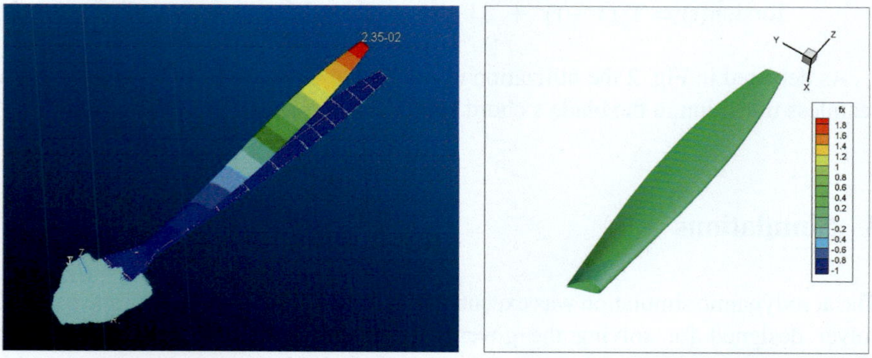

(a) Deformation (b) Load distribution

Fig. 4 Deformation and distribution figure

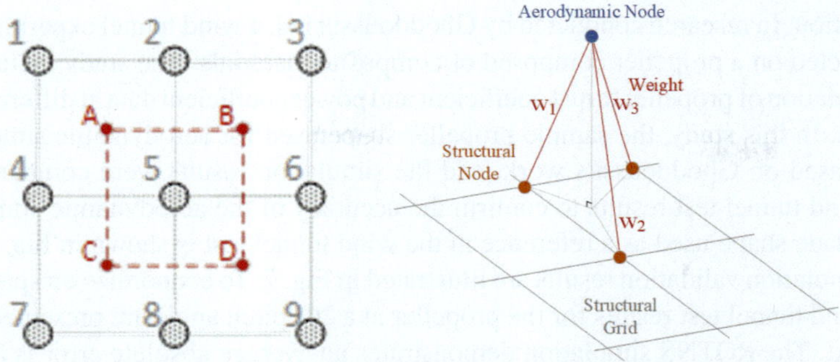

(a) Panel method for conversion (b) Mapping algorithm for distribution

Fig. 5 Load conversion and distribution

4 Conversion and Distribution Algorithm

The ROTNS program generates an output file encompassing pressure values at all nodes within the structured blade aerodynamic grid. For the determination of the vectorial force at a specific target node, the program transforms the aerodynamic pressure into an equivalent force vector. Figure 5a illustrates an illustrative instance of this procedure, featuring the target node designated as "node 5" and its adjacent nodes identified as "node 1 to 9." To ascertain the coordinates of the geometric centers (A to D) for rectangles 1245, 2356, 4578, and 5689, we assume that points A, B, C, and D exist within the same plane. The pressure, denoted as 'p,' at node 5 is considered to act upon the rectangle ABCD. By computing the normal vector and the area of the rectangle ABCD, we can derive the equivalent vectorial force.

Equivalent vectorial forces acting on the blade's aerodynamic grids are translated into loads applied to the structural grids via custom in-house programs developed in C++. Or each aerodynamic node, the program systematically scans through all structural nodes to pinpoint a subset with the smallest spatial separation. From this subset, three nodes are chosen for the distribution of aerodynamic loads. Figure 5b provides a visual representation of how the aerodynamic equivalent vectorial force is decomposed and allocated to the structural nodes, taking into account their relative positions in relation to the aerodynamic node.

5 Experiments

In order to validate the accuracy of the fluid–structure interaction solver, it is necessary to examine the simulation performance of both the aerodynamic simulation and blade deformation simulation. To achieve this, a series of tests were conducted, including wind tunnel testing, ground static thrust testing, and structural deformation testing. The wind tunnel test was conducted specifically for aerodynamic simulation validation. In research conducted by Ghoddoussi [14], a wind tunnel experiment was conducted on a propeller composed of composite materials. The study included the presentation of propeller thrust coefficient and power coefficient data at different pitch angles. In this study, the sample propeller shape used for aerodynamic simulations was based on Ghoddoussi's work, and the simulation results were compared with the wind tunnel test results to confirm the accuracy of the aerodynamic simulation. The blade shape used as a reference in the wind tunnel test is shown in Fig. 6.

Simulation validation results are illustrated in Fig. 7. To economize on space, only the wind tunnel test results for the propeller at a 20° pitch angle are presented in this context. The ROTNS simulation demonstrates an average absolute error is no more than 0.003 for the thrust coefficient and the power coefficient.

The accuracy of fluid–structure interaction simulation was evaluated by conducting two tests: the ground static thrust test and the structural deformation test. Figure 8 illustrates the pictures of the tests.

A representative propeller was employed for the ground static thrust evaluation. The testing ground was situated at an altitude of 0.5 km, featuring a laminar wind speed of 0 m/s. A six-component force balance was utilized to measure the propeller's

Fig. 6 Propeller shape figure ($\beta\_0.75 = 25°$)

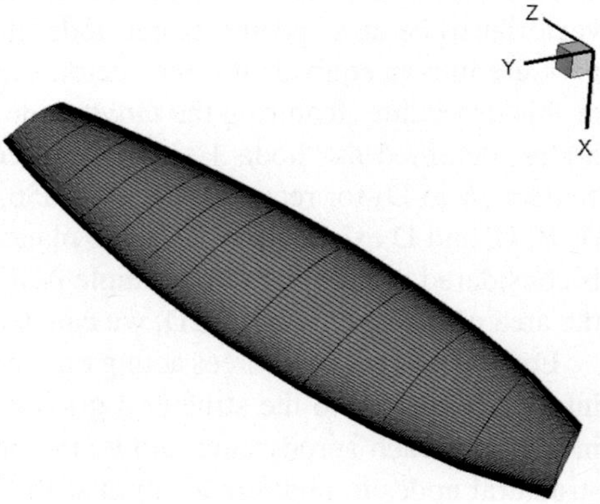

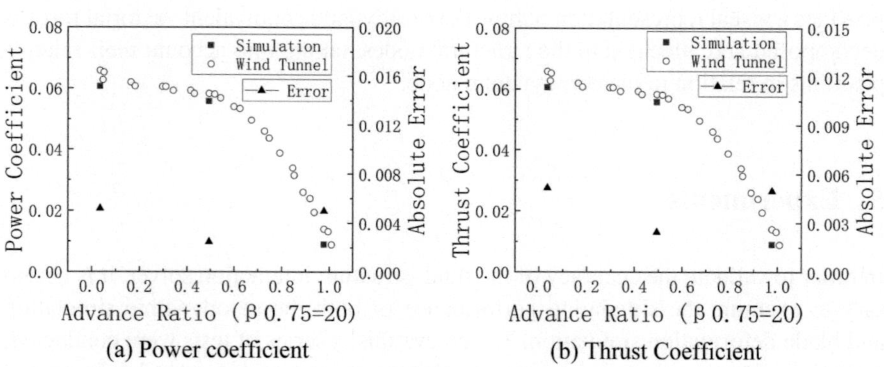

Fig. 7 Result figure of wind tunnel test

(a) Ground static thrust test (b) Structural deformation test

Fig. 8 Pictures of ground static thrust test and structural deformation test

thrust. The static thrust examination was conducted in triplicate to compute the average resultant thrust. At a rotational speed of 140 revolutions per minute (r/min), the propeller yielded an average output thrust of 1355 Newtons (N). In comparison, the simulation outcome recorded 1304 N, reflecting a relative error of 3.8%.

After the ground static thrust test, the same propeller was subjected to the structural deformation test. Sandbags were used to simulate aerodynamic load. The simulated load corresponded to 1350 N at the 75% blade radius. A high accuracy laser rangefinder was used to measure the blade tip deflection. The structural deformation test was performed 3 times to calculate the average blade tip deflection. The experimental blade tip deflection was 25.3 mm. The simulation result was 23.5 mm, with an absolute error of 1.8 mm. Therefore, the calculated results showed good agreement with the experimental results, which confirmed that the accuracy of the solver was good enough for propulsion optimal design. Following the ground static thrust test, the structural deformation test was conducted using the same propeller. Sandbags were employed to simulate aerodynamic load of 1350 N on the 75% blade radius. A high accuracy laser rangefinder measured the blade tip deflection. The structural deformation test was repeated 3 times to obtain the average blade tip deflection. The calculated results demonstrated good agreement with the experimental results. The measured blade tip deflection was 25.3 mm, while the simulation result was 23.5 mm, yielding an absolute error of 1.8 mm.

A comparison of the calculation accuracy of the commercial software solver [6] with the propeller fluid-structural interaction solver developed in this paper revealed that the latter had slightly higher accuracy in aerodynamic simulation but slightly lower in structural simulation, which validated that the accuracy of the solver was sufficient for propulsion optimal design.

References

1. Liu F, Du M, Yue L, Zhang Q (2017) High efficiency propeller design at low-Reynolds number and high altitude. In: Chinese aeronautics science and technology conference, p 7 (in Chinese)
2. Lissaman PBS (1983) Low-Reynolds-number airfoils. Annu Rev Fluid Mech 15(1):223–239
3. García-Gutiérrez A, Gonzalo J, López D et al (2020) Stochastic design of high altitude propellers. Aerosp Sci Technol 107:106283
4. Hubner B, Walhorn E, Dinkier D (2001) Strongly coupled analysis of fluid-structure interaction using space-time finite elements. In: 2nd European conference on computational mechanics
5. Qu Z, Yang Y (2018) Aeroelastic analysis for the high-altitude propeller by using fluid-structure interaction method. Trans Jpn Soc Aeronaut Space Sci, Aerosp Technol Jpn 16(3):236–241
6. Chen G (2014) Study on performance of composite propellers based on fluid and structure interaction. Harbin Engineering University
7. Xu J-H et al (2017) Investigation on improving efficiency of high-altitude propeller with tandem configuration. In: 35th AIAA applied aerodynamics conference
8. Xu J, Han Z, Song W et al (2020) Efficient aerodynamic optimization of propeller using hierarchical kriging models. J Phys: Conf Ser 1519(1):012019
9. Xu J, Song W, Yang X et al (2019) Aerodynamic performance of variable-pitch propellers for high-altitude UAVs. IOP Conf Ser: Mater Sci Eng 686(1):012019
10. Dongchen W, Bifeng S, Jun J, et al (2023) High-altitude airship variable-pitch propeller performance analysis at multiple altitudes. Proc Inst Mech Eng, Part G: J Aerosp Eng 09544100231167710
11. Sederberg TW (2012) Computer aided geometric design. Brigham Young University, Provo, UT
12. Choi J-W, Curry R, Elkaim G (2008) Path planning based on Bézier curve for autonomous ground vehicles. In: Advances in electrical and electronics engineering-IAENG special edition of the world congress on engineering and computer science 2008. IEEE
13. Baydas S, Karakas B (2019) Defining a curve as a Bezier curve. J Taibah Univ Sci 13(1):522–528
14. Ghoddoussi A, Miller LS (2016) A more comprehensive database for low Reynolds number propeller performance validations. In: 34th AIAA applied aerodynamics conference

Experimental Investigation on Frost Density and Thermal Conductivity Under Cryogenic Condition

Shi Shangguan, Lei Wang, Peijie Sun, Li Yan, Bowen Liu, and Yanzhong Li

Abstract Cryogenic frosting is a common phenomenon in the field of aerospace industry. However, for the case that the cryogenic surface is cooled from room temperature, such as a cryogenic propellant fueling to an uninsulated cryogenic tank, the frost formation process could be unique. In this paper, the density and thermal conductivity of frost deposited on the cold surface cooled from room temperature to cryogenic temperature were experimentally investigated under forced condition. Weighting method was adopted to measure the final frost density at the end of each experiment, and frost thermal conductivity distribution inside the frost layer was obtained by analyzing the temperature profile measured at different heights. The results show that the final frost density increases with air velocity and air absolute humidity and varies from 54.7 to 556.7 kg/m^3 in this study. The cooling rate in the initial cooling process exerts a certain effect on the frost structure, which could be observed by the variations of frost density. The frost thermal conductivity is layered along the thickness direction of the frost layer, but the thermal conductivity value, which has been formed at a certain height, does not change significantly with time. Moreover, correlations of frost density and frost thermal conductivity are established with ambient temperature, ambient humidity, and incoming air velocity as variables. Compared to the experimental data, the deviations of the predicted frost density and frost thermal conductivity are both within 10%. The present study could deepen the understanding on frost features in cryogenic conditions, and the established correlations provide reliable tools of estimating frost properties, which are beneficial to the design of cryogenic propellant fueling process.

Keywords Frost properties · Cryogenic frosting · Initial cooling process · Forced convection

---

S. Shangguan · L. Wang (✉) · B. Liu · Y. Li
Institute of Refrigeration and Cryogenic Engineering, Xi'an Jiaotong University, Xi'an 710049, China
e-mail: wanglei-epe@xjtu.edu.cn

P. Sun · L. Yan
Shanghai Institute of Aerospace System Engineering, Shanghai 201109, China

Z. Liu et al. (eds.), *Advances and Challenges in Advanced Unmanned Aerial Systems*, Springer Aerospace Technology, https://doi.org/10.1007/978-981-99-8045-1_40

Nomenclature

Fo	Fourier number
Nu	Nusselt number
Pr	Prandtl number
Re	Renolds number
R	Gas constant
T	Temperature (K)
h	Heat transfer coefficient ($W/m^2 \cdot K^{-1}$)
k	Thermal conductivity ($W/m \cdot K^{-1}$)
m	Mass flux to frost ($kg/m^2 \cdot s^{-1}$)
p	Pressure (kPa)
q	Heat flux (W/m^2)
t	Time (min)
v	Air velocity (m/s)
w	Absolute humidity (g/kg)
δ	Thickness (mm)
ρ	Density (kg/m^3)
σ	Stefan-Boltzmann constant = 5.67×10^{-8} $W/m^2 \cdot K^{-4}$
ε	Essimmity

1 Introduction

Frosting on a surface with varied temperatures is a common and interesting phenomenon. For the fueling process of a cryogenic propellant tank without thermal insulation, the tank outer surface is cooled from room temperature to ordinary-low temperature, and finally cryogenic temperature. The temperature at the tank outer surface is varied and affected by the probable propellant boiling and convection during the cooling process. In this case, the frost formation could be quite complicated, and the cryogenic propellant supply could be affected adversely by frosting. Therefore, the frosting performance and its properties in this process are necessary for the design of cryogenic propellant fueling system.

Frost properties are key factors in the study of frost layer growth, and four factors, including cold surface temperature, ambient temperature, air absolute humidity, and air flow rate are usually considered in the research on frost properties [1, 2]. Hayashi et al. [3] conducted frosting experiments in virtue of a general-low temperature cold plate. It was found that the frost density growth rate could be accelerated by a high air velocity, and the highest frost density approached 500 kg/m^3 when the air flow rate was 6 m/s. Byun et al. [4] analyzed the frost density at ultra-low temperatures under forced convection conditions. It was pointed out that the absolute humidity was the main factor contributing to the frost density. By contrast, the frost density did not change significantly with other factors like the cold surface temperature. Jeong et al.

[5] studied the influence of cold surface temperature and air flow rate on frost density as well. The results showed that the frost density decreased with the cold surface temperature but increased with the air flow rate. Although scholars had studied the effects of the cold surface temperature, which was frequently maintained constant in their experiments, and few researchers focused on cryogenic frosting involving the initial cooling process. A cryogenic frosting experiment with initial cooling process was conducted by Choi and Kim [6], and the density of frost deposited under natural convection was also measured. It was found that under natural convection, the frost formed on a surface with cryogenic temperature was usually light and porous, and the frost density was lower than 100 kg/m^3. Comparatively, a higher frost density varied from 300 to 500 kg/m^3 was found in the case of frosting at moderately low temperatures under forced convection. With the frosting time and modified Jakob number but not frost surface temperature as variables, a semi-empirical frost density correlation was proposed by Hermes et al. [7], and the deviation of the predicted results was within 10%.

Another key factor affecting frost growth is the frost thermal conductivity. Since the frost could be considered as a mixture of ice and air, the frost structure could be reflected by the frost density. Therefore, many correlations of frost thermal conductivity were expressed as a function of frost density [8, 9]. Tong et al. [10] modified the random mixture model by considering the effect of latent thermolysis on steam condensation, and the predicted frost thermal conductivity showed good agreement with experimental data. With the porosity expressed by the frost density as variable, Auracher [11] established a theoretical correlation of frost thermal conductivity. For the measurement of frost thermal conductivity in experiments, it could be indirectly obtained by the Fourier equation and the direct measurement of the temperature difference and thickness, and the heat flux could be measured with heat flux meters or calculated based on energy balance [12]. Xie et al. [13] experimentally studied the frost thermal conductivity under natural convection condition, and the frost thermal conductivity was fit as a function of the distance along tangent direction of the vertical cryogenic plate with least square method. Brian et al. [14] conducted frosting experiments at moderately low temperature and cryogenic temperature and proposed a correlation of frost thermal conductivity. In addition to frost density, the cold surface temperature was considered as a factor in the correlation. However, the cold surface temperature was regarded as a constant in the correlation, and the initial cooling process was not included.

It could be found that few scholars focus on the properties of frost deposited on a cold surface with various temperature. Therefore, in this work, a series of frosting experiments were carried out on a horizontal cold surface cooled from room temperature to cryogenic temperature under convection condition. Frost properties, including frost density and frost thermal conductivity, were measured, and related correlations were proposed based on experimental data. This work provides a reference for the design of cryogenic propellant fueling process, also for the establishment of cryogenic frosting model.

2 Experimental Apparatus and Methods

Figure 1 shows the experimental apparatus of cryogenic frost deposition. An air duct with a total length of 2.5 m and the square cross section of 45 × 45 cm was adopted. The whole air duct was placed in the enthalpy difference chamber, which could control the ambient temperature from 0 to 50 °C with an accuracy of ±0.5 °C, and the relative humidity from 35 to 85% with an accuracy of ±2.0%. The air flow was generated by a speed frequency controlled axial blower, and the adjustment range was from 0 to 12 m/s. A homogeneous arrangement of air flow over the test cold plate was obtained with two air flow rectifiers equipped inside the wind channel respectively in the front and latter metallic channel parts. Sensors for measuring air temperature and air humidity were installed above the cold plate and behind the latter air flow rectifier to monitor the experiment environment stability. The air velocity was measured by EE650 wind speed sensors with an accuracy of ±0.5 m/s, which were installed above the cold plate and in the latter part of the wind channel. Therefore, a forced convection condition with controlled ambient environment could be realized.

Two 20 × 20 cm aluminum plates with grooves on the back, which were fixed with stainless steel parallel pipeline of DN10, were fabricated to assemble the test cold plate. The schematic diagram of the cold plate structure is shown in Fig. 2. To keep a stable liquid nitrogen supply, a pressurized liquid nitrogen tank was used, and a high-pressure nitrogen gas bottle was adopted to provide pressurization in the liquid

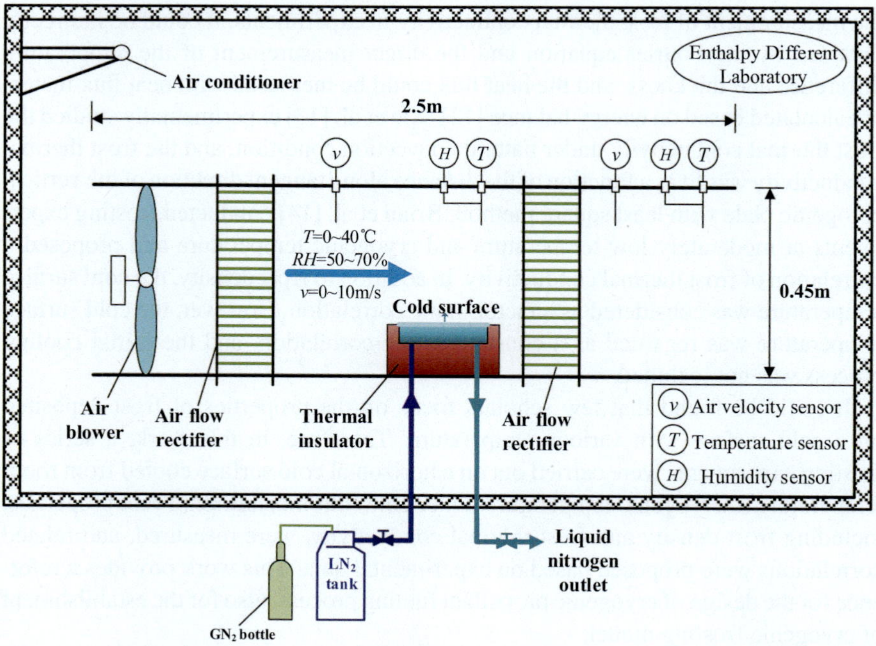

Fig. 1 Schematic diagram of frost deposition test apparatus

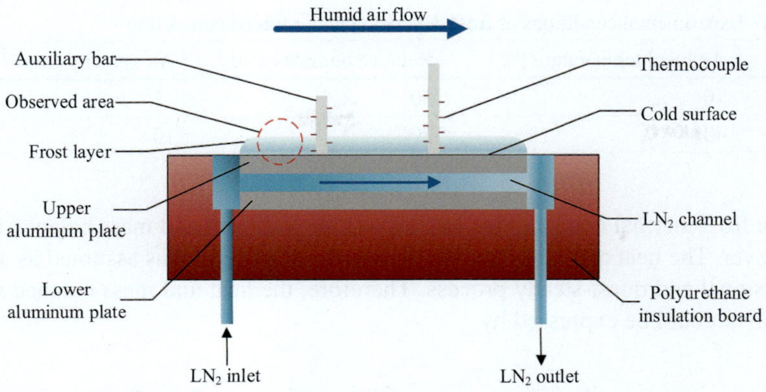

Fig. 2 Schematic diagram of cold plate arrangements

nitrogen tank. Polyurethane insulation board and polyurethane foam were adopted as the thermal insulation and support materials, and aluminum foil insulation cotton was used as the thermal insulation material of liquid nitrogen inlet pipeline.

The cold plate surface temperature was measured by five T-type thermocouples, which were evenly buried inside the upper aluminum plate. Besides, two auxiliary polyester hollow bars, with holes spaced 2 mm and 3 mm respectively, were connected to the cold surface in the form of a thread for the purpose of measuring the temperature distribution inside the probable frost layer as well as the adjacent air temperature. The temperature measuring points of thermocouples were passed through the inside of the bars and out of these holes, and the protruding parts were kept parallel to the cold plate surface to avoid the adverse effect of frost formation on the temperature measurement. Measurements of cold plate surface temperature and temperature distribution above the cold plate were realized by the Omega TT-T-30-SLE thermocouple with an accuracy of ±0.5 °C.

The frost thickness was obtained with a projection method, and a cold light source was set in front of the CCD camera. When the frost layer gradually grew, the proportion of the shadow part in the picture captured by the CCD camera also gradually increased. After processing with the image software, the frost thickness could be calculated with an uncertainty of 2.85%. When the frosting time was over, the liquid nitrogen inlet valve was firstly closed, then the measurement of the frost density was carried out. The frost deposited on the cold plate surface was scraped and placed into a paper cup with a stainless steel ruler. Then it was weighed by the MTQ500D electronic balance with an accuracy of ±0.001 g. This operation was repeated twice for better accuracy. With measured frost mass, removal area and frost thickness, the frost density could be calculated. To minimize disturbance, the frost layer not affected by the turbulence effect of the auxiliary bars was considered as observed area, as shown in Fig. 2. Also, the frost used to calculate frost density was scraped from this area.

Table 1 Experimental conditions of frost deposition under forced convection

	Ambient temperature (°C)	Relative humidity (%)	Air velocity (m·s^{-1})
Min	0	50	5
Max	40	70	10

The frost thermal conductivity was calculated with heat and mass balance in the frost layer. The heat and mass transfer within the frost layer was assumed as a one-dimensional and quasi-steady process. Therefore, the heat and mass balance in the frost layer could be expressed by

$$q_{total} = k_f \frac{T_{sur} - T_w}{\delta_f} = \varepsilon\sigma(T_a^4 - T_{sur}^4) + h(T_a - T_{sur}) + (2.88 \times 10^6 - 195T_{sur})m \quad (1)$$

The convective heat transfer coefficient h was determined by

$$\overline{Nu} = \frac{hL}{k_a} = 0.664\mathrm{Re}^{1/2}\mathrm{Pr}^{1/3} \quad (2)$$

The deposition mass flux is obtained from

$$m = h_m(\frac{RHp_{sat,a}}{RT_a} - \frac{p_{sat,f}}{RT_f}) = \frac{d(\delta_f \rho_f)}{dt} \quad (3)$$

With frost density and calculated parameters, the effective frost thermal conductivity could be obtained with heat and mass balance. Also, the frost thermal conductivity at different heights from the cold plate surface could be calculated with the total heat flux q_{total} and the measured temperatures.

The experimental conditions, including ambient temperature, relative humidity and air velocity, were listed in Table 1. The experimental cases were labelled under different experimental conditions. The case labelling AT20RH50V5, for example, represents the experimental condition with the ambient temperature of 20 °C, the air relative humidity (*RH*) of 50% and the air velocity of 5 m/s.

3 Results and Discussion

3.1 Frost Density

Figure 3 shows the frost layer structure development at different ambient temperatures when the air relative humidity is 70% and the incoming air velocity is 5 m/s.

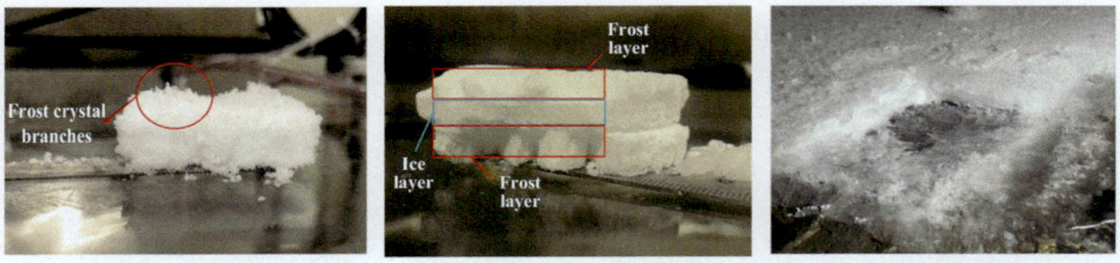

(a) Frost layer (T_a = 10 °C, RH = 70%, v = 5m/s) (b) Frost-ice-frost layer (T_a=20 °C, RH = 70%, v = 5m/s) (c) Ice layer (T_a =40 °C, RH = 70%, v = 5m/s)

Fig. 3 Frost structure variation with ambient temperature

It could be seen that when the ambient temperature was relatively low, the frost deposited on the cold surface is light, and the frost crystal branches on the frost surface could be easily observed, as shown in Fig. 3a. When the ambient temperature increases to 20 °C, the upper and lower frost is still light frost, but the middle layer frost is transformed to a hard ice layer, as shown in Fig. 3b. With the ambient temperature further increases, the frost layer is almost transformed to a hard ice layer, and only sporadic dots of frost spread on the frost surface, as shown in Fig. 3c. This phenomenon could be explained as the result of the increase of heat transfer intensity on the air side. When the ambient temperature increases, the frost at the frost layer surface experiences a strengthened repeated melting-solidification process. As a result, a special frost structure of frost-ice-frost is formed. When the ambient temperature is 40 °C, the heat exchange between air and frost surface becomes more intense, so that a more frequent melt-solidify is caused, and the frost layer is transformed into the ice layer.

Figures 4 and 5 display the effects of ambient conditions, including ambient temperature, air humidity and air velocity, on frost density. According to the frost layer structural changes, the frost layer density could be divided into three regions, including frost area, transition area and ice area. In the frost area, the frost layer is light and the frost porosity is relatively high. In the transition area, the frost consists of light frost and hard ice. In the ice area, the frost layer is almost hard ice. When the frost layer structure changes, the frost density also changes obviously, which could be reflected by the fact that the growth rate of frost density across different regions is higher than that of frost density in the same region. Besides, it could be found that the frost density increases with the ambient temperature, air humidity and air velocity. According to the theory of Choi [5], the heat and mass transfer in the frosting process consists of two mechanisms, involving the increase of frost density and the growth of frost thickness. With the increase of ambient temperature, air humidity and air velocity, the stronger phase change driving force also results in a higher frost layer density.

The test cases and the corresponding frost density results are summarized in Table 2. For the cases AT40RH50V10, AT40RH70V5 and AT40RH70V10, the cold plate surface temperatures did not decrease to the liquid nitrogen temperature over the entire experimental period. This is because the heat exchange on the air side was quite intense.

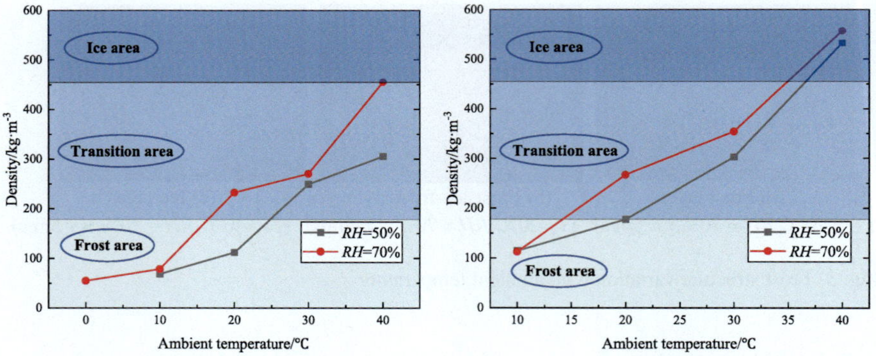

(a) $v = 5$ m/s (b) $v = 10$ m/s

Fig. 4 Frost density variations with air humidity and ambient temperature

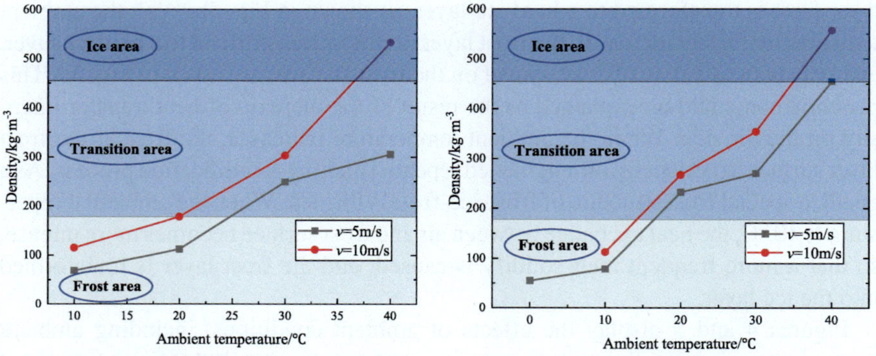

(a) $RH = 50\%$ (b) $RH = 70\%$

Fig. 5 Frost density variations with air velocity and ambient temperature

3.2 Frost Thermal Conductivity

Given the fact that the frost layer is a mixture of air and ice, the variation of frost thermal conductivity could also indicate the structure change of deposited frost layer. Figure 6 shows the variations of frost thermal conductivity with time at different frost thicknesses. It could be found that the frost thermal conductivity nearly keeps constant with time. This may indicate that once the frost forms at the cold surface, the frost structure would not change obviously, especially for the frost formed in the early frosting period as well as in initial cooling period. Since the value of frost thermal conductivity is relatively small in this period, it could be speculated that for the frost layer formed in the initial cooling period is light, and the frost porosity is relatively high.

Moreover, it could be found that the thermal conductivity of frost layer is significantly stratified in the direction of frost layer thickness, which means the frost

Table 2 Summary of experimental results

Case labelling	Cold surface temperature (K)	Absolute humidity (g·kg^{-1})	Cooling time (min)	Frost density (kg·m^{-3})	Remarks
AT0RH70V5	80.65	2.64	19	54.7	Non-freezing
AT10RH50V5	83.15	3.79	17	68.5	Non-freezing
AT10RH50V10	79.15	3.79	23	114.5	Non-freezing
AT10RH70V5	79.15	5.32	20	78.2	Non-freezing
AT10RH70V10	78.65	5.32	28	112.3	Non-freezing
AT20RH50V5	80.15	7.26	20	111.8	Non-freezing
AT20RH50V10	79.15	7.26	26	177.4	Freezing
AT20RH70V5	78.15	10.21	21	232.1	Freezing
AT20RH70V10	79.65	10.21	22	266.5	Freezing
AT30RH50V5	78.65	13.31	21	248.9	Freezing
AT30RH50V10	78.65	13.31	26	302.5	Freezing
AT30RH70V5	78.65	18.80	39	269.8	Freezing
AT30RH70V10	78.15	18.80	40	353.8	Freezing
AT40RH50V5	78.65	23.52	41	305.2	Freezing
AT40RH50V10	211.65	23.52	–	532.3	Freezing
AT40RH70V5	193.15	33.43	–	454.5	Freezing
AT40RH70V10	213.65	33.43	–	556.7	Freezing

structure is not uniform during the whole frosting period. When the cold surface is cooled to liquid nitrogen temperature, the thermal conductivity of frost formed at this time is higher than that in the initial cooling period. Especially, when the ambient temperature rises to 30 °C, the frost layer is almost transformed into ice layer, and the thermal conductivity of frost formed at the thickness of 9–15 mm far outstrips that at other thickness. However, the thermal conductivity of frost formed at other thickness variation is not significant. Therefore, it could be speculated that repeated melting-solidification occurring at frost surface exerts few effects on the lower frost layer. Generally, the effective frost thermal conductivity increases with time and with ambient temperature. The increase rate of the effective frost thermal conductivity slows down as the ambient temperature increases, which means the effective frost thermal conductivity could be considered as a constant at a relatively high ambient temperature.

Figure 7 illustrates the variation of frost thermal conductivity with air humidity and air velocity. It could be found that increasing air humidity or air velocity could both accelerate the frost thermal conductivity. Comparatively, the influence of air humidity on frost thermal conductivity is more significant.

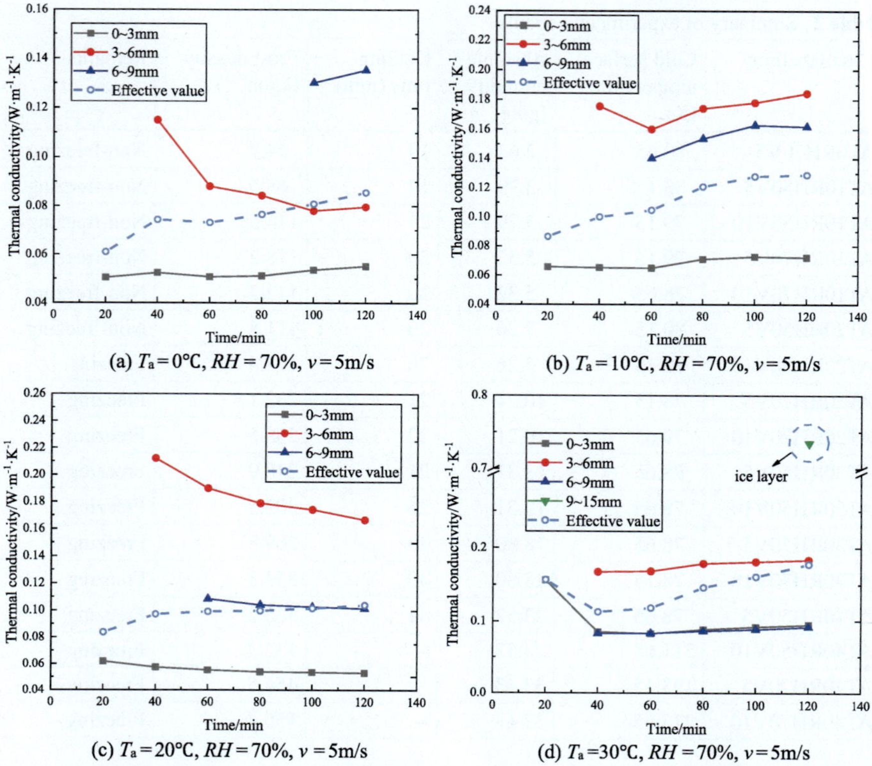

Fig. 6 Frost thermal conductivity variations with ambient temperature and time

Fig. 7 Frost thermal conductivity variations with different ambient conditions

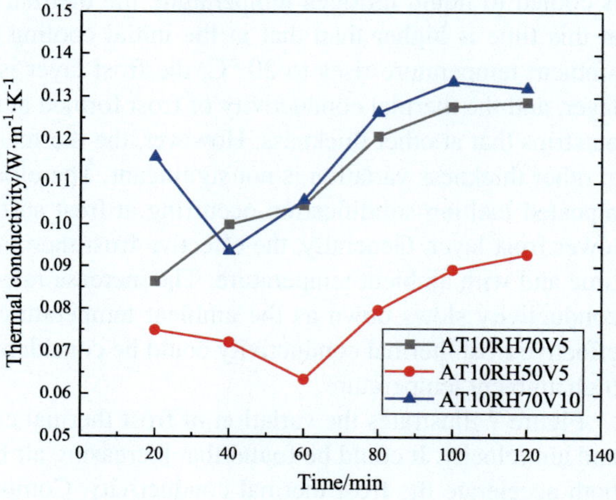

4 Correlations

4.1 Frost Density

Based on the controlled experimental conditions, the frost density is expressed as a correlation of absolute humidity, cooling time, total frosting time, ambient temperature, triple point of water, Reynolds number, and final cold surface temperature. This correlation is valid for 0 °C < T_a < 40 °C, 6 × 104 < Re < 1.5 × 105, 2.64 g/kg < w_a < 23.52 g/kg and 0 min < t_{cool} < 40 min when the cold wall temperature is cooled to about 80 K. Table 3 displays the values of correlation coefficients. The operation for measuring frost density during frosting will bring instability to the experimental environment, which could interfere the frost formation. Therefore, the change of frost density over time is not considered in the correlation, so that the predicted value by this correlation is the frost density when t_{total} is 120 min.

$$\frac{\rho_f}{\rho_i} = a w_a^b \mathrm{Re}^c \left(\frac{t_{cool}}{t_{total}}\right)^d \left(\exp\left(\frac{T_a - T_{tp}}{T_a - T_w}\right)\right)^e \tag{4}$$

Figure 8 shows the comparison between the measured frost density and correlated frost density. It could be seen that the predicted frost density shows good agreement with experimental results, and the relative deviations are within 10%.

Table 3 Coefficients of correlated frost density

Limited range of w_a	a	b	c	d	e
[2.64, 10)	0.08665	−0.51955	0.18226	1.11277	14.36430
[10, 23.52]	4.3765e−4	0.80028	0.36232	−0.30371	−1.50487

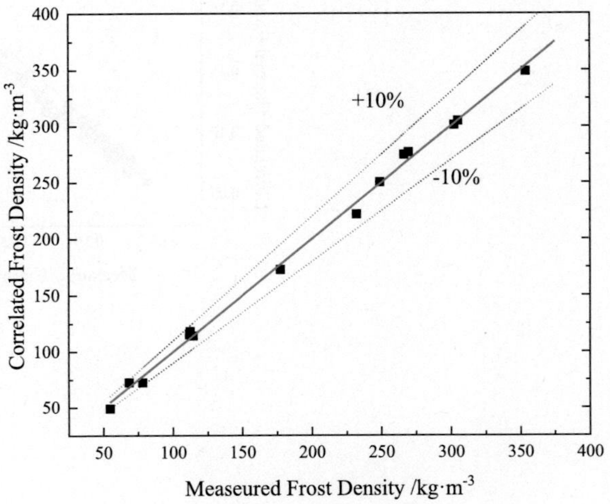

Fig. 8 Comparison of measured and correlated frost densities

4.2 Effective Frost Thermal Conductivity

Based on the experimental results, a correlation for the effective frost thermal conductivity is established, which is expressed as

$$\frac{k_f}{k_i} = aw_a^b(\ln \mathrm{Re})^c \mathrm{Fo}^d \left(\exp\left(\frac{T_a - T_{tp}}{T_a - T_w(t)}\right)\right)^e \tag{5}$$

Table 4 displays the coefficients of the correlation. Compared to the correlation of frost density, frosting time is considered as a variable in this correlation, which is reflected by the Fourier number and the variation of the cold wall temperature. Figure 9 shows the comparison of the correlation and the experimental data. The average error of the predicted effective frost thermal conductivity is approximately within ±10%.

Table 4 Coefficients of correlated frost thermal conductivity

Limited range of w_a	a	b	c	d	e
[2.64, 7)	3.97566e−4	0.76866	1.48323	0.16058	−2.59784
[7, 10)	0.74846	1.29872	−2.68259	0.17162	11.56064
[10, 12)	0.60086	−4.55851	−1.82191	−0.08142	133.47843
[12, 18.8]	4.39829e−9	1.16887	5.11519	0.26734	4.00672

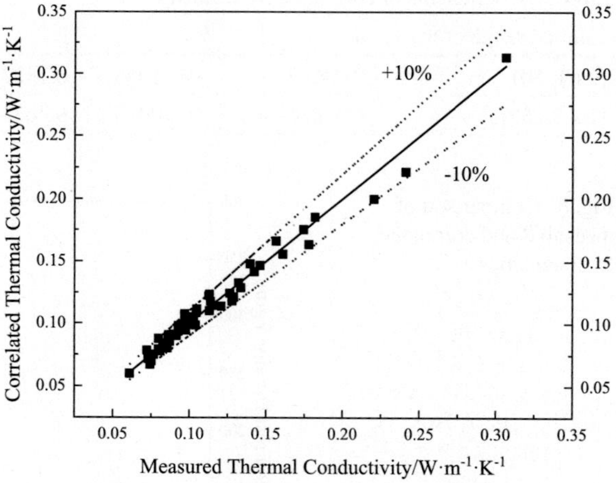

Fig. 9 Comparison of measured and correlated effective frost thermal conductivities

5 Conclusions

In this work, frost density and frost thermal conductivity were experimentally investigated in virtue of an air duct under cryogenic frosting, and the influence of initial cooling process was also involved. Several conclusions may be drawn as follows:

(1) The frost density increases with both of air absolute humidity and air velocity, and the variation range is about from 54.7 to 556.7 kg/m^3. As the frost density increases, the proportion of ice in the frost layer also increases. As a result, the frost composition changes with ambient conditions from light frost to frost-ice-frost structure and finally to ice.
(2) The increase of ambient temperature, air humidity and air velocity promote the increase of frost thermal conductivity, and the variations of frost thermal conductivity could also indicate the change of frost structure. The thermal conductivity of frost layer is significantly stratified in the direction of frost layer thickness, and the thermal conductivity of the formed frost layer does not change significantly with time.
(3) The correlations for predicting the frost properties are established by considering various parameters, including ambient conditions, cooling time cost and frosting time, and the predicted frost density and frost thermal conductivity show well agreement with the experimental results within a maximum error of 10%.

Generally, this study significantly expands the existing experimental database of cryogenic frosting and provides a significant reference for the design of cryogenic propellant fueling process, which is also helpful for developing theoretical models.

Acknowledgements This work is supported by National Natural Science Foundation of China (No. 51976151).

References

1. Byun S, Jeong H, Son H et al (2022) Frost formation from general-low to ultra-low temperatures: a review. Int J Heat Mass Transf 195:123164
2. Mengjie S, Chaobin D (2018) Review on the measurement and calculation of frost characteristics. Int J Heat Mass Transf 124:586–614
3. Hayashi Y, Aoki A, Adachi S et al (1977) Study of frost properties correlating with frost formation types. J Heat Transfer 99(2):239–245
4. Byun S, Jeong H, Hyeon K et al (2021) Quantitative analysis of frosting characteristics at ultra-low temperatures under forced convection conditions. Energy Build 248:111186
5. Jeong H, Byun S, Kim D et al (2021) Frost growth mechanism and its behavior under ultra-low temperature conditions. Int J Heat Mass Transf 169:120941
6. Choi S, Kin S (2017) Effect of initial cooling on heat and mass transfer at the cryogenic surface under natural convection condition. Int J Heat Mass Transf 112:850–861
7. Hermes C, Loyola F, Nascimento V (2014) A semi-empirical correlation for the frost density. Int J Refrig 46:100–104

8. Shin J, Tikhonov A, Kim C et al (2003) Experimental study on frost structure on surfaces with different hydrophilicity: density and thermal conductivity. J Heat Transfer 125:84–94
9. Yang D, Lee K (2004) Dimensionless correlations of frost properties on a cold plate. Int J Refrig 27:89–96
10. Tong J, Yang Z, Fan Y (1996) Research on comprehensive thermal conductivity of frost layer. Cryogenics 5:46–50
11. Auracher H (1986) Effective thermal conductivity of frost. In: Proceedings of the international symposium on heat and mass transfer in refrigeration cryogenics, Dubrovnik, Yugoslavia, pp 285–301
12. Qu K, Komori S, Jiang Y (2006) Local variation of frost layer thickness and morphology. Int J Therm Sci 45(2):116–123
13. Xie F, Chen S, Yao S, et al (2012) Experimental study on frost layer's thermal conductivity of cryogenic vertical plate under natural convection condition. Cryogenics 40(5):9–13
14. Brian P, Reid R, Shah Y (1970) Frost deposition on cold surfaces. Ind Eng Chem Fundam 9(3):375–380

Study on Effect of Single-Tooth Radial Forging Process Parameters on Spline Shaft Forming

Kun Li, Shengdun Zhao, Dawei Zhang, and Chao Zhang

Abstract Higher requirement of the performance of spline shaft has been put forward with the development of aerospace technology. As a precision advanced manufacturing technology, radial forging technology is suitable for the manufacturing of high-performance spline shaft. Aiming at the influence of process parameters on plastic forming in single-tooth radial forging (STRF) process of spline shaft, the effects of process parameters such as initial forging temperature of the blank, forging speed of the dies, and friction condition on the forming load, and distribution of equivalent stress and equivalent strain of the workpiece were investigated. Firstly, the introduction to STRF process of spline shaft was given. And then a finite element model was established based on the FORGE software, and the range of each process parameter was determined. Through the finite element analysis, the influence laws of various parameters of STRF process on spline shaft forming were obtained. The simulation results show that the forming load and equivalent stress decrease with the increasing of the initial forging temperature of the blank, and increase with the increasing of the forging speed. Within the friction coefficient range, the effect of friction conditions on die load and equivalent stress is not significant. The equivalent strain and the cumulative plastic deformation degree of the material are almost unchanged under different parameter condition. The obtained results in this paper can provide reference for the application and development of advanced manufacturing process of high-performance gear parts used in aerospace.

Keywords Single-tooth radial forging · Process parameters · High-performance spline shaft · Finite element analysis

K. Li (✉) · S. Zhao · D. Zhang · C. Zhang
School of Mechanical Engineering, Xi'an Jiaotong University, Xi'an 710049, Shannxi, China
e-mail: 1950220460@qq.com

S. Zhao
e-mail: sdzhao@mail.xjtu.edu.cn

D. Zhang
e-mail: Zhangdawei2000@mail.xjtu.edu.cn

Z. Liu et al. (eds.), *Advances and Challenges in Advanced Unmanned Aerial Systems*,
Springer Aerospace Technology, https://doi.org/10.1007/978-981-99-8045-1_41

1 Introduction

As a typical mechanical part, high-performance spline shafts are widely used in aerospace, modern tanks, heavy-duty equipment, and other fields due to their reliable performance and strong bearing capacity [1]. Spline shaft forming methods are mainly divided into two types: one is cutting process, and the other is plastic deformation process. Currently, most gear manufacturers produce spline shafts mainly by cutting process [2]. There are many shortcomings with the process of tooth shaped parts by cutting: reducing the performance of the part with the fibrous tissue of the material cutting off, low material utilization and high production costs, and low manufacturing efficiency [3, 4].

The plastic forming process of spline shaft has the advantages of material saving, reliable part performance, and high forming efficiency, and is increasingly being valued. The plastic forming methods of spline shaft mainly include rolling, extrusion, and other processes [5, 6]. The rolling process is mainly used to produce small module tooth shaped parts with high efficiency, but the die phase adjustment is complex and prone to defects such as disordered or misaligned teeth. The extrusion process has high forming accuracy and is mainly used to form small module parts. However, there are shortcomings such as high load and insufficient filling of teeth. The above-mentioned plastic forming process has not been well promoted in actual production and manufacturing due to its large forming force and complex die phase adjustment.

With the rapid development of aerospace technology, higher requirement is put forward for the performance of spline shaft. Therefore, the development of advanced manufacturing processes of spline shaft used in aerospace is becoming increasingly urgent [7]. As a precision advanced forging and forming technology, radial forging technology has gained more attention than before and gradually used in the production of high-performance shafts and tube parts. Radial forging process is a high-end precision forging process with 2–8 forging dies uniformly arranged along the circumference of the billet. Due to its advantages such as low forming force, high forming efficiency, and good performance of forged parts, it has gradually been valued by scholars. As a new plastic forming process of spline shaft, STRF process is based on the characteristics of radial forging process, such as high forging frequency and small deformation, and has many advantages such as low forming force, high material utilization rate, continuous material fiber structure, and high mechanical properties of the formed parts. Therefore, it is suitable for the plastic forming of high-performance spline shaft [8].

In order to further study the influence of process parameters on the forging load, equivalent stress and equivalent strain in STRF process of spline shaft, a finite element model of STRF process of spline shaft was established based on the FORGE software, and the finite element simulation was conducted under different process parameters such as initial forging temperature, forging speed, and friction conditions. After analysing the distribution of equivalent stress and equivalent strain, and forging load, the influence of various process parameters on the radial forging process of the spline shaft was obtained. It can provide an important basis for the subsequent practical plastic forming manufacturing and application of high-performance spline shaft used in aerospace.

2 Principle of STRF Process of Spline Shaft

Radial forging technology is a precision forging technology using forming the shaft and tube parts [9]. It is generally formed by high forging frequency of four dies uniformly arranged along the circumferential direction of the blank. In the striking gap of the forging dies, the blank is fed axially and rotated about its own axis until the radial forging forming of the entire workpiece is completed [10, 11].

Based on the radial forging process, STRF process of spline shaft is shown in Fig. 1. The four forging dies are uniformly distributed in the circumferential direction of the blank, and each forging die has the structure of only one internal tooth and the tooth shape is consistent with that of the spline shaft to be formed [12]. The forming process of STRF process is shown in Fig. 2. The blank is forged by the four forging dies with single tooth under high-frequency. Four evenly distributed teeth are formed on the blank after each time forging. During the forging gap, the manipulator holds the blank and rotates along its own axis, and the blank is feeding along the axial direction synchronously until the entire spline shaft is formed completely.

During STRF process of spline shaft, only one tooth is formed with each forging die at a time, which belongs to the plastic deformation under local loading, and the forming load can be effectively reduced. In addition, the material in the tooth shape area of spline shaft formed by STRF process will not be cut off, and the material structure is continuous. The performance of the formed spline shaft can be improved accordingly. Meanwhile, STRF process belongs to the near net forming process resulting in high material utilization rate, and the production efficiency of spline shaft is higher with high forging frequency of the forging dies.

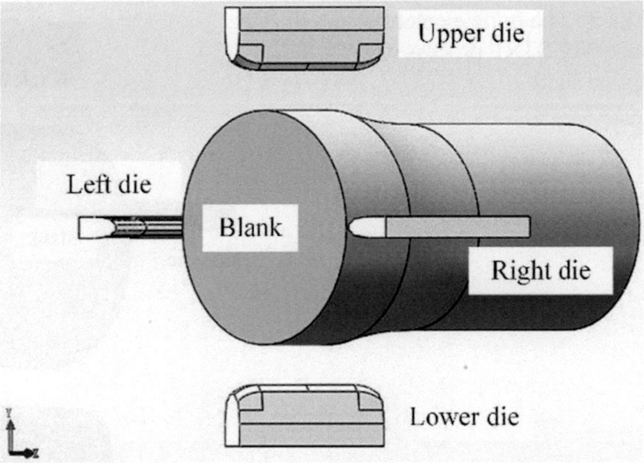

Fig. 1 STRF process of spline shaft

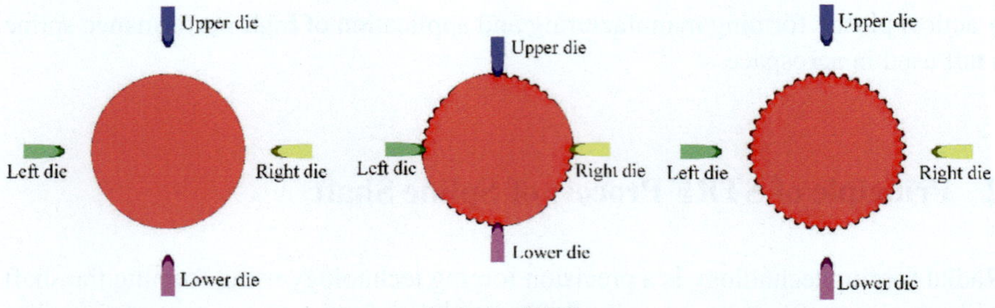

(a) Before STRF process (b) During STRF process (c) After STRF process

Fig. 2 The forming process of spline shaft with STRF process

3 Establishment of Finite Element Model

In order to further study the influence of STRF process parameters on the formation of spline shaft teeth, a finite element analysis model was established based on the commercial finite element software FORGE NxT 1.0 from French company TRANSVALOR. The tooth shape parameters were as follows: modulus was 1.25 mm, number of teeth was 40, pressure angle was 30°, and displacement coefficient was +0.25. The spline shaft CAD three-dimensional model was converted and imported into the FORGE software. The model was meshed and basic parameters were set. The grid was refined locally in the tooth forming area of the blank and the finite element model was shown in Fig. 3. Each time the forging die forged, the blank was rotated by 9°. In order to improve the calculation efficiency, the blank was only rotated without feeding.

The blank material is 45 steel, which is commonly used for spline shaft. The density of 45 steel is 7850 kg/m^3, and the specific heat is 778. The hot compressive tests were conducted at temperatures of 750, 800 and 1000 °C, and the strain rates of

Fig. 3 The finite element model of STRF process of spline shaft

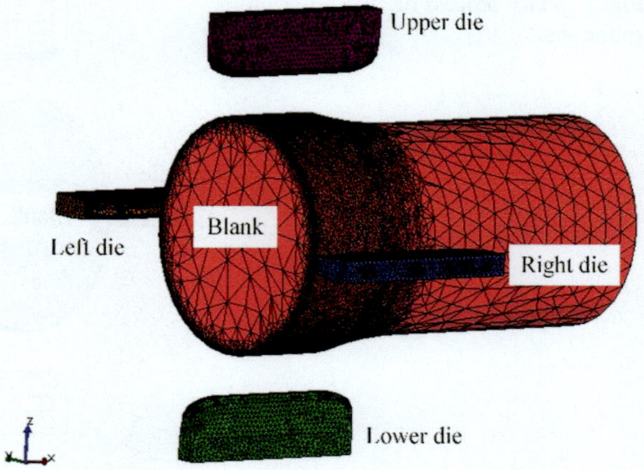

0.1 s^{-1} and 1 s^{-1}, respectively. The test data were processed and analyzed, and then the constitutive model of 45 steel is established, as shown in Eq. (1) [12]. According to literature [12], the deformation resistance of 45 steel decreases with the increase of forming temperature and increases with the increase of strain rate. And then, Eq. (1) is imported into the FORGE software.

$$\sigma = 427.398\exp^{-0.00130T}\varepsilon^{0.139}\exp^{-0.000135/\varepsilon}\dot{\varepsilon}^{0.157} \tag{1}$$

where σ is the true stress, ε is the true strain, T is the temperature, and $\dot{\varepsilon}$ is the strain rate. Using the control variable method, the simulation of STRF process of spline shaft was carried out under different process parameters, which were the initial forging temperature of the blank was 950, 1050 and 1150 °C, the forging speed was 50, 100 and 200 mm/s, and the friction coefficient was 0.1, 0.3 and 0.5. The effects of the above process parameters on the forming load, equivalent stress and equivalent strain would be obtained.

4 Analysis of Finite Element Simulation Results

4.1 The Effect of Initial Forging Temperature on the Forming Process

Figure 4 shows the forging load during STRF process under different initial forging temperature. The maximum load of the forging die under the initial forging temperature of 950 °C is reached 35.5 kN. When the initial forging temperature of the blank increasing, the forging load is decreased, and the maximum load at 1150 °C is decreased to 21.1 kN. As the initial forging temperature of the blank increases, the softening effect of the material is more obvious, and the deformation resistance of the material during forming is decreased, therefore the forging load is decreased with increasing of the initial forging temperature.

Figure 5 shows the stress distribution in the tooth shape region of single tooth die forging at different initial forging temperature. As can be seen from Fig. 5, the average equivalent stress value in tooth profile area is decreased with the increase of

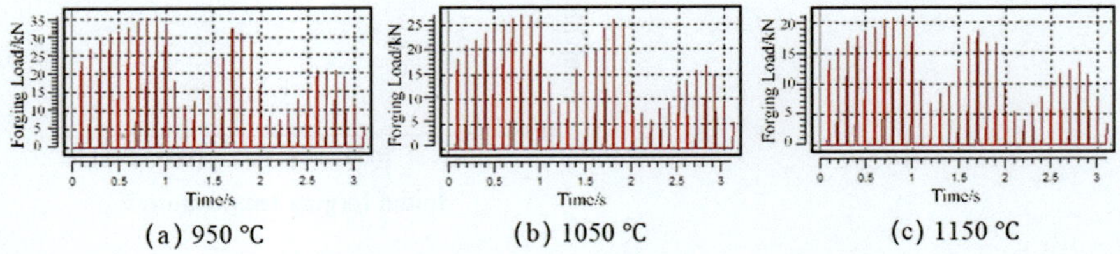

Fig. 4 Forging load under different initial forging temperature

Fig. 5 Equivalent stress in tooth profile area under different initial forging temperature

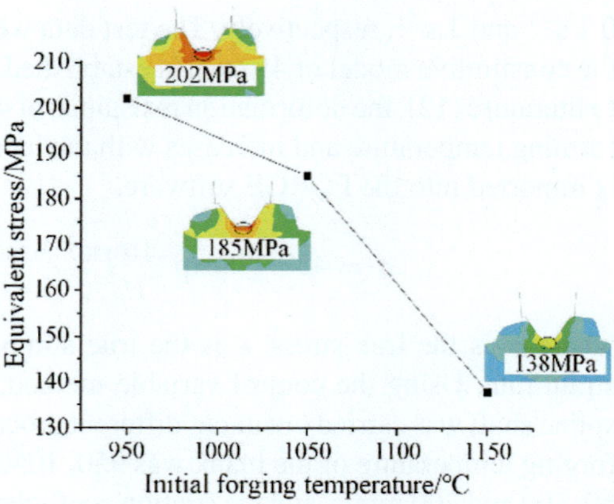

the initial forging temperature of the blank. The equivalent stress value is the smallest at 1150 °C, which is 138 MPa. This result is consistent with the above rule of forging load variation, which is caused by the softening of the material as the temperature increases.

The equivalent strain can reflect the degree of cumulative plastic deformation of the material. Figure 6 shows the distribution of equivalent strain in tooth profile area after radial forging at different initial temperature. As can be seen from Fig. 6, under different initial forging temperature, the equivalent strain values in the tooth shape region are basically the same. This is because regardless of the initial forging temperature being high or low, the reduction of single tooth shaped forging dies during radial forging is the same, resulting in the equivalent strain values in the tooth shape deformation region after radial forging not significantly different and the cumulative plastic deformation degree of the material basically the same.

Fig. 6 Equivalent strain in tooth profile area under different initial forging temperature

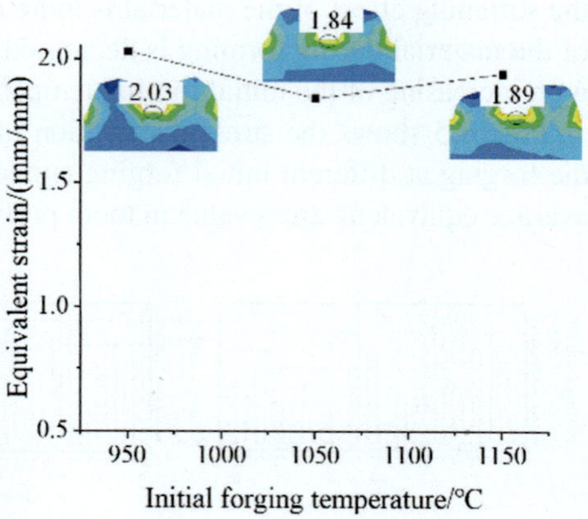

4.2 The Effect of Forging Speed on the Forming Process

Figure 7 shows the variation of forging load during STRF process of spline shaft at different forging speed. The maximum load of the forging die at each forging speed is 25.3 kN, 27.2 kN and 30.1 kN, respectively. The forging load is increased with the increase of the forging speed, which is determined by the characteristics of the material itself. From the true stress–strain curve of 45 steel, it can be seen that at the same temperature the true stress is increased with the increase of strain rate. As the forging speed increasing, the strain rate is correspondingly increased, so the deformation resistance of the material is increased, resulting in a corresponding increase of the load on the forging die.

Figure 8 shows the variation curves of the equivalent stress value and equivalent strain value with different forging speed. It can be seen that the equivalent stress value is increased with the increase of the forging speed, consistent with the change rule of the forging load with the forging speed. However, the variation of the equivalent strain value under different forging speed is not significant, the reason is that the reduction of the forging die is the same at different forging speed, and the cumulative plastic deformation degree of the material is basically the same.

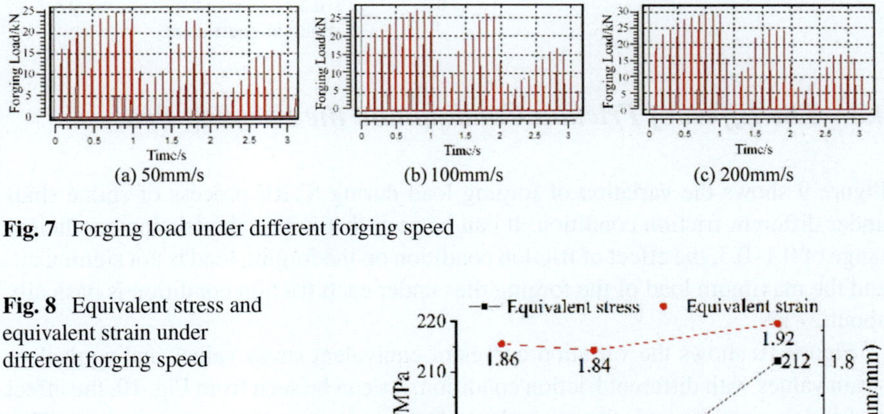

Fig. 7 Forging load under different forging speed

Fig. 8 Equivalent stress and equivalent strain under different forging speed

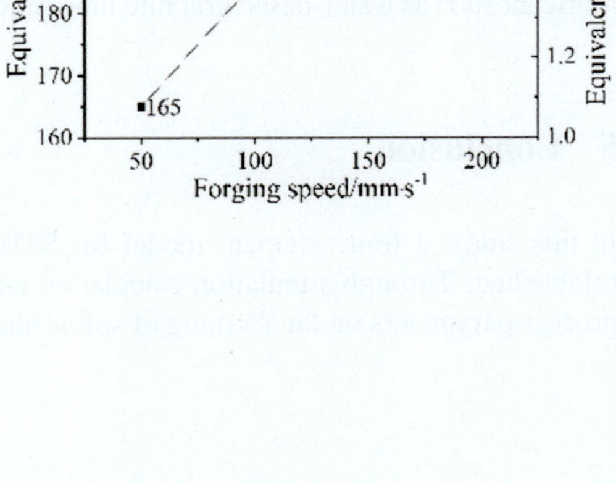

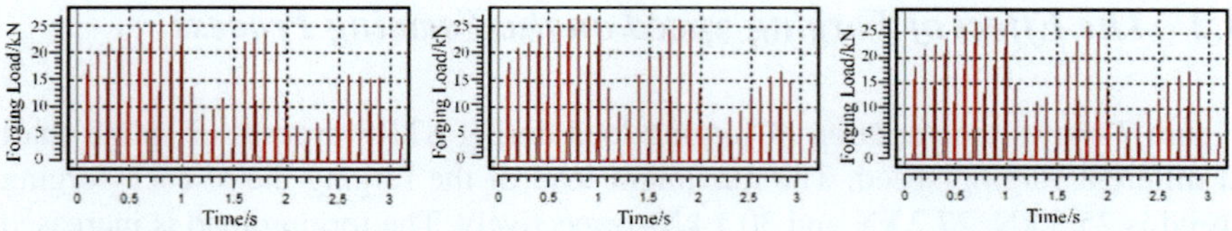

(a) Friction coefficient is 0.1 (b) Friction coefficient is 0.3 (c) Friction coefficient is 0.5

Fig. 9 Forging load under different friction condition

Fig. 10 Equivalent stress and equivalent strain under different friction condition

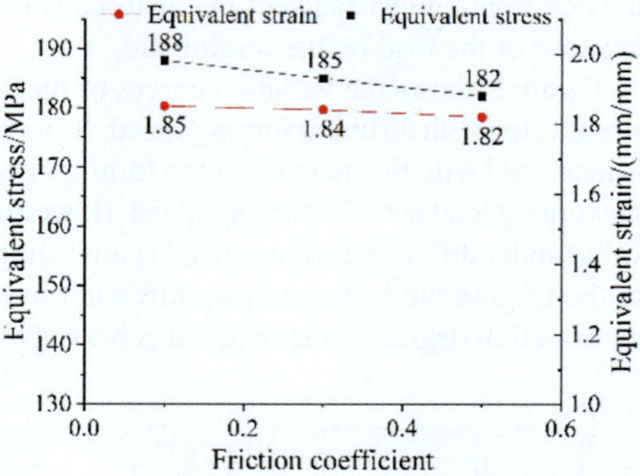

4.3 *The Effect of Friction Condition on the Forming Process*

Figure 9 shows the variation of forging load during STRF process of spline shaft under different friction condition. It can be seen that within the friction coefficient range of 0.1–0.5, the effect of friction condition on the forging load is not significant, and the maximum load of the forging dies under each friction condition is basically about 27 kN.

Figure 10 shows the variation curves of equivalent stress values and equivalent strain values with different friction condition. As can be seen from Fig. 10, the effect of friction condition on the equivalent stress and equivalent strain during STRF process of spline shaft is not significant. There are no special requirements for lubrication condition during STRF process of spline shaft, so conventional hot forging lubricant such as water-based graphite lubrication can be used during STRF process.

5 Conclusion

In this study, a finite element model for STRF process of spline shaft has been established. Through simulation calculation and analysis, the influence of various process parameters on the forming of spline shaft has been obtained, as follows:

1. During STRF process of spline shaft, the forging load and equivalent stress values are decreased with the increase of the initial forging temperature of the blank. When the initial forging temperature is 950 °C, 1050 °C, and 1150 °C, the corresponding maximum forging load is 35.5 kN, 27.2 kN, and 21.1 kN respectively, and the average equivalent stress in tooth profile area is about 202 MPa, 185 MPa, and 138 MPa respectively.
2. The forging load and equivalent stress are increased with the increase of forging speed. When the forging speed is respectively 50 mm/s, 100 mm/s, and 200 mm/s, the corresponding maximum forging load is 25.3 kN, 27.2 kN, and 30.1 kN and the average equivalent stress in tooth profile area is about 165 MPa, 185 MPa, and 212 MPa, respectively.
3. In the range of friction coefficient from 0.1 to 0.5, the effects of friction condition on forging load and equivalent stress are not significant. Therefore, conventional hot forging lubricants such as water-based graphite lubrication can be selected for STRF process of spline shaft.

References

1. Li Y, Zhao S, Zhao Y et al (2014) Analysis and experiment on the effect of system parameters on the axial-infeed rolling forming process of spline shaft. J Mech Eng 50(22):50–56 (in Chinese)
2. Luo Y, Ma Z, Fang X et al (2021) Influencing factors of effective tooth depth in the cylindrical gear rolling process with axial infeed. Int J Adv Manuf Technol 118:497–510
3. Fu X, Wang B, Wang T (2021) Experimental and numerical investigations on the deformation defects by gear rolling process with local induction heating. Int J Adv Manuf Technol 117:835–844
4. Hou H, Zhao Y, Cui M et al (2021) Experimental study on the axial-infeed incremental warm rolling process for spline shaft production. Int J Adv Manuf Technol 112:2177–2186
5. Wang Z et al (2008) Numerical simulation of involutes spline shaft in cold rolling forming. J Cent South Univ Technol 15(S2):278–283
6. Zhang D, Zhao S, Wu S et al (2015) Phase characteristic between dies before rolling for thread and spline synchronous rolling process. Int J Adv Manuf Technol 81:513–528
7. Lu Q, Yu D et al (2020) Analysis on fracture of output spline shaft for a certain type of turbine stater. Aviat Maint Eng (10):109–112
8. Li K, Zhao S et al (2021) Discussion of near net forming process for complex curved tooth profile using radial forging. Heavy Mach (05):19–23
9. Simonov MY, Shaimanov GS et al (2016) Effect of structure on the dynamic crack resistance and special features of the micromechanism of crack growth in steel 35Kh after cold radial forging. Met Sci Heat Treat 58(1–2):82–90
10. Darki S, Raskatov EY (2020) Analysis of the hot radial forging process according to the finite element method. Int J Adv Manuf Technol 110:1061–1070
11. Panov D, Pertsev A, Smirnov A et al (2019) Metastable austenitic steel structure and mechanical properties evolution in the process of cold radial forging. Materials (Basel) 12(2058):1–18
12. Li K, Zhao S, Zhang C et al (2022) Deformation behavior during single-tooth radial forging of large-scale spline shaft. J Plast Eng 29(02):69–75 (in Chinese)

Influence of Friction on Deformation Characteristics in Incremental Sheet Forming Process

Guang-Can Yang and Da-Wei Zhang

Abstract Incremental sheet forming (ISF) process is a flexible forming technology that utilizes a simple tool to act on the sheet metal point by point and form the target part through plastic accumulation. It has the potential to provide strong technical support for rapid response prototype manufacturing at low-cost. The friction between the tool and sheet is an important parameter affecting the formability and surface quality of the process, and understanding its influence on the deformation characteristics of the sheet is an essential basis for reasonably optimizing the friction parameters and exploring forming mechanisms. Therefore, this work establishes a numerical analysis model for ISF under different friction conditions, extracts the equivalent strain and stress data during the whole forming process of arbitrary element in the contact area and shear stress of the element, and compares and analyses the average equivalent strain and stress values with the calculation results of the analytical model. It is found that the friction condition hardly affects the equivalent strain and stress values of the sheet in the contact deformation area during ISF process, but it has a significant impact on the shear stress.

Keywords Incremental sheet forming · Friction · Equivalent strain and stress

1 Introduction

Incremental sheet forming (ISF) is a flexible process in which a simple tool is used to act on the surface of the sheet point by point [1], which has excellent potential application value in the fields of aerospace and biomedicine [2]. However, the small contact area and the high pressure resulted in poor friction conditions. Under the action of the adhesion mechanism, the surface quality of the formed part can be

G.-C. Yang · D.-W. Zhang (✉)
School of Mechanical Engineering, Xi'an Jiaotong University, Xi'an, China
e-mail: zhangdawei2000@mail.xjtu.edu.cn

G.-C. Yang
e-mail: yanggc1995@stu.xjtu.edu.cn

Z. Liu et al. (eds.), *Advances and Challenges in Advanced Unmanned Aerial Systems*, Springer Aerospace Technology, https://doi.org/10.1007/978-981-99-8045-1_42

damaged [3], and even the "orange peel" effect can occur to form waste parts [4]. Therefore, the forming quality of ISF largely depends on friction [5]. Based on the importance of friction in the forming process, an in-depth study of its influence on the deformation characteristics of sheet metal is an essential basis for understanding forming mechanism and optimizing friction parameters of ISF process.

The deformation of sheet during ISF is related to friction. Lu [6] pointed out that increasing friction would improve the forming stability and inhibit necking, but would increase the stress triaxiality and reduce formability. Yang [7] proposed an analytical model for calculating the twist angle, revealing the mechanism of the twist phenomenon. Jackson et al. [8] reduced thickness-shear by reducing friction, and significantly reducing wear between the tool and the part. Kim [9] pointed out that friction can increase strain and pressure, reduce stress, delay the development of cracks, and improve formability. Although these studies have investigated the effects of friction on stability, formability, and thickness-shear during ISF, there are still many outstanding questions to be answered regarding the influence of friction on the equivalent strain and stress in the contact area during the forming process.

This work establishes a numerical analysis model for ISF under different friction conditions and extracts the equivalent strain, stress, and shear stress of an arbitrary element in the contact area during the whole forming process. The average equivalent strain and stress are compared with the calculation results of the analytical model to analyze the influence of friction on the deformation characteristics in ISF.

2 Finite Element Analysis Model

The numerical analysis model shown in Fig. 1 is established through the dynamic explicit analysis module of ABAQUS. The upper and lower pressing plates can be considered rigid bodies. Compared to the performance of sheet, the hardness of tool is relatively high, and it can be set as a rigid body with an end radius of $r = 7$ mm. The size of the sheet is 200 × 200 mm, thickness $t_o = 0.8$ mm. Considering that the length and width are much larger than the thickness, the variation of stress in the thickness direction can be ignored, so the shell element is assigned to the sheet, and its material mechanical property is shown in Eq. (1). Refining the mesh of possible contact areas with a size of 0.5 mm. The friction coefficients are 0 when there is no friction, 0.1 when the lubrication condition is good, 0.2 when the lubrication condition is poor, and 0.3 when there is no lubrication. The feed speed is 1000 mm/min, the rotational speed is 500 rpm, the step size is $z = 0.5$ mm, and the wall angle is $\alpha = 45°$. The geometric shape of the formed part is shown in Fig. 1, with a maximum diameter of D = 70 mm and forming depth of 20 mm.

$$\sigma_e = 359.2582\varepsilon_e^{0.24} \tag{1}$$

The tool only compresses the sheet on a small area that continuously changes as the moves. And an element begins to deform when it first enters the deformation zone,

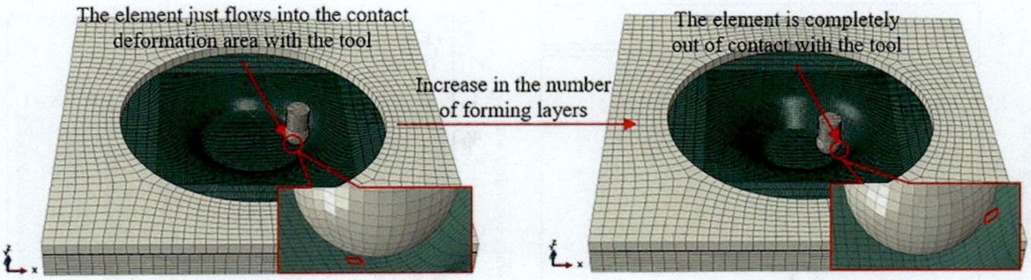

Fig. 1 Schematic diagram of finite element analysis model

it flows out of the deformation zone after the cumulative deformation of multiple forming layers, as shown in Fig. 1. The influence of friction on the deformation characteristics could be reflected by the deformation characteristics of the whole deformation process of an arbitrary element under different friction conditions.

3 Results and Discussion

3.1 Influence of Friction on Equivalent Strain

According to the deformation process of ISF, the equivalent strain of an arbitrary element remains basically unchanged after the instant of contact with the tool increases, as shown in Fig. 2a. Then the tool may continue to contact the element after entering the next forming layer, causing the instant increase of the equivalent strain of the element, and then maintaining stability for a period of time. When the tool nears the element again, it may still cause continued accumulation of strain, so the equivalent strain has a tendency to continue to increase. This cyclic accumulation of strain increases until the element completely disengages from the deformation region, and the equivalent strain value remains stable. According to the variation of the equivalent strain during the whole forming process of an arbitrary element, it can be seen that the changing trend of the equivalent strain is consistent under different friction conditions, with a high degree of data coincidence, and there is no noticeable change with the friction conditions. It is preliminarily speculated that the friction effect has almost no influence on the equivalent strain of the sheet during ISF.

When $\mu = 0$, the average equivalent strain $\varepsilon_e = 0.3638$, when $\mu = 0.1$, $\varepsilon_e = 0.35605$, when $\mu = 0.2$, $\varepsilon_e = 0.35744$, and when $\mu = 0.3$, $\varepsilon_e = 0.36561$, while the equivalent strain model Eq. (2) [10] can obtain $\varepsilon_e = 0.3544$ under the same forming parameters, as shown in Fig. 2b. Therefore, compared to the analytical value, the simulation values at $\mu = 0, 0.1, 0.2$ and 0.3 change by 2.6524%, 0.4656%, 0.8578%, and 3.1631%, respectively. It can be seen that the deviations between the two are relatively small. That is, there is no significant change and regularity in the equivalent strain value when the friction coefficient changes. Therefore, from the above analysis

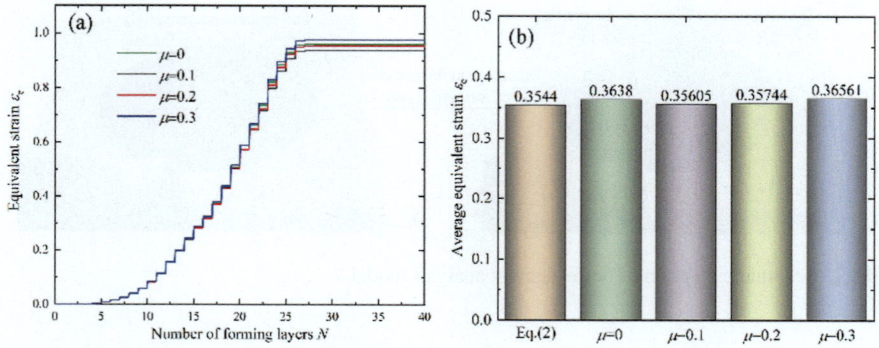

Fig. 2 Equivalent strain under different friction conditions: **a** equivalent strain; **b** average equivalent strain

and comparison, it can be determined that the effect of friction on the equivalent strain of any element in ISF process is not significant.

$$\varepsilon_e = \frac{2}{\sqrt{3}} \ln\left(\frac{rt_o}{t_o \cos\alpha\left(r + \frac{t_o \cos\alpha}{2}\right)}\right) \quad (2)$$

3.2 *Influence of Friction on Equivalent Stress and Shear Stress*

Similarly, when the tool is away from the element, the element stress is small, and as it gradually approaches, the element stress gradually increases, as shown in Fig. 3a. When the tool contact element deforms it, the equivalent stress momentarily increases to a specific value, and after the element disengages from contact with the tool, the stress in the element rapidly decreases. The tool repeatedly extrudes the element along the forming path until the forming process of the element is completed. The variation law of the equivalent stress under different friction conditions is consistent, and there is almost no significant difference in the numerical value. After the element entirely flows out of the deformation zone, the stress basically remains unchanged. It can be seen that the friction condition hardly affects the equivalent stress of an arbitrary element in the processing region during ISF.

According to Eqs. (1) and (2), the equivalent stress is calculated to be $\sigma_e =$ 280.0834 MPa. The equivalent stress $\sigma_e = 269.103$, 268.6, 268.407 and 269.068 MPa when $\mu = 0$, 0.1, 0.2 and 0.3 calculated by the simulation model, as shown in Fig. 3b. The deviation of the equivalent stress between the calculated values of the analytical model and the simulation model under different friction conditions are 3.9204%, 4.2214%, 4.1689% and 3.9329%, respectively. It can be seen that there

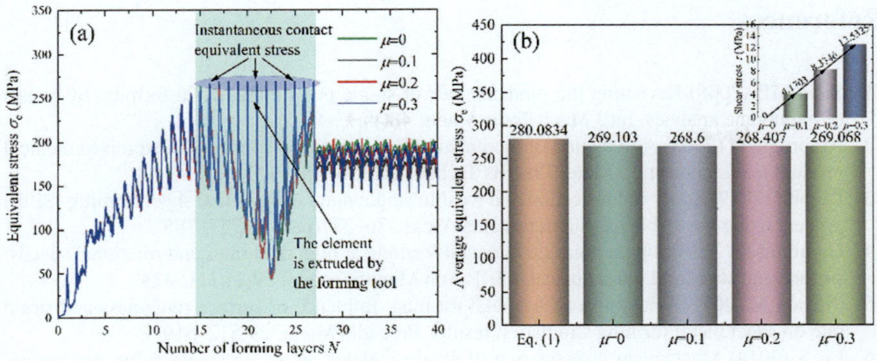

Fig. 3 Equivalent stress under different friction conditions: **a** equivalent stress of the whole forming process; **b** average equivalent stress calculated by analytically and numerically

is a good consistency between the two. Then, by analyzing the influence of friction conditions on the equivalent stress, it is found that the equivalent stress values under different friction coefficients are very similar, and it can be basically determined that the influence of changes in friction conditions on the equivalent stress is not significant. The shear stress $\tau = 0$, 4.1703, 8.3346 and 12.5325 MPa when $\mu = 0$, 0.1, 0.2 and 0.3, that is, the values of shear stress are relatively small, but friction conditions can have an evident impact on them.

In summary, the influence of friction effects on the equivalent stress and strain in the contact deformation region can be ignored, but it has a relatively obvious impact on the shear stress.

4 Conclusions

In this work, the numerical analysis models for ISF under different friction conditions were established, and the equivalent stress, strain and shear stress during the whole forming process of an arbitrary element were extracted. By comparing the average values of equivalent stress and strain with the calculated results of the analytical model, and analyzing the shear stress under friction conditions. It is found that compared to the deformation characteristics without friction, the maximum variation of the equivalent strain under other friction conditions is only 3.163%, and the maximum variation of the equivalent stress is only 4.221%, indicating that the equivalent stress and strain are unaffected by the friction conditions during ISF. However, friction has a noticeable impact on shear stress.

References

1. Silva MB (2008) Revisiting the fundamentals of single point incremental forming by means of membrane analysis. Int J Mach Tools Manuf 48(1):73–83
2. Ambrogio G (2005) Application of incremental forming process for high customised medical product manufacturing. J Mater Process Technol 162(10):156–162
3. Diabb J (2017) Study of lubrication and wear in single point incremental sheet forming (SPIF) process using vegetable oil nanolubricants. Wear 376–377(part A):777–785
4. Hamilton K (2010) Single point incremental forming at high feed rates and rotational speeds: surface and structural consequence. CIRP Ann Manuf Technol 59(1):311–314
5. Sigvant M (2019) Friction in sheet metal forming: influence of surface roughness and strain rate on sheet metal forming simulation results. Procedia Manuf 29:512–519
6. Lu B (2014) Mechanism investigation of friction-related effects in single point incremental forming using a developed oblique roller-ball tool. Int J Mach Tools Manuf 85:14–29
7. Yang Z (2020) Mechanism of twist in incremental sheet forming of thermoplastic polymer. Mater Des 195:108997
8. Jackson K (2009) The mechanics of incremental sheet forming. J Mater Process Technol 209(3):1158–1174
9. Kim YH (2002) Effect of process parameters on formability in incremental forming of sheet metal. J Mater Process Technol 130–131:42–46
10. Fang Y (2014) Analytical and experimental investigations on deformation mechanism and fracture behavior in single point incremental forming. J Mater Process Technol 214(8):1503–1515

GPSR Compliance

The European Union's (EU) General Product Safety Regulation (GPSR) is a set of rules that requires consumer products to be safe and our obligations to ensure this.

If you have any concerns about our products, you can contact us on ProductSafety@springernature.com

In case Publisher is established outside the EU, the EU authorized representative is:

Springer Nature Customer Service Center GmbH
Europaplatz 3
69115 Heidelberg, Germany

Batch number: 10731143

Printed by Printforce, the Netherlands